Primate Parasite Ecology

The Dynamics and Study of Host–Parasite Relationships

Edited by
MICHAEL A. HUFFMAN
Kyoto University, Japan

COLIN A. CHAPMAN
McGill University, Montréal

CAMBRIDGE
UNIVERSITY PRESS

CAMBRIDGE
UNIVERSITY PRESS

University Printing House, Cambridge CB2 8BS, United Kingdom

One Liberty Plaza, 20th Floor, New York, NY 10006, USA

477 Williamstown Road, Port Melbourne, VIC 3207, Australia

314-321, 3rd Floor, Plot 3, Splendor Forum, Jasola District Centre, New Delhi - 110025, India

79 Anson Road, #06-04/06, Singapore 079906

Cambridge University Press is part of the University of Cambridge.

It furthers the University's mission by disseminating knowledge in the pursuit of education, learning and research at the highest international levels of excellence.

www.cambridge.org
Information on this title: www.cambridge.org/9781108829403

First published 2009
First paperback edition 2020

A catalogue record for this publication is available from the British Library

Library of Congress Cataloging in Publication data
Primate parasite ecology : the dynamics and study of host–parasite relationships / edited by Michael A. Huffman, Colin A. Chapman.
p. ; cm.
Includes bibliographical references and index.
ISBN 978-0-521-87246-1 (hardback)
1. Parasites – Ecology. 2. Parasitology. 3. Host-parasite relationships.
4. Primates. I. Huffman, Michael A. II. Chapman, Colin A.
[DNLM: 1. Host-Parasite Interactions. 2. Primates – parasitology.
3. Communicable Diseases – transmission. 4. Ecosystem. 5. Parasitology.
QX 45 P952 2009]
QL757.P743 2009
571.9′99198 – dc22 2008038887

ISBN 978-0-521-87246-1 Hardback
ISBN 978-1-108-82940-3 Paperback

Cambridge Studies in Biological and Evolutionary Anthropology 57

Primate Parasite Ecology

The Dynamics and Study of Host–Parasite Relationships

Anyone who has spent an extended period in the tropics has an idea, through caring for others or first-hand experience, just what it is like to be a primate parasite host. Monkeys and apes often share parasites with humans, for example the HIV viruses which evolved from related viruses of chimpanzees and sooty mangabeys, and so understanding the ecology of infectious diseases in non-human primates is of paramount importance. Furthermore, there is accumulating evidence that environmental change may promote contact between humans and non-human primates and increase the possibility of sharing infectious disease. Written for graduate students and academic researchers, this book addresses these issues and provides up-to-date information on the methods of study, natural history, and ecology/theory of the exciting field of primate parasite ecology.

MICHAEL A. HUFFMAN is an Associate Professor, and the first tenured non-Japanese faculty member, at Kyoto University's Primate Research Institute, Japan. He is currently an editor for the *American Journal of Primatology*. His research on host–parasite relationships and primate self-medication has involved multi-disciplinary international collaborations on species around the world, spanning over 15 countries.

COLIN A. CHAPMAN is a Professor in the Department of Anthropology and McGill School of Environment at McGill University and a Canada Research Chair in Primate Ecology and Conservation. He has been an associate scientist with the Wildlife Conservation Society since 1995 and for the last 18 years has conducted research in the Kibale National Park, Uganda.

Cambridge Studies in Biological and Evolutionary Anthropology

Series editors

Also avilable in the series

39 *Methods in Human Growth Research* Roland C. Hauspie, Noel Cameron & Luciano Molinari (eds.) 0 521 82050 2
40 *Shaping Primate Evolution* Fred Anapol, Rebecca L. German & Nina G. Jablonski (eds.) 0 521 81107 4
41 *Macaque Societies: A Model for the Study of Social Organization* Bernard Thierry, Mewa Singh & Werner Kaumanns (eds.) 0 521 81847 8
42 *Simulating Human Origins and Evolution* Ken Wessen 0 521 84399 5
43 *Bioarchaeology of Southeast Asia* Marc Oxenham & Nancy Tayles (eds.) 0 521 82580 6
44 *Seasonality in Primates* Diane K. Brockman & Carel P. van Schaik 0 521 82069 3
45 *Human Biology of Afro-Caribbean Populations* Lorena Madrigal 0 521 81931 8
46 *Primate and Human Evolution* Susan Cachel 0 521 82942 9
47 *The First Boat People* Steve Webb 0 521 85656 6
48 *Feeding Ecology in Apes and Other Primates* Gottfried Hohmann, Martha Robbins & Christophe Boesch (eds.) 0 521 85837 2
49 *Measuring Stress in Humans: A Practical Guide for the Field* Gillian Ice & Gary James (eds.) 0 521 84479 7
50 *The Bioarchaeology of Children: Perspectives from Biological and Forensic Anthropology* Mary Lewis 0 521 83602 6
51 *Monkeys of the Ta'i Forest* W. Scott McGraw, Klaus Zuberbühler & Ronald Noe (eds.) 0 521 81633 5
52 *Health Change in the Asia-Pacific Region: Biocultural and Epidemiological Approaches* Ryutaro Ohtsuka & Stanley I. Ulijaszek (eds.) 978 0 521 83792 7
53 *Technique and Application in Dental Anthropology* Joel D. Irish & Greg C. Nelson (eds.) 978 0 521 87061 0
54 *Western Diseases: An Evolutionary Perspective* Tessa M. Pollard 978 0 521 61737 6
55 *Spider Monkeys: Behavior, Ecology and Evolution of the Genus* Ateles Christina J. Campbell (ed.) 978 0 521 86750 4
56 *Between Ziology and Culture* Holger Schutkowski (ed.) 978 0 521 85936 3

Contents

Contributors

JULIE M. ALLEN, Florida Museum of Natural History, and Department of Zoology, University of Florida, Gainesville, FL 32611, USA

SUMMER J. ARRIGO-NELSON, Centre ValBio, BP 33 – Ranomafana, 312 Ifanadiana Madagascar and Department of Anthropology, University of Notre Dame, Notre Dame, IN 46556, USA

FRANCISCO J. AYALA, Department of Ecology and Evolutionary Biology, University of California, Irvine, Irvine, CA 92697, USA

CHRIS BAKUNEETA, Department of Zoology, Makerere University, Kampala, Uganda

BRIAN BANNON, Department of Anthropology, Stony Brook University, Stony Brook, NY 11794, USA

MASSIMO BARDI, Department of Psychology, Marshall University, Huntington, WV, USA

MAGDALENA BERMEJO, Department of Animal Biology, Faculty of Biology, University of Barcelona, Av. Diagonal 645, ES-08028, Barcelona, Spain and Programme de conservation et utilisation rationnelle des Ecosystèmes Forestiers en Afrique Centrale (ECOFAC), BP 15115 Libreville, Gabon

DWIGHT D. BOWMAN, Department of Veterinary Microbiology and Immunology, College of Veterinary Medicine, Cornell University, Ithaca, NY 14853, USA

COLIN A. CHAPMAN, Department of Anthropology and McGill School of Environment, McGill University, 855 Sherbrooke St. West, Montreal, Quebec, H3A 2T7, Canada and Wildlife Conservation Society, 2300 Southern Boulevard, Bronx, NY 10460, USA

ANTHONY DI FIORE, Center for the Study of Human Origins, Department of Anthropology, New York University, 25 Waverly Place, New York, NY 10003, USA

TODD DISOTELL, Center for the Study of Human Origins, Department of Anthropology, New York University, 25 Waverly Place, New York, NY 10003, USA

JOZEF DUPAIN, African Wildlife Foundation, Boulevard du 30 juin n 2515, BP 2396, Kinshasa, Gombe, RDC.

SHELLY FLANNIGAN, Florida Museum of Natural History, University of Florida, Gainesville, FL 32611, USA

SHIHO FUJITA, Department of Veterinary Medicine, Faculty of Agriculture,Yamaguchi University, Yoshida1677–1, Yamaguchi-shi, Yamaguchi, 753–8515, Japan

PASCAL GAGNEUX, Glycobiology Research and Training Center, Cellular and Molecular Medicine East, University of California San Diego, 9500 Gilman Drive, La Jolla, CA 92093, USA

PAOLA GARCIA, Departamento de Parasitología, Facultad de Farmacia, Universidad Complutense de Madrid, Plaza Ramón y Cajal s/n, 28040 Madrid, Spain

ROBIN B. GASSER, Department of Veterinary Science, The University of Melbourne, 250 Princes Highway, Werribee, Victoria 3030, Australia

TONY L. GOLDBERG, University of Illinois Department of Pathobiology, 2001 South Lincoln Avenue, Urbana, IL 61802, USA

FRANCISCO PONCE GORDO, Departamento de Parasitología, Facultad de Farmacia, Universidad Complutense de Madrid, Plaza Ramón y Cajal s/n, 28040 Madrid, Spain

SHUNJI GOTOH, Wild Animal Research Clinic, Tatsugo, Amami, Japan

ELLIS C. GREINER, Department of Infectious Diseases and Pathology, College of Veterinary Medicine, University of Florida, Gainesville, FL 32610, USA

JOHANNA M. DE GRUIJTER, Department of Veterinary Science, The University of Melbourne, 250 Princes Highway, Werribee, Victoria 3030, Australia and Department of Parasitology, Leiden University Medical Center, Leiden, PO Box 9600, 2300 RC Leiden, The Netherlands

A. L. HARIVELO, Department of Paleontology and Biological Anthropology, University of Antananarivo, Antananarivo, Madagascar

HIDEO HASEGAWA, Department of Infectious Diseases (Biology), Faculty of Medicine, Oita University, Hasama, Yufu, Oita 879–5593, Japan

ALEXANDER D. HERNANDEZ, Department of Ecology and Social Behavior, Primate Research Institute, Kyoto University, Inuyama, Aichi, 484–8506, Japan

STACEY A. M. HODDER, Department of Anthropology, McGill University, 855 Sherbrooke St. West, Montreal, Quebec, H3A 2T7, Canada

KRISTINA L. HOGG, College of Veterinary Medicine, Cornell University, Ithaca, NY 14853, USA

MICHAEL A. HUFFMAN, Department of Ecology and Social Behavior, Primate Research Institute, Kyoto University, 41–2 Kanrin, Inuyama, Aichi, 484–8506, Japan

TAKASHI KAGEYAMA, Center for Human Evolution Modeling Research, Primate Research Institute, Kyoto University, 41–2 Kanrin, Inuyama, Aichi 484–8506, Japan

TARANJIT KAUR, Department of Biomedical Sciences and Pathobiology, VA-MD Regional College of Veterinary Medicine, Virginia Tech, Duck Pond Drive (0442), Blacksburg, VA 24061, USA

NATALIE P. LEO, Center for Human Evolution Modeling Research, Primate Research Institute, Kyoto University, Inuyama, Aichi 484–8506, Japan

JESSICA E. LIGHT, Florida Museum of Natural History, University of Florida, Gainesville, FL 32611, USA

ANDREW J. MACINTOSH, Department of Ecology and Social Behavior, Primate Research Institute, Kyoto University, Inuyama, Aichi, 484–8506, Japan

ANTOINETTE MCINTOSH, Department of Infectious Diseases and Pathology, College of Veterinary Medicine, University of Florida, Gainesville, FL 32610, USA

DAVID MODRÝ, Department of Parasitology, University of Veterinary and Pharmaceutical Sciences, Palackého 1–3, 612 42 Brno, Czech Republic and Institute of Parasitology, Biological Centre, Academy of Sciences of the Czech Republic, Branišovská 31, 370 05 České Budějovice, Czech Republic

TONI LYN MORELLI, Centre ValBio, BP 33 – Ranomafana, 312 Ifanadiana Madagascar and Department of Ecology and Evolution, Stony Brook University, Stony Brook, NY 11794 USA

MICHAEL P. MUEHLENBEIN, Department of Anthropology, Evolutionary Physiology and Ecology Laboratory, Indiana University, 701 E. Kirkwood Ave., Student Building 130, Bloomington, IN 47405, USA

CARLOS NELL, Departamento de Parasitología, Facultad de Farmacia, Universidad Complutense de Madrid, Plaza Ramón y Cajal s/n, 28040 Madrid, Spain

CHARLES L. NUNN, Max Planck Institute for Evolutionary Anthropology, D-04103 Leipzig, Germany, and University of California, Department of Integrative Biology, Berkeley, CA 94720–3140, USA

ASAMI OGASAWARA, Center for Human Evolution Modeling Research, Primate Research Institute, Kyoto University, Inuyama, Aichi 484–8506, Japan

PAULA PEBSWORTH, Roots & Shoots, The Jane Goodall Institute, 4245 N. Fairfax Drive, Suite 600, Arlington, VA 22203, USA

ALICE N. PELL, Department of Animal Science, Cornell University, Ithaca, NY, 14853, USA

KLÁRA JUDITA PETRŽELKOVÁ, Institute of Vertebrate Biology, Academy of Sciences of the Czech Republic, Květná 8, 603 00 Brno, Czech Republic and Zoo Liberec, Masarykova 1347/31, 460 01 Liberec, Czech Republic

ANTON M. POLDERMAN, Department of Parasitology, Leiden University Medical Center, Leiden, PO Box 9600, 2300 RC Leiden, The Netherlands

FELIX RATELOLAHY, Department of Paleontology and Biological Anthropology, University of Antananarivo, Antananarivo, Madagascar

DAVID L. REED, Florida Museum of Natural History, University of Florida, Gainesville, FL 32611, USA

JOSÉ DOMINGO RODRÍGUEZ-TEIJEIRO, Department of Animal Biology, Faculty of Biology, University of Barcelona, Av. Diagonal 645, ES-08028, Barcelona, Spain

JESSICA M. ROTHMAN, Department of Biology and McGill School of Environment, McGill University, 3534 University Avenue, Montreal, Quebec, H3A 2A7, Canada

SADIE J. RYAN, Department of Anthropology and McGill School of Environment, McGill University, Montreal, Quebec, Canada H3A 2T7 and Department of Anthropological Sciences, Building 360, Stanford University, Stanford, CA 94305, USA

RAJA SENGUPTA, Department of Geography, McGill University, Montreal, Quebec, H3A 2K6, Canada

JATINDER SINGH, Department of Biomedical Sciences and Pathobiology, VA-MD Regional College of Veterinary Medicine, Virginia Tech, Duck Pond Drive (0442), Blacksburg, VA 24061, USA

MICHAEL V. K. SUKHDEO, Department of Ecology, Evolution and Natural Resources, Rutgers University, NJ 088903, USA

SUZANNE C. SUKHDEO, Department of Ecology, Evolution and Natural Resources, Rutgers University, NJ 088903, USA

WILLIAM M. SWITZER, Laboratory Branch, Division of HIV/AIDS Prevention, National Center for HIV, Hepatitis, STD, and TB Prevention, Centers for Disease Control and Prevention MS G-45, Atlanta, GA 30333, USA

MELISSA A. TOUPS, Florida Museum of Natural History and Department of Zoology, University of Florida, Gainesville, FL 32611, USA

KAZUNARI USHIDA, Laboratory of Animal Science, Kyoto Prefectural University, Shimogamo, Kyoto 606–8522, Japan

SYLVIA K. VITAZKOVA, University of the Virgin Islands, VI-EPSCoR, Division of Science and Mathematics, and Center for Marine and Environmental Studies, #2 John Brewer's Bay, St. Thomas, US Virgin Islands 00802

PETER D. WALSH, Max Planck Institute for Evolutionary Anthropology, Deutscher Platz 6, 01203 Leipzig, Germany

ANNA H. WEYHER, Department of Anthropology, Washington University in St. Louis, St. Louis, MO 63130, USA

NATHAN D. WOLFE, Department of Epidemiology, School of Public Health, University of California Los Angeles, 650 Charles E. Young Drive South, CHS 71–279B, Box 177220, Los Angeles, CA 90095–1772, USA

PATRICIA C. WRIGHT, Department of Anthropology, Stony Brook University, Stony Brook, NY 11794 USA and Centre ValBio, BP 33 – Ranomafana, 312 Ifanadiana, Madagascar, and Institute of Biotechnology, and Department of Ecology and Systematics, University of Helsinki, FIN-00014, Finland

JEFFREY WYATT, Centre ValBio, BP 33 – Ranomafana, 312 Ifanadiana Madagascar and Department of Comparative Medicine, University of Rochester Medical Center, Rochester, NY 14627 USA

Preface

Anyone who has spent any extended period living and researching in the tropics has an idea, whether by second-hand experience caring for others or by being infected themselves with malaria, amebic dysentery, sand fleas, or some of the other local milieu of parasites, just what it is like to be a primate parasite host. The effects of parasitism can be serious or even deadly, warranting that all precautionary measures be taken. However, for some like ourselves who have had the experience more than once, it can lead to an interest to understanding the nature of host–parasite relationships and the effect parasites can have on the host. For both of us, the study of primate parasite ecology is truly infectious, and it is our wish that this enthusiasm is transmitted to you the reader!

The sudden appearance of diseases like SARS (Severe Acute Respiratory Syndrome) and bird flu or the devastating impacts that diseases like Ebola have had on both human and wildlife communities, and the immense social and economic costs created by viruses like HIV underscore our need to understand the ecology of infectious diseases. Given that monkeys and apes often share parasites with humans, understanding the ecology of infectious diseases in non-human primates is of paramount importance. This is well illustrated by the HIV viruses, the causative agents of human AIDS, which evolved recently from related viruses of chimpanzees (*Pan troglodytes*) and sooty mangabeys (*Lophocebus atys*) and the outbreaks of Ebola virus, which trace their origins to zoonotic transmissions from local apes. A consideration of how environmental change may promote contact between humans and non-human primates and increase the possibility of sharing infectious disease detrimental to humans or non-human primates is now critical to both conservation and human health planning.

Such emerging diseases and the impact that they have had on humans and wildlife has stimulated a considerable amount of recent research and it is clear that the field of primate disease ecology has recently been gaining momentum. The study of disease adds a new and important dimension to primatology, as most previous research has focused on predation and resource competition, with almost no research on infectious disease as an ecological force. The relevance of issues of disease ecology is wide with an expected impact on a diverse set of researchers including the veterinary sciences, conservation, zoonotic diseases, zoology, and evolutionary biology. This is a very young field, but the time is right to gather together what information has recently become available on primate disease ecology to clearly illustrate the "state of the art" and to also point the way forward.

This book covers a diversity of aspects of host–parasite relationships integrating laboratory methodology, field research, and theory. In general, the chapters fall into three broad categories: (1) Methods to study primate–parasite interactions, (2) The natural history of primate–parasite interactions, and (3) The ecology of primate–parasite interactions. Within this general framework chapters in the section on field research cover a variety of primate species ranging from tropical to temperate habitats. They cover host–parasite, pathogen interactions of both internal and external parasites. Authors address the dynamic nature of host–parasite relationships and look at such aspects as host behavioral counter-measures in response to infection, inter- and intra-species difference in parasite prevalence as a consequence of climatic and environmental variation, habitat fragmentation, and seasonality. Chapters include original research papers, reviews, methodology, theory on various aspects of host–parasite ecology research, and resources for species identification.

This book would not have come to fruition had it not been for the enthusiasm and efforts of all the authors and colleagues who offered their time and assistance in preparing and reviewing the manuscripts. To all of you we give our hearty appreciation.

Part I

Methods to study primate–parasite interactions

1 *Collection methods and diagnostic procedures for primate parasitology*

ELLIS C. GREINER AND ANTOINETTE McINTOSH

Photograph by Jessica Rothman

Introduction

A great deal of energy has been expended on studying parasites of free-ranging primates. Many of these studies have been confined to fecal surveys as it is both difficult and not practical to gain knowledge of parasites in these hosts by using the older and normal method of studying the parasites that are recovered at

Primate Parasite Ecology. The Dynamics and Study of Host–Parasite Relationships, ed. Michael A. Huffman and Colin A. Chapman. Published by Cambridge University Press.

necropsy. Thus, this chapter is designed to aid researchers in being as productive and efficient as possible in gathering information that is most useful for the conservation of these interesting creatures. The type of information obtainable depends upon the restrictions placed on the collection and preservation of the samples to be examined. If the primates are darted and anesthetized, then feces, blood, and ectoparasites may be collected. If the primates are not alive, then parasites and specimens could be recovered from various organs and tissues at necropsy. We need to be opportunistic with such endeavors as we cannot make the contributions with only feces that we can from recovery of parasites at necropsy. The eggs or worms found at necropsy have not been matched to the adult worm identifications in most cases as they have been with domestic animals species. Feces, therefore, is the only practical sample that can be collected and examined with reference to parasites in free-ranging primates for which capture is not an option. Some general references that might be useful for aiding in the detection and identification of primate parasites include those by Melvin & Brooke (1974), Anon. (1979), Zajac & Conboy (2006), and Garcia (2007). A flow chart of potential diagnostic procedures is depicted in Figure 1.1.

Collection of fecal and blood samples precautions

Because primate blood and feces are potential sources of organisms infectious to humans, personal safety of individuals collecting these samples in the field is imperative and can be achieved by following "universal precautions." "Universal precautions," as defined by the Centers for Disease Control (CDC), "are a set of precautions designed to prevent transmission of human immunodeficiency virus (HIV), hepatitis B virus (HBV), and other bloodborne pathogens when providing first aid or healthcare. Under universal precautions, blood and certain body fluids of all patients are considered potentially infectious for HIV, HBV and other bloodborne pathogens" (CDC website).

Basically, this means wearing gloves and protective clothing, such as disposable aprons, thorough hand washing, proper disposal of needles, scalpels, syringes and any items contaminated with blood, mucous, feces, and other body fluids in appropriate leak-proof and puncture-resistant containers for disposal. Refer to OSHA (Occupational Safety and Health Act) regulations for specifics on "universal precautions" and recommendations and procedures to follow in case of animal bites, contaminated needle sticks, or wounds or cuts becoming exposed to any of these fluids.

Some points to consider in maintaining sample integrity for optimal parasite detection are:

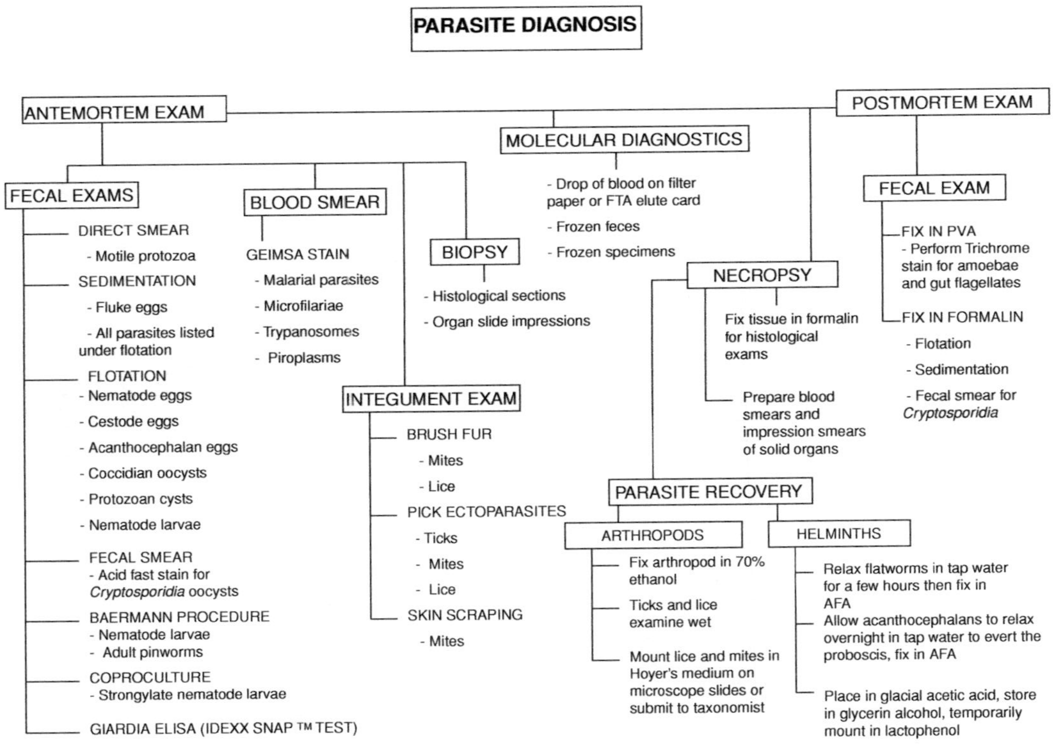

PARASITE DIAGNOSIS
ANTEMORTEM EXAM
POSTMORTEM EXAM
MOLECULAR DIAGNOSTICS
- Drop of blood on filter paper or FTA elute card
- Frozen feces
- Frozen specimens
FECAL EXAMS
DIRECT SMEAR
- Motile protozoa
SEDIMENTATION
- Fluke eggs
- All parasites listed under flotation
FLOTATION
- Nematode eggs
- Cestode eggs
- Acanthocephalan eggs
- Coccidian oocysts
- Protozoan cysts
- Nematode larvae
FECAL SMEAR
- Acid fast stain for *Cryptosporidia* oocysts
BAERMANN PROCEDURE
- Nematode larvae
- Adult pinworms
COPROCULTURE
- Strongylate nematode larvae
GIARDIA ELISA (IDEXX SNAP ™ TEST)
BLOOD SMEAR
GEIMSA STAIN
- Malarial parasites
- Microfilariae
- Trypanosomes
- Piroplasms
BIOPSY
- Histological sections
- Organ slide impressions
INTEGUMENT EXAM
BRUSH FUR
- Mites
- Lice
PICK ECTOPARASITES
- Ticks
- Mites
- Lice
SKIN SCRAPING
- Mites
NECROPSY
Fix tissue in formalin for histological exams
Prepare blood smears and impression smears of solid organs
PARASITE RECOVERY
ARTHROPODS
Fix arthropod in 70% ethanol
Ticks and lice examine wet
Mount lice and mites in Hoyer's medium on microscope slides or submit to taxonomist
HELMINTHS
Relax flatworms in tap water for a few hours then fix in AFA
Allow acanthocephalans to relax overnight in tap water to evert the proboscis, fix in AFA
Place in glacial acetic acid, store in glycerin alcohol, temporarily mount in lactophenol
FECAL EXAM
FIX IN PVA
- Perform Trichrome stain for amoebae and gut flagellates
FIX IN FORMALIN
- Flotation
- Sedimentation
- Fecal smear for *Cryptosporidia*

Figure 1.1. Flow chart of diagnostic procedures.

1. Avoid contamination of sample with water or urine "because water can be contaminated with free-living organisms that can be mistaken for (human) parasites" (Garcia, 2007).
2. Loss of motility of protozoa may occur if sample is contaminated by urine.
3. Time frames for examination or fixation (recommended by Garcia, 2007):
   a. Liquid feces: Examine or preserve within 30 minutes of passage (trophozoites).
   b. Soft feces: Examine or preserve within 1 hour of passage (trophozoites and cysts).
   c. Formed feces: Examine or preserve within 24 hours of passage (cysts).
4. If quantity allows for attempt at larval culture, do not refrigerate this portion of sample.

Fecal exams

Examination of feces is useful when attempting to gain information of the parasite fauna in a given primate. There are a number of procedures that could be used to detect the different groups of parasites. The number of exams that can be performed on one sample depends somewhat on the volume of feces obtained from each host. If at least 2 g of feces can be obtained, then the following procedures could be done: fecal flotation, fecal sedimentation, direct smear, a *trichrome* or *iron hematoxylin* stain for protozoa from fecal smears prepared after appropriate fixation, acid-fast stain for *Cryptosporidium* sp. on unfixed smears, and Baermann procedures for larval or tiny nematode recovery. How the samples are collected and fixed will depend on whether the examinations will be conducted in the field or sent away to a reference laboratory for processing or taken back into a laboratory setting. Because of the potential for the presence of infectious organisms, sample containers should be placed in plastic leak-proof bags before transport and the shipping container needs to be able to withstand the rigors of postal systems.

Solutions for fecal procedures

Saturated sodium nitrate

Nitrate of soda (commercial grade fertilizer) 5 lb
Hot tap water 1 gallon

OR
Sodium nitrate 400 g
Hot water 1000 ml

Stir ingredients in appropriate-sized container until dissolved. Specific gravity should be 1.20–1.25. The actual specific gravity should be measured with a hydrometer and recorded on the container when each batch of flotation solution is prepared.

Sheather's sugar

Granulated sugar 454 g (1 lb)
Tap water 355 ml (12 fluid oz)
Liquefied phenol crystals or formaldehyde 6.7 ml (a preservative and mold inhibitor or simply store in the refrigerator)

Dissolve sugar in hot water by stirring over a heat plate. After sugar is dissolved and the solution has cooled to room temperature, add liquefied phenol (or formaldehyde) or store in the refrigerator. The *specific gravity should be 1.27* and this should be checked with a hydrometer and recorded on the container when each batch of flotation medium is prepared.

Detergent solution for simple sedimentation

Add 5 ml of liquid dish detergent into 4 liters of tap water. Avoid formation of bubbles when in use.

Zinc sulfate for detection of* Giardia *sp. cysts

$ZnSO_4 \cdot 7H_2O$ 336 g
Distilled water 1000 ml

Mix and adjust solution to specific gravity of 1.18. While some parasitologists think they must use zinc sulfate in order to float cysts of this flagellate, sodium nitrate can be used with good results.

0.85% sodium chloride (normal saline)

NaCl 8.50 g
Distilled water 1000 ml

This is used only for direct smear preparations looking for motility of gut inhabiting protozoa.

Saturated sodium chloride

Water 1000 ml
NaCl table grade

Dissolve NaCl in water by stirring until there is a salt residue in the bottom of the container that will not go into solution.

Procedures

Fecal flotation

Simple sodium nitrate flotation: for recovery of nematode and tapeworm eggs, coccidian oocysts, mites, and larval nematodes.

Materials:

Saturated sodium nitrate or **Fecasol™ specific gravity 1.2
15 ml conical centrifuge tubes and tube rack to hold them vertically (or Fecalyzer™)
Wooden tongue depressors or applicator sticks
50 ml paper cups or small disposable plastic containers
Glass microscope slides and 22 × 22 mm coverslips
Paper towels
Compound microscope.

Procedure using 15 ml centrifuge tubes:

1. Conduct steps 2 through 6 over paper towels.
2. Place 1–3 g of feces into the 50 ml container.
3. Add 15–20 ml sodium nitrate solution and *mix well, thoroughly breaking up feces.*
4. Place one layer of cheesecloth or gauze over the top of the container. Pour and strain the mixture into a 15 ml conical centrifuge tube (Figure 1.2) until you have formed a slight positive meniscus at the top. Set cover slip on top. The liquid should just touch the coverslip, but not spill over. Many eggs that have already floated to the top will be lost if you overfill the tube and spillage occurs when the coverslip is placed on the tube.
5. Let stand vertically for a minimum of 10 minutes (Figure 1.3).
6. Gently lift coverslip straight up without losing any fluid, and place it on a glass slide.

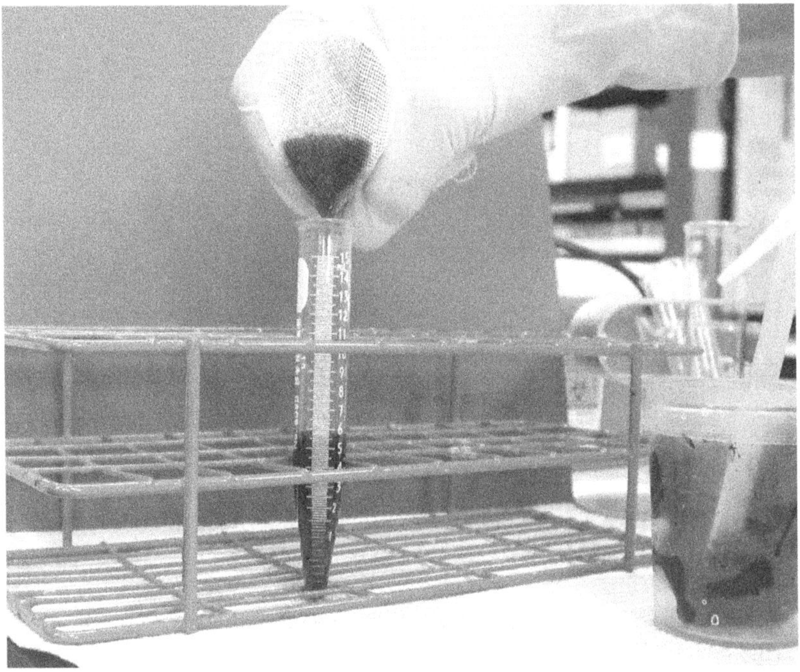

Figure 1.2. Setting up a flotation.

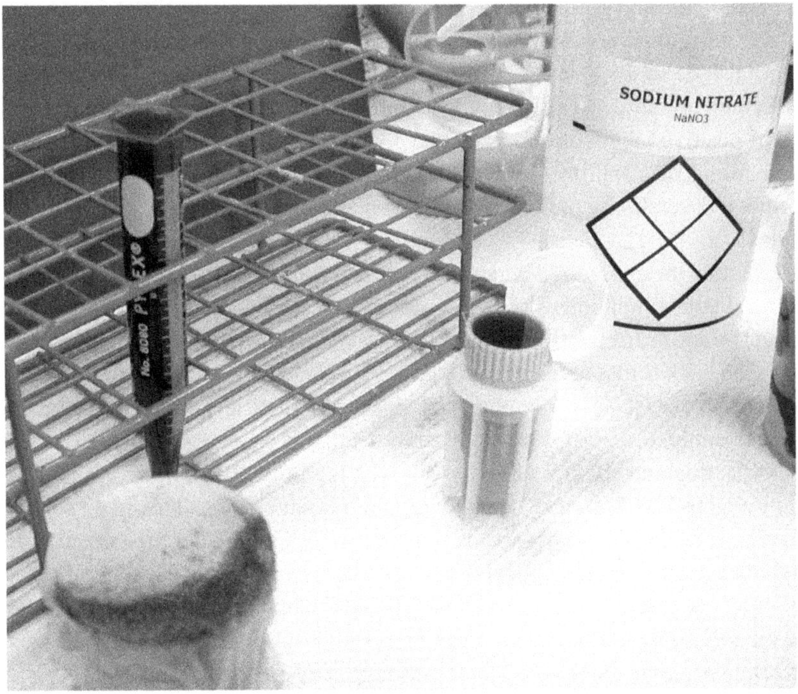

Figure 1.3. Flotation running.

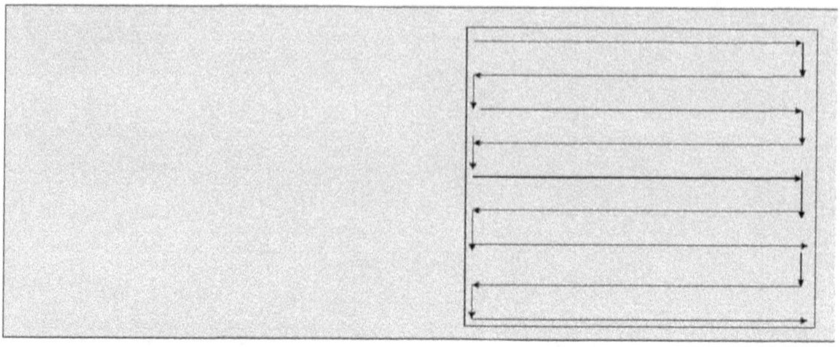

Figure 1.4. How to scan a slide systematically.

Procedure using Fecalyzer™ (strictly follow the directions that come with the Fecalyzer™)

The following is a basic summary of the procedure.

This method assumes you have at least 2–4 g of feces to work with or you will be picking up any freshly dropped fecal samples. Using the apparatus, pick up a portion of the fecal sample in the bottom of the white center portion of the apparatus, and place it in the outer container that is half full of Fecasol™ or sodium nitrate solution. Mix the sample well by turning and twisting the internal tube in the solution. Then press the white central portion down with sufficient pressure to seal the edges and fill the tube to the top forming a slight positive meniscus and place the coverslip on top being careful to avoid over filling and spilling. Let stand a minimum of 10 minutes (Figure 1.3). Remove coverslip by picking it straight up and place it on a glass microscope slide.

Examine the *entire area* under the coverslip using a low power (10×) objective and adjust light intensity to give good contrast. Begin at one corner of the coverslip and scan cross to the opposite edge, then move down one field of view and scan back across to the opposite edge. Continue this method until the entire area under the coverslip is observed (Figure 1.4). If any suspicious objects are encountered or you want to observe details of eggs, cysts, or oocysts for identification and confirmation, move to the high-dry 40× objective (increase light slightly) and observe details and take measurements. This higher magnification is required to observe the characteristics of smaller protozoan cysts and oocysts in order to confirm their presence. Do not allow the slide to stand too long because the salt will begin to crystallize on the edges of the coverslip making the observation of any eggs or cysts in those areas difficult and may affect an accurate diagnosis if their presence is overlooked.

Fecal sedimentation

This method is used primarily to recover trematode and acanthocephalan eggs, but other helminth eggs, nematode larvae, mites, protozoan cysts, and oocysts will sediment as well.

Materials:

Detergent solution, cheesecloth, small plastic beaker, plastic 50 ml conical centrifuge tubes, appropriate size tube rack to hold tubes vertically, pipettes, glass slides, and coverslips. (Optional: 5% methyl green or 0.1% methylene blue stains if available), compound microscope, dissecting microscope if available.

Procedure:

1. In a small beaker or cup, break up and mix 2–4 g of feces in a small amount of sedimentation solution and stir until uniform mixture is achieved (no large chunks). Continue adding detergent solution to an approximate volume of 50 ml avoiding the formation of bubbles.
2. Place one layer of cheesecloth over the beaker and pour mixture through cheesecloth into the 50 ml tube. If the tube is not full at this point, add more detergent solution to mixture by pouring it through the cheesecloth that is still covering the cup (the material trapped in the cheesecloth will be rinsed back into cup and possibly release more eggs into the solution), swirl to mix then continue by pouring through the cheesecloth from the cup into the tube until it is full or contains approximately 50 ml of filtered mixture.
3. Place the filled tube in rack and allow sedimentation to proceed for 5 minutes in a vertical position.
4. After a sedimentation time of 10 minutes, decant supernatant carefully and return to vertical position in one movement before any sediment reaches the lip of the tube *or* aspirate approximately 75% of the supernatant with a pipette to avoid mixing or losing the sediment at the bottom of the tube which contains the eggs.
5. Resuspend the sediment collected at the bottom of the tube by swirling or carefully tapping the bottom of the tube and refill tube with tap water by letting the water gently flow down the inside wall of the tube while avoiding the formation of bubbles.
   Note: Avoid formation of bubbles when adding water because eggs can be trapped in or adhere to detergent bubbles and will not descend during the next step and will be lost in the decanting process.
6. Allow to stand and sediment vertically for 5 minutes and decant as in step 4.

7. Repeat steps 5 and 6 until supernatant remains clear after 5 minutes standing time.
8. When supernatant is clear after standing 5 minutes, decant or pipette off supernatant down to remaining sediment.
9. *Optional step:* Add 1 drop of methyl green or methylene blue stain to sediment before next step. The dye will stain the background debris, but will not stain parasite eggs. The eggs will stand out with their natural color against the stained background, making them easier to detect.
10. Pipette a well mixed and unstained or stained drop of sediment onto a glass slide and apply coverslip. With the microscope set with the low power 10× objective, systematically examine entire area under coverslip (as in Figure 1.4) for eggs, larvae, etc. Do not overload the slide as eggs will be hidden in debris. To observe all contents of the sediment, making several slides will be required. If a dissecting microscope is available, then the sediment may be placed in a small Petri dish and systematically scanned. Again do not overload the dish as eggs will be lost amongst the debris.

Sequential flotation and sedimentation

If sample size is not adequate to perform several independent examinations, a sequential flotation/sedimentation may be employed. First perform the simple fecal flotation exam as described above. Decant about half of the solution from the flotation tube and then add sedimentation solution, treat it as in the sedimentation procedure just outlined as there might still be a lot of fine material in the sediment. To then observe for trematode eggs which do not float in flotation media, pipette an aliquot of the pellet (material that has sedimented to the bottom of the tube) being careful not to disturb or re-mix the material back into solution. Place one or two drops on slide, apply coverslip (take care not to make the preparation too thick) and observe for trematode ova using the systematic approach described in above procedures, making several slides until all sediment has been examined (if a dissecting microscope is available, larger aliquots, still in shallow layers, can be placed in Petri dishes and observed for eggs). Stained preparations make finding eggs at lower powers of magnification easier. Eggs found in this manner must then be collected out with a fine pipette and placed on a slide with coverslip to be measured and identified.

Direct smears

The direct smear is a routine exam used primarily to detect the presence of intestinal protozoa by observing their trophozoites or motile stages (although other helminth eggs and larvae may be detected in a direct smear, they would have to be present in very large numbers to be found with this method). It is not prudent to rely on a direct exam to diagnose the presence of helminths. Because the sample volume used for direct exam is minute compared with that used in flotation or sedimentation, the chances of finding those eggs present in low numbers is significantly reduced. Unlike the flotation and sedimentation procedures that are methods used to concentrate and cleanse eggs present in the sample, a direct smear is thus an inferior choice for detecting eggs.

Refer to "Collection of fecal and blood samples" section for recommended times for direct examination or preservation. Direct smear exams should be performed on a fresh, unfixed sample optimally within 30 minutes of passage because loss of motility can occur quickly with some protozoa.

Materials:

Normal saline (0.85% NaCl), applicator stick, glass slide, coverslip, newspaper, pipette, and compound microscope.

Procedure:

1. Prepare on or over newspaper as soon as possible after collecting feces.
2. Pipette 1 drop of normal saline onto a labeled glass slide.
3. With the tip of an applicator stick, pick up a very small amount of feces, blood-tinged mucous on feces, genital washing, oral cavity scraping or mucous. Mix with the saline on glass slide with a circular motion until the specimen covers an approximately 1 × 2 cm area.
4. Pick out any bits large enough to interfere with proper placement of the coverslip.
   Hint: You should be able to read newsprint through the preparation, if you cannot, it is too thick and you will not be able to observe motility through the debris or focus through the layers.
5. Apply coverslip and observe for motility by systematic examination of entire area under the coverslip (see Figure 1.4) using low magnification 10× objective. If you discover a parasite, move to the high dry 40× objective to observe details for identification. If no parasites are found at low power, scan again at 40× for small protozoa before calling the sample negative. Scan slide slowly because

the formation of pseudopodia in amoeba is very slow and easily overlooked.

Polyvinyl alcohol preservation

Protozoan cysts, oocysts, and trophozoites are fragile and will deteriorate and lose the morphological characteristics that facilitate their accurate identification. Therefore, it is important to fix or preserve a fresh sample as soon after collection as possible. The method of preservation and fixing feces in PVA (polyvinyl alcohol) is used when it is not convenient to perform a direct smear for protozoa and you wish to send samples to a laboratory for detection and identification of any motile protozoa present, especially amoebae and *Giardia* sp., by use of the Trichrome stain. There are two methods of preservation that can be employed depending on supplies on hand and convenience.

Materials:

PVA or Zn-PVA fixative, slides, applicator sticks, small unbreakable plastic vials or tubes with tight fitting *screw-on caps*. Zn-PVA is more environmentally friendly because it does not contain mercury as does regular PVA.

Method 1:

Thoroughly mix 1 part feces in 3 parts Zn-PVA in a small tube or container until a uniformly liquid mixture is achieved. Do not simply drop a pellet or formed ball of feces in fixative without mixing. Screw cap on tightly to prevent spillage of contents in transport. The receiving laboratory will prepare slides for staining upon receipt.

Method 2:

If you have slide boxes or carriers and wish to prepare your own slides on site, samples may be fixed and slides prepared simultaneously at site of collection. Place 3 drops of PVA or Zn-PVA fixative on a slide and mix in a small amount of feces picked up with the tip of a wooden applicator stick. When the sample is well mixed, smear it over 1/3 of the glass surface to obtain a *thin layer* across the slide. To facilitate this, you can hold the applicator stick horizontally across the slide and "roll" or push it across the length of the slide spreading the mixture evenly making a nice, confluent thin smear (Figure 1.5). This will also push large chunks of debris away that could make eventual coverslip placement and reading of the slide difficult. Allow the slides to dry undisturbed overnight at room temperature in a dust-free area to ensure thorough drying before staining.

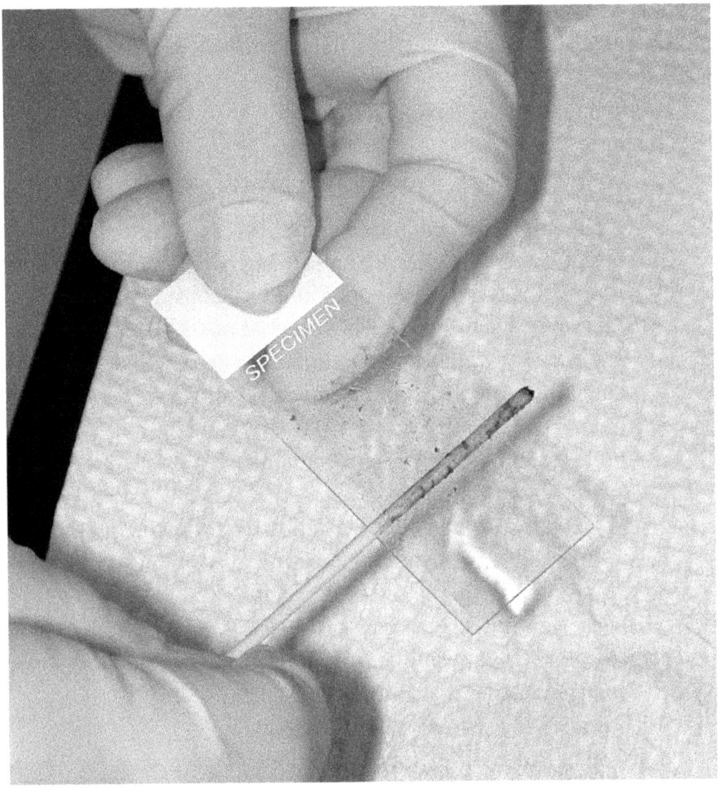

Figure 1.5. Making a PVC fecal smear.

Acid-fast stain

Cryptosporidium oocysts are much smaller than those of other coccidia, but they are similar in size and shape to other fecal components, especially some yeast. Therefore, to avoid confusion, differential staining techniques are more desirable than concentration methods such as Sheather's sugar flotation. Any acid-fast stain (such as Kinyoun's, and Ziehl-Neelsen available through Becton Dickinson/BBL) is the most favored method to achieve this purpose. The oocysts stain bright red, whereas yeast, bacteria, and other fecal debris only take up the counterstain.

This stain can also be used with various other samples when looking for the presence of acid-fast bacilli. (If acid-fast stain kits are unavailable, steps 1 and 2, preparation of smears, can be done for later submission to a laboratory.)

Materials:

Kinyoun's acid-fast stain*, glass slides, coverslips, applicator stick, staining rack.

Compound microscope equipped with a 100× oil immersion lens.

Procedure:

1. Both fresh and fixed (in 10% formalin) fecal samples can be processed for staining. If using fixed sample, drain the formalin off before picking up sample to prepare smear. Many laboratories recommend that samples be fixed because of biohazard considerations.
2. Pick up a small amount of feces (about 3–4 mm in diameter) with applicator stick and spread a thin layer across the slide removing large chunks as you push and roll the stick across the slide.
3. Let slide air dry.
4. Follow stain manufacturers recommended procedure.
   OR
   Good results can be achieved by the following method.
   Place slide on a staining rack, flood with Kinyoun Carbol Fuchsin stain and allow to stand for 5 minutes.
5. Rinse gently with a stream of water.
6. Decolorize with Kinyoun Decolorizer (acid alcohol) until no more stain appears in the washing.
7. Rinse gently with a stream of water.
8. Counterstain with Kinyoun's Brilliant Green stain for 30 seconds.
9. Rinse gently with a stream of water and allow to air dry.
10. Add 2 or 3 drops of microscope immersion oil to the smear and cover with coverslip and scan under high power 40× objective looking for small (4–7 microns), bright pink oocysts.

To measure and confirm a diagnosis of *Cryptosporidium*, viewing with the 100× oil objective (usually has a black ring around the lens tube) is necessary. Add a drop or two of immersion oil to the dry smear, add a coverslip, and add a drop of oil over coverslip and observe for bright pink oocysts measuring 4–7 microns (Figure 1.6).

Coproculture

This procedure is used to provide another look at the types of strongylate nematodes present in an animal. Coproculture allows the eggs found in the feces to develop into the infective L_3 larvae. This facilitates differentiation of nematodes with morphologically similar eggs.

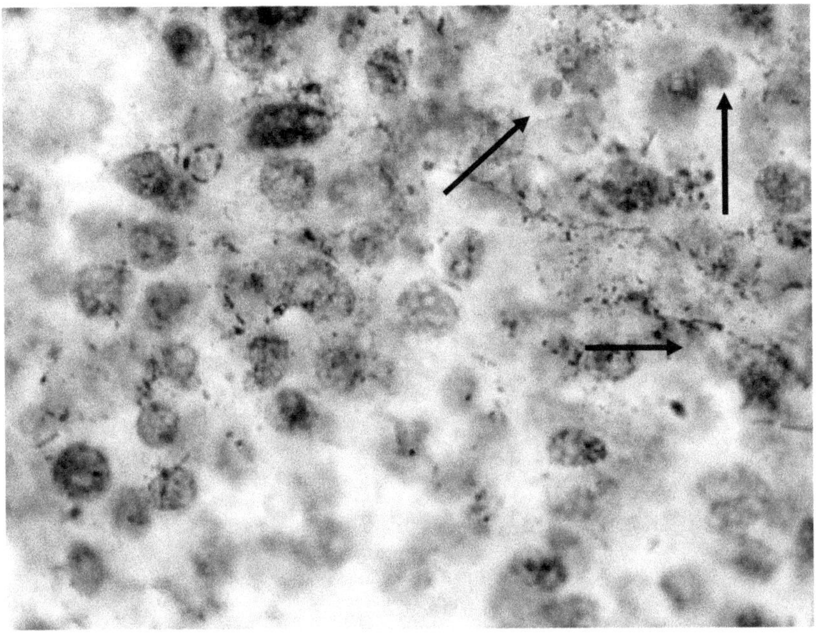

Figure 1.6. *Cryptosporidium* oocysts stained with Kinyoun's acid-fast procedure.

Materials:

Medium-sized container or jar with a screw-on lid, gauze or cheesecloth, string, vermiculite (used to help aerate soil for potting plants), compound microscope, slides, coverslips, dissecting microscope if available.

Procedure:

1. Mix 5 g of feces with equal parts of vermiculite and moisten the mixture.
2. The mixture is enclosed in two layers of gauze and the corners of the gauze are tied together with the string to make a ball of the mixture.
3. Water is placed in the jar and this gauze packet is suspended about 1 mm above the water by holding the string across the rim of the jar and screwing the lid over it. The string is caught between the lid and the container to hold the fecal ball in position at the surface of the water in the bottom of the container (Figure 1.7).
4. Place in a dark cabinet or drawer. Check daily for evaporation and add water to maintain the position of the ball close to the surface of the water as needed. The packet should remain moist. If it seems dry,

Figure 1.7. Coproculture incubating.

remoisten by gently dripping water over the packet. Do not let all of the water evaporate from the container.

5. After 7–10 days, open the container and check the water for the presence of motile nematode larvae. This is accomplished by pouring the water into a centrifuge tube and allowing it to stand for 10–15 minutes. The sediment may be transferred to a microscope slide, cover slip added and the entire coverslip scanned as in Figure 1.1. Or the fecal ball may be placed into a Baermann apparatus (see next procedure) and the larvae harvested in that manner. Or place the fluid in a Petri dish and use a dissecting microscope to systematically scan at 25×. If no larvae, place the fecal ball into a Baermann apparatus (see next procedure). Compare the larvae with those in the references listed at the end of this chapter.

An alternate procedure for larval cultures is the Harada–Mori filter paper culture technique (see Chapter 2, this volume by Hideo Hasegawa). This present procedure may offer a little more sensitivity than the Harada–Mori method

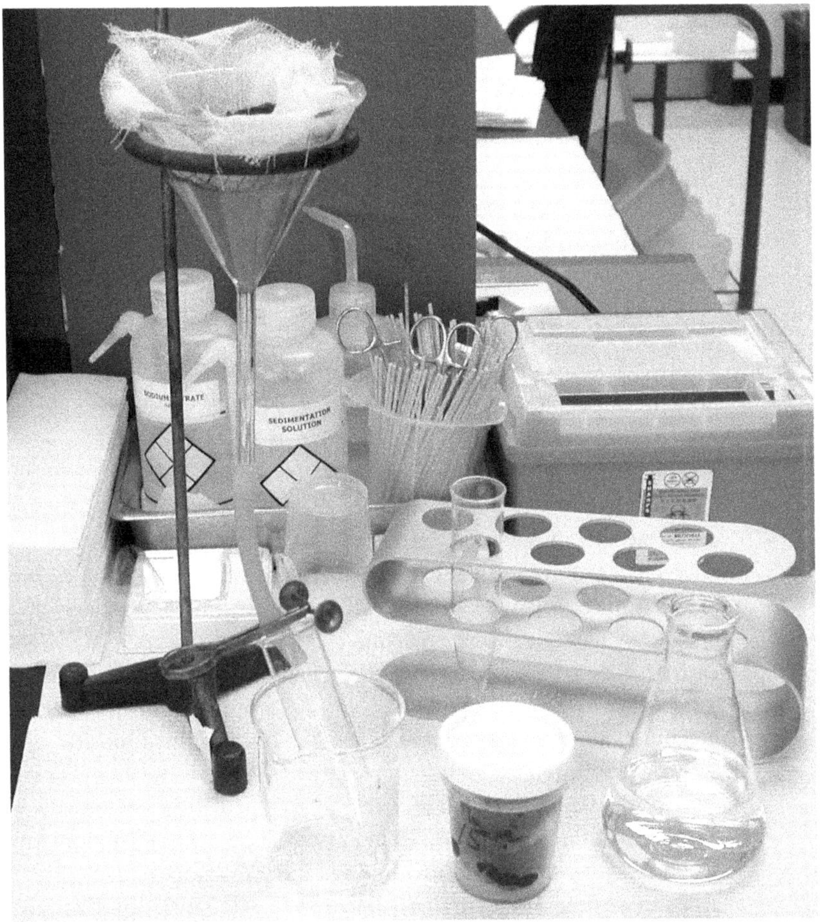

Figure 1.8. Baermann apparatus set up.

in that the larger sample size may increase the chances of larval recovery in infections with fewer eggs.

Baermann procedure

This procedure is used to recover living nematode larvae or tiny adult nematodes from feces, vomit, minced host tissue, and coproculture (Figure 1.8).

Materials:

Funnel with tight-fitting rubber tubing on stem, pipette tip, clamp on

Rubber tubing (as illustrated), funnel stand, wire mesh or tea strainer, gauze/cheesecloth.

Petri dish, centrifuge tube, centrifuge, Lugol's iodine, compound microscope and dissecting microscope (if available).

Procedure:

1. Firmly support the funnel in rack or stand.
2. Attach pinchcock as far distally on rubber hosing as possible and make certain that the tube is closed off completely at the distal end.
3. Fill the funnel reservoir with warm water (about 37–40 °C).
4. Release clamp to remove any trapped air; close clamp while water is still in stem of funnel or squeeze rubber tubing to force air out of stem while the stem is full of water.
5. Place approximately 10 g of feces (field collections would rarely provide this quantity so use as much as possible), fecal culture, or minced tissue on top of 3 layers of gauze. Wrap gauze around sample and place this into warm water in funnel reservoir.
6. Cover material with warm water by gently pouring additional water down the side of the funnel and not through sample.
7. Allow to stand overnight. Gaining experience with the nematodes you are collecting, you may find they will be down within a few hours and then you can modify your procedure.
8. Draw off 2–5 ml of sediment into small Petri dish and scan bottom of dish systematically with dissecting scope for parasites at 25× magnification. If negative, collect another 10–15 ml and transfer to a centrifuge tube; concentration of parasites is achieved by allowing the tube to stand upright for 10 minutes.
9. Examine sediment under dissecting microscope or compound microscope. For measuring and identifying larvae recovered, place 2 to 3 drops of sediment onto a glass slide, cover with coverslip. Examine with 10× objective to locate, switch to 40× objective to identify. If motile nematode larvae are thrashing back and forth, add a small drop of Lugol's iodine solution to edge of coverslip and this will diffuse under the coverslip, killing and fixing the larvae and staining them so the esophagus may be more clearly seen.

If you have an anesthetized primate in hand...

Blood smears

This is the most common method used in diagnosis of blood protozoa such as *Plasmodium* sp., *Hepatocystis* sp., *Trypanosoma* sp. piroplasms or microfilariae. Smears should be made from whole anti-coagulated blood collected

(a)

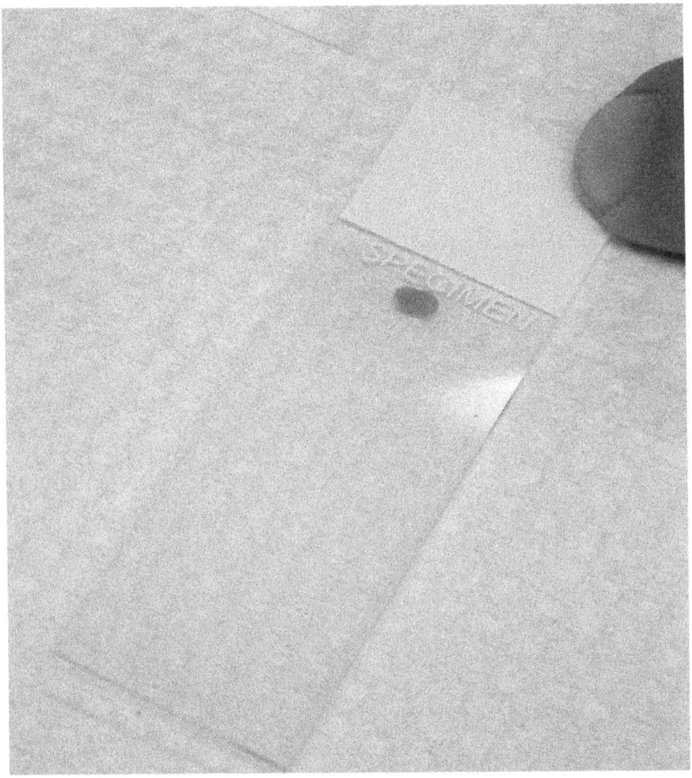

Figure 1.9. a, b, and c. Making a blood smear.

in EDTA or heparinized vacutainer tubes or a drop of fresh unclotted blood prepared *immediately* at collection. *This procedure requires some beforehand practice on spare or discarded blood samples as it requires good consistent technique and coordination to prepare useable blood films.*

Materials:

Giemsa stain stock solution, stain dishes or jars, graduated cylinder, absolute methyl alcohol, pipette, distilled water (buffered to pH 7.0), clean glass slides, coverslips, and compound microscope.

Procedure:

1. Place one drop of blood on one end of a slide (Figure 1.9a).
2. Place the end of a second slide so it backs up and just touches the drop of blood facing the opposite end from which it was placed, and held at an angle of 20–40 degrees (Figure 1.9b). Keeping both slides

(b)

Figure 1.9. (*cont.*)

in contact throughout the process, touch the drop with the "spreader slide" and allow the blood drop to spread across the width of the edge of the spreader slide (Figure 1.9b), then push the spreader slide in one smooth movement with light pressure across the other slide towards the opposite end drawing the blood out into a thin film (Figure 1.9c).

3. Air dry the film by waving it back and forth. Do not blow on it.
4. Fill a staining dish or jar with absolute methyl alcohol. Immerse smear for 3 minutes to fix it.
5. Air dry slide. If you can stain the smears in the field, do so. If not, the slides should be kept dry in a closed container.
6. Fill second staining dish or jar with working solution of stain prepared by manufacturer's directions (usually 2 ml stock solution to 48 ml buffered water).
7. Immerse slides in the stain solution for amount of time recommended by manufacturer (usually 20–30 minutes).

(c)

Figure 1.9. (*cont.*)

8. Rinse stained slides in tap water gently. Air dry.
9. Gently smear a thin coat of immersion oil across film as this will enhance resolution.
10. Systematically scan smear for at least 5 minutes with high dry 40× objective. Any parasitized cells should be examined more closely with 100× oil immersion lens.
11. If no parasitized cells are found after 5 minutes, change to oil immersion objective and scan systematically for 10 minutes more before calling sample negative.

Oil may be removed by adding a few drops of xylene on the slide and gently wiping dry. Another common stain is the Wright's stain, which is relatively inexpensive and easier to use. The stain is commercially available in liquid form (ready for use) and also as a powder (which must be dissolved in methyl alcohol, before use). The fixing solution, methyl alcohol, is combined with the stain, so you are not required to fix the blood smear, just add 10–20 drops of the stain onto the blood smear and let it stand for 1 minute, then add an equal volume of distilled water, let stain 2–5 minutes, wash in distilled water, air dry.

Scotch tape method for pinworm detection

Pinworm eggs are rarely shed in feces. Ideally this procedure is conducted early in the morning, as soon as the animal awakens or after rest, as the female pinworm migrates to the anus to deposit eggs while the animal is asleep. In wild caught animals, this can only be used on animals that have been anesthetized, as would be the case with blood collections.

Materials:

Glass slides, scotch tape (approximately the length of the slide), compound microscope.

Procedure:

1. Hold the folded piece of tape between thumb and forefinger with sticky side out and press against the exposed anus allowing any pinworm ova present to adhere to tape.
2. Place the tape, sticky side down, on the slide.
3. Scan slide for presence of pinworm ova under 10× objective, switching to 40× objective for confirmation. View is enhanced by adding water under the tape.

Collection and fixation of ectoparasites

Materials:

Small vials or tubes with screw caps, 70% ethyl alcohol, small strips of paper cut to fit inside the vials or tubes, pencil.

Procedure:

1. Collect as many ectoparasites as possible. Because identification depends largely on morphological characteristics of the head, mouthparts and number of legs, care should be taken when removing to keep the parasite as intact as possible.
2. Place in vials containing 70% ethyl alcohol. Use a pencil to label the strips of paper with the animal identification number, site in which it was found and the species of the primate. Place the strip in the appropriate vial with each type of parasite collected. (Each different

type of ectoparasite should be placed in a separate vial even if taken from the same primate.)

If worms can be recovered at necropsy or via biopsy

Fecal surveys are not the complete picture. Often the identification of the eggs is speculative. To be sure which strongylate nematodes are present, one needs to see the adults. The same may be said for the eggs of flukes and tapeworms. Thus we must be opportunistic and when a primate is killed by some accident or is found dead, necropsies need to be performed and the helminths present need to be matched with the eggs in the feces. It is important to help your odds for correct helminth identification, and the following information will assist you. Always keep the specimens wet-immersed in water or saline until in the proper fixative. Whenever possible, initially place worms from different organs into different labeled vials containing physiological saline (0.85% NaCl) before removing saline and fixing worms. Each label should indicate host species, accession or necropsy number, organ from which worm was removed, and collector's name. If there are many specimens, provide many specimens. Nematode infections may be comprised of multiple species and you are trying to provide both sexes of as many species as are present. Fluke infections also may be a mixture of species. If you provide a large number of specimens, the chances you provided specimens in good condition will increase. Also, depending upon the case, what your final objectives are and the species of the host will determine how detailed you might need the identification, i.e. to superfamily, genus, species. Do not request species identifications if they are not needed, as in exotic species, this can result in hours of work for the parasitologist.

Trematodes and cestodes should be placed in a dish containing either tap water or physiological saline and allowed to relax for 30–60 minutes. They should then be fixed in AFA (85 ml of 85% ethanol, 10 ml commercial formalin, 5 ml glacial acetic acid). Specimens may be stored and sent to the parasitologist in this solution. It is best to allow large tapeworms to flatten as much as possible and not pack them too tightly, as representative sections need to be mounted flat to see structures necessary for identification. Be sure you have included scoleces (tapeworm hold-fast organs) with tapeworms as these structures are often lost, resulting in non-identification in many cases. Flukes should be fixed as flat as possible and this can be accomplished by fixing specimens between glass slides with very light pressure applied to the upper slide. Being fixed flat is not "squashing" as in a "squash prep" of a mature cestode proglottid to observe for presence of eggs for tentative diagnosis.

Nematodes should be dipped in concentrated glacial acetic acid or hot 70% ethanol to fix them in as straight a posture as possible. After they have stopped writhing, transfer them into glycerin-alcohol (90 ml 70% ethanol, 10 ml glycerin). They may be stored indefinitely in this solution. They can be cleared for identification by adding glycerin to the dish containing the specimens and allowing the alcohol to evaporate leaving the worms in glycerin. A short-term, quick alternative is to clear roundworms in lactophenol (2 parts glycerin, 1 part distilled water, 1 part melted phenol crystals and 1 part liquid lactic acid). After identifying the specimens, remove the specimens from the lactophenol and return them to glycerin-alcohol.

The acanthocephalans or spinyheaded worms require special attention due to their unique attachment to the gut wall by the hold-fast organ. Most "acanths" cannot be simply pulled loose of the gut wall without destroying the main taxonomic structure – the proboscis. These worms should be either carefully dissected free of the host tissue or be presented to the parasitologist fixed while still attached to the gut wall. Sometimes "acanths" lose their turgidity and pass in the feces. They resemble wrinkled worms in this condition. If separated from host tissue, then place worms in tap water overnight to allow proboscis eversion, then fix in AFA or 70% ethanol.

If sufficient numbers of parasites are present, fix a representative sampling of these taxa in 95% ethanol as these could then be used more easily by molecular parasitologists.

Glossary of terms

Eggs per gram of feces (EPG)

A means of determining the level of environmental contamination by an infected host or a potential indication of reduction in numbers of worms left following drug administration. This measure does not equate across the board with the number of worms producing the eggs as there are a number of factors that influence the number of eggs being shed. These undoubtedly include age of host, age of worms, quality of host diet, host immune system function, and types of parasites present.

The following definitions are standard terms used in survey work and are from Bush *et al.* (1997).

Prevalence is the number of hosts infected/infested with a particular parasite divided by the number of hosts examined and is commonly multiplied by 100 and expressed as a percentage.

Incidence is the number of new hosts that become infected or infested with a particular parasite during a specific time interval divided by the number of uninfected/uninfested hosts at the start of the interval.

Intensity (of infection) is the number of individuals of a particular parasite in a single infected or infested host. This deals only with those hosts infected or infested, and does not include a measure of the uninfected hosts.

Abundance is the number of individuals of a parasite in or on a single host including those that are not infected or infested.

Density is the number of individuals of a particular parasite per some measured sampling unit taken from a host or habitat. This might be the number of trypanosomes per ml of blood.

References/Bibliography

Anderson, R. C. (2000). *Nematode Parasites of Vertebrates: Their Development and Transmission*, 2nd edn. New York, NY: CABI Publishing.

*Anderson, R. C., Chabaud, A. G. & Willmott, S. (1974). *CIH Keys to the Nematode Parasites of Vertebrates*, 1st edn. Farnham Royal, UK: Commonwealth Agricultural Bureau.

Anonymous (1979). *Manual of Veterinary Parasitological Laboratory Techniques*. London: Ministry of Agriculture, Fisheries and Food Technical Bulletin No. 18. Her Majesty's Stationery Office.

Bush, A. O., Lafferty, K. D., Lotz, J. M. & Shostak, A. W. (1997). Parasitology meets ecology on its own terms: Margolis *et al.* revisited. *Journal of Parasitology* 83, 575–583.

Garcia, L. S. (2007). *Diagnostic Medical Parasitology*, 5th edn. Washington, DC: American Society for Microbiology.

*Gibson, D. I., Jones, A. & Bray, R. A. (eds.) (2002). *Keys to the Trematoda, Vol. 1*. London: CAB International and The Natural History Museum.

*Jones, A., Gibson, D. I. & Bray, R. A. (eds.) (2005). *Keys to the Trematoda, Vol. 2*. London: CAB International and The Natural History Museum.

*Khalil, L. F., Jones, A. & Bray, R. A. (eds.) (1994). *CIH Key to the Cestode Parasites of Vertebrates*, 1st edn. Wallingford, UK: CAB International.

Melvin, D. M. & Brooke, M. M. (1974). *Laboratory Procedures for the Diagnosis of Intestinal Parasites*. Atlanta, GA: U.S. Department of Health Education and Welfare Publ. No. (CDC)75–8282.

*Price, R. D., Hellenthal, R. A., Palma, R. L., Johnson, J. P. & Clayton, D. H. (2003). *The Chewing Lice: World Checklist and Biological Overview.* Champaign, IL: Illinois Natural History Survey Special Publication 24.

*Schmidt, G. D. (1986). *Handbook of Tapeworm Identification*. Boca Raton, FL: CRC Press, Inc.

Zajac, A. M. & Conboy, G. A. (2006). *Veterinary Clinical Parasitology*, 7th edn. Ames, IA: Blackwell Publishing.

*References that will help with identification at least to genus.

2 Methods of collection and identification of minute nematodes from the feces of primates, with special application to coevolutionary study of pinworms

HIDEO HASEGAWA

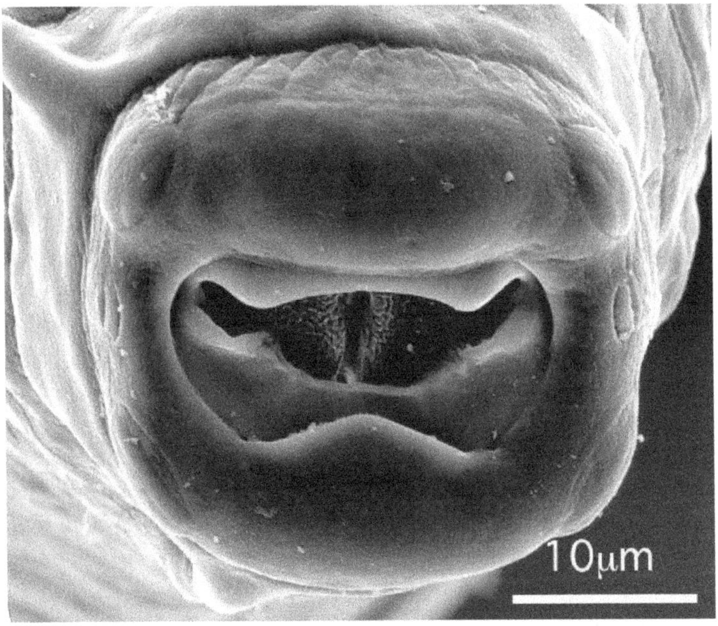

SEM by Hideo Hasegawa (*Trypanoxyuris (Buckleyenterobius) atelis*, host: *Ateles geoffroyi*)

Primate Parasite Ecology. The Dynamics and Study of Host–Parasite Relationships, ed. Michael A. Huffman and Colin A. Chapman. Published by Cambridge University Press.

Introduction

The improvement of methods used in the collection and identification of minute nematode species opens new avenues for field research into primate parasite ecology and evolution. In field research, fecal samples are often collected and used for diet analysis, as well as parasitological examination. When worms in the feces are visible to the naked eye, they may be picked up and fixed for later identification. However, with this method, only worms with relatively large size (e.g. *Oesophagostomum* spp. or *Streptopharagus* spp.) are collected, while minute worms such as male pinworms usually go unnoticed. Actually, two new *Enterobius* species recently reported from Asian monkeys were erected based on females only (Hasegawa *et al.*, 2002, 2003). Much more attention needs to be given to tiny nematodes at fecal examination. A simple method described herein may be effective to recover minute worms from fecal samples. Special remarks are also made with regards to observation methods of nematode species with particular reference to their significance in the study of coevolution of pinworms and their hosts.

The idea of coevolution of pinworms and their primate hosts was first presented by Cameron (1929). He found that one pinworm species restricts itself to one genus of host, and suggested that the speciation of pinworm lags behind the host (delayed cospeciation). More than a half-century later, his empirical insight was tested by objective analyses. Brooks & Glen (1982) first applied cladistic analysis on 13 pinworm species of primates and found that the phylogenetic trees of pinworms and host primates have the same gross topology. More recently, Hugot (1999) also applied a cladistic analysis on 45 species of primate pinworms and compared the phylogenetic tree reconstructed with that of the host primates. His results supported the Cameron hypothesis generally, but also suggested host switching and speciation of pinworm without speciation of host (duplication). These cladistic analyses were based on morphological characteristics, some of which have not been known adequately in many species. Today, combination of detailed morphological observation and DNA analysis is essentially necessary to understand the coevolutionary history of pinworms further.

DNA sequencing analysis has become increasingly easy to perform, and use of DNA nucleotide bases as characters for phylogenetic analysis is now regarded to bring much more promising results. In this chapter I describe methods and some results from our recent efforts of collection, identification, and DNA analysis of pinworms and *Strongyloides* to illustrate several methods of value for research into primate parasite ecology.

Figure 2.1. "Gauze-washing" method to recover minute nematodes. The surgical gauze used for filtration of feces in formalin-ether concentration (A) is washed on a fine strainer with mesh aperture size of 100 μm (B), put the feces aside by sprinkling tap water (C), and the residues left on the strainer are transferred to a beaker by pouring water from the back of the strainer (D).

Recovery of parasitic worms from fecal material of primates

For tiny nematodes like pinworms, there are two recommended methods for collecting the worms present, depending particularly upon the condition and quantity of the fecal samples. For fresh feces, the Baermann procedure is advantageous to concentrate living nematodes effectively (see Greiner & McIntosh, Chapter 1 in this volume for details of the procedure). However, it is often difficult to apply the Baermann method in field conditions, and only a small amount of feces are fixed in 10% formalin for later examination by a qualitative method such as the formalin-ether sedimentation at the laboratory.

For the sedimentation procedure, fixed feces are strained with one layer of surgical gauze. The used gauze is then always discarded. However, small nematodes may effectively be trapped by fibers of the gauze. Such worms can be recovered by washing the gauze with running tap water over a fine strainer (mesh aperture size 0.1 mm or less) (Figure 2.1). The residues left on the strainer are transferred to a Petri dish and observed under a stereomicroscope. In our experience of working with chimpanzee fecal material collected from Rubondo Island in Lake Victoria, Tanzania, 25 out of 179 fecal samples, each containing only several grams of feces fixed in formalin, were found to be pinworm-positive (Hasegawa *et al.*, 2005 and unpublished observation). Of these, only nine samples were found to be egg-positive using formalin-ether sedimentation, while 19 were found worm-positive by using the above-described "gauze-washing."

Modified Harada–Mori fecal culture

Fecal culture is advantageous for the detection of Strongylida nematodes and *Strongyloides* spp. The egg morphology of members of Strongylida closely resemble each other, making genus level identification difficult. Genus and sometimes species identification of Strongylida nematodes is made easier by using the morphology of filariform larvae rather than the more common use of eggs. If free-living adults of *Strongyloides* can be recovered, species identification is possible. Several methods have been developed for fecal culture, and the modified Harada–Mori's filter paper culture may be the most convenient and suitable for most field surveys.

Before departure to the field, the following material should be prepared for fecal cultures (Figure 2.2). (1) Polyethylene tube (approx. 180 × 25 mm): These home-made tubes can be fabricated from polyethylene sheets or bags using an electric heat sealer. Using relatively thick polyethylene sheets (e.g. thickness of 0.05 mm) is recommended. A thinner polyethylene sheet can be used, but sometimes perforation may occur causing leakage and loss of culture liquid. If possible, the end should be made obliquely to make later observation easier. (2) Filter paper (Whatman #1: approx. 140 × 40 mm) and wooden applicators: The applicator is used to spread feces onto the filter paper. Wood or bamboo sticks such as disposable chopsticks or spits are suitable. However, you can even find appropriate applicators from natural materials in the field. (3) Water: Tap water or drinking water is usable. If only well water or river water is available, the water should be boiled before use to kill free-living nematodes in the water. (4) Magnifying glass: More than 10× magnification is needed. The ocular glass (10×) of a microscope can also be used as a magnifying glass. (5) Pipette: Disposable plastic pipettes are convenient. Fine glass pipettes such as the Pasteur pipette are fragile and not suitable for field surveys. (6) Fixative: 5% formalin is recommended for morphological examination. 95% ethanol is used for samples undergoing DNA analysis. Heat-fixation or Lugol's iodine

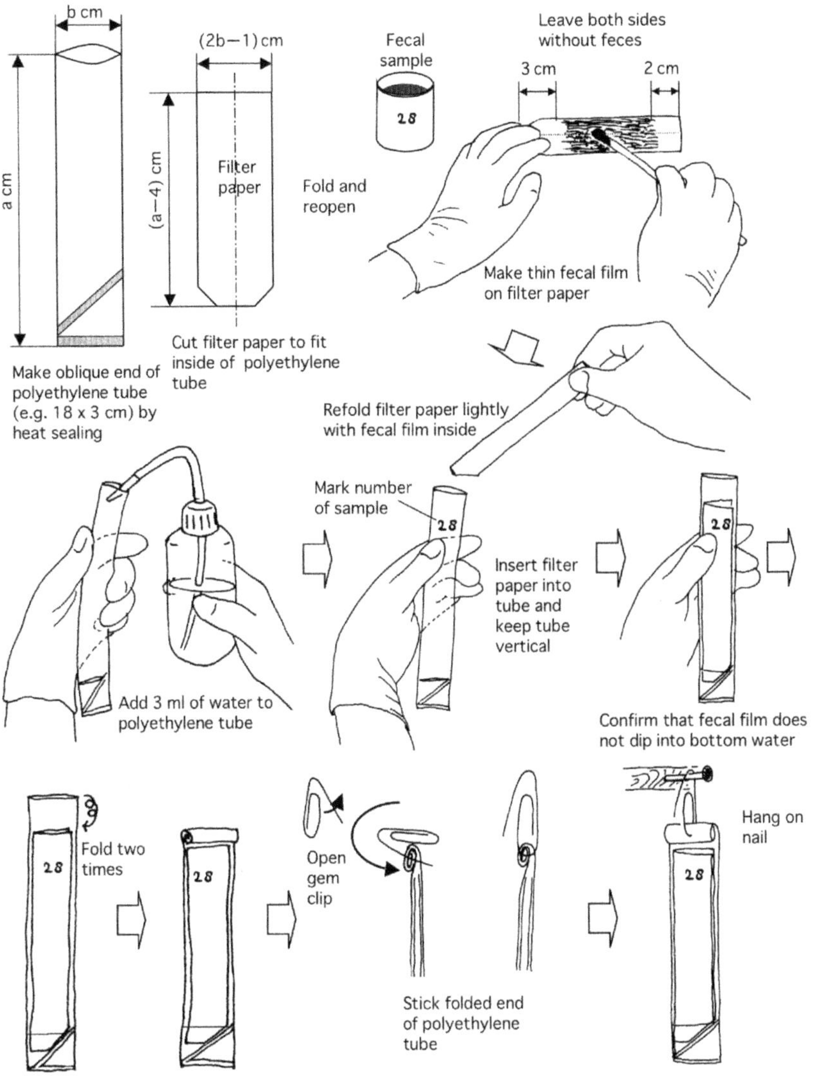

Figure 2.2. Modified Harada–Mori fecal culture.

solution staining are suitable methods for temporary sample preparation if microscopic examination on-site is possible. (7) Scissors, forceps, disposable grooves, paper sheet such as used newspaper. Gem paper clip is useful for hanging the culture tube.

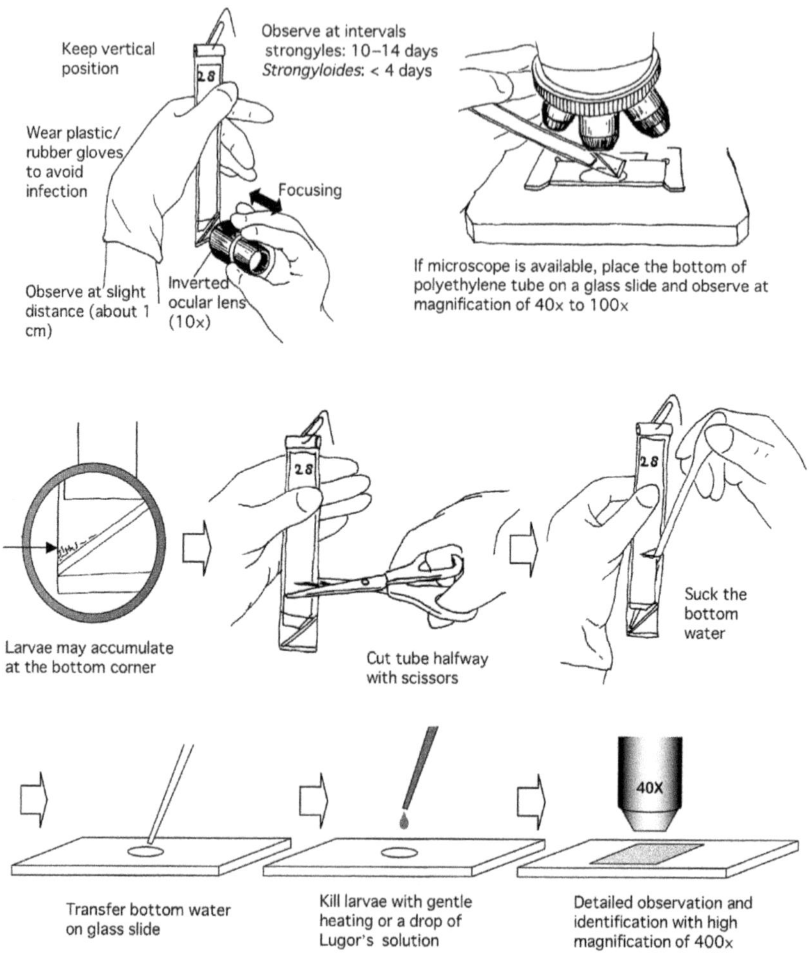

Figure 2.2. (*cont.*)

Procedure

Fold the filter paper along its longitudinal axis, and then reopen. Spread feces thinly on the inner side of the filter paper except for about 2–3 cm on both ends. Too thick a fecal layer may cause contamination of water with bacteria and fungi. Keep the polyethylene tube straight vertically and add water to about 2 cm depth. After this time, keep the tube vertically positioned throughout the culture, examination, and handling process. Insert the lightly refolded filter paper into the polyethylene tube. Confirm that the lower end of the filter paper is immersed in the water, but do not allow the fecal film to make contact with the water. Turn down the top of the tube twice, and hang it on an appropriately improvised rack device such as a row of nails on board. Place the culture in the dark avoiding sunlight. If the water level evaporates to below the end of the filter paper, carefully add water into the system without contacting the fecal smear.

Check the water at the bottom of the tube at regular intervals. If eggs are present in the feces, the hatching larvae should begin to migrate and appear in the water. The incubation time depends on the nematode species and temperature. *Strongyloides* spp. may appear in the bottom layer of water within 4 days of culture. Filariform larvae of Strongylida nematodes may be found in water about 7 days to 2 weeks later. Filariform larvae of *Strongyloides* may swim in the water actively, whereas larvae of Strongylida nematodes remain at the bottom and slightly wave about. Keep the tube positioned vertically and observe the bottom with a magnification glass. If the ocular lens of a microscope is used, look from the reversed side of the lens.

To collect the larvae, cut the tube slightly above the water level, then insert a pipette to aspirate the bottom layer of water. If microscopic examination is made on-site, place this water on a glass slide, place a glass coverslip on top, heat gently with a cigarette lighter or add a drop of Lugor's iodine solution to kill the larvae. If examination cannot be made on-site, fix the larvae with 10% formalin and store in appropriate vials for later examination. Fixation with ethanol with a concentration of over 95% ethanol is preferred for DNA analysis of the larvae.

The used polyethylene tubes, filter papers, and fecal applicators should be burnt to kill all pathogens. Scissors or any tools touched by the filter paper or bottom water can be sterilized with heat or ethanol. Wear disposable plastic or rubber gloves to prevent infection.

Identification of nematodes appearing in the culture needs practical experience. Presence or absence of sheath on body, shape of pharyngeal spear, esophagus, lumen of intestine, and tail are key characteristics for identification. Useful pictorial keys for genera of filariform larvae of zoonotic nematodes found in the culture are provided by Little (1981), and are reproduced in some textbooks (e.g. Beaver *et al.*, 1984). Caution should be paid to distinguish parasitic nematode larva from free-living nematodes, which may often contaminate feces that are not fresh enough. Free-living nematodes usually appear with various developmental stages, and often with an adult stage, whereas only *Strongyloides* species of parasitic nematodes develop to adults in the culture.

Morphological observation of nematodes

Nematodes collected from the feces should be cleared (made transparent) before observation. Glycerol-ethanol solution and lactophenol solution are usually used for this purpose. However, the latter may damage smaller sized worms such as pinworms. Transfer the worms into a small Petri dish containing ethanol-glycerol solution and leave the dish without cover. Within a few days, ethanol evaporates leaving the cleared worm in glycerol. Then, place the worm into a drop of 50% glycerol solution on a glass slide and put a coverslip over it. A fine needle used to mount insects on a board or a small piece of nylon fishing line of appropriate diameter placed beside the worms may prevent flattening of the worm by the pressure of the coverslip. Some researchers place a fixed worm directly on a glass slide and mount with glycerol. However, such a procedure may cause irreversible shrinkage of the worm obscuring essential structures.

En face examination of the head is essential in nematode systematics. The simplest technique is as follows. Place the cleared worm on a glass slide, and remove excess glycerol from around the worm carefully. Presence of excess fluid may prevent accurate cutting. Under a stereomicroscope, decapitate the worm with a razor blade. Then, add a drop of 50% glycerol solution on the worm and put a coverslip over it. After confirming the decapitated head under the stereomicroscope, observe it under a light microscope. If the direction of the head is not right, gently move the coverslip rim with fine forceps to rotate the head to give right en face view. Various modifications of the en-face technique have been proposed (cf. Hasegawa *et al.*, 2004).

Observation using an oil-immersion lens focusing at many planes through the head is necessary for understanding the three-dimensional structure of minute nematodes such as pinworms (Figure 2.3A–E). Scanning electron microscopy (SEM) observation is suitable for understanding surface topography, especially cephalic and caudal structures of nematodes. However, SEM observation often gives an image different from that of light microscopy because with the latter, inner structures are seen in addition to surface structures. The dehydration procedure in processing the material for SEM observation also causes shrinkage, giving a distorted image. Hence, it is essential to observe the material by light microscopy even if SEM observation is applied (Figure 2.3E–H).

For morphological identification of nematodes, CIH keys to the nematode parasites of vertebrates (Anderson *et al.*, 1974–1983) are useful. Anderson (2000) is also indispensable to understand life histories of nematode parasites. The overall review of primate parasites by Yamashita (1963) is still useful for study.

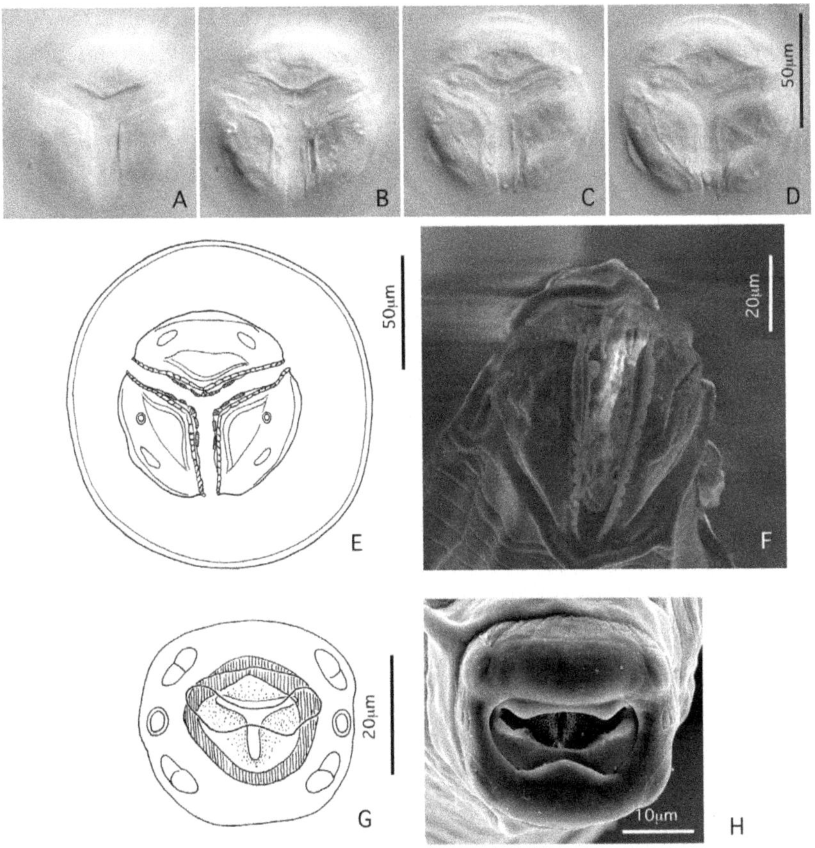

Figure 2.3. Light microscopic and SEM observation of cephalic end of pinworms. A-F. Female of *Enterobius (Colobenterobius) serratus* (host: *Nasalis larvatus*), G-H. Male of *Trypanoxyuris (Buckleyenterobius) atelis* (host: *Ateles geoffroyi*). Observation of different planes of the head (A–D) is necessary to understand the three-dimensional structure (E). Scanning electron microscopy (SEM) often gives a different image due to dehydration process (F, H). E and G: Hasegawa *et al.* (2002, 2004), reproduced with permission of the Helminthological Society of Washington.

Pinworms of primates and their significance in the study of parasite–primates coevolution

Pinworms parasitic in primates are divided into three large groups, which correspond to three genera in the subfamily Enterobiinae, *Enterobius*, *Trypanoxyuris,* and *Lemuricola.* Members of *Enterobius* are parasites of catarrhine primates, namely Cercopithecidae and Hominidae of the Old World; *Trypanoxyuris* pinworms are parasites of platyrrhine primates, Cebidae and Callithricidae, of the

New World; *Lemuricola* is known from Cheirogaleidae, Lemuridae, Daubentoniidae, Indridae, and Lorisidae (cf. Hugot *et al.*, 1996; Hugot, 1999). A useful key to the genera of Enterobiinae is proposed by Hugot *et al.* (1996).

Cephalic morphology

The cephalic structure of pinworms is very important for identification. In some species, cephalic appearance is different by sex. Four cephalic papillae and two amphidial pores are invariably observed (Figure 2.4A–J). Inner labial papillae are hardly discernible. There are various shapes of the mouth. Principally, the mouth is surrounded by three lips, forming a triradiate appearance. In some pinworms in New World monkeys, the mouth is elongated laterally and surrounded by dorsal and ventral lips (Figure 2.3G, H; Figure 2.4F). In most species the head is bilaterally symmetrical, but a few species show asymmetrical structure (Figure 2.4D, E).

Lateral alae

Sexual dimorphism is often observed. In general, males have single-crested lateral alae, while females possess double lateral alae or double-crested lateral alae. In females of many species the cervical region has single lateral alae, which are discontinuous becoming double-crested lateral alae in the more posterior body (Figure 2.5A, B, G). In females of some species such as *Enterobius (Enterobius) vermicularis* and *Trypanoxyuris (Buckleyenterobius) atelis*, the lateral alae are single-crested throughout (Figure 2.5E, F, H). Commencing and ending levels of the lateral alae are also characteristic to each species.

Male caudal structure

Precloacal ventral thickenings are prominent in pinworms parasitic in prosimians, such as *Lemuricola* spp. Similar thickenings called mamelons are present in pinworms of Syphaciinae parasitic in rodents. In pinworms of Old World monkeys and apes, these ornamentations are not clearly manifested.

Number of papillae and position of phasmidial pores: It has been claimed that the number of papillae pairs in primate pinworms is four (Hugot *et al.*, 1996). However, some researchers have described five pairs of genital papillae in the pinworms of prosimians (cf. Dollfus & Chabaud, 1955; Chabaud *et al.*, 1965). We also observed five pairs of papillae in *Lemuricola* of Madagascar

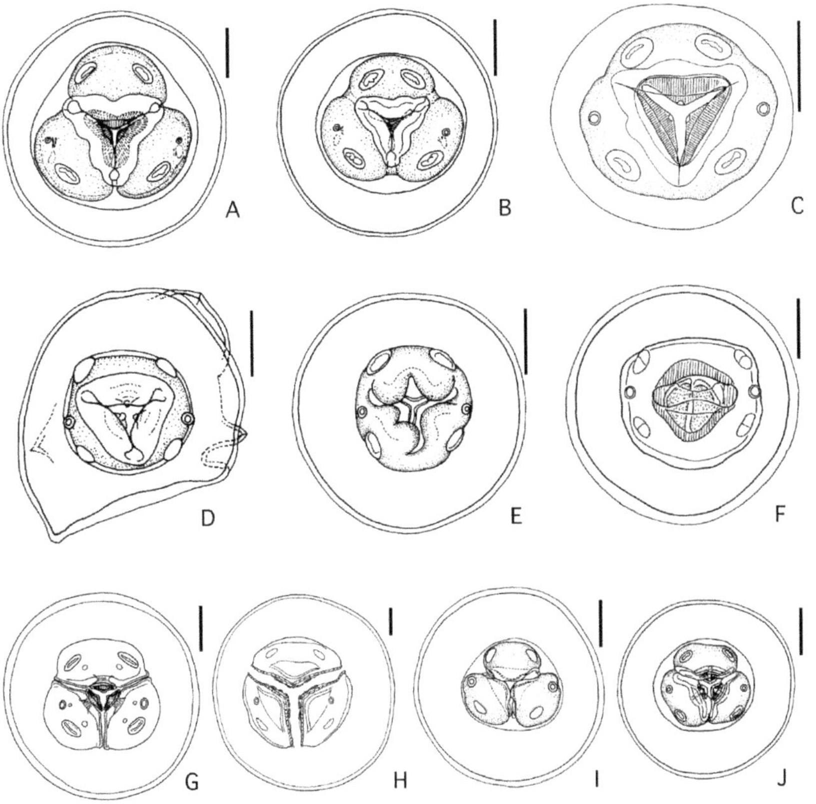

Figure 2.4. En-face views of cephalic ends of female pinworms parasitic in prosimians (A–C), New World monkeys (D–F) and Old World monkeys and apes (G–J) (scale bar: 20 μm). A. *Lemuricola (Madoxyuris) vauceli* (host: *Eulemur fulvus*). B. *Lemuricola (Madoxyuris) bauchoti* (host: *Eulemur fulvus*). C. *Lemuricola (Protenterobius) nycticebi* (host: *Nycticebus coucang*). D. *Trypanoxyuris (Trypanoxyuris) microon* (host: *Aotus azarae*). E. *Trypanoxyuris (Trypanoxyuris) sceleratus* (host: *Saimiri sciureus*). F. *Trypanoxyuris (Buckleyenterobius) atelis* (host: *Ateles geoffroyi*). G. *Enterobius (Colobenterobius) pygatrichus* (host: *Pygathrix roxellana*). H. *Enterobius (Colobenterobius) serratus* (host: *Nasalis larvatus*). I. *Enterobius (Enterobius) anthropopitheci* (host: *Pan troglodytes*). J. *Enterobius (Enterobius) vermicularis* (host: *Pan troglodytes*). F, G, and H: Hasegawa *et al.* (2002, 2003, 2004) reproduced with permission from the Helminthological Society of Washington.

lemurs and *Trypanoxyuris* in monkeys of South America. The "fifth" papillae are located lateral or laterodorsal of the 2nd and 3rd papillae. More interestingly, the "fifth" papillae are thin, fiber-like, and connected to the cuticular surface in *Lemuricola (Madoxyuris) bauchoti* and protruded on surface in *L. (Madoxyuris) vauceli* (Figure 2.6A). Meanwhile, the 'fifth' papillae in *Trypanoxyuris*

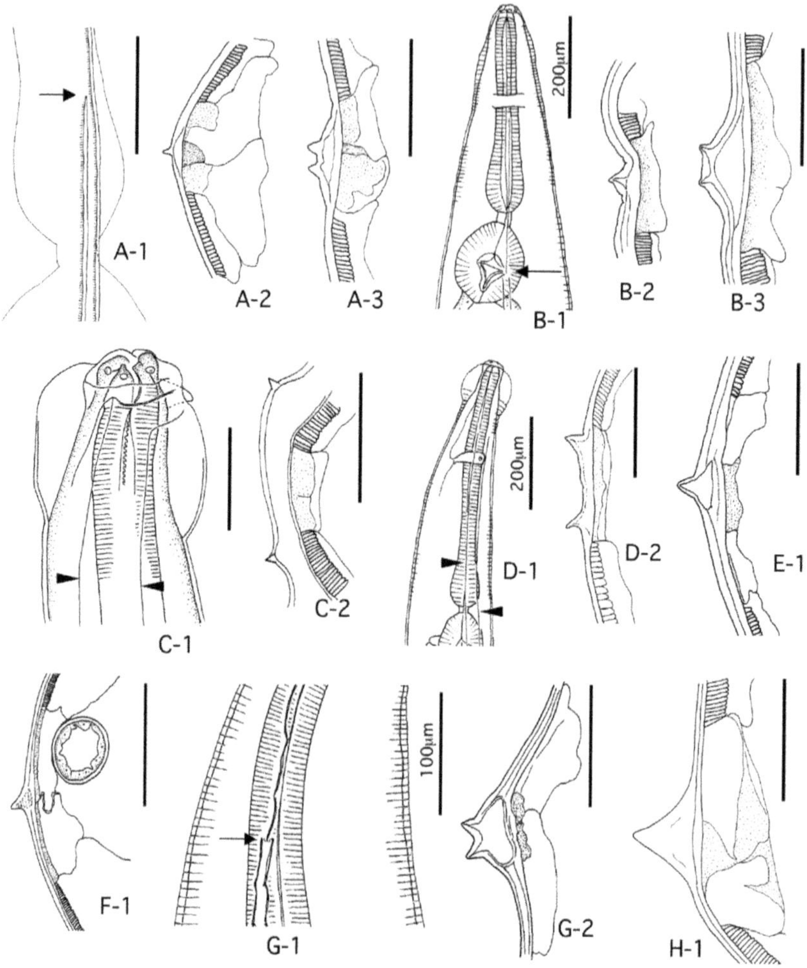

Figure 2.5. Lateral alae of female pinworms parasitic in prosimians (A, B), New World monkeys (C–E) and Old World monkeys and apes (F–H) (A-1, B-1, C-1, D-1, G-1 in lateral views; A-2 and B-2 in cross-sections through esophageal corpus; others in cross-sections of midbody. Scale bar: 50 μm unless otherwise specified). A. *Lemuricola (Madoxyuris) bauchoti* (host: *Eulemur fulvus*). B. *Lemuricola (Protenterobius) nycticebi* (host: *Nycticebus coucang*). C. *Trypanoxyuris (Trypanoxyuris) sceleratus* (host: *Saimiri sciureus*). D. *Trypanoxyuris (Trypanoxyuris) microon* (host: *Aotus azarae*). E. *Trypanoxyuris (Buckleyenterobius) atelis* (host: *Ateles geoffroyi*). F. *Enterobius (Colobenterobius) serratus* (host: *Nasalis larvatus*). G. *Enterobius (Enterobius) anthropopitheci* (host: *Pan troglodytes*). H. *Enterobius (Enterobius) vermicularis* (host: *Pan troglodytes*). Arrowheads and arrows in figures of lateral views indicate lateral alae and transition point from single-crested to double-crested alae, respectively. (E-1, F-1, G-1, G-2: Hasegawa *et al.* (2003, 2004, 2005), reproduced with permission from the Helminthological Society of Washington and American Society of Parasitologists.

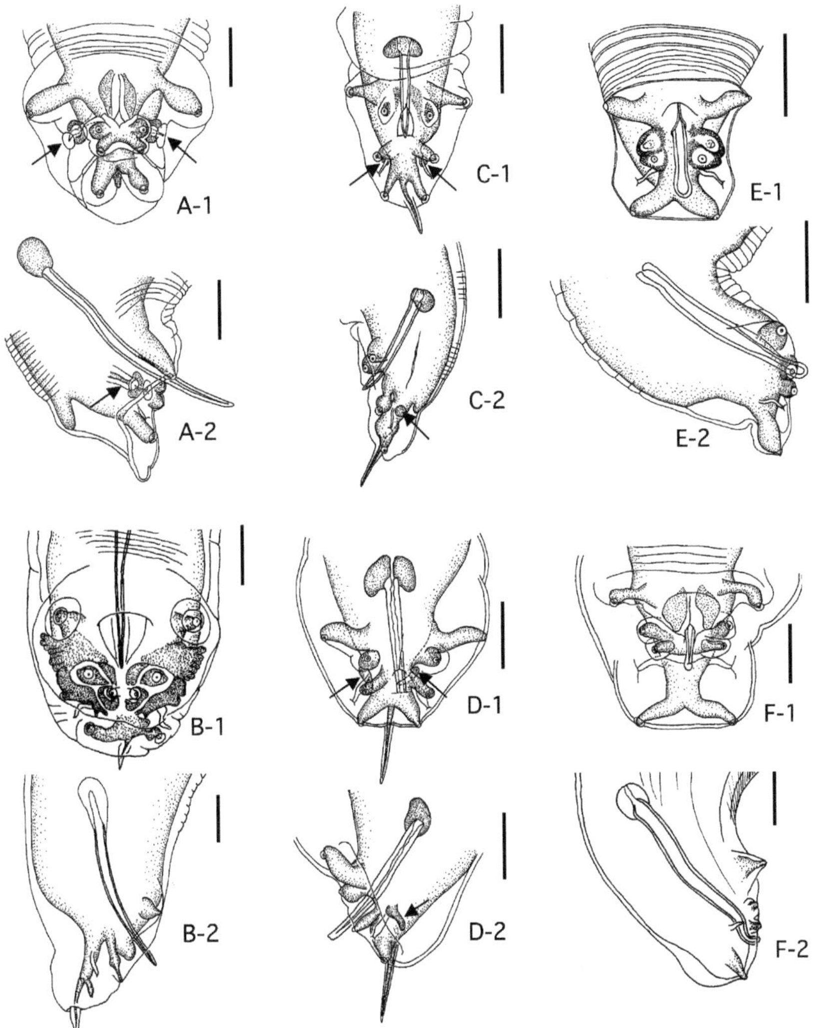

Figure 2.6. Ventral and lateral views of caudal end of male pinworms parasitic in prosimians (A, B), New World monkeys (C, D), and Old World monkeys and apes (E, F) (scale bar: 20 μm). A. *Lemuricola (Madoxyuris) bauchoti* (host: *Eulemur fulvus*). B. *Lemuricola (Protenterobius) nycticebi* (host: *Nycticebus coucang*). C. *Trypanoxyuris (Trypanoxyuris) sceleratus* (host: *Saimiri sciureus*). D. *Trypanoxyuris (Trypanoxyuris) microon* (host: *Aotus azarae*). E. *Enterobius (Enterobius) anthropopitheci* (host: *Pan troglodytes*). F. *Enterobius (Enterobius) vermicularis* (host: *Pan troglodytes*). Arrows indicate "fifth" papillae (see text). (E-1 and E-2: Hasegawa *et al.* (2005), reproduced with permission from the American Society of Parasitologists.

(Buckleyenterobius) atelis, T. (Trypanoxyuris) microon, T. (Trypanoxyuris) sceleratus, all parasitic in South American monkeys, end in a space beneath the cuticle and do not reach the cuticular surface (Hasegawa *et al.*, 2004) (Figure 2.6C, D). These "fifth" papillae are unclear in *Lemuricola (Protenterobius) nycticebi*, a pinworm of the slow loris in South-East Asia (Figure 2.6B). They are not seen in *Enterobius (Enterobius) bipapillatus, E. (E.) anthropopitheci*, and *E. (E.) vermicularis* (Hasegawa *et al.*, 2005) (Figure 2.6E, F).

Spicule and its appendix/formations

Spicule is a stable chitinized structure with shape characteristic to species (Figure 2.6A–F). The spicule of human pinworm, *E. (E.) vermicularis*, has a curved distal end, while that of chimpanzee pinworm, *E. (E.) anthropopitheci*, has a rather straight distal end (Figure 2.6E-2, F-2). The spicule of chimpanzee pinworm has a membranous formation on the ventral side, whereas such formation is not found in human pinworm. Spicules have additional structures especially in the proximal end. In human pinworm, the proximal formation develops during growth, changing its shape (Hasegawa *et al.*, 1998). Enough caution should be paid to judge that the shape of such appendices is a species-specific stable character.

The tail process is not observed in members of *Enterobius* in Old World primates, but is present in *Trypanoxyuris* of New World monkeys and some *Lemuricola* of prosimians. The perianal cuticular thickenings are often prominent. However, this area is also often difficult to delimitate clearly. It is also pointed out that the degree of sclerotization may vary according to growth stage (cf. Hasegawa *et al.*, 1998).

Female reproductive organs

Position of the vulva and direction of vagina are important features to be observed. The ovejecter is divided into three portions, thick-muscular vagina, vagina uterina lined with cuboidal epithelial cell, which is separated by a cellular wall (so-called diaphragm) from the third portion of which the wall structure is identical with that of the uterus. The relative position of the cellular wall is an important characteristic (Hugot *et al.*, 1996).

Egg morphology

The eggs of Enterobiinae are usually ellipsoidal, often asymmetrical with one side being more convex. In cross-section, the eggs are of more or less triangular

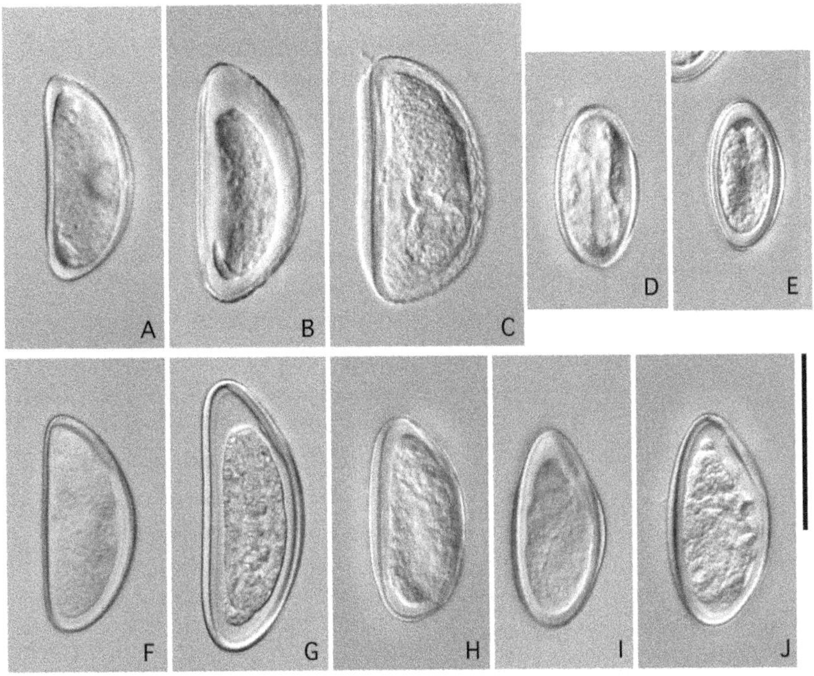

Figure 2.7. Egg morphology of pinworms parasitic in prosimians (A–C), New World monkeys (D, E), and Old World monkeys and apes (F–J) (scale bar: 50 μm). A. *Lemuricola (Madoxyuris) vauceli* (host: *Eulemur fulvus*). B. *Lemuricola (Madoxyuris) bauchoti* (host: *Eulemur fulvus*). C. *Lemuricola (Protenterobius) nycticebi* (host: *Nycticebus coucang*). D. *Trypanoxyuris (Trypanoxyuris) sceleratus* (host: *Saimiri sciureus*). E. *Trypanoxyuris (Buckleyenterobius) atelis* (host: *Ateles geoffroyi*). F. *Enterobius (Colobenterobius) pygatrichus* (host: *Pygathrix roxellana*). G. *Enterobius (Colobenterobius) serratus* (host: *Nasalis larvatus*). H. *Enterobius (Colobenterobius) bipapillatus* (host: *Miopithecus talapoin*). I. *Enterobius (Enterobius) anthropopitheci* (host: *Pan troglodytes*). J. *Enterobius (Enterobius) vermicularis* (host: *Pan troglodytes*).

appearance. The operculum is rudimentary or absent. The shell is often pitted at various extents. As shown in Figure 2.7, the eggs possess morphological features and size characteristic to species. However, species identification on only eggs may be often difficult.

DNA analysis of small nematodes

DNA analysis is possible when nematodes are fixed in ethanol. Formalin-fixed worms are not suitable for DNA extraction. DNA can and always should be

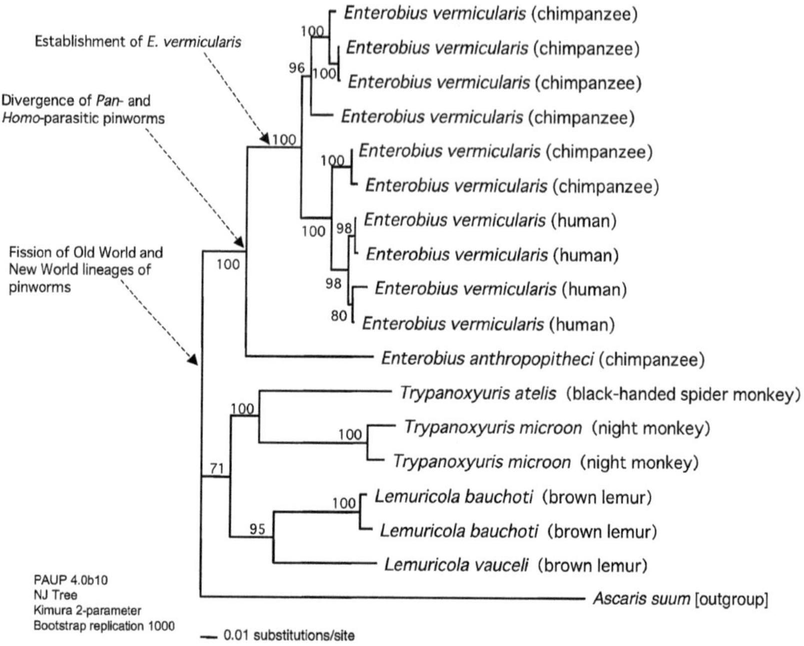

Figure 2.8. Phylogenetic tree of pinworms parasitic in primates reconstructed by neighbor-joining method based on 761 nucleotide sequence of cox1 DNA, and evolutionary events suggested by this tree. The values at the nodes represent bootstrap confidence level based on 1000 resampling.

amplified using only one individual of minute nematode stages such as the rhabditoid or filariform larvae of *Strongyloides*. DNA can be extracted by conventional phenol-chloroform extraction or commercially available extraction kits. Mitochondrial cytochrome C oxidase subunit 1 (cox1) and internal transcribed spacers (ITS) of nuclear ribosomal DNA (rDNA) are often analyzed because they usually give important information on evolutionary relationships and species identification. In general, mitochondrial genes diversify more rapidly than rDNA. Cox1 gene is useful for detecting "cryptic" species, while ITS regions are useful for "diagnosis" of species identification (Blouin, 2002). Nuclear 5S, 5.8S, 18S, or 28S rDNA are usually stable among congeners. However, it was recently found that a portion 18S rDNA forming a stem and loop is a promising marker for species identification of *Strongyloides* spp. (Hasegawa *et al.*, 2006).

It was recently found that a phylogenetic tree reconstructed from the cox1 gene of pinworm parasites was generally congruent with host phylogeny of Anthropoidea (Nakano *et al.*, 2006), but failed to make a significant tree when prosimian pinworms were included (Figure 2.8). Probably, the substitution of nucleotide bases has reached "saturation" during the period of primate evolution, obscuring real phylogeny. More stable regions, such as rDNA, are useful to examine the coevolution of pinworms and primates.

The phylogenetic analysis of DNA sequences of pinworms in extant primates may not only demonstrate the coevolutionary process between them but also suggest participation of extinct primates in the process. Reed *et al.* (2004) analyzed cox1 and cytb sequences of human louse, *Pediculus humanus*, and demonstrated the presence of two lineages, one of which was assigned to have evolved with *Homo erectus*, and secondarily host-switched to *Homo sapiens*. Nakano *et al.* (2006) found that the cox1 gene of the human pinworm, *Enterobius (Enterobius) vermicularis*, forms three clusters in their phylogenetic analysis. Because pinworm species of primates are often host genus-specific, such clusters might have been formed during evolution and dispersal of *Homo*.

The pinworms are very minute nematodes and may be often overlooked or neglected as less pathogenic or trifling parasites in field surveys. However, they possess very important clues for understanding evolutionary processes in primates. Host switching by the parasites might have occurred during ecological interactions between host primates (see Chapters 10 and 11, this volume for related discussions of host–parasite evolution and specificity in lice). It is expected that collaborations between ecologists, parasite taxonomists, and molecular systematists could open many new areas of fruitful research in the evolutionary biology of primates.

References

Anderson, R. C. (2000). *Nematode Parasites of Vertebrates. Their Development and Transmission*, 2nd edn. Wallingford, Oxon: CABI Publishing.

Anderson, R. C., Chabaud, A. G. & Willmott, S. (1974–1983). *CIH Keys to Nematode Parasites of Vertebrates*. Nos. 1–10. Farnham Royal, Buckinghamshire: Commonwealth Agricultural Bureaux.

Beaver, P. C., Jung, R. C. & Cupp, E. W. (1984). *Clinical Parasitology*, 9th edn. Philadelphia, PA: Lea & Febiger.

Blouin, M. S. (2002). Molecular prospecting for cryptic species of nematodes: mitochondrial DNA versus internal transcribed spacer. *International Journal for Parasitology*, **32**, 527–531.

Brooks, D. R. & Glen, D. R. (1982). Pinworms and primates: a case study in coevolution. *Proceedings of the Helminthological Society of Washington*, **49**, 76–85.

Cameron, T. W. M. (1929). The species of *Enterobius* Leach in primates. *Journal of Helminthology*, **7**, 161–182.

Chabaud, A. G., Brygoo, E. R. & Petter, A. J. (1965). Les nématodes parasites de lémurien malgaches. VI. Description de six espèces nouvelles et conclusions générales. *Annales de Parasitologie Humaine et Compareé*, **40**, 181–214.

Dollfus, R. P. & Chabaud, A. G. (1955). Cinq espèces de nématodes chez un atèle [*Ateles ater* (G. Cuvier 1823)] mort a la ménagerie du muséum. *Archives du Muséum National d'Histoire Naturelle, Paris*, **3**, 27–40.

Hasegawa, H., Takao, Y., Nakao, M. *et al.* (1998). Is *Enterobius gregorii* Hugot, 1983 (Nematoda: Oxyuridae) a distinct species? *Journal of Parasitology*, **84**, 131–134.

Hasegawa, H., Murata, K. & Asakawa, M. (2002). *Enterobius (Colobenterobius) pygatrichus* sp. n. (Nematoda: Oxyuridae) collected from a golden monkey *Pygathrix roxellana* (Milne-Edwards, 1870) (Primates: Cercopithecidae: Colobinae). *Comparative Parasitology*, **69**, 62–65.

Hasegawa, H., Matsuo, K. & Onuma, M. (2003). *Enterobius (Colobenterobius) serratus* sp. n. (Nematoda: Oxyuridae) collected from the proboscis monkey, *Nasalis larvatus* (Wurmb,1787) (Primates: Cercopithecidae: Colobinae), in Sarawak, Borneo, Malaysia. *Comparative Parasitology*, **70**, 128–131.

Hasegawa, H., Ikeda, Y., Diaz-Aquino, J. J. & Fukui, D. (2004). Redescription of two pinworms from the black-handed spider monkey, *Ateles geoffroyi*, with reestablishment of *Oxyuronema* and *Buckleuenterobius* (Nematoda: Oxyuroidea). *Comparative Parasitology*, **71**, 166–174.

Hasegawa, H., Ikeda, Y., Fujisaki, A. *et al.* (2005). Morphology of chimpanzee pinworms, *Enterobius (Enterobius) anthropopitheci* (Gedoelst, 1916) (Nematoda: Oxyuridae), collected from chimpanzees, *Pan troglodytes*, on Rubondo Island, Tanzania. *Journal of Parasitology*, **91**, 1314–1317.

Hasegawa, H., Hayashida, S., Ikeda, Y. & Sato, H. (2006). 18S rDNA sequence as a possible marker for species-specific diagnosis of strongyloidiasis. *Proceedings of the 75th Annual Meeting of the Japanese Society of Parasitology*, p. 92.

Hugot, J. P. (1999). Primates and their pinworm parasites: the Cameron hypothesis revisited. *Systematic Biology*, **48**, 523–546.

Hugot, J. P., Gardner, S. L. & Morand, S. (1996). The Enterobiinae subfam. nov. (Nematoda, Oxyurida) pinworm parasites of primates and rodents. *International Journal for Parasitology*, **26**, 147–159.

Little, M. D. (1981). Differentiation of nematode larvae in coprocultures. Guidelines for routine practice in medical laboratories. *W.H.O. Technical Report Series*, **666**, 144–150.

Nakano, T., Okamoto, M., Ikeda, Y. & Hasegawa, H. (2006). Mitochondrial cytochrome *c* oxidase subunit 1 gene and nuclear rDNA regions of *Enterobius vermicularis* parasitic in captive chimpanzees with special reference to its relationship with pinworms in humans. *Parasitology Research*, **100**, 51–57.

Reed, D. L., Smith, V. S., Rogers, A. R., Hammond, S. L. & Clayton, D. H. (2004). Molecular genetic analysis of human lice supports direct contact between modern and archaic humans. *Public Library of Science – Biology*, **2**, e340.

Yamashita, J. (1963). Ecological relationships between parasites and primates. I. Helminth parasites and primates. *Primates*, **4**, 1–96.

3 *The utility of molecular methods for elucidating primate–pathogen relationships – the* Oesophagostomum bifurcum *example*

ROBIN B. GASSER, JOHANNA M. DE GRUIJTER, AND ANTON M. POLDERMAN

SEM by Thushara Chandrasiri (*Oesophagostomum stephanostomum*, oral cavity view (scale bar 500 μm), host *Pan troglodytes schweinfurthii*)

Primate Parasite Ecology. The Dynamics and Study of Host–Parasite Relationships, ed. Michael A. Huffman and Colin A. Chapman. Published by Cambridge University Press.

Introduction

Investigating the transmission of particular pathogens of different primate species is of fundamental biological and ecological interest and also of major significance for the prevention and control of infectious diseases of humans. Various studies of non-human primates, employing coproscopic examination procedures, have reported parasitic nematode stages consistent with *Trichuris trichiura*, *Strongyloides fuelleborni*, and *Oesophagostomum* from humans (Muriuki *et al.*, 1998; Rothman & Bowman, 2003; Jones-Engel *et al.*, 2004; Legesse & Erko, 2004; Phillips *et al.*, 2004) and suggested that parasite transmission may be occurring between non-human primates and humans, thereby having human health implications. Such interpretations can be misleading because, although parasites are frequently identified and distinguished on the basis of morphological features, the host they infect, their pathological effect(s) on the host or/and their geographical origin, these criteria are frequently insufficient for specific identification and diagnosis (Gasser, 2006). The limitations of traditional approaches and advantages of molecular tools are illustrated in the present chapter, using *Oesophagostomum bifurcum* (Nematoda: Strongylida: Oesophagostominae) as an example.

Numerous studies have shown that the infection of humans with the nodule worm, *O. bifurcum* is common in northern Togo and Ghana (reviewed by Polderman *et al.*, 1999) (Figure 3.1). The infection commonly causes pathological effects (i.e. nodules in the intestine and surrounding tissues) in humans, resulting in serious clinical disease (= oesophagostomiasis) (Gigase *et al.*, 1987; Storey *et al.*, 2000). In spite of the human health importance of *O. bifurcum*, there have been significant gaps in the knowledge of its epidemiology and ecology (cf. Polderman & Blotkamp, 1995). It had been postulated that some species of non-human primates may act as reservoir hosts for human oesophagostomiasis (Stewart & Gasbarre, 1989). However, Blotkamp *et al.* (1993) indicated some morphological variability between the *O. bifurcum* adults from human and those from non-human primates, stimulating molecular investigations into the epidemiology and genetic make-up of *O. bifurcum* populations from these primate hosts. In particular, molecular methods, including polymerase chain reaction (PCR)-based mutation scanning, sequencing, and fingerprinting (reviewed by Gasser, 2006), have allowed the genetic characterization and/or specific identification of *O. bifurcum*, providing insights into the primate–parasite relationship. This chapter briefly describes the state of knowledge of *O. bifurcum*, its biology and epidemiology, the disease it causes, and the approaches employed for the diagnosis of infection. It illustrates, using the primate–*O. bifurcum* system as the example, how the application of molecular methods has enhanced our understanding of the epidemiology of

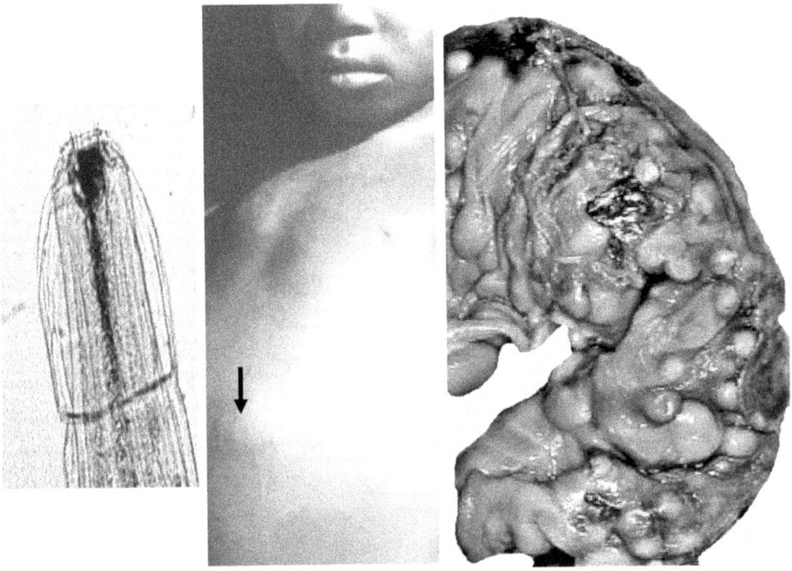

A B C

Figure 3.1. *Oesophagostomum bifurcum* (A – anterior part of adult worm) is endemic in the northern parts of Ghana and Togo in Africa and causes clinical oesophagostomiasis in humans (B – "Dapaong tumor"; arrowed) as a consequence of granulomatous nodules forming around encysted juvenile nematode stages in the intestinal wall and/or surrounding tissues. Nodules are visible in a section of large intestine removed surgically from a human patient (C).

the parasite and has led to the discovery of multiple, genetically distinct variants of *O. bifurcum*, inferred to have particular primate host affiliations and transmission.

Oesophagostomiasis and its relevance in humans and other primates

Species of *Oesophagostomum* are parasitic nematodes of the order Strongylida. They belong to the subfamily Oesophagostominae within the family Chabertiidae (Durette-Desset *et al.*, 1994). They are called "nodule worms," because (usually) their larvae cause small, nodular granulomata and/or abscesses mainly in the large intestines of the infected animals (characteristically in the submucosa or muscularis) and less frequently in ectopic sites (Orihel, 1970). The infection of humans with *Oesophagostomum* was first recognized by Brumpt (1902) in southern Ethiopia, and three years later, the description of this case was published (Railliet & Henry, 1905). The authors concluded that this parasite was a new species and called it *Oesophagostomum brumpti*. Subsequently, more cases of human oesophagostomiasis were reported, including infections with *O. apiostomum* (Nigeria) (Leiper, 1911), *O. stephanostomum* (East Africa) (Thomas, 1910), *O. stephanostomum var. thomasi* (Brazil) (Railliet & Henry, 1910), *O. aculeatum* (Indonesia) (Lie Kian Joe, 1949), and *O. bifurcum* (Ghana) (Haaf & van Soest, 1964). Some of the species infecting humans were considered as synonymous (Travassos & Vogelsang, 1932), and Chabaud & Larivière (1958) proposed, based on a morphological study, that three main species were responsible for human oesophagostomiasis, namely *O. stephanostomum* (e.g. in Brazil), *O. aculeatum* (in South-East Asia), and *O. bifurcum* (in Africa). These species were commonly found to infect non-human primates, considered to be their usual hosts. In 1987, Gigase and coworkers reported 54 cases of human oesophagostomiasis in northern Togo. This and another study in northern Ghana (Haaf & van Soest, 1964) stimulated Polderman and coworkers to conduct research into the occurrence and prevalence of human oesophagostomiasis in these countries, known also to be endemic for hookworm disease. Interestingly, the strongylid eggs detected in fecal samples from humans in northern Ghana and Togo were misidentified microscopically as "hookworm eggs," but were shown (by coproculture) to represent nematodes of the genus *Oesophagostomum*. In a morphological study, Blotkamp *et al.* (1993) proposed that *O. bifurcum* was involved in these countries. Thereafter, various epidemiological and clinical studies of *O. bifurcum* were undertaken (Krepel *et al.*, 1992a; Blotkamp *et al.*, 1993; Polderman & Blotkamp, 1995; Gasser *et al.*, 1999; Pit *et al.*, 1999b; Polderman *et al.*, 1999; Storey *et al.*, 2000, 2001a), and the efficacy of anthelmintic treatment against the parasite was investigated (Krepel *et al.*, 1993; Storey *et al*, 2001b; Ziem *et al.*, 2004).

Also the life cycle of *O. bifurcum* in humans received attention. *Oesophagostomum bifurcum* is a dioecious nematode. Each adult female produces ~5000 eggs per day (Krepel & Polderman, 1992b); the eggs are excreted in the feces and subsequently develop into the first-, second-, and third-stage larvae (L1, L2, L3, respectively) in the environment. The development from egg to L3 takes 4–7 days, depending on conditions. Importantly, the L3s of *Oesophagostomum* spp. are resistant to environmental stresses. Even after long periods of desiccation (several months) or freezing, some of them can remain viable for extended periods of time (Spindler, 1936; Rose & Small, 1980; Barger *et al.*, 1984; Polderman & Blotkamp, 1995; Pit *et al.*, 2000). When desiccated, the L3s shrink within their sheath, and such "dormant" L3s can "revive" after rehydration (cf. Polderman & Blotkamp, 1995). The infection of humans by *O. bifurcum* is considered to occur via the ingestion of the L3 in contaminated water, food, soil, or dust, but percutaneous infection cannot yet be excluded as a mode of transmission (Krepel, 1994; Polderman & Blotkamp, 1995; Eberhard *et al.*, 2001). After ingestion, the L3s penetrate the intestinal wall and form tiny nodules, in which their development continues. Some larvae remain in these nodules and develop through to juvenile worms, whereas others enter the intestinal lumen to develop into male and female adults. After the copulation with males, the females start to produce eggs. The tissue-dwelling juvenile worms and the surrounding inflammation cause the pathological effects.

Although *O. bifurcum* infection in humans is often asymptomatic, two distinct types of clinical disease are defined. (1) The uninodular disease, also referred to as the "Dapaong tumor," presents as a painful, abdominal mass with a diameter of 2–11 cm, formed around a single or a small cluster of encapsulated juvenile worms, frequently adhering to the abdominal wall (cf. Polderman *et al.*, 1999). Usually, patients do not suffer from the effects of this manifestation, but intestinal occlusion and abscessation can occur, leading to considerable discomfort and abdominal pain. In such cases, surgery may be needed to remove the masses to avoid rupture and peritonitis. (2) The much less common multinodular disease is characterized by hundreds of pea-sized, granulomatous nodules within the colonic wall and other intra-abdominal structures, together with gross thickening and edema of the wall. This type of clinical presentation is often associated with considerable abdominal pain, persistent diarrhea, and weight loss (Storey *et al.*, 2000). Multinodular disease can lead to progressive destruction of the colonic wall, in which case, total or partial colonectomy is indicated (Gigase *et al.*, 1987; Storey *et al.*, 2000). In non-human primates, *O. bifurcum* infection can also be pathogenic, particularly in captive primates with considerable burdens, and suffering from stresses of confinement and following transportation (Stewart & Gasbarre, 1989). Heavy infestations can cause diarrhea, weight loss, and abdominal adhesions. Lesions caused by this parasite are small, dark nodules (diameter of $\leq$ 8 mm), located from the submucosa to the serosal surface of the caecum and colon, and occasionally in the colonic mesentery (Stewart & Gasbarre, 1989).

Diagnosis of oesophagostomiasis using traditional, coproscopic approaches

The specific identification of members of the genus *Oesophagostomum* has been based primarily on morphological features of the adult worms (Skrjabin *et al.*, 1991). However, the relatively simple body plan and an overlap in some morphological characters between members of this genus have caused some controversies. Between 1905 and the early 1980s, there had been confusion about the taxonomic status of *Oesophagostomum* species infecting primates.

For example, specimens referred to as *O. apiostomum* by Leiper (1911) are now classified as either *O. bifurcum* or *O. aculeatum* (Chabaud & Larivière, 1958), and *O. bifurcum* and *O. brumpti*, considered to be synonymous by Chabaud & Larivière (1958), were described as two different species (Glen & Brooks, 1985). There had been controversy about the species status of *O. bifurcum* infecting primates in northern Togo and Ghana. Also, it was not clear whether the parasite causing human oesophagostomiasis was the same species as that infecting non-human primates. To tackle this question, some morphological (Blotkamp, unpublished) and genetic (Gasser *et al.*, 1999) studies were conducted initially but were inconclusive.

Traditionally, the specific diagnosis of oesophagostomiasis has been based on the detection of typical strongylid-type eggs in host feces. However, the eggs of many species of strongylids infecting primates are (on an individual egg basis) morphologically indistinguishable, which represents a considerable diagnostic limitation. For example, the eggs of *Oesophagostomum* spp., hookworms, *Trichostrongylus* spp., and *Ternidens deminutus* are morphologically similar (Goldsmid, 1991; Polderman & Blotkamp, 1995). To overcome this problem, coproculture is used to allow eggs to develop into the L3s, which can then be identified microscopically to the genus level (Polderman *et al.*, 1991; Blotkamp *et al.*, 1993). However, coproculture is laborious and time consuming, and requires personnel skilled in identifying and differentiating the larvae microscopically.

Epidemiological aspects of human oesophagostomiasis

Oesophagostomum bifurcum infection in humans in northern Togo and Ghana occurs in almost every village, with the highest prevalence (up to 90%) being in the rural areas (Krepel *et al.*, 1992a; Pit *et al.*, 1999b; Storey *et al.*, 2000; Yelifari *et al.*, 2005). Krepel *et al.* (1992) reported that the prevalence in northern Togo and Ghana is usually lower in children of < 5 years of age, although heavy infections in children of 3–4 years have been reported in a more recent survey (Pit *et al.*, 1999b). There is a significant increase in the prevalence in children between the ages of 2 and 10 years, which indicates an increased rate of transmission of the parasite in this age group. Females (particularly adults) are usually more frequently infected with *O. bifurcum* than males in the same age group (Ziem *et al.*, 2006). However, the differences in prevalence between human females and males cannot yet be explained satisfactorily. It is possible that the difference is caused by the daily activities of females, such as cooking, washing, and the fetching of water, which may imply a higher frequency of contact with L3s in contaminated water, soil, or dust. It is also possible that the higher prevalence relates to differences in the immune response(s) or susceptibility in females. In all villages endemic for *O. bifurcum*, co-infection with the hookworm *Necator americanus* occurs, whereas other species of gastrointestinal helminths appear to be rare. Krepel *et al.* (1992a) showed that there was a correlation between the infection with *O. bifurcum* and *N. americanus* in that most persons were either infected with both parasites or not at all, and suggested explanations for this observation. Firstly, similar factors could be associated with the risk of infection with both parasites, such as poor hygiene, agricultural practices, and/or the relative lack of potable water. Secondly, there could be a similarity in transmission, although this seems to be unlikely as transmission of *N. americanus* occurs percutaneously (Hotez *et al.*, 2005) and *O. bifurcum* transmission is considered to occur via the oral route (Polderman & Blotkamp, 1995). In contrast, transmission of the human hookworm *Ancylostoma duodenale* can occur percutaneously and orally (Hotez *et al.*, 2005). Until recently, it was unclear whether *N. americanus* is the only species of hookworm infecting humans in Togo and Ghana, but *A. duodenale* has been demonstrated (de Gruijter *et al.*, 2005b).

Human oesophagostomiasis appears to be localized mainly to several foci in northern Togo and Ghana; the prevalence of the disease decreases rapidly toward the south of these countries (Krepel *et al.*, 1992a; Pit *et al.*, 1999b; Yelifari *et al.*, 2005). In the north, the number of non-human primates has decreased significantly in the last decades which may suggest that humans have become a preferred host for *O. bifurcum* (Polderman, unpublished). In the south, there are locations (e.g. Baobeng-Fiema or Mole National Park) where *O. bifurcum* is commonly found in the mona monkey (*Cercopithecus mona*), patas monkey (*Erythrocebus patas*), and olive baboon (*Papio anubis*), but not in humans, although they live in close association with these non-human primates. These observations raise a number of important questions regarding the epidemiology of human oesophagostomiasis, such as which factors prevent the parasite from infecting humans in the south of Togo and Ghana and whether the parasite causing human oesophagostomiasis in northern Togo and Ghana represents a different species or variant compared with that found in monkeys, thereby having a different host preference and/or geographical distribution.

The establishment and application of molecular-diagnostic methods for studying the epidemiology of *O. bifurcum*

Central to the development of a DNA-based approach for the identification of parasites to species is the choice of one or more appropriate target regions (genetic markers) (Gasser, 2001; Gasser & Chilton, 2001). Various studies of strongylid nematodes (to which nodule worms belong) have consistently shown that the internal transcribed spacers (ITS) of nuclear ribosomal DNA provide genetic markers for the identification of species (reviewed by Gasser *et al.*, 1999; Gasser, 2006). The definition of specific markers in the second ITS (ITS-2) of *O. bifurcum* has had important implications for diagnosis and for studying the epidemiology of oesophagostomiasis. Based on the research of *O. bifurcum* in humans in Togo (Romstad *et al.*, 1997), Verweij *et al.* (2000) established a PCR for the specific amplification of the ITS-2 DNA from *O. bifurcum* (fg) from human feces. Using a panel of >150 well-defined fecal and DNA samples, the diagnostic assay achieved high sensitivity (~ 95%) and specificity (100%) (cf. Verweij *et al.*, 2000). Subsequently, Verweij *et al.* (2001) employed this PCR assay to estimate the prevalence of both *O. bifurcum* and *N. americanus* infections in humans in northern Ghana. The *Oesophagostomum bifurcum*-PCR was positive in 57 of 61 (93%) fecal samples known to contain *O. bifurcum* L3s in coproculture. The *Necator americanus*-PCR was positive in 137 of 146 (94%) fecal samples known to contain *N. americanus* L3 larvae in coproculture. A PCR assay also detected 26 additional *O. bifurcum* cases in 72 samples from *O. bifurcum*-endemic villages for which no *O. bifurcum* L3s were found and 45 *N. americanus* cases in 78 samples from which no *N. americanus* L3s were obtained after larval culture. No *O. bifurcum* DNA was detected in 91 fecal samples from individuals from two non-endemic villages. Clearly, these results demonstrated the usefulness and advantages of the PCR assays (Verweij *et al.*, 2000, 2001) as epidemiological tools to estimate the prevalence and distribution of *O. bifurcum* in humans and for the differentiation of *O. bifurcum* from *N. americanus* infection. The development of PCR-based copro-diagnostic methods represented a significant improvement and overcame the limitations of traditional coproscopic methods.

Since non-human primates were considered to represent a potential zoonotic reservoir for human oesophagostomiasis (Polderman *et al.*, 1999), van Lieshout *et al.* (2005) undertook a study of *O. bifurcum* in different primate host species. These authors tested, using both microscopic and species-specific PCR (cf. Verweij *et al.*, 2000, 2001) methods, fecal samples (n = 349) from olive baboon, mona monkey, and western black-and-white colobus monkey (*Colobus polykomos*) from two areas external to that in Ghana endemic for human *O. bifurcum*. A high percentage (75–99%) of samples from the olive baboon and mona monkeys were shown to contain *O. bifurcum*. Most test-positive samples contained a large number of L3 (>100) after coproculture, whereas no *O. bifurcum* was detected in the feces from the black-and-white colobus monkeys. Studying the behavior of the non-human primates, focusing on defecation, food consumption, and physical contact with humans, indicated favorable conditions for zoonotic transmission.

Interestingly, larger numbers of L3s were found in feces from mona monkey and olive baboon (van Lieshout *et al.*, 2005) than those reported for human feces (Pit *et al.*, 1999a; Yelifari *et al.*, 2005), indicating a higher intensity of infection and effective transmission for non-human primates. In contrast, there appeared to be no risk of infection in the aboreal (i.e. tree-living) black-and-white colobus monkeys. Interestingly, to date, no cases of oesophagostomiasis have been reported in the two study areas. Also, the majority of inhabitants (n = ~700) living in the Mole National Park (northern region of Ghana) and a selection of individuals (n = ~100) from the Baobeng Fiema Monkey Sanctuary (Brong-Ahafo region) were tested on multiple occasions by copro-culture and/or PCR for the presence of *O. bifurcum*. None of the human stool samples were shown to be test-positive for *O. bifurcum*, whereas *N. americanus* infection was common (prevalence of ~30%). The apparent absence of *Oesophagostomum* from humans in both regions is remarkable for multiple reasons. Firstly, the ecological conditions for transmission seem to be favorable; the estimates of the prevalence and the intensities of infection in non-arboreal, non-human primates are high, with no major seasonal fluctuations. This indicates that the free-living, potentially infective, larval stages successfully develop throughout the year. Secondly, the behavioral study indicated a close relationship and interaction between human and non-human primates in the same habitat. Thirdly, humans are suitable hosts for *O. bifurcum*, as demonstrated by the high prevalence of human infection several hundred kilometers to the north of the study areas. Thus, it is still unclear why the parasite does not infect humans in the two study areas. Given that *O. bifurcum* infection was not detected in humans in either study area, the findings from the study by van Lieshout *et al.* (2005) supported the proposal that *O. bifurcum* from humans in the north of Ghana is biologically distinct from *O. bifurcum* from mona monkey and/or olive baboon hosts.

The application of DNA fingerprinting methods reveals genetic substructuring within *O. bifurcum* according to primate species

Since the magnitude of nucleotide variation in selected nuclear ribosomal and mitochondrial DNA loci was not adequate to establish genetic substructuring within *O. bifurcum* according to primate host species (Gasser *et al.*, 1999; de Gruijter *et al.*, 2002), further studies were undertaken using DNA fingerprinting methods (de Gruijter *et al.*, 2004, 2005a). The method of RAPD or arbitrarily primed-polymerase chain reaction (AP-PCR) (Welsh & McClelland, 1990; Williams *et al.*, 1990) relies on the PCR-based amplification of genomic DNA fragments using (mainly) single primers (~10-mers) of arbitrary sequence, and

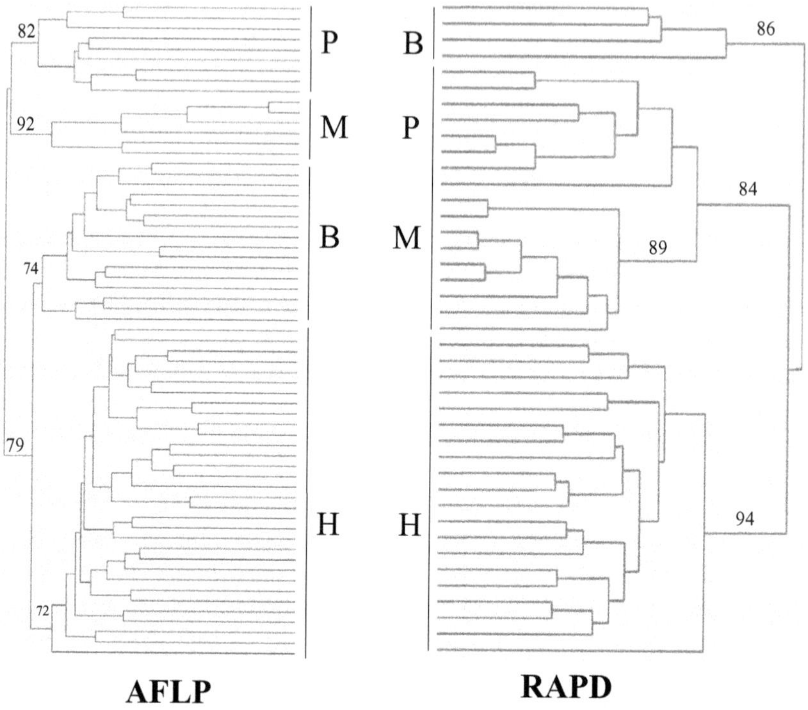

Figure 3.2. Simplified trees based on cluster analyses of amplified fragment length polymorphism (AFLP) or random amplified polymorphic DNA (RAPD) data, indicating the genetic variation among specimens of *Oesophagostomum bifurcum* from humans (H), mona monkey (*Cercopithecus mona*) (M), patas monkey (*Erythrocebus patas*) (P), and olive baboon (*Papio anubis*) (B) (de Gruijter *et al.*, 2004, 2005). Totals of 63 and 41 adult worms of *O. bifurcum* were employed in the AFLP and RAPD analyses, respectively. Numbers refer to bootstrap (RAPD) or co-phenetic (AFLP) values.

subsequent electrophoretic separation of the products. Applied to *O. bifurcum*, RAPD analyses (conducted at high stringency) using selected primers revealed a considerable polymorphism (several hundreds of polymorphic bands) among individuals from different species of primate from Ghana. Cluster analysis of the profile data revealed that *O. bifurcum* represented three distinct genetic groups, namely those from humans, those from the patas or the mona monkey, and those from the olive baboon (Figure 3.2). Hence, population genetic substructuring within *O. bifurcum* was associated with host species, and *O. bifurcum* from human and non-human primates represented genetically distinct groups. That *O. bifurcum* from humans and from the patas monkey (from the Bolgatanga–Bawku region) grouped into different clusters showed that there was no apparent association between *O. bifurcum* variant and the geographical origin of the host species based on the RAPD data set. This was also indicated for *O. bifurcum* from the patas monkey and from the olive baboon (from the Tamale region) which were divided into distinct clusters. In another study, de Gruijter *et al.* (2004) subjected >60 *O. bifurcum* adults from human, patas monkey, mona monkey, and olive baboon hosts from Ghana to amplified fragment length polymorphism (AFLP) analysis (cf. Vos *et al.*, 1995). The cluster analysis of the data set also revealed multiple genetically distinct groups (Figure 3.2), namely *O. bifurcum* from the patas monkey (1), from the mona monkey (2), from humans (3), and from the olive baboon (4). Hence, these findings were in accordance with those achieved using the RAPD analysis and supported the proposal for genetic substructuring within *O. bifurcum* according to host species. Although the sample sizes were small in these studies (de Gruijter *et al.*, 2004, 2005a), the findings revealed genetically distinct, host species-affiliated variants of *O. bifurcum* in Ghana, suggesting distinct transmission patterns for *O. bifurcum* for different primate hosts and raising questions about their epidemiology and ecology.

Concluding remarks

Recent studies have demonstrated the utility of molecular tools for the specific diagnosis of oesophagostomiasis and for studying the genetic make-up of *O. bifurcum* from different primate host species, giving new insights into the primate–parasite relationships. Importantly, molecular evidence suggests that *O. bifurcum* in humans is genetically distinct from that harbored by some non-human primates, that these genetic variants have distinct transmission patterns, and that non-human primates are not a source of human oesophagostomiasis. This evidence also supports the proposal that *O. bifurcum* from humans in the north of Ghana is biologically distinct from *O. bifurcum* from mona monkey and/or olive baboon hosts living in close association with humans (van Lieshout *et al.*, 2005). These findings and interpretations therefore have important implications for the prevention and control of human oesophagostomiasis, because the non-human primates examined thus far do not seem to harbor a parasite which is readily transmissible to humans. With this information, future work can now focus on studying the fundamental details of the ecology of these different host-affiliated genetic variants of *O. bifurcum* in Africa, providing a foundation for conducting similar studies in a range of countries where oesophagostomiasis is endemic in primates. For instance, it would be interesting to elucidate the genetic make-up and the transmission patterns of *Oesophagostomum* spp. in East Africa (e.g. Uganda) where it is common in colobus monkeys and chimpanzees that interact with humans (Muehlenbein, 2005; Chapman *et al.*, 2006). Developing real-time PCR assays will also be useful for estimating the intensity and incidence of *O. bifurcum* infections. Such assays may also find application for monitoring the effectiveness of mass treatment of humans (with albendazole) against oesophagostomiasis. Also, the DNA fingerprinting methods, particularly AFLP, should be useful for investigating the population genetics and systematics of a range of other pathogens and their association with and transmission among different species of primates (Muriuki *et al.*, 1998; Rothman & Bowman, 2003; Jones-Engel *et al.*, 2004; Legesse & Erko, 2004; Phillips *et al.*, 2004). This is of particular importance in cases where there is uncertainty about the specific identity of an infectious agent or the genetic make-up of its populations. Clearly, such molecular methods assist in studying the geographical spread of parasites and/or for investigating possible sources and routes of infection. While the focus of this chapter has been on *O. bifurcum*, molecular methods will find broad applicability to other pathogens of human and non-human primate hosts and their host–pathogen relationships.

Acknowledgements

Funding was provided by the Dutch Foundation for the Advancement of Tropical Research (WOTRO-NWO), the Australian Research Council, the Australian Academy of Science, Elchrom Scientific AG, and the Collaborative Research Program of the University of Melbourne. J. K. van der Reijden (Department of Infectious Diseases, Leiden University Medical Center, The Netherlands), E. E. C. Agbo (Division of Animal Sciences, Institute for Animal Science and Health, ID-Lelystad, The Netherlands), Paul Janssen (Center for Molecular Design, Janssen Pharmaceutica, Vosselaar, Belgium), and Guus Simons (Keygene, Wageningen, The Netherlands) are thanked for assistance and discussions. Thanks are also due to the following persons who provided, or assisted in providing, some of the samples used in this study: J. Ziem (University for Development Studies, Tamale, Ghana), M. Adu-Nsiah (Wildlife Division, Accra, Ghana), V. Asigri (Parasitic Diseases Research Laboratory, Tamale, Ghana), M. Hassel (Mole National Park, Ghana), D. Laar and S. Amponsah (Ghana), J. Blotkamp and E. Brienen (The Netherlands).

References

Barger, L. A., Lewis, R. J. & Brown, G. F. (1984). Survival of infective larvae of nematode parasites of cattle during drought. *Veterinary Parasitology*, **14**, 143–152.

Blotkamp, J., Krepel, K. P., Kumar, V. *et al.* (1993). Observations on the morphology of adults and larval stages of *Oesophagostomum* sp. isolated from man in northern Togo and Ghana. *Journal of Helminthology*, **67**, 49–61.

Chabaud, A. G. & Larivière, M. (1958). Sur les oesophagostomes parasites de l'homme. *Bulletin de la Societé de Pathologie Exotique*, **51**, 384–393.

Chapman, C. A., Speirs, M. L., Gillespie, T. R., Holland, T. & Austad, K. M. (2006). Life on the edge: gastrointestinal parasites from the forest edge and interior primate groups. *American Journal of Primatology*, **68**, 397–409.

de Gruijter, J. M., Polderman, A. M., Zhu, X. Q. & Gasser, R. B. (2002). Screening for haplotypic variability within *Oesophagostomum bifurcum* (Nematoda) employing a single-strand conformation polymorphism approach. *Molecular and Cellular Probes*, **16**, 183–188.

de Gruijter, J. M., Ziem, J., Verweij, J. J., Polderman, A. M. & Gasser, R. B. (2004). Genetic substructuring within *Oesophagostomum bifurcum* (Nematoda) from human and non-human primates from Ghana based on random amplification of polymorphic DNA analysis. *American Journal of Tropical Medicine and Hygiene*, **71**, 227–233.

de Gruijter, J. M., Gasser, R. B., Polderman, A. M., Asigri, V. & Dijkshoorn, L. (2005a). High resolution DNA fingerprinting by AFLP to study the genetic variation among *Oesophagostomum bifurcum* (Nematoda) from human and non-human primates in Ghana. *Parasitology*, **130**, 229–237.

de Gruijter, J. M., van Lieshout, L., Gasser, R. B. *et al.* (2005b). PCR-based differential diagnosis of *Ancylostoma duodenale* and *Necator americanus* in humans from northern Ghana. *Tropical Medicine and International Health*, **10**, 574–580.

Durette-Desset, M. C., Beveridge, I. & Spratt, D. M. (1994). The origins and evolutionary expansion of the Strongylida (Nematoda). *International Journal for Parasitology*, **24**, 1139–1165.

Eberhard, M. L., Kovacs-Nace, E., Blotkamp, J. *et al.* (2001). Experimental *Oesophagostomum bifurcum* in monkeys. *Journal of Helminthology*, **75**, 51–56.

Haaf, E. & van Soest, A. H. (1964). Oesophagostomiasis in man in North Ghana. *Tropical and Geographical Medicine*, **16**, 49–53.

Gasser, R. B. (2001). Identification of parasitic nematodes and study of genetic variation using PCR approaches. In *Parasitic Nematodes: Molecular Biology, Biochemistry and Immunology*, ed. M. W. Kennedy & W. Harnett. Oxon and New York: CABI Publishing, pp. 53–82.

Gasser, R. B. (2006). Molecular tools – advances, opportunities and prospects. *Veterinary Parasitology*, **136**, 69–89.

Gasser, R. B. & Chilton, N. B. (2001). Applications of single-strand conformation polymorphism (SSCP) to taxonomy, diagnosis, population genetics and molecular evolution of parasitic nematodes. *Veterinary Parasitology*, **101**, 201–213.

Gasser, R. B., Woods, W. G., Blotkamp, J. *et al.* (1999). Screening for nucleotide variations in ribosomal DNA arrays of *Oesophagostomum bifurcum* by polymerase chain reaction-coupled single-strand conformation polymorphism. *Electrophoresis*, **20**, 1486–1491.

Gigase, P., Baeta, S., Kumar, V. & Brandt, J. (1987). *Helminth Zoonosis*, ed. S. Geerts, V. Kumar & J. Brandt. Dordrecht: Martinus Nijhoff, pp. 233–236.

Glen, D. R. & Brooks, R. (1985). Phylogenetic relationships of some Strongylate nematodes of primates. *Proceedings of the Helminthological Society of Washington*, **52**, 227–236.

Goldsmid, J. M. (1991). The African hookworm problem. In *Parasitic Helminths and Zoonoses in Africa*, ed. C. N. L. MacPherson & P. S. Craig. London: Unwin Hyman, pp. 101–137.

Hotez, P. J., Bethony, J., Bottazzi, M. E., Brooker, S. & Buss, P. (2005). Hookworm: "The Great Infection of Mankind." *PLoS Medicine*, **2**, e67.

Jones-Engel, L., Engel, G. A., Schillact, M. A. *et al.* (2004). Prevalence of enteric parasites in pet macaques in Sulawesi, Indonesia. *American Journal of Primatology*, **62**, 71–82.

Krepel, H. P. (1994). *Oesophagostomum bifurcum* infection in man: a study on the taxonomy, diagnosis, epidemiology and drug treatment of *Oesophagostomum bifurcum* in northern Togo and Ghana. PhD thesis. Leiden, The Netherlands.

Krepel, H. P., Baeta, S. & Polderman, A. M. (1992a). Human *Oesophagostomum* infection in northern Togo and Ghana: epidemiological aspects. *Annals of Tropical Medicine and Parasitology*, **86**, 289–300.

Krepel, H. P. & Polderman, A. M. (1992b). Egg production of *Oesophagostomum bifurcum*, a locally common parasite of humans in Togo. *American Journal of Tropical Medicine and Hygiene*, **46**, 469–472.

Krepel, H. P., Haring, T., Baeta, S. & Polderman, A. M. (1993). Treatment of mixed *Oesophagostomum* and hookworm infection: effect of albendazole, pyrantel pamoate, levamisole and thiabendazole. *Transactions of the Royal Society of Tropical Medicine and Hygiene*, **87**, 87–89.

Legesse, M. & Erko, B. (2004). Zoonotic intestinal parasites in *Papio anubis* (baboon) and *Cercopithecus aethiops* (vervet) from four localities in Ethiopia. *Acta Tropica*, **90**, 231–236.

Leiper, R. T. (1911). The occurence of *Oesophagostomum apiostomum* as an intestinal parasite of man in Nigeria. *Journal of Tropical Medicine and Hygiene*, **14**, 116–118.

Lie Kian Joe (1949). Helminthiasis of the intestinal wall caused by *Oesophagostomum apiostomum* (Willach, 1891) Railliet and Henry, 1905. *Documenta Neerlandia et Indonesia de Morbis Tropicis*, **1**, 75–80.

Muehlenbein, M. P. (2005). Parasitological analyses of the male chimpanzees (*Pan troglodytes schweinfurthii*) at Ngogo, Kibale National Park, Uganda. *American Journal of Primatology*, **65**, 167–179.

Muriuki, S. M., Murugu, R. K., Munene, E., Karere, G. M. & Chai, D. C. (1998). Some gastro-intestinal parasites of zoonotic (public health) importance commonly observed in old world non-human primates in Kenya. *Acta Tropica*, **71**, 73–82.

Orihel, T. C. (1970). The helminth parasites of non-human primates and man. *Laboratory Animal Care*, **20**, 395–401.

Phillips, K. A., Hass, M. E., Grafton, B. W. & Yrivarren, M. (2004). Survey of the gastrointestinal parasites of the primate community at Tambopata National Reserve, Peru. *Journal of Zoology* (*London*), **264**, 149–151.

Polderman, A. M., Krepel, H. P., Baeta, S., Blotkamp, J. & Gigase, P. (1991). Oesophagostomiasis, a common infection of man in northern Togo and Ghana. *American Journal of Tropical Medicine and Hygiene*, **44**, 336–344.

Polderman, A. M., Anemana, S. D. & Asigri, V. (1999). Human oesophagostomiasis: a regional public health problem in Africa. *Parasitology Today*, **15**, 129–130.

Pit, D. S., de Graaf, W., Snoek, H., de Vlas, S. J., Baeta, S. M. & Polderman, A. M. (1999a). Diagnosis of *Oesophagostomum bifurcum* and hookworm infection in humans: day-to-day and within-specimen variation of larval counts. *Parasitology*, **118**, 283–288.

Pit, D. S., Rijcken, F. E., Raspoort, E. C., Baeta, S. M. & Polderman, A. M. (1999b). Geographic distribution and epidemiology of *Oesophagostomum bifurcum* and hookworm infections in humans in Togo. *American Journal of Tropical Medicine and Hygiene*, **61**, 951–955.

Pit, D. S., Blotkamp, J., Polderman, A. M., Baeta, S. & Eberhard, M. L. (2000). The capacity of the third-stage larvae of *Oesophagostomum bifurcum* to survive adverse conditions. *Annals of Tropical Medicine and Parasitology*, **94**, 165–171.

Polderman, A. M. & Blotkamp, J. (1995). *Oesophagostomum* infections in humans. *Parasitology Today*, **11**, 451–460.

Railliet, A. & Henry, A. (1905). Encore un nouveau sclerostomien (*Oesophagostomum brumpti* nov. sp.) parasite de l'homme. *Comptes rendus des séances de la Société de Biologie et de ses filiales*, **58**, 643–645.

Railliet, A. & Henry, A. (1910). Etude zoologique de l'oesophagostome de Thomas. *Annals of Tropical Medicine and Parasitology*, **4**, 89–94.

Romstad, A., Gasser, R. B., Monti, J. R. *et al.* (1997). Differentiation of *Oesophagostomum bifurcum* from *Necator americanus* by PCR using genetic markers in spacer ribosomal DNA. *Molecular and Cellular Probes*, **11**, 169–176.

Rose, J. H. & Small, A. J. (1980). Observations on the development and survival of the free-living stages of *Oesophagostomum dentatum* both in their natural environments out-of-doors and under controlled conditions in the laboratory. *Parasitology*, **81**, 507–517.

Rothman, J. & Bowman, D. D. (2003). A review of endoparasites of mountain gorillas. In *Companion and Exotic Animal Parasitology*, ed. D. D. Bowman. Ithaca, NY: International Veterinary Information Service.

Skrjabin, K. I., Shikhobalova, N. P., Schulz, R. S. *et al.* (1991). Strongylata. In *Keys to Parasitic Nematodes*, Vol. 3, ed. K. I. Skrjabin *[English Translation of Opredelitel' Paraziticheskikh Nematod. Izdatel'stvo Akademii Nauk SSSR, Moscow]*. New Delhi: Amerind Publishing Co., pp. 230–247.

Spindler, L. A. (1936). Effects of various physical factors on survival of eggs and infective larvae of the swine nodular worm *Oesophagostomum dentatum*. *Parasitology*, **22**, 259.

Stewart, T. B. & Gasbarre, L. C. (1989). The veterinary importance of nodular worms (*Oesophagostomum* spp.). *Parasitology Today*, **5**, 209–213.

Storey, P. A., Faile, G., Hewitt, E. *et al.* (2000). Clinical epidemiology and classification of human oesophagostomiasis. *Transactions of the Royal Society of Tropical Medicine and Hygiene*, **94**, 177–182.

Storey, P. A., Steenhard, N. R., van Lieshout, L. *et al.* (2001a). Natural progression of *Oesophagostomum bifurcum* pathology and infection in a rural community of northern Ghana. *Transactions of the Royal Society of Tropical Medicine and Hygiene*, **95**, 295–299.

Storey, P. A., Bugri, S., Magnussen, P. & Polderman, A. M. (2001b). The effect of albendazole on *Oesophagostomum bifurcum* infection and pathology in children from rural northern Ghana. *Annals of Tropical Medicine and Parasitology*, **95**, 87–95.

Thomas, H. W. (1910). The pathological report of a case of oesophagostomiasis in man. *Annals of Tropical Medicine and Parasitology*, **4**, 57–88.

Travassos, L. P. & Vogelsang, E. G. (1932). Pesquizas helminthologicas realisadas em Hamburgo. X. Contribuição ao conhecimento das especies de *Oesophagostomum* dos primates. *Mem. Inst. Oswaldo Cruz*, **26**, 251–328.

van Lieshout, L., de Gruijter, J. M., Adu-Nsiah, M. *et al.* (2005). *Oesophagostomum bifurcum* in non-human primates is not a potential reservoir for human infection in Ghana. *Tropical Medicine and International Health*, **10**, 1315–1320.

Verweij, J. J., Polderman, A. M., Wimmenhove, M. C. & Gasser, R. B. (2000). PCR assay for the specific amplification of *Oesophagostomum bifurcum* DNA from human faeces. *International Journal for Parasitology*, **30**, 137–142.

Verweij, J. J., Pit, D. S., van Lieshout, L. *et al.* (2001). Determining the prevalence of *Oesophagostomum bifurcum* and *Necator americanus* infections using specific PCR amplification of DNA from faecal samples. *Tropical Medicine and International Health*, **6**, 726–731.

Welsh, J. & McClelland, M. (1990). Fingerprinting genomes using PCR with arbitrary primers. *Nucleic Acids Research*, **18**, 7213–7218.

Williams, J. G. K., Kubelik, A. R., Livak, K. J., Rafalski, J. A. & Tingey, S. V. (1990). DNA polymorphisms amplified by arbitrary primers are useful as genetic markers. *Nucleic Acids Research*, **8**, 6531–6535.

Yelifari, L., Bloch, P., Magnussen, P. *et al.* (2005) Distribution of human *Oesophagostomum bifurcum*, hookworm and *Strongyloides stercoralis* infections in northern Ghana. *Transactions of the Royal Society of Tropical Medicine and Hygiene*, **99**, 32–38.

Ziem, J. B., Kettenis, I. M., Bayita, A. *et al.* (2004). The short-term impact of albendazole treatment on *Oesophagostomum bifurcum* and hookworm infections in northern Ghana. *Annals of Tropical Medicine and Parasitology*, **98**, 385–390.

Ziem, J. B., Spannbrucker, N., Magnussen, P. *et al.* (2005). *Oesophagostomum bifurcum*-induced nodular pathology in a highly endemic area of Northern Ghana. *Transactions of the Royal Society of Tropical Medicine and Hygiene*, **99**, 417–422.

Ziem, J. B., Olsen, A., Magnussen, P. *et al.* (2006). Distribution and clustering of *Oesophagostomum bifurcum* and hookworm infections in Northern Ghana. *Parasitology*, **132**, 525–534.

4 *The application of endocrine measures in primate parasite ecology*

MICHAEL P. MUEHLENBEIN

Photograph by Michael Muehlenbein. (*Pongo pygmaeus*)

Primate Parasite Ecology. The Dynamics and Study of Host–Parasite Relationships, ed. Michael A. Huffman and Colin A. Chapman. Published by Cambridge University Press.

Introduction

The past decade has witnessed a tremendous amount of growth in the research, development, and use of endocrinological analyses in field primatology (Whitten *et al.*, 1998; Wasser *et al.*, 2000; Mohle *et al.*, 2002; Hodges & Heistermann, 2003; Ziegler & Wittwer, 2005). New, non-invasive techniques allow monitoring of endocrine function in these animals under natural conditions, without disrupting normal animal activities. These methods have been used most frequently to describe endocrine correlates of behavior (Muehlenbein *et al.*, 2004), development (Robbins & Czekala, 1997), and reproduction (Strier & Ziegler, 2005) in non-human primates (referred to as "primates" throughout the remainder of this chapter). The use of endocrine measures in primate parasite ecology is relatively rare, although phylogenetic relationships with humans warrant such work. In fact, immune–endocrine interactions have only once been investigated in apes under natural infection conditions (Muehlenbein, 2006).

The primary purpose of this chapter is to illustrate how endocrinological considerations may further facilitate studies of primate ecological parasitology. That is, endocrine and immune (i.e. parasite) measures can complement each other in the interpretations of studies on social relationships, anthropogenic disturbance (habituation and tourism), and nutrition in primates. Various ecological and social pressures in wild primates are described throughout this chapter, with particular reference to the wealth of information that can be obtained through combined analysis of endocrine and parasite data. This chapter begins with a basic introduction to endocrinology and methodological considerations, and concludes with a discussion on the benefits of utilizing endocrine measures in future studies of primate parasite ecology.

Basic endocrinology and methodological considerations

Basic endocrinology

A complete review of endocrine physiology is provided by Griffin & Ojeda (2000). In brief, the endocrine system is a coordinated system of tissues that utilize chemical signals (hormones) to regulate the function of various other tissues and cells to elicit some necessary function, including metabolism, reproduction, development, behavior, and the maintenance of homeostasis. The primary endocrine organs include the hypothalamus, pituitary, thyroids, adrenals, testes, ovaries, pineal, placenta, and pancreas, although other tissues/organs also secrete and respond to hormones. The hypothalamus is the coordinating neuroendocrine center that integrates sensory information to elicit stimulating and inhibiting hormone responses. The different endocrine "axes" include the hypothalamic-pituitary gonadal (HPG), hypothalamic-pituitary-adrenal (HPA), and the hypothalamic-pituitary-thyroid (HPT). The release of small amounts of hormone from the hypothalamus causes chemical signals to become amplified through the stimulation of the pituitary, and eventually the target cells (gonads, adrenals, thyroid, etc.).

Important hormones produced by the hypothalamus include gonadotropin-releasing hormone (GnRH), corticotropin-releasing hormone (CRH), thyrotropin-releasing hormone (TRH), growth hormone-releasing hormone, dopamine, and prolactin-releasing factor. The anterior pituitary gland responds to these stimuli to produce a number of other hormones, including growth hormone, prolactin, thyroid-stimulating hormone (TSH), luteinizing hormone (LH), follicle-stimulating hormone (FSH), and adrenocorticotropin (ACTH). Primary hormones of the posterior pituitary include vasopressin and oxytocin.

After stimulation from TSH from the anterior pituitary, the thyroid and parathyroid produce triiodothyronine (T3) and thyroxine (T4). Luteinizing hormone from the anterior pituitary stimulates the production of testosterone, dihydrotestosterone (DHT), androstenedione and other androgens from the testes, as well as estradiol, estrone, and other estrogens from the ovaries. Follicle-stimulating hormone supports sperm and follicle development, and CRH from the anterior pituitary stimulates production of cortisol, corticosterone, and dihydroepiandrosterone (DHEA) from the adrenals. Other hormones include progestins from the corpus luteum, chorionic gonadotropin from the placenta, melatonin from the pineal gland, insulin from the pancreas, leptin and adiponectin from adipose tissue, ghrelin from the gastrointestinal tract, as well as epinephrine (adrenaline), norepinephrine, dopamine, serotonin, ß-endorphins, and other neuroendocrine transmitters from the adrenals and neural tissue. Importantly, this brief list is nowhere near exhaustive, as such a complete discussion of each steroid, protein, monoamine, and fatty-acid derived hormone is beyond the scope of this chapter.

For those readers considering using endocrine measures in their work, this should be done in direct consultation with an endocrinologist so that one may choose the most appropriate hormones and methods to adequately address the specific research questions. That said, a few things should be kept in mind for now. Firstly, all hormones exhibit pluralistic effects, including synergistic and antagonistic ones, and several hormones often regulate a singular function. Secondly, circulating hormone levels are determined not just through production and secretion, but also through clearance (target cell uptake, metabolic degradation, and excretion) and feedback loops (e.g. high testosterone levels can signal the hypothalamus to produce less LH). Thirdly, the effects of hormones on the body do not just depend on absolute levels in circulation, but also the presence of co-activators and the number and affinity of available receptors. Finally, it should be remembered that investigations into endocrine correlates of infection and immunity in wild primates cannot resolve the issue regarding causation between these factors. The relationships between hormones and disease are reciprocal ones and questions regarding causation will be best answered under carefully controlled conditions, as would be the case with captive animals exposed to known amounts of hormone or infection.

It is also imperative to note that variation in hormone levels within and between individuals is a natural aspect of organismal phenotypic plasticity. Reporting average hormone levels for an individual, sex, or species may ignore important inter-individual variation. Each animal will have a different homeostatic "set point" around which hormone levels can vary in response to different ecological and social stimuli. The ability to appropriately adjust hormone levels in response to these stimuli is an adaptive response that facilitates the allocation of metabolic resources towards different functions, like growth or reproduction, tissue anabolism or catabolism, etc. (Muehlenbein & Bribiescas, 2005). Considered in this light, comparing the variation in hormone level within an individual animal before, during, and after various stimuli, like parasite infection, may be more insightful than comparing average hormone levels between different animals.

Methodological considerations

Choosing your sample

Blood can yield data on all types of hormones. However, darting, trapping, and otherwise invasively sampling primates can be logistically and ethically challenging, particularly for endangered species (like many primates). In a captive setting with habituated animals, blood draws and spinal taps are much more feasible. It must be kept in mind that invasive sampling will, in almost every circumstance, result in activation of the stress responses and result in altered endocrine parameters. In addition, steroid hormones typically circulate in the blood bound to either albumin or sex-hormone binding globulin to prevent deactivation by enzymes and promote solubility. Some assays are designed to measure the "free" or "bioavailable" hormone not bound to these factors, while other assays are not.

Urine samples can yield steroid hormones (estrogens and androgens), progestins, catecholamine (serotonin) metabolites, and protein hormones (luteinizing hormone and follicle-stimulating hormone), whereas only steroids can currently be recovered from feces (Whitten *et al.*, 1998). Hormones excreted in urine and feces are also frequently conjugated as glucuronides or sulfates by the liver in order to increase solubility, reduce activity, and facilitate excretion, which can make assays for urine and feces more complex (Ziegler & Wittwer, 2005). For urinary assays (not feces), sample water content must also be controlled for by determining creatinine levels and expressing hormone results per unit of creatinine.

Urinary steroids are excreted relatively rapidly (Ziegler *et al.*, 1989; Mohle *et al.*, 2002) compared to fecal steroids. Because they are not as influenced by minor rapid fluctuations in the hypothalamic-pituitary-gonadal axis of the endocrine system, hormone values from fecal samples represent an average value over a given length of time. This time is dependent upon gut retention and transit, which is longer in larger-bodied animals (e.g. approximately 48 hours in chimpanzees) (Milton & Demment, 1988; Whitten *et al.*, 1998).

While fecal samples have the benefit of yielding information about chronic endocrine activity, the long excretion time means that there will be a time lag between any stimulus and a subsequent endocrine response measurable in feces. Neurotransmitter levels can change within seconds of a stimulus, and these changes can be detected immediately in cerebrospinal fluid. On the other hand, steroid hormones might change 5–15 minutes after a stimulus, and be detectable in feces only 24 hours later. In fact, cortisol in yellow baboons has a fecal excretion lag time of 26 hours (Wasser *et al.*, 1993). While urine also represents a cumulative index of endocrine function, urinary hormones are excreted more rapidly and thus are more conducive for identifying hormonal correlates of singular stressful events.

Collection

Hormone levels can vary by animal age, sex, and reproductive status (Bercovitch & Clarke, 1995), and therefore samples should not be randomly collected, but rather only collected from known animals. In the case of urine and fecal samples, it is best to witness the defecation event to corroborate the sample with its owner. Because hormone levels vary on a moment-to-moment basis, multiple samples must be also collected from each individual to generate accurate baseline levels. Hormone levels also vary by season and time of day (Sousa & Ziegler, 1998; Lynch *et al.*, 2002). To avoid spurious results due to diurnal effects, it is advised to collect urine and fecal samples only at one time of day (morning or afternoon) (Sousa & Ziegler, 1998), and to record time of sample collection.

Samples can be collected off vegetation, a tarpaulin, or similar collection device. Care must be taken to avoid contamination with water, soil, or excrement from other animals. A few milliliters of urine or a gram of feces are usually more than adequate, although higher volume/mass may be necessary to run many different assays. For urine, it is often convenient to collect the first morning void shortly after the animal awakes. For feces, a wooden applicator stick can be used to take a sample from the center of the fecal mass, avoiding the abovementioned contaminants in addition to the same animal's urine. Samples can be placed in 15 ml polypropylene conical-bottom centrifuge tubes, immediately sealed and labeled. Samples should be frozen or extracted (see below) as soon as possible so as to avoid oxidation and bacterial metabolism of hormone metabolites (Washburn & Millspaugh, 2002).

Storage

Freezing the samples is the gold standard for preservation. However, this can be difficult in the field in the absence of electricity or liquid nitrogen, and frozen samples can be difficult to transport. If both organic solvents and electricity are available in the field, solid-phase extraction (SPE) is preferred (Ziegler & Wittwer, 2005). This involves extracting analytes from the material and adhering them to a sorbent matrix inside of an SPE cartridge that can then be stored and shipped at ambient temperature. By extracting the steroids in the field using SPE techniques, the transportation of alcohol and infectious agents is eliminated (Ziegler & Wittwer, 2005). This also eliminates the need for an import permit from the Centers for Disease Control and Prevention.

The SPE technique used will depend upon the hormone of interest as well as the species in question. For example, New World monkeys have higher glucocorticoid levels than many Old World primates (Coe *et al.*, 1992), and thus may require less fecal material for the assay. Furthermore, while storing the loaded SPE cartridge at ambient temperatures (cool, dry place out of direct sunlight) in the field may be feasible for several weeks or months, the cartridges should then be frozen whenever possible to prevent potential breakdown of the hormones.

To provide an example of SPE, the technique that M. Muehlenbein and colleagues use with wild orangutans is as follows. Firstly, the sample is mixed within the tube, and 0.2 g of stool are weighed and combined with 2.5 ml of distilled water and 2.5 ml of 95% ethanol in a clean tube. The steroid is extracted from the sample by vortexing the tube for 5 minutes. It is very important that each sample is vortexed for exactly the same amount of time to avoid over- or under-extraction relative to other samples. The sample is then centrifuged at 3300 rpm for 10 minutes, the supernatant (containing the extracted hormone) is kept and the fecal pellet discarded. An Alltech Prevail SPE cartridge (500 mg bed weight) is primed with 2 ml of distilled water using a disposable syringe. Next, 2 ml of the fecal extract is loaded into the cartridge using a clean syringe, followed by another 2 ml of distilled water. Both ends of the cartridge are capped, and the cartridge is labeled and stored in a zip-lock bag at ambient temperature. The loaded cartridges are shipped to the laboratory where they are stored at −20 °C until assayed.

In the absence of electricity or organic solvents, fecal samples can be dehydrated in the field using a portable oven placed atop a kerosene stove (Muehlenbein *et al.*, 2004), although this may alter hormone levels in the sample. Blood, urine, and feces (solubilized) can be aliquoted and dried onto filter paper, which is stable at room temperature for short periods of time (less than one month) (Worthman & Stallings, 1997). Feces can be preserved in ethanol or isopropyl alcohol and stored at room temperature for short periods of time, although this method is no longer preferred due to restrictions of combustible material on airplanes.

Analysis

Hormone levels can be determined via radioimmunoassay, enzyme immunoassay, or high performance liquid chromatography with either ultraviolet detection or mass spectrometry (Whitten *et al.*, 1998; Ziegler & Wittwer, 2005). The majority of commercial immunoassay kits are designed for use with human samples and are generally not validated for primates. Different species will metabolize hormones differently (Bahr *et al.*, 2000; Wasser *et al.*, 2000; Heistermann *et al.*, 2006), and because of this, assays are often custom designed. For example, M. Muehlenbein, T. Ziegler, and D. Wittwer have developed and validated an orangutan fecal cortisol assay using material from wild orangutans from Sabah, Malaysia and captive orangutans from Cleveland Metroparks Zoo. For this assay, the loaded SPE cartridges are washed with 1 ml of 20% methanol and then eluted with 2 ml of 100% methanol using a vacuum manifold. Using a nitrogen evaporator, the methanol is evaporated and the sample rehydrated with 1 ml of 100% ethanol. Solvolysis (to hydrolyze conjugated steroids) is performed by incubating the samples overnight with ethyl acetate, a sulfuric acid solution, and a saturated sodium chloride solution. After the incubation, water is added, the tubes centrifuged, and the ethyl acetate fraction aspirated and evaporated until dryness. The samples are rehydrated with 100% ethanol and are then run in an enzyme immunoassay (cortisol antibody and horseradish peroxidase conjugate obtained from C. Munro) with eight standards and two controls. More detailed reviews of the methods employed by other investigators for measuring endocrine responses in captive and wild primates are given elsewhere (Whitten *et al.*, 1998; Hodges & Heistermann, 2003; Ziegler & Wittwer, 2005), and the reader is encouraged to refer to these and other works.

Endocrine correlates with ecological and social stressors in primates

Combined analysis of endocrine and infection data can produce a wealth of information in primate studies, yet immune–endocrine interactions have only once been investigated in apes under natural infection conditions (Muehlenbein, 2006). Projects investigating the effects of social and ecological stressors on primates are particularly suited for integrating such endocrine and infection measures. "Stress" is broadly defined here as the disruption of physiological or psychological homeostasis, and stress responses are those mechanisms activated to re-establish homeostasis (or "allostasis"). Glucocorticoids (mainly cortisol in primates) are released from the adrenal cortex within minutes following activation of the hypothalamic-pituitary-adrenal axis, and epinephrine (adrenaline) is released within seconds from the adrenal medulla following activation of the sympathetic nervous system. Acute stress responses and glucocortocoids are necessary to prepare the body to cope with crisis (Romero, 2004). However, chronic exposure to stressors and overactivation of the stress responses can produce pathological effects (i.e. "allostatic load"), including impaired cognition, growth, and immunity (McEwen, 1998). For example, Alberts and others (1992) observed an inverse association between cortisol and total lymphocyte levels in wild female baboons, and Sapolsky & Spencer (1997) found an inverse association between cortisol and insulin-like growth factor I in wild male baboons. Elevated glucocorticoid levels can also contribute to suppression of reproductive function in males and females (Bambino & Hsueh, 1981; Sapolsky, 1985).

Other important hormones have pluralistic effects on the reproductive and immune systems. For example, testosterone exhibits some suppressive effects on immune functions (Grossman, 1995; Muehlenbein & Bribiescas, 2005). High testosterone levels can also compromise survivorship by increasing energetic costs and the risk of negative energy balance (Ketterson *et al.*, 1992; Marler *et al.*, 1995; Bribiescas, 2001), increasing the risk of prostate cancer (Soronen *et al.*, 2004), facilitating the production of oxygen radicals (Zirkin & Chen, 2000), increasing risk of injury due to hormonally augmented behaviors such as aggression, violence, and risk taking (Wilson & Daly, 1985; Dabbs, 1996), and reducing tissue (especially adipose) and organ maintenance (Bribiescas, 2001; Muehlenbein & Bribiescas, 2005). In contrast, estradiol and other estrogens appear to be immunostimulating. For example, estrogens likely upregulate the production of antioxidant enzymes (Vina *et al.*, 2006) which may decrease oxidative damage of mitochondrial DNA in females (Borras *et al.*, 2007) and protect against the oxygen radicals produced by inflammatory stress (Asaba *et al.*, 2004). At the same time, estrogens may contribute to increased risk of heart disease, reproductive cancers, and stroke (Anderson *et al.*, 2004) as well as increased incidence of autoimmune diseases in women (Cutolo *et al.*, 2006). Although the incorporation of endocrine measures in primate projects involving estrogens, progestins, prostaglandins, oxytocin, prolactin, and other hormones will certainly be insightful for primate parasite ecology (e.g. altered infection status during pregnancy and lactation), for purposes of brevity I will limit the following discussion to the relationships between infection level, cortisol, and testosterone during the maintenance of social relationships, anthropogenic disturbance, and under-nutrition in wild primates.

Social status, cortisol, testosterone, and disease

Members of group living species often form dominance hierarchies and put considerable effort into attaining and maintaining high rank to increase access to nutritional resources and mating opportunities. Endocrinological correlates of dominance rank have been sought, with two of the most frequent assumptions being that high ranking (dominant) animals have high testosterone levels, and that low ranking (subordinate) animals have high cortisol levels (Abbott *et al.*, 2003; Muehlenbein *et al.*, 2004). With regards to the former, high testosterone levels would theoretically augment aggressive behaviors which would facilitate the acquisition of high rank. However, correlations between testosterone and aggression are largely inconsistent. It is likely the case that testosterone is most strongly associated with aggression (and subsequently dominance rank) in male animals only during situations of social instability, such as during challenges by conspecific males for territory or access to mates, the establishment of territorial boundaries, or the presence of receptive females (Wingfield *et al.*, 1990).

It is also frequently assumed that subordinate animals are subjected to the chronic stress of failed agonistic interactions and subsequently exhibit high cortisol levels. However, correlations between cortisol and subordinance have also proved largely inconsistent (Abbott *et al.*, 2003). In some species, cortisol levels are higher in subordinate than dominant animals, whereas in other species the opposite is true (Sapolsky, 2005). The directionality of these associations may depend largely on the types of stressors faced by dominant animals versus subordinates, the costs of attaining and maintaining dominance, and access to social support systems (Abbott *et al.*, 2003; Goymann & Wingfield, 2004; Sapolsky, 2005).

The relationships between parasite infection, social status, and hormones are certainly complex. Muller-Graf *et al.* (1996) found no association between helminth infection and dominance rank in olive baboons, whereas Hausfater & Watson (1976) did demonstrate higher intestinal helminth infection in higher ranking yellow baboons. Eley *et al.* (1989) also demonstrated increased louse prevalence in lower ranking olive baboons. Hormone levels were not determined in any of these studies.

Using the largest habituated population of wild chimpanzees in the world, Muehlenbein (2006) has identified significant positive associations between testosterone, cortisol, and intestinal parasite richness (the number of unique intestinal parasite species recovered from hosts' fecal samples). New analyses also indicate that intestinal parasite richness is directly associated with both testosterone level and dominance rank (unpublished data). In this case, testosterone could facilitate dominance status by mediating aggressive or other behaviors, while simultaneously imposing a cost on dominance through immunosuppression and increased likelihood of acquiring multiple infections. Clearly, attaining and maintaining dominance is not an easy job, and increased intestinal infection may just be one of the risks involved. Further investigating the relationships between infection status, social variables, and endocrine measures in other primate species will be interesting.

Habituation, tourism, cortisol, and disease

Human disturbance of wildlife is obviously very complicated, and anthropogenic disturbances in general may affect animal physiology adversely (Walker *et al.*, 2005). The impact that humans have on primate health has typically focused on disease transmission potential (Homsey, 1999; Wallis & Lee, 1999; Adams *et al.*, 2001; Woodford *et al.*, 2002), a vital consideration given the increasing demand from tourists to experience direct encounters with these animals. Ecotourism is a potential tool to assist conservationist efforts in preserving populations of these wild animals, particularly great apes. It is therefore important to produce definitive guidelines that will protect visitors from possible risks as well as ensure long-term well-being of the animals, and this begins with monitoring the effects of habituation on animal physiology.

Habituation consists of a waning response following repeated stimulation without reinforcement. The effects of habituation and tourism on animal physiology have only rarely been investigated even though such stressors can theoretically cause immunosuppression, increasing susceptibility to infectious diseases, and decreasing reproductive success. Among wild adult Magellanic penguins (*Spheniscus magellanicus*), unhabituated animals exhibit elevated plasma corticosterone levels in response to tourist visitation compared with habituated animals, although this response diminishes quickly during the habitation process (Walker *et al.*, 2006). More importantly, habituation may permanently alter adrenocortical tissue function as evidenced by blunted corticosterone responses following capture and restraint as well as blunted responses following exogenous ACTH treatment. Similarly, tourist-exposed marine iguanas in the Galápagos exhibit reduced stress responses to capture and restraint compared with more isolated animals (Romero & Wikelski, 2002). Attenuated acute stress responses in habituated animals could be potentially detrimental because acute responses are necessary for normal "fight-or-flight" reactions (Sapolsky *et al.*, 2000; Romero, 2004).

Unfortunately, endocrine and infection measures have never been combined in any study on habituation or ecotourism. In one case, intestinal parasite infection was higher in habituated versus non-habituated gorillas in Bwindi Impenetrable National Park, Uganda (Kalema, 1995). This could be due to infections transmitted from human contact, from the stress of constant human presence, or other ecological factors. Amongst the Karisoke mountain gorilla population, Czekala & Robbins (2001) report that fecal samples have been collected from the night nests of habituated and unhabituated silverbacks for the purpose of cortisol analysis, although the results are not yet published. A logical hypothesis for any species under investigation would be that habituation and altered frequency of human presence/contact is associated with changes in glucocorticoid and infection levels.

Behavioral responses to habituation and tourism have been evaluated in the Kanyanchu chimpanzee group of Kibale National Park, Uganda, and reactions of animals included fleeing, charging, vocalizations, and hiding, which would indicate fear and stress (Johns, 1996). Endocrine stress responses to habituation and tourism have not been evaluated in this population, although investigators with Makerere University and the Jane Goodall Institute are presently conducting such studies in other primate communities in the region (M. Housholder, personal communication, 2006). Similarly, researchers with the Dian Fossey Gorilla Foundation International have been investigating endocrine stress responses in gorillas visited by tourists on a daily basis compared with those gorillas previously habituated but not currently used for ecotourism (T. Stoinski, personal communication, 2006). It would be most fruitful to pair detailed parasitological analyses with these and other similar studies.

M. Muehlenbein, M. Ancrenaz, and colleagues have been undertaking one such study in the lower Kinabatangan Wildlife Sanctuary, Sabah, Malaysia. Fecal samples are collected from orangutans involved with the Kinabatangan Orangutan Conservation Project and Red Ape Encounters ecotourism program and analyzed for both intestinal parasites and cortisol before, during, and after controlled visits from tourist groups of varying sizes. Results of this study are pending, but it is hopeful that such studies can help produce guidelines that will ensure the long-term well-being of these animals.

Nutrition, cortisol, and disease

Food availability and distribution are limiting factors for animal group size (Sterk *et al.*, 1997; Chapman *et al.*, 2004). Nutrition also limits an animal's reproductive and immunological capabilities. For example, prolonged energy restriction can lead to immune suppression and increased susceptibility to infection (Klasing, 1998; Koski & Scott, 2001). Strenuous exercise or participation in energetically demanding tasks can also compromise immune function (Shephard *et al.*, 1998). Changes in metabolism following infection as well as in vivo assessments of energy consumption by the immune system suggest that immunocompetence is energetically expensive (Newsholme & Newsholme, 1989; Lockmiller & Deerenberg, 2000). Taken together, evidence suggests that development and activation of immune responses imposes an encompassing stress that can be described as an energetic burden subject to allocation mechanisms (Sheldon & Verhulst, 1996; Raberg *et al.*, 1998; Schmid-Hempel, 2003; Muehlenbein & Bribiescas, 2005). In addition, the activation of steroid hormones that promote tissue anabolism (e.g. testosterone) and catabolism (e.g. cortisol) essentially prevent energy and nutrients from being used for other purposes, like immunocompetence (Wedekind & Folstad, 1994; Sheldon & Verhulst, 1996; Muehlenbein & Bribiescas, 2005).

The relationships between nutrition, hormones, disease risk, and wildlife abundance are interesting and complicated. Chapman *et al.* (2006) have been the first to investigate such relationships by analyzing associations between food availability (i.e. forest fragment size), gastrointestinal parasite infection, and population size of red colobus monkeys (*Procolobus rufomitratus tephrosceles*) in Kibale National Park, Uganda. Forest decline was associated with increased parasitic infection in the animals, and fecal cortisol levels were directly associated with nematode infection, although cortisol was not related to population change. With decreased nutrient availability, immunity may be impaired, facilitating host colonization. Gastrointestinal parasite infections can also impair host nutrient absorption due to gut damage (Lunn & Northrop-Clewes, 1993). Nutritional demands of the host are also highest during infection, and so the problem becomes amplified in diseased animals (Coop & Holmes, 1996).

The endocrine results from this and similar studies may be difficult to interpret because it is unclear if elevated cortisol levels are a cause or consequence of parasite infection. High cortisol levels could increase susceptibility to infection in some animals. Cortisol levels could also be elevated to facilitate tissue catabolism in nutritionally stressed animals, regardless of infection level. Elevated cortisol levels during infection or starvation could also function to suppress the reproductive system so that the host can forego current reproductive events for future ones during more opportune times (Muehlenbein & Bribiescas, 2005). Further elucidating the intricate relationships between nutrition, disease, and stress will be valuable for conservation biologists.

Future directions

The benefits of utilizing endocrine measures in studies of primate parasite ecology are tremendous. For behavioral ecologists, this will allow for further detailed clarification of the hormonal and disease correlates with dominance rank and social stress. For conservation biologists, incorporation of endocrine measures may benefit wildlife management strategies by detailing the relationships between nutrition, disease risk, and physiological stress as well as begin to describe any possibly detrimental physiological effects of habituation and ecotourism. Others will also benefit from the further development of non-invasive measures of endocrine and immune functions, particularly physiological ecologists attempting to clarify immune–endocrine interactions in wild primates as well as those researchers involved in primate psychoneuroimmunology.

As detailed above, non-invasive methods of sample collection and hormone analysis have been used to monitor reproductive condition as well as physiological stress responses to environmental and social stimuli in primates. These non-invasive methods can yield information about baseline physiological status in wild animals that cannot otherwise be gleaned using more invasive methods. However, in the absence of blood samples, we are left with relatively few measures of immune function. Parasite egg/cyst/larvae abundance in any given fecal sample may not directly correlate with the number of parasites in a host at any given time. Importantly, it is unknown whether or not parasite excretion reflects the immune status of a host. It is, however, possible that parasite richness (the number of unique intestinal parasite species recovered from a host's fecal sample) may reflect the ability of the host to control infections with multiple parasites at any given time (Muehlenbein, 2006). Other future non-invasive measures in primates may include the quantification of secretory IgA in feces (Hau *et al.*, 2001) as well as β-2 microglobulin, neopterin, and C-reactive protein in urine. Paired with endocrinological assays, these and other non-invasive measures will surely be useful in primate parasite ecology.

A single biological sample can now bear information about hormone levels, infection status, and genotype, all of which can complement current methods in primatology. One future avenue of research to utilize such multifactorial measures will include psychoneuroimmunology, or the study of complex relationships between the brain, particularly behavioral aspects of emotion and personality, and immune function, and how these relationships are mediated via hormones and other chemical messengers (Moynihan & Ader, 1996; Segerstrom *et al.*, 2006). For example, personality or "behavioral style" correlates with various physiological measures associated with survival in adult male rhesus macaques infected with simian immunodeficiency virus, and personality factors (particularly sociability, or the tendency to engage in affiliative interactions) may likely mediate disease outcomes (Capitanio *et al.*, 1999). Similarly, antibody responses to tetanus toxoid booster immunization are elevated in "high-sociable" male rhesus macaques following social separation compared to "low-sociable" animals (Maninger *et al.*, 2003). Behavioral styles also correlate with hormone levels in primates (Virgin & Sapolsky, 1997). How behavioral styles translate into disease susceptibility, as mediated through endocrine responses, will be interesting to investigate, particularly in wild primates. Behavioral endocrinology in general will benefit greatly from the incorporation of immune measures.

Physiological ecologists will also benefit from the utilization of the methodologies discussed here, particularly when it comes to further detailed description of immune–endocrine interactions in wild primate populations. In particular, the "immunocompetence handicap hypothesis" is yet to be evaluated in primates. This hypothesis states that testosterone may balance the competing demands of increased reproductive success afforded by exaggerated secondary sexual characteristics with increased susceptibility to infection (Folstad & Karter, 1992). That is, the immunosuppressive actions of testosterone should make secondary sexual characteristics (which rely on testosterone) costly and honest signals that only certain males can maintain. Maintaining high testosterone levels in order to bolster reproductive effort may also theoretically reduce the amount of energy and/or nutrients available for energetically expensive immune responses (Wedekind & Folstad, 1994; Sheldon & Verhulst, 1996; Muehlenbein & Bribiescas, 2005).

Testosterone is associated with the coloration of sexual skin in male rhesus macaques and mandrills (Vandenbergh, 1965; Setchell & Dixson, 2001), however no studies to date have investigated whether or not colorations in male primates are honest signals of immunocompetence. It may be that these are reliable indicators of pathogen resistance (Hamilton & Zuk, 1982) which are maintained via testosterone's antagonist pleiotropic interactions with the immune and reproductive systems (Folstad & Karter, 1992). The utilization of both endocrinological and immunological measures will be useful in evaluating these and other aspects of primate life histories.

References

Abbott, D. H., Kevernem, E. B., Bercovitch, F. B. *et al.* (2003). Are subordinates always stressed? A comparative analysis of rank differences in cortisol levels among primates. *Hormones and Behavior*, **43**, 67–82.

Adams, H. R., Sleeman, J. M., Rwego, I. & New, J. C. (2001). Self-reported medical history survey of humans as a measure of health risk to the chimpanzees (*Pan troglodytes schweinfurthii*) of Kibale National Park, Uganda. *Oryx*, **35**, 308–312.

Alberts, S. C., Sapolsky, R. M. & Altmann, J. (1992). Behavioral, endocrine, and immunological correlates of immigration by an aggressive male into a natural primate group. *Hormones and Behavior*, **26**, 167–178.
Anderson, G. L., Limacher, M. and the Women's Health Initiative Steering Committee. (2004). Effects of conjugated equine estrogen in postmenopausal women with hysterectomy. *Journal of the American Medical Association*, **291**, 1701–1712.
Asaba, K., Iwasaki, Y., Yoshida, M. *et al.* (2004). Attenuation by reactive oxygen species of glucocorticoid suppression on proopiomelanocortin gene expression in pituitary corticotroph cells. *Endocrinology*, **145**, 39–42.
Bahr, N. I., Palme, R., Mohle, U., Hodges, J. K. & Heistermann, M. (2000). Comparative aspects of the metabolism of cortisol in three individual nonhuman primates. *General and Comparative Endocrinology*, **117**, 427–438.
Bambino, T. H. & Hsueh, A. J. (1981). Direct inhibitory effect of glucocorticoids upon testicular luteinizing hormone receptor and steroidogenesis in vivo and in vitro. *Endocrinology*, **108**, 2142–2148.
Bercovitch, F. B. & Clarke, A. S. (1995). Dominance ranks, cortisol concentrations, and reproductive maturation in male rhesus macaques. *Physiology and Behavior*, **58**, 215–221.
Borras, C., Gambini, J. & Vina, J. (2007). Mitochondrial oxidant generation is involved in determining why females live longer than males. *Frontiers in Bioscience*, **12**, 1008–1013.
Bribiescas, R. (2001). Reproductive ecology and life history of the human male. *Yearbook of Physical Anthropology*, **44**, 148–176.
Capitanio, J. P., Mendoza, S. P. & Baroncelli, S. (1999). The relationship of personality dimensions in adult male rhesus macaques to progression of simian immunodeficiency virus disease. *Brain, Behavior, and Immunity*, **13**, 138–154.
Chapman, C. A., Chapman, L. J., Naughton-Treves, L., Lawes, M. J. & McDowell, L. R. (2004). Predicting folivorous primate abundance: validation of a nutritional model. *American Journal of Primatology*, **62**, 55–69.
Chapman, C. A., Wasserman, M. D., Gillespie, T. R. *et al.* (2006). Do food availability, parasitism, and stress have synergistic effects on red colobus populations living in forest fragments? *American Journal of Physical Anthropology*, **131**, 525–534.
Coe, C. L., Savage, A. & Bromley, L. J. (1992). Phylogenetic influences on hormone levels across the primate order. *American Journal of Primatology*, **28**, 81–100.
Coop, R. L. & Holmes, P. H. (1996). Nutrition and parasite interaction. *International Journal of Parasitology*, **26**, 951–962.
Cutolo, M., Capellino, S., Sulli, A. *et al.* (2006). Estrogens and autoimmune diseases. *Annals of the New York Academy of Sciences*, **1089**, 538–547.
Czekala, N. & Robbins, M. M. (2001). Assessment of reproduction and stress through hormone analysis in gorillas. In *Mountain Gorillas: Three Decades of Research at Karisoke*, ed. M. M. Robbins, P. Sicotte & K. J. Stewart. New York, NY: Cambridge University Press, pp. 317–339.
Dabbs, J. M. (1996). Testosterone, aggression, and delinquency. In *Pharmacology, Biology, and Clinical Applications of Androgens*, ed. S. Bhasin, H. L. Gabelnick, J. M. Spieler *et al.* New York, NY: Wiley-Liss, pp. 179–190.

Eley, R. M., Strum, S. C., Muchemi, G. & Reid, G. D. F. (1989). Nutrition, body condition, activity patterns, and parasitism of free-ranging troops of olive baboons (*Papio anubis*) in Kenya. *American Journal of Primatology*, **18**, 209–219.

Folstad, I. & Karter, A. J. (1992). Parasites, bright males and the immunocompetence handicap. *American Naturalist*, **139**, 603–622.

Goymann, W. & Wingfield, J. C. (2004). Allostatic load, social status and stress hormones: the costs of social status matter. *Animal Behaviour*, **67**, 591–602.

Griffin, J. E. & Ojeda, S. R. (eds.) (2000). *Textbook of Endocrine Physiology*. New York, NY: Oxford University Press.

Grossman, C. J. (ed.) (1995). *Bilateral Communication Between the Endocrine and Immune Systems*. New York, NY: Springer-Verlag.

Hamilton, W. D. & Zuk, M. (1982). Heritable true fitness and bright birds: a role for parasites? *Science*, **218**, 384–387.

Hau, J., Andersson, E. & Carlsson, H. E. (2001). Development and validation of a sensitive ELISA for quantification of secretory IgA in rat saliva and faeces. *Lab Animal*, **35**, 301–306.

Hausfater, G. & Watson, D. F. (1976). Social and reproductive correlates of parasite ova emissions by baboons. *Nature*, **262**, 688–689.

Heistermann, M., Palme, R. & Ganswindt, A. (2006). Comparison of different enzyme immunoassays for assessment of adrenocortical activity in primates based on fecal analysis. *American Journal of Primatology*, **68**, 257–273.

Hodges, J. K. & Heistermann, M. (2003). Field endocrinology: monitoring hormonal changes in free-ranging primates. In *Field and Laboratory Methods in Primatology: A Practical Guide*, ed. J. M. Setchell & D. J. Curtis. New York, NY: Cambridge University Press, pp. 282–294.

Homsey, J. (1999). Ape tourism and human diseases: how close should we get? Report for the International Gorilla Conservation Programme Regional Meeting, Rwanda.

Johns, B. G. (1996). Responses of chimpanzees to habituation and tourism in the Kibale Forest, Uganda. *Biological Conservation*, **78**, 257–262.

Kalema, G. (1995). Epidemiology of the intestinal parasite burden of mountain gorillas in Bwindi Impenetrable National Park, S.W. Uganda. *Newsletter of the British Veterinary Zoological Society*, Autumn, 18–34.

Ketterson, E. D., Nolan, V. Jr., Wolf, L. & Ziegenfus, C. (1992). Testosterone and avian life histories: effects of experimentally elevated testosterone on behavior and correlates of fitness in the darkeyed junco (*Junco hyemalis*). *American Naturalist*, **140**, 980–999.

Klasing, K. C. (1998). Nutritional modulation of resistance to infectious diseases. *Poultry Science*, **77**, 1119–1125.

Koski, K. G. & Scott, M. E. (2001). Gastrointestinal nematodes, nutrition and immunity: breaking the negative spiral. *Annual Review of Nutrition*, **21**, 297–321.

Lockmiller, R. L. & Deerenberg, C. (2000). Trade-offs in evolutionary immunology: just what is the cost of immunity? *Oikos*, **88**, 87–98.

Lunn, P. G. & Northrop-Clewes, C. A. (1993). The impact of gastrointestinal parasites on protein-energy malnutrition in man. *Proceedings of the Nutrition Society*, **52**, 101–111.

Lynch, J. W., Ziegler, T. E. & Strier, K. B. (2002). Individual and seasonal variation in fecal testosterone and cortisol levels of wild male tufted capuchin monkeys, *Cebus paella nigritus*. *Hormones and Behavior*, **41**, 275–287.

Maninger, N., Capitanio, J. P., Mendoza, S. P. & Mason, W. A. (2003). Personality influences tetanus-specific antibody response in adult male rhesus macaques after removal from natal group and housing relocation. *American Journal of Primatology*, **61**, 73–83.

Marler, C. A., Walsberg, G., White, M. L. & Moore, M. C. (1995). Increased energy expenditure due to the increased territorial defense in male lizards after phenotypic manipulation. *Behavioral Ecology and Sociobiology*, **37**, 225–231.

McEwen, B. S. (1998). Stress, adaptation, and disease: allostasis and allostatic load. *Annals of the New York Academy of Sciences*, **840**, 33–44.

Milton, K. & Demment, M. W. (1988). Digestion and passage kinetics of chimpanzees fed high and low fiber diets and comparison with human data. *Journal of Nutrition*, **118**, 1082–1088.

Mohle, U., Heistermann, M., Palme, R. & Hodges, J. K. (2002). Characterization of urinary and fecal metabolites of testosterone and their measurement for assessing gonadal endocrine function in male nonhuman primates. *General and Comparative Endocrinology*, **129**, 135–145.

Moynihan, J. A. & Ader, R. (1996). Psychoneuroimmunology: animal models of disease. *Psychosomatic Medicine*, **58**, 546–558.

Muehlenbein, M. P. (2006). Intestinal parasite infections and fecal steroid levels in wild chimpanzees. *American Journal of Physical Anthropology*, **130**, 546–550.

Muehlenbein, M. P. & Bribiescas, R. G. (2005). Testosterone-mediated immune functions and male life histories. *American Journal of Human Biology*, **17**, 527–558.

Muehlenbein, M. P., Watts, D. P. & Whitten, P. L. (2004). Dominance rank and fecal testosterone levels in adult male chimpanzees (*Pan troglodytes schweinfurthii*) at Ngogo, Kibale National Park, Uganda. *American Journal of Primatology*, **64**, 71–82.

Muller-Graf, C. D. M., Collins, D. A. & Woolhouse, M. E. J. (1996). Intestinal parasite burden in five troops of olive baboons (*Papio cynocephalus anubis*) in Gombe Stream National Park, Tanzania. *Parasitology*, **112**, 489–497.

Newsholme, P. & Newsholme, E. A. (1989). Rates of utilization of glucose, glutamine and oleate and formation of end-products by mouse peritoneal macrophages in culture. *Biochemistry*, **261**, 211–218.

Raberg, L., Grahn, M., Hasselquist, D. & Svensson, E. (1998). On the adaptive significance of stress-induced immunosuppression. *Proceedings of the Royal Society of London, Series B, Biological Sciences*, **265**, 1637–1641.

Robbins, M. M. & Czekala, N. M. (1997). A preliminary investigation of urinary testosterone and cortisol levels in wild male mountain gorillas. *American Journal of Primatology*, **43**, 51–64.

Romero, L. M. (2004). Physiological stress in ecology: lessons from biomedical research. *Trends in Ecology and Evolution*, **19**, 249–255.

Romero, L. M. & Wikelski, M. (2002). Exposure to tourism reduces stress-induced corticosterone levels in Galápagos marine iguanas. *Biological Conservation*, **108**, 371–374.

Sapolsky, R. M. (1985). Stress-induced suppression of testicular function in wild baboons: role of glucocorticoids. *Endocrinology*, **116**, 2273–2278.

Sapolsky, R. M. (2005). The influence of social hierarchy on primate health. *Science*, **308**, 648–652.

Sapolsky, R. M. & Spencer, E. M. (1997). Insulin-like growth factor I is suppressed in socially subordinate male baboons. *American Journal of Physiology: Regulatory, Integrative and Comparative Physiology*, **273**, R1346–R1351.

Sapolsky, R. M., Romero, L. M. & Munck, A. U. (2000). How do glucocorticoids influence stress-responses? Integrating permissive, suppressive, stimulatory, and adaptive actions. *Endocrine Reviews*, **21**, 55–89.

Schmid-Hempel, P. (2003). Variation in immune defence as a question of evolutionary ecology. *Proceedings of the Royal Society of London, Series B, Biological Sciences*, **270**, 357–366.

Segerstrom, S. C., Lubach, G. R. & Coe, C. L. (2006). Identifying immune traits and biobehavioral correlates: generalizability and reliability of immune responses in rhesus macaques. *Brain, Behavior and Immunity*, **20**, 349–358.

Setchell, J. M. & Dixson, A. F. (2001). Changes in the secondary sexual adornments of male mandrills (*Mandrillus sphinx*) are associated with gain and loss of alpha status. *Hormones and Behavior*, **39**, 177–184.

Sheldon, B. C. & Verhulst, S. (1996). Ecological immunology: costly parasite defenses and trade-offs in evolutionary ecology. *Trends in Ecology and Evolution*, **11**, 317–321.

Shephard, R. J., Castellani, J. W. & Shek, P. N. (1998). Immune deficits induced by strenuous exertion under adverse environmental conditions: manifestations and countermeasures. *Critical Reviews in Immunology*, **18**, 545–568.

Soronen, P., Laiti, M., Torn, S. *et al.* (2004). Sex steroid hormone metabolism and prostate cancer. *Journal of Steroid Biochemistry and Molecular Biology*, **92**, 281–286.

Sousa, M. B. C. & Ziegler, T. E. (1998). Diurnal variation on the excretion patterns of fecal steroids in common marmoset (*Callithrix jacchus*) females. *American Journal of Primatology*, **46**, 105–117.

Sterk, E. M. H., Watts, D. P. & van Schaik, C. P. (1997). The evolution of female social relationships in nonhuman primates. *Behavioral Ecology and Sociobiology*, **41**, 291–309.

Strier, K. B. & Ziegler, T. E. (2005). Variation in the resumption of cycling and conception by fecal androgen and estradiol levels in female northern muriquis (*Brachyteles hypoxanthus*). *American Journal of Primatology*, **67**, 69–81.

Vandenbergh, J. G. (1965). Hormonal basis of sex skin in male rhesus monkeys. *General and Comparative Endocrinology*, **56**, 31–34.

Vina, J., Sastre, J., Pallardo, F. V., Gambini, J. & Borras, C. (2006) Role of mitochondrial oxidative stress to explain the different longevity between genders: protective effect of estrogens. *Free Radical Research*, **40**, 1359–1365.

Virgin, C. E. Jr. & Sapolsky, R. M. (1997). Styles of male social behavior and their endocrine correlates among low-ranking baboons. *American Journal of Primatology*, **42**, 25–39.

Walker, B. G., Boersma, P. D. & Wingfield, J. C. (2005). Field endocrinology and conservation biology. *Integrative and Comparative Biology*, **45**, 12–18.

Walker, B. G., Boersma, P. D. & Wingfield, J. C. (2006). Habituation of adult Magellanic penguins to human visitation as expressed through behavior and corticosterone secretion. *Conservation Biology*, **20**, 146–154.

Wallis, J. & Lee, D. R. (1999). Primate conservation: the prevention of disease transmission. *International Journal of Primatology*, **20**, 803–826.

Washburn, B. E. & Millspaugh, J. J. (2002). Effects of simulated environmental conditions on glucocorticoid metabolite measurements in white-tailed deer feces. *General and Comparative Endocrinology*, **127**, 217–222.

Wasser, S. K., Thomas, R., Lair, P. P. *et al.* (1993). Effects of dietary fibre on faecal steroid measurements in baboons (*Papio cynocephalus cynocephalus*). *Journal of Reproduction and Fertility*, **97**, 569–574.

Wasser, S. K., Hunt, K. E., Brown, J. L. *et al.* (2000). A generalized fecal glucocorticoid assay for use in a diverse array of nondomestic mammalian and avian species. *General and Comparative Endocrinology*, **120**, 260–275.

Wedekind, C. & Folstad, I. (1994). Adaptive or nonadaptive immunosuppression by sex-hormones. *American Naturalist*, **143**, 936–938.

Whitten, P. L., Brockman, D. K. & Stavisky, R. C. (1998). Recent advances in noninvasive techniques to monitor hormone-behavior interactions. *Yearbook of Physical Anthropology*, **41**, 1–23.

Wilson, M. & Daly, M. (1985). Competitiveness, risk taking, and violence: the young male syndrome. *Ethology and Sociobiology*, **6**, 59–73.

Woodford, M. H., Butynski, T. M. & Karesh, W. B. (2002). Habituating the great apes: the disease risks. *Oryx*, **36**, 153–160.

Wingfield, J. C., Hegner, R. E., Dufty, A. M. & Ball, G. F. (1990). The "challenge hypothesis": theoretical implications for patterns of testosterone secretion, mating systems, and breeding strategies. *American Naturalist*, **136**, 829–846.

Worthman, C. M. & Stallings, J. F. (1997). Hormone measures in finger-prick blood spot samples: new field methods for reproductive endocrinology. *American Journal of Physical Anthropology*, **104**, 1–22.

Ziegler, T. E. & Wittwer, D. J. (2005). Fecal steroid research in the field and laboratory: improved methods for storage, transport, processing, and analysis. *American Journal of Primatology*, **67**, 159–174.

Ziegler, T. E., Sholl, S. A., Scheffler, G., Hagerty, M. A. & Lasley, B. L. (1989). Excretion of estrone, estradiol, and progesterone in urine and feces of the female cotton top tamarin (*Saguinus oedipus oedipus*). *American Journal of Primatology*, **17**, 185–195.

Zirkin, B. R. & Chen, H. (2000). Regulation of Leydig cell steroidogenic function during aging. *Biology of Reproduction*, **63**, 977–981.

5 *Using agent-based models to investigate primate disease ecology*

CHARLES L. NUNN

Photograph by Tara Harris (*Colobus guereza*)

Introduction

Infectious disease has played a major role in human history and evolution, and a wide diversity of parasites continues to impact human health around the world. Infectious disease also plays a major role in the lives of non-human primates (Nunn & Altizer, 2006). In fact, wild primates harbor many of the parasites and

Primate Parasite Ecology. The Dynamics and Study of Host–Parasite Relationships, ed. Michael A. Huffman and Colin A. Chapman. Published by Cambridge University Press.

pathogens that also infect humans (Pedersen *et al.*, 2005), such as yellow fever virus, *Schistosoma*, and the simian equivalent of human immunodeficiency virus. Many of these infectious diseases can have tremendous negative impacts on wild primate populations, as seen, for example, in the large die-offs of African apes due to Ebola infection (Walsh *et al.*, 2003; Leroy *et al.*, 2004; Bermejo *et al.*, 2006). Understanding the dynamics of infectious diseases in primates can provide key insights to the origins of human diseases, their spread through human populations, and the evolutionary impacts of pathogens on host behavior. Moreover, knowledge of infectious agents and their effects could be increasingly important for primate conservation efforts (Nunn & Altizer, 2006), especially in the context of global climate change (Chapman *et al.*, 2005; Nunn *et al.*, 2005).

Investigating primate disease ecology can be accomplished using two main approaches. One approach combines field and laboratory methods to build "baseline" data on infectious diseases that occur naturally in wild primate populations (Leendertz *et al.*, 2006). With such data, it becomes possible to pinpoint emerging infectious agents amidst the background of endemic parasites and pathogens (Leendertz *et al.*, 2006), and to understand the factors that lead to differences in parasitism among individuals, populations, and even species (e.g. McGrew *et al.*, 1989a, b; Muller-Graf *et al.*, 1996, 1997; Moore & Wilson, 2002). Baseline data also provide a means to understand whether parasites are shared among host species (De Gruijter *et al.*, 2004, 2005), although it is not always possible to obtain baseline data before a conservation threat is identified. The results from such studies are often largely descriptive, but their value should not be underestimated, and collecting these data requires diligent and time-consuming efforts in the field and the laboratory.

The second set of tools is more abstract. These methods make use of statistical procedures that synthesize parasite data from multiple locations to identify broad-scale ecological and evolutionary patterns, and development of theoretical models to investigate the factors that can lead to parasite spread. One example includes phylogeny-based comparative methods that allow biologists to investigate the evolutionary drivers of parasitism (Poulin, 1995, 1997; Morand & Harvey, 2000). These methods have been used in primates to study associations between host and parasite traits (Sorci *et al.*, 1997), to examine whether parasitism correlates with host extinction risk (Altizer *et al.*, 2007; Nunn *et al.*, 2007), to test whether parasite richness correlates with host population density (Nunn *et al.*, 2003), and to investigate whether primate species that live near the equator have more parasites (Nunn *et al.*, 2005). Other tools allow an investigator to study the geographical distribution of parasites, for example to identify gaps in our knowledge of parasites (Hopkins & Nunn, 2007), or to investigate parasitism in relation to host geographic range overlap and rates of host diversification (Nunn *et al.*, 2004). These approaches are beginning to reveal the factors that account for broad patterns of parasitism among host lineages (Poulin & Morand, 2004; Nunn & Altizer, 2006). However, we are still in the infancy of understanding the geographic and phylogenetic drivers of disease risk in primates, and replicating tests in other mammalian groups will be critical for understanding the generality of patterns found in primates (e.g. Ezenwa *et al.*, 2006; Lindenfors *et al.*, 2007).

Another set of theoretical approaches has been used to investigate the dynamics of parasites within host populations. These tools provide a means to investigate whether particular host characteristics impact the spread of disease. For example, sexually transmitted diseases (STDs) might spread more readily in populations of promiscuous individuals, as compared with populations of monogamous hosts. Models can also investigate how characteristics of the parasites interact with host traits. For example, interactions between parasite infectious period and host dispersal are critically important in socially structured populations, such as primates: when host dispersal rates are low, parasites will need a longer infectious period to become established in the population (Cross *et al.*, 2005). Importantly, these models can be spatially explicit and parameterized with real-world characteristics, including characteristics of geographic areas where disease outbreaks occur, such as rivers and mountain ranges (Smith *et al.*, 2002; Ostfeld *et al.*, 2005).

A variety of modeling approaches have been developed to study the spread of infectious diseases in humans and wildlife (Anderson & May, 1991; Hudson *et al.*, 2002). These models have been used, for example, to investigate patterns of cowpox transmission in wild rodent populations (Begon *et al.*, 1999) and to study the links between population density and brucellosis infection in bison (Dobson & Meagher, 1996). A primary goal of many modeling studies is to calculate the basic reproductive rate of the pathogen, or R_0. When R_0 exceeds 1, the pathogen is expected to persist in a population; when R_0 is less than 1, the pathogen is expected to go extinct (Anderson & May, 1991). Many analytical models make simplistic assumptions about host social organization, variation in contact among individuals in groups, and the spatial configuration of groups. For example, a model might assume that individuals contact all other individuals at an equal rate, or that dispersal can occur from one group to any other group in the population (essentially, a non-spatial model). Nunn & Altizer (2006) provide an overview of the basics of epidemiological modeling as applied to primates.

In this chapter, I focus on a particular type of modeling approach that offers great promise for modeling the spread of infectious diseases in primate social systems. This approach involves agent-based modeling, which has increased in prominence as a tool for modeling complex systems in ecology (Grimm & Railsback, 2005; Grimm *et al.*, 2005). A major advantage of agent-based models is that they overcome many of the simplifying assumptions that are often

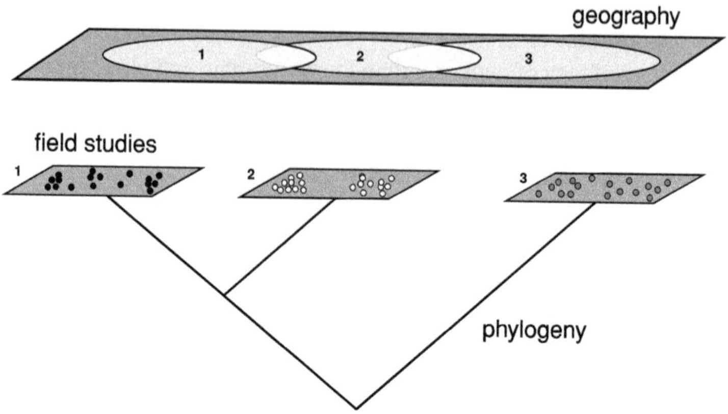

Figure 5.1. Agent-based models as a link between field research and broad-scale ecological and evolutionary research. Numbers refer to the geographical and phylogenetic relationships of three different species. Field research is revealing important roles of infectious disease in primate ecology, but we lack an understanding of how processes within and between primate groups scale up to larger patterns at the population and species level. Modeling approaches – including agent-based models – provide a means to understand these processes and thus the drivers of broader geographical and phylogenetic patterns. Modeling therefore serves as a crucial component in an integrative approach to disease ecology (Nunn & Altizer, 2006).

required of analytical models; indeed, agent-based models are increasingly used in primate ecology and behavior, for example in the context of gorilla population dynamics and dispersal (Robbins & Robbins, 2004, 2005), fission-fusion social systems in spider monkeys (Ramos-Fernandez *et al.*, 2006), and the dynamics of ape sociality (te Boekhorst & Hogeweg, 1994). A limitation of agent-based models is that they can be cumbersome to describe fully, making it difficult for others to understand all the assumptions, to make sense of output, and to replicate the study. Moreover, if the developer of a model is not careful to include a wide range of values for each of the input parameters, agent-based models will provide less generality than an analytical model. Dunbar (2002) provides an overview of modeling approaches – including agent-based models – in studies of primate behavioral ecology.

Agent-based modeling can be used in an integrative framework to better understand comparative patterns and results from the field. As noted above, comparative and broad-scale ecological studies provide key insights to the factors that lead to different numbers and types of parasites in different evolutionary lineages. These studies rely on parasite data from the field. What is lacking, however, is an understanding of the processes that generate and maintain the broad evolutionary patterns uncovered in comparative studies (Figure 5.1).

The patterns revealed in these broad-scale studies are ultimately generated through ecological and evolutionary processes operating within species (Brown, 1995). By focusing on the processes that generate variation in parasitism within populations, agent-based models can therefore provide greater linkage between field and comparative approaches, and such models can lead to new predictions for field and comparative research.

In what follows, I begin with three examples of how agent-based modeling has been applied to investigate the role of infectious disease in animals. I then briefly discuss some of the practicalities of agent-based modeling, including programming strategies and tools; the goal is to sketch a road map that primatologists and parasitologists can follow to initiate their own agent-based models of disease dynamics. I conclude by identifying several outstanding questions to be addressed in future studies of primate disease ecology, including an example of how to use agent-based models to generate testable predictions. Throughout, it is important to keep in mind that the art of model building is to balance simplification with the complexity of the real world; a model that is too simplified relative to the phenomena in question can fail to satisfactorily answer the major questions, while a model that is too complex is no easier to study than reality.

Agent-based modeling and disease ecology: background and three examples

Background and terminology

An agent-based model makes use of individual "agents" that are generated and live in computer memory. These agents can exhibit heterogeneities in sex, age, dominance rank, and other traits that are of interest in the context of disease dynamics. The model will typically have a time unit over which the agents interact, ranging from a single day to months, and hereafter called a "time step." These interactions therefore have a time component, and if the model is spatially explicit, the interactions can also be restricted in space. Agent-based models address questions numerically, i.e. by using simulations, rather than seeking an analytical solution, such as through use of equations to identify equilibrium points. Thus, the interactions among agents often have a stochastic component, meaning that a particular type of interaction, such as mating, occurs with some probability in each time step, rather than occurring deterministically.

After iteratively running the simulation multiple times, it is possible to identify general patterns that emerge from individual interactions in the simulation. These "emergent properties" include features such as equilibrium prevalence, rates of population decline, and even the degree of host sociality, all of which occur at the population level rather than the individual level. Thus, an important aspect of an agent-based model is that individual interactions lead to emergent properties that cannot be predicted in advance of running the simulations.

It is important to consider what type of modeling approach to use to address a particular question (Dunbar, 2002). Given some of the limitations of agent-based modeling mentioned above, why should a researcher develop an agent-based model, as compared with other types of models? One reason is that primate social systems violate many of the assumptions of the standard modeling approaches. For example, most epidemiological models become difficult (if not impossible) to solve analytically when one has to incorporate spatial structure, where structure involves both contact patterns of individuals within social groups and the spatial proximity of groups in the population. In addition, interactions can be asymmetrical within groups; a dominant male is more likely to have mating access to more females than a subordinate male, for example, and therefore is more likely to host a wider diversity of STDs (Graves & Duvall, 1995; Thrall *et al.*, 2000; Nunn & Altizer, 2004). Decisions might also hinge on the desired outcome of a model; if one is interested in the properties that emerge from simple behavioral rules, then an agent-based model is likely to be most appropriate (Grimm & Railsback, 2005). A final consideration when selecting a modeling approach involves its flexibility – is it relatively easy to alter the model to study new questions? Agent-based models, if built with foresight for extension and flexibility, can often be used as a platform for further model development, specifically by adding new variables to an existing "core" of social and ecological interactions (Grimm & Railsback, 2005). In addition, new questions can often be asked of existing sets of simulated data, or the data can be investigated in more than one way (e.g. by using different statistical methods).

Example 1: Spread of sexually transmitted diseases in animal mating systems

A wide variety of sexually transmitted diseases can be found in primates, including papillomaviruses, herpesviruses, and retroviruses, such as simian immunodeficiency virus (SIV) and simian T-lymphotropic virus (STLV). Little is known about the effects of these STDs on primate hosts, but a syphilis-like genital infection led to severe disfigurement and deaths in a troop of baboons at Gombe (Wallis & Lee, 1999), and STDs in wild primates are likely to lead to increased rates of infertility and vertical transmission to offspring (Lockhart *et al.*, 1996). In non-primates, STDs have even been the source of conservation concern (e.g. koalas: McColl *et al.*, 1984; Canfield *et al.*, 1991; Augustine, 1998).

Could STDs act as a selective force on the evolution of animal mating behavior? Female choice, male–male competition, and alternative mating tactics can influence mating success, and through their effects on mating, these factors should also impact the spread of STDs. For example, if females in a social group choose to mate with the same male, we might expect that this male will be more likely to be exposed to (and thus infected with) the STDs that some females in the group carry, which were presumably acquired from males that were previously preferred but now are no longer in the group or have declined in rank (Graves & Duvall, 1995; Thrall *et al.*, 2000; Nunn & Altizer, 2004). Thus, from the standpoint of selection on female choice, this male is likely to serve as the primary conduit for uninfected females to acquire STDs. If the STD is harmful to female reproductive success, these costs could offset any benefits, such as good genes, obtained from mating with the preferred male (Kokko *et al.*, 2002).

Other social factors can also influence the spread of STDs and should therefore be included in a model. For example, immigrants could be the source of new infections, including STDs, provided that immigrants have mating experience in their previous groups and remain infected after successful transfer to a new group. This and other considerations led Freeland (1976) to suggest that primates might be selected to "challenge" outsiders who are attempting to immigrate, specifically as a way to determine if they are carrying any infectious organisms. More generally, the risk from immigrants predicts that higher rates of dispersal increase the probability that an STD will become established in a population. Similarly, mortality will tend to result in the elimination of sexually transmitted organisms because the pathogen will die when the host dies (indeed, background mortality is a key parameter in most epidemiological models; Anderson & May, 1991). Thus, in addition to mating behavior itself, sexual transmission in primates involves demographic patterns, such as dispersal, and life history traits, such as mortality.

Thrall *et al.* (2000) investigated many of these variables in an agent-based model that simulated the spread of STDs in a polygynous mating system. The model was initiated by creating an equal number of male and female agents (n = 250 members of each sex). Males were assigned an attractivity code, which was pulled from a lognormal distribution, where the variance of this distribution reflected the degree of polygyny to be simulated. When this variance was high, only a few males were assigned high codes, and only these males attracted females, resulting in extreme polygyny. By contrast, when the variance was low, all males tended to attract an equal number of females, resulting in a system more akin to monogamy (since there was an equal number of male and female agents in the model). Thus, a single parameter – variance in the lognormal distribution – resulted in either high skew (extreme polygyny) or low skew (monogamy), with the mechanism related to either female choice or male–male competition (even though use of the term "attractivity code" implies a mechanism based on female choice).

Once groups were formed in the Thrall *et al.* (2000) model, a single male and a single female were randomly selected as sources of the initial infection. Mating occurred within groups, and females dispersed between groups, with simulations lasting 500 generations (each time step was assumed to equal one mating season). The infection caused sterility in infected individuals, but did not increase the death rate. When an individual died due to background mortality, he or she was replaced with an individual of the same sex in a randomly selected group. This was done to maintain constant population size over time, under the assumption that deaths are balanced by births in a population at demographic equilibrium; reproductive success was assessed by counting, for each agent, the number of mating seasons in which both the focal agent and its partner were healthy. It was assumed that individuals do not recover from the STD, consistent with the long infectious period of infectious agents with sexual transmission (Lockhart *et al.*, 1996).

The simulation model revealed a number of interesting dynamics that provided new insights to the links between mating systems and the spread of STDs in polygynous groups. Among these results, the authors showed that increasing female dispersal tended to favor the establishment of an STD (Thrall *et al.*, 2000). High levels of dispersal are essentially equivalent to increased promiscuity, and at extremely high levels of dispersal, individuals effectively change partners every mating season, resulting in high prevalence of infection in the simulated populations. Another result involved variation in mating success, or reproductive skew. Reproductive skew exhibited a major effect in the simulations, with the prevalence of infection increasing more rapidly for females than for males as skew increased.

The authors also investigated whether introduction of an STD could act as a "brake" on the evolution of polygynous mating systems. Simulations revealed that more attractive males were more likely to become infected with an STD. Because "successful" males also had larger harems, they tended to infect more females, lowering female reproductive success in these groups (due to the effects of sterility caused by the STD). Indeed, the simulations revealed that females mating with more attractive males tended to have lower reproductive success. This effect could therefore limit the evolution of polygyny, depending on the details of host ecology, the mating system, and the effects of the pathogen on host reproduction (see Thrall *et al.*, 2000). However, the situation was different for males, where an STD tended to reduce the reproductive success of dominant males compared to the absence of an STD, but more dominant individuals still benefited from monopolizing more matings. Thus, the presence of an STD could result in a conflict between males and females, with selection on females tending to favor reduced reproductive skew, and selection on (successful) males favoring increased skew.

Another study also investigated whether the presence of an STD could alter mate choice dynamics in a socially structured animal population. Kokko *et al.* (2002) developed both analytical and agent-based approaches to investigate whether STDs impact female choice. In the agent-based model, males varied in quality, with a fraction identified as high quality, and the remaining males considered to be low quality (producing two classes of males: high and low quality). Similarly, females varied in their faithfulness, with a fraction seeking extra-pair copulations when paired with a low-quality male (thus producing two classes of females: faithful and unfaithful). Extra-pair copulations by unfaithful females take place only with high-quality males. The model is spatially explicit and initialized with an equal number of males and females. Females give birth, with the litter size determined as a function of both infection status of the social parents and quality of the male biological parent; infection reduced litter size, while high male quality increased litter size. Infected individuals remained infected for life.

As expected, the prevalence of the STD increased with increasing levels of extra-pair mating. A more surprising result is that the simulations revealed two equilibria in Kokko *et al.*'s (2002) model. If the proportion of unfaithful females is initially low (close to zero), the benefits of extra-pair mating are great, due to the lower prevalence in the population. This leads to selection for extra-pair mating until the fitness benefits of unfaithfulness and fitness costs of the STD are balanced, producing an equilibrium of approximately 15% unfaithful females in the simulated population. If the simulation is initialized with most females being unfaithful, however, the overall prevalence of the STD in the population increased rapidly, resulting in a different equilibrium. In this case, females are likely to become infected by their social partner, and therefore experience fewer costs when seeking matings with more attractive males. Thus, the proportion of unfaithful females approaches 100%. The probability of transmission interacts strongly with the proportion of unfaithful females to determine the resulting equilibrium point (or points); thus, the model provided a means to investigate the complex and unexpected dynamics that can emerge from individual interactions.

In summary, the innovative studies by Thrall *et al.* (2000) and Kokko *et al.* (2002) provide new insights to the links between mating systems, life history, and the spread of STDs. Future work on primate behavior and STDs could focus on three main areas. First, it would be interesting to examine how different probabilities of transmission (male-to-female versus female-to-male) influence the spread of an STD, as many STDs in humans appear to spread more efficiently from male-to-female (e.g. in the case of HIV; Alexander, 1990; Padian *et al.*, 1997). Second, primate social groups are composed of multiple males, whereas Thrall *et al.*'s (2000) model focused on polygynous groups; multiple males could be included in social groups to investigate how this changes infection dynamics. Finally, promiscuity in primates is often driven by infanticide avoidance. Thus, a useful extension of the Kokko *et al.* (2002) framework for primates would consider the benefits of promiscuity to reduce the risk of infanticide, rather than only genetic benefits of mating with a high-quality male. Unexpected dynamics are likely to emerge when an STD is added to the mix of interactions involving group size, composition, and infanticide risk.

Example 2: Spread of emerging infectious diseases in wild primates

Many infectious diseases have no readily discernible effects on host populations, while others, such as Ebola and yellow fever, have caused major die-offs in wild primates (Nunn & Altizer, 2006). For example, a recent study estimates that 5000 gorillas have died of Ebola in the Republic of Congo (Bermejo *et al.*, 2006). Similar effects have been noted for population extinctions of mantled howler monkeys due to yellow fever (Galindo & Srihongse, 1967), while a population of red howlers underwent an 85% decline over a 4-year period due to an unknown infectious disease (Pope, 1998).

Understanding the conditions under which an infectious disease can cause population declines is critically important for primate conservation (Chapman *et al.*, 2005; Nunn & Altizer, 2006) and also for public health (Rouquet *et al.*, 2005). The factors that underlie population declines include both host and parasite traits. For example, we might expect that an STD is less able to establish in a population of monogamous hosts, or that high rates of host dispersal can lead to a "tipping point" at which a highly lethal infectious agent can spread to multiple groups, causing significant mortality. Similarly, a longer incubation period increases the chances that an individual will immigrate into another group before pathology and infectiousness occurs, enabling the infectious agent to overcome limits on disease spread when populations are subdivided into stable social groups (Cross *et al.*, 2005; Loehle, 1995). Finally, when disease-induced mortality (virulence) is low, an infectious agent might spread more widely, while higher mortality can cause the pathogen to "burn out" before many individuals are infected, due to the fact that the pathogen will die when the host dies. In many cases, it is difficult to disentangle whether a parameter relates to the host or the parasite; the value for incubation period, for example, will represent the interaction of many biological factors involving parasite developmental rates, host immune responses to the parasite, and environmental stressors experienced by the host (e.g. food shortages or inclement weather).

In collaboration with P. Thrall, K. Stewart, and A. Harcourt, I developed a spatially explicit, agent-based model to investigate the factors that lead a highly pathogenic disease, such as Ebola, to become established in a primate population (Nunn *et al.*, 2008). The model was initiated by forming groups of males and females on a square lattice (i.e. a matrix), with the number of males and females per group pulled from a Poisson distribution characterized by user-defined values. Once groups were formed, a female was selected randomly as infected. The infection spread to other individuals within the female's group, and could also spread to other groups through dispersal of infected individuals. Importantly, the infectious agent was assumed to cause high mortality, with the peak probability of death over a 9-day infectious period occurring on day 5 and disease-related mortality normally distributed over this infectious period. We also investigated the effects of several infection characteristics, including incubation period, disease-induced mortality, and the probability of transmission within groups, as these will interact with host behavioral patterns to determine whether a pathogen can persist when a population is divided into social groups (Cross *et al.*, 2005).

As with all modeling approaches, it is helpful to have a specific hypothesis in mind when building an agent-based model; this hypothesis can then be evaluated in a framework that is as general as possible with respect to the characteristics of organisms to which the model applies. Often, the hypothesis makes intuitive sense, but it is difficult to know with certainty whether it really works as one thinks, and more importantly, the conditions under which the mechanisms involved will influence disease dynamics. In this case, my coauthors and I tested the hypothesis that a highly pathogenic disease could increase rates of dispersal in polygynous systems, thus leading to increased disease spread to other groups in the population. We called this pathogen-mediated dispersal, or PMD. Our reasoning was based on reports from the literature and our own knowledge of primate biology, and can be summarized as follows. In a polygynous system, such as gorillas, females will disperse if the resident male dies from disease. These dispersing females could therefore carry the disease to other groups (see also Thrall *et al.*, 2000). In polygynandrous systems, such as multi-male, multi-female communities of chimpanzees, some males might die, but it is unlikely that all the males would acquire the disease and die. Thus, a pathogenic disease might cause high mortality in one multi-male, multi-female group, but it is not as likely to increase rates of dispersal from the group, and thus less likely to spread between groups. With so many parameters – group size and composition, rates of dispersal, background mortality, disease-related mortality, and incubation period – it is impossible to comprehend the conditions under which our intuition actually works (if indeed the PMD hypothesis is supported theoretically). We therefore designed a

Table 5.1 *Parameters varied in agent-based model of Nunn* et al. *(2008)*

Definition	Numerical range
Incubation period (individuals are infected but non-infectious)	1 to 10 days
Average number of females in groups (integer values)	1 to 10
Average number of males in groups (integer values)	1 to 5
Baseline probability of female dispersal per day	0 to 0.005
Baseline probability of male dispersal per day	0 to 0.005
Per-contact transmission probability	0 to 0.8
Baseline disease-independent mortality rate per day	0 to 0.001
Mortality multiplier for an infectious individual in a group	0 to 1

Table provides values for eight parameters that were varied randomly in an agent-based model for the spread of a pathogenic disease in a socially structured population (Nunn *et al.*, 2008). The infectious period was held constant at 9 days, and all simulations were run with an initial population size of 2000 individuals. A single female was randomly selected as infected at the start of the simulation, with all subsequent infections arising from host-to-host transmission. Individuals that died from disease were not replaced.

model to investigate these parameters and how they interact to influence disease dynamics.

After developing the model and refining the question, we ran 30 000 simulations under a spatially explicit model, with dispersing individuals moving in a random walk across the lattice until they found a group with one or more members of the opposite sex. Thus, the spread of the infection tended to be local because dispersal was local, radiating out from the initially infected individual. Table 5.1 summarizes the parameters that we varied. Simulations were run with different sets of parameters, and a simulation run was terminated when the prevalence of infection hit zero (which it always did, due to the high disease-related mortality rates used in the simulations). We analyzed the simulation output using regression trees (De'ath & Fabricius, 2000), which provide a means to investigate interactions among traits (and are therefore a useful approach for investigating simulation output, see below).

The simulation revealed several interesting and unexpected patterns. First, we were surprised at how rarely a single infection leads to a massive die-off in the simulated populations. The infection was sustained for an average of only 20.4 time steps before the pathogen went extinct in the simulated population (a time step was equivalent to one day). In no cases did a single infection lead to the complete extinction of the host population. Given the massive die-offs that have been observed in wild populations, this result suggests that most pathogenic emerging diseases require multiple spillover events to cause substantial depressions in host populations. Alternatively, it could be that rates of contact between groups are higher, for example in overlap zones (Walsh

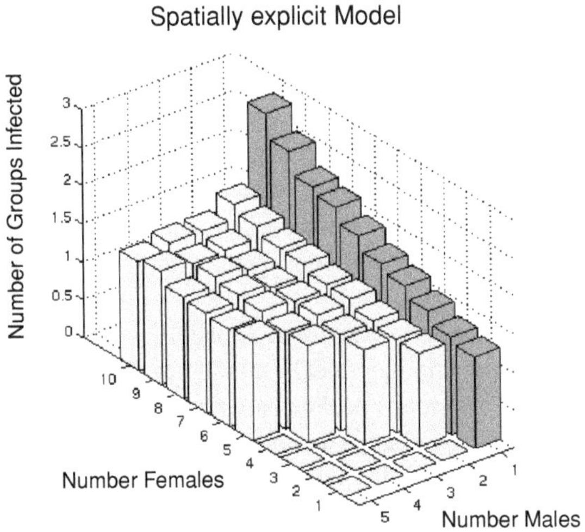

Figure 5.2. Effect of disease in different social systems in a spatially explicit simulation model. Plot shows total number of groups infected. Simulations of a single-male (polygynous) system are shown as dark bars, simulations of non-polygynous systems as light bars. Polyandrous groups were not simulated. Other variables were varied using a random-sampling technique, with ranges for variables given in Table 5.1. Notice that for a given group size, the number of infections was higher in polygynous than non-polygynous groups. From Nunn *et al.* (2008).

et al., 2007), which would increase rates of disease spread among groups. In our model, contact only occurs through dispersal of infected individuals; this assumption could be relaxed in future extensions of the model.

Second, in terms of host parameters, group size had an effect on the number of infections within a single infection run. A statistical analysis of the simulation output revealed that an average of 10.2 infections occur following an initial infection when there are fewer than 6.5 females per group, but 20.9 infections occur when the average number of females is greater than 6.5 (based on a regression tree analysis; see Nunn *et al.*, 2008). This effect makes intuitive sense because most contacts occur within groups, with few opportunities for the infection to move to other groups. Thus, a larger group should lead to more infections simply because most individuals in the group will become infected. A longer incubation period and higher rates of dispersal further increased the number of infections.

We also examined the number of groups that were infected. Analyses revealed that more groups were infected in simulations of single-male than multi-male groups, with an added effect of the number of females in single-male groups (Figure 5.2). We used a regression tree model to make predictions for how mating system and group size interact to influence the number of groups that are infected. These analyses revealed that 1.38 groups were infected in multiple male-groups, 1.51 were infected in single-male groups with fewer than 6.5 females, and 2.34 were infected in single-male groups with more than 6.5 females. As expected under PMD, the effect of polygyny was driven by dispersal of infected individuals when the single male of a polygynous group died, with more instances of dispersal by infected individuals observed in simulations of single-male groups, and with more infected individuals dispersing as group size increased.

The simulations just described were run in a spatially explicit model, based on the reasoning that dispersing individuals should have more difficulty finding groups as the density of groups declines after a disease outbreak, thereby slowing the spread of disease. Indeed, a number of previous studies have shown that spatial structure can significantly impact disease dynamics and patterns of virulence, as well as longer-term evolutionary processes (Thrall & Antonovics, 1995; Gandon *et al.*, 1996; Boots & Sasaki, 1999; Roy & Kirchner, 2000; Carlsson-Graner & Thrall, 2002; O'Keefe & Antonovics, 2002). For comparison, we therefore also ran simulations in a non-spatial model, with dispersing individuals transported to any of the groups in the population regardless of distance from the original group. The differences in the results were striking (Figure 5.3). Specifically, we found that the number of infections increased by two-fold in the non-spatial model (15.6 vs. 36) and the number of group-extinctions increased, on average, by a similar amount (1.49 vs. 3.19 in a non-spatial model).

Finally, we applied the model to study the spread of Ebola in chimpanzees and gorillas, with the model parameterized for seven different ape populations. Based on PMD, we expected that a simulated outbreak of Ebola would result in more gorilla groups being infected than chimpanzee groups, based on the fact that gorillas are generally polygynous while chimpanzees are polygynandrous. We found support for this prediction (Figure 5.4). Evidence thus far available from the field supports the possibility that gorillas suffer more from Ebola than chimpanzees (Bermejo *et al.*, 2006), although quantitative tests are not yet possible. Thus, this example shows how a modeling approach can be taken to understand the social conditions under which a disease can impact host populations. In this case, it provided a way to understand the conditions under which PMD can occur (see also Thrall *et al.*, 2000).

Example 3: Modeling the spatial spread of infectious disease

Spatial context is emerging as a major area of research in epidemiological studies, resulting in the new field of "spatial epidemiology" (Ostfeld *et al.*, 2005). The previous example provided a hint of why spatial context is important

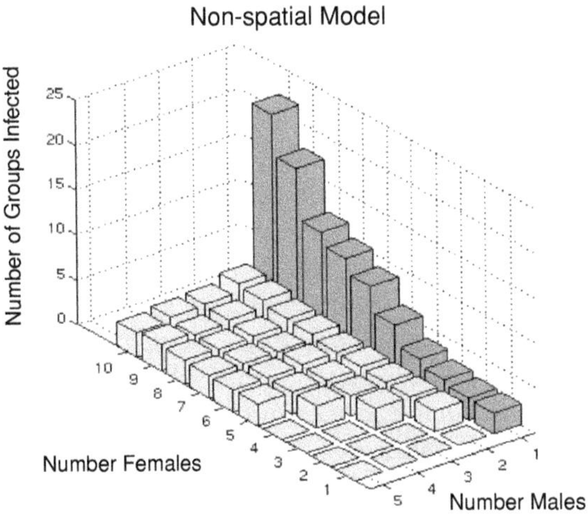

Figure 5.3. Effect of disease in different social systems in a non-spatially explicit simulation model. Plot shows total number of groups infected and is provided for comparison to the spatially explicit model in Figure 5.2. Simulations of a single-male (polygynous) system are shown as dark bars, simulations of non-polygynous systems as light bars. Polyandrous groups were not simulated. Other variables were varied using a random-sampling technique, with ranges for variables given in Table 5.1. As compared with the spatial model, the same conclusions would be reached, but the support is greater in the non-spatial model, due to the fact that disease spreads to more groups when dispersal is not constrained to be local.

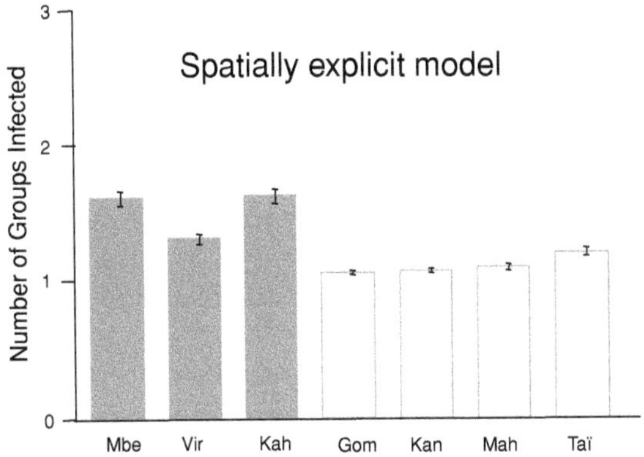

Figure 5.4. Simulations parameterized for chimpanzees (open bars) and gorillas (filled bars). Plots show results (mean values + 1 SE) for number of groups infected. Population codes: Mbe: Mbeli; Vir: Virungas; Kah: Kahuzi; Gom: Gombe; Kan: Kanyawara; Mah: Mahale; Taï: Taï National Park. From Nunn *et al.* (2008).

when investigating the spread of infectious disease, but it used space in only a general sense. The last modeling example in this chapter reinforces the importance of linking epidemiological models with specific spatial features, e.g. in a geographical information system (GIS). The example involves the spread of parapoxvirus in squirrels (Rushton *et al.*, 2000). This virus is moving rapidly through squirrel populations in Britain, and seems to be spreading from introduced gray squirrels (*Sciurus carolinensis*) to the endemic red squirrels (*Sciurus vulgaris*), possibly hastening the decline of the latter species. The disease causes death in nearly 100% of infected red squirrels, but most gray squirrels show no signs of infection or increased mortality, possibly because they can mount an efficient immune response but still carry the disease. The disease is thought to spread through contact with lesions on infected individuals (for more information, see Rushton *et al.*, 2000, 2006; Sainsbury *et al.*, 2000; Tompkins *et al.*, 2003; Gurnell *et al.*, 2006).

Rushton *et al.* (2000) investigated the factors that might lead to parasite-mediated competition between red and gray squirrels using an agent-based modeling approach (see also Gurnell *et al.*, 2006; Rushton *et al.*, 2006). The model used real-world data on the distribution of squirrels of both species in the county of Norfolk, England (Reynolds, 1985). In the fragmented populations that were used to parameterize the model, animals moved between populations through dispersal. Population dynamics were individually based and involved user-defined probabilities of mortality, fecundity, and dispersal. In a GIS, habitat was identified as suitable for reproduction (i.e. contains food needed to produce offspring) or suitable for dispersal (available for movement but not reproduction) using 25 m × 25 m grid cells and a land cover map. Areas of suitable habitat were identified as habitat blocks, and the simulation kept track of population sizes and dynamics in each block (which were modeled as populations linked through dispersal). Further details on the processes of birth, mortality, and dispersal are given in Rushton *et al.* (2000). On these geographically defined populations, a disease model was overlaid, with inputs involving population densities of the two species of squirrels, encounter rates between individual squirrels, disease-related mortality, and length of the infective period.

The main output from the model involved the densities of healthy and infected individuals at the end of the simulation run. Data were analyzed using standard parametric statistics. The results revealed that the primary variables influencing red squirrel declines were encounter rate with other squirrels and the infection rate; as these parameters increased, red squirrels declined more rapidly. Other variables were also significant predictors of red squirrel survival, including gray squirrel mortality rates. In 378 of 500 simulations, gray squirrels were able to successfully invade Norfolk County, and in 83 of these runs, gray squirrels invaded all the available habitats. In terms of infectious disease, gray squirrel infection rate had a negative impact on red squirrel persistence: with increasing prevalence in gray squirrels, red squirrels were more likely to go extinct. The results from the model were compared to real-world data on the population sizes of squirrels over time (Reynolds, 1985). Although some spatial and temporal differences were found between the observed and model results, the model produced generally high congruence with the observed population declines in Norfolk County.

Rushton *et al.*'s (2000) model revealed that parapoxvirus could be a cause of red squirrel declines, and it showed how spatial data could be integrated with individual-based modeling, including real-world data on the spatial and temporal spread of disease. It also illustrates how agent-based models can be used to address conservation concerns involving invasive species and the spread of disease across species boundaries. The agent-based model used by Rushton *et al.* (2000) also reveals a weakness of agent-based models: it is remarkably difficult to build a simple model – and to describe its internal procedures clearly enough – so that readers can make sense of the output. Rushton *et al.* (2000) actually pair two models, and with complicated simulation rules and parameterization, the resulting dynamics are not always obvious or intuitive. In contrast, a later model by Tompkins *et al.* (2003) is much easier to understand. This model is not agent-based but instead examines the connections among habitat patches and dynamics of infection within patches using differential equations. In fact, the Tompkins *et al.* (2003) model provided a closer match to patterns of decline documented by Reynolds (1985). This example therefore reveals the importance of making a model as simple as possible, with an easier to understand model sometimes providing more insights than a more complicated model.

Building an agent-based model

Just as artists must overcome the blank canvas, and scientific manuscripts start with a blank computer screen, it can be daunting to know how to start a modeling project. Agent-based modeling is no different in this respect from other types of models. In addition, building an agent-based model typically takes substantially longer than one expects when starting a project; a span of 2 years can easily pass from the start of building a model to analyzing the output. What are the skills, equipment, and background that are needed to build an agent-based model to study disease ecology? How does one start a model, and what are the key host and parasite parameters to include in the model? Here I give a very brief overview of this topic; further information on general issues in agent-based modeling can be found in Dunbar (2002) and Grimm & Railsback (2005).

The first ingredient for starting a modeling project is patience and preparation, but new computer packages are making programming substantially easier than in the past. The computer package MATLAB, for example, provides toolboxes that can do the statistics or graphics – code that a budding programmer once had to write for him or herself, often with poorer quality results and greater potential to introduce errors. While a number of hard-core programmers use C++ to program agent-based models, a number of other packages designed for agent-based modeling offer a great deal of flexibility without the pain of learning to program in more general-purpose languages. A particularly useful package in this regard is NetLogo (Wilensky, 1999). This package runs on multiple operating systems, comes with many wonderful code examples, has a very appealing graphical interface, and is freely available for download on the internet (http://ccl.northwestern.edu/netlogo/).

In terms of computer hardware, it helps to have a computer with a fast processor, since the models usually iterate through procedures thousands (or even millions) of times. Agent-based modeling does not require a high-level of programming experience – provided that the computer is fast enough to make code efficiency less of a constraint than it used to be – and there are many good books and courses available for learning how to program, depending on the computer language used. Several papers have been published on this and related topics in primates, or have applied agent-based approaches to address important questions in primate behavior (te Boekhorst & Hogeweg, 1994; Robbins & Robbins, 2004, 2005; Ramos-Fernandez *et al.*, 2006). Reading these papers can provide insights to the general process of building an agent-based model.

Once the computer is booted up and the software has loaded, the first step in agent-based modeling is to create the host population in which to unleash the disease. Depending on the question at hand, the user has to make decisions regarding group size, group composition, age structure of the population, and details on how agents disperse between groups. The spatial organization of groups can also be important, for example regarding how many neighbors each group has, and whether any geographic or social barriers to disease spread exist in the population. In most cases it is important to run the simulation in a spatial context, with the groups distributed on a matrix (lattice) and with agents more likely to disperse to close groups than to groups that are farther away. Finally, life history factors are important, particularly mortality. Animals that die from natural causes can be replaced by births in the same or different group, thus holding population size constant. Disease-related deaths are a different story, however, with deaths of these individuals tending to depress population size, at least in the short-term. This is a realistic assumption for introduced diseases, and is important to include, as population declines can impact subsequent spread of disease.

Equally important are decisions about which variables to hold constant and which to vary. For example, one might be interested in the evolution of host or parasite traits, such as group size or the duration of infection. It would be possible to do this by allowing these parameters to evolve over time. However, it is usually best to leave these experiments for later, once the basic dynamics of the system with static variables are understood. Indeed, many coevolutionary simulations can produce chaotic dynamics, and it is often impossible to make sense of these without first understanding emergent patterns and disease dynamics in more static simulations. Of course, these chaotic dynamics can be of interest in their own right, especially in systems that allow host behavior to coevolve with parasite characteristics.

After virtual primates have populated the computer-generated spatial arena, it is time to add a disease to the system. A modeler might choose to have a certain initial percentage of the population infected, or could randomly select one or several individuals to be infected. The latter option might be appropriate, for example, for the introduction of a novel disease to a susceptible host population. If regular spillover is of interest, it might be worthwhile to regularly "seed" the computer landscape with infected individuals. This can also help to ensure that the disease persists when stochastic events cause extinctions of disease early in the simulation, or when allowing the evolution of host or parasite traits in the context of regular disease spillover.

The order of events in a simulation can affect the outcome, and it is therefore essential to think carefully about the relative timing of procedures in the simulation. For example, should background mortality take place before or after disease-related mortality? Do animals disperse before or after procedures for disease transmission take place? Equally relevant is whether events happen simultaneously or sequentially (Grimm & Railsback, 2005). In the case of transmission, it is often preferable to first determine how many transmission events occur, and once all individuals have been examined, make the changes to infection status simultaneously for all individuals. Otherwise, individuals who are examined later in a simulation time step will be more likely to be infected by individuals who were infected earlier in the same time step. When in doubt, it is usually easy to reorganize the events in the simulation model to see how different arrangements impact the simulation results.

Often it is difficult to parameterize the model, given uncertainty or variation in values of real-world parameters included in the simulation. This is especially true for some infection parameters such as incubation period or disease-related mortality, where solid numbers are difficult to come by for wild animals.

In addition, a modeler might be interested in assessing how variation in a parameter affects the results. Ways of dealing with variation and uncertainty are possible, for example by selecting values randomly from the range of likely values. A more statistically and computationally powerful method is to use Latin hypercube sampling, or LHS (McKay *et al.*, 1979). Latin hypercube sampling can be considered an extension of Latin square sampling and is basically a randomization procedure without replacement; it provides a way to more efficiently sample parameter space and has been applied successfully to a variety of questions in epidemiology (Seaholm *et al.*, 1988; Blower & Dowlatabadi, 1994; Rushton *et al.*, 2000).

Equally important is consideration of the variables to measure from the simulation model. The builder of a simulation model is essentially "god," in that he or she has omniscient knowledge of each detail that takes place in the simulation. Nonetheless, it makes the task of analyzing the simulation output easier when only the essential output parameters are recorded. Moreover, keeping track of variables and writing data to disk for later analysis is very time consuming at a computational level; retaining too many output parameters will usually slow the program considerably. With these issues in mind, what are the main parameters that might be important for analyzing disease spread in an agent-based model? One key parameter involves patterns of prevalence over time, as this can be used to assess whether an equilibrium has been reached. Other parameters involve prevalence broken down by age and sex, the percentage of groups that are infected, and the number of deaths due to disease. It is also useful to have information on the actual spread of disease over time. Who acquired the infection from whom, and what were the characteristics of the source individuals, such as dominance rank, age, and sex? Collectively, these different sources of information provide the keys to uncovering the mechanisms that drive global patterns of disease spread.

Finally, it is easy to underestimate how long it will take to build the program and, especially, to debug and double-check the code prior to running simulations. Usually it is best to check the code by running subsets of it, to be sure that it is performing as expected. It is also important to fully document the code using comments, i.e. to write out exactly what each line of code does (or should do!), and then to print out all the code and take a fresh look at it on paper, rather than on the computer screen. To save headaches later, these checks should be taken very seriously prior to running the simulation; depending on the number of parameters that are varied, it could take weeks or even months to finish the simulations, and discovery of a bug at a later stage can be extremely costly in terms of lost computing time. While all computer programs have bugs, the user can limit the number of bugs substantially by careful proofing, by closely examining values of variables in subsets of code, and by comparing output to expected results.

Summary and conclusions

We often have notions for how particular factors influence the spread of disease in primate populations, but only rarely are these questions pressing enough to secure the financial resources to investigate these processes in real-time in real populations. Theoretical models provide a means to assess whether our intuition is correct and therefore can be an important means to investigate the drivers of disease dynamics in primate populations. In addition, theoretical models often produce unexpected results, thus generating new insights. Indeed, by providing theoretical support for an intuition about factors that impact the spread of disease, one can use the results of the simulation as "initial results" in a grant application. Finally, agent-based models can be used to investigate phenomena that operate over large spatial and temporal scales that are impractical (or impossible) to address using field data.

Agent-based models can also be used to generate predictions that can be tested in other ways, both in the field and comparatively. In previous work, for example, Sonia Altizer and I used Thrall *et al.*'s (2000) model to generate predictions for comparative tests of STDs in primates (Nunn & Altizer, 2004). Specifically, we predicted that female prevalence of an STD would be greater than male prevalence across non-monogamous primates, given that this sex difference in Thrall *et al.* (2000) was generally consistent across simulations that used different levels of dispersal and mortality. Indeed, we found support for this prediction (Figure 5.5). We are now working with Peter Thrall to extend his model to apply more directly to multi-male–multi-female primate groups. Indeed, this project illustrates a basic principle of iterative model development, with an initial model used to generate simple predictions, use of empirical data to test those predictions, and then further model development based on the empirical results.

Here, I focused on agent-based models, but it is important to remember that other modeling approaches can be used to study primate behavior (Dunbar, 2002) or the spread of disease (Hudson *et al.*, 2002). There are many reasons for using agent-based models to study primate disease ecology, including the ability to incorporate spatial structure within and between groups, and the relative ease of generating such a model. An agent-based model can often aid the developer of such a model in understanding the underlying mechanisms. It is essential to keep in mind, however, that models are most successful when

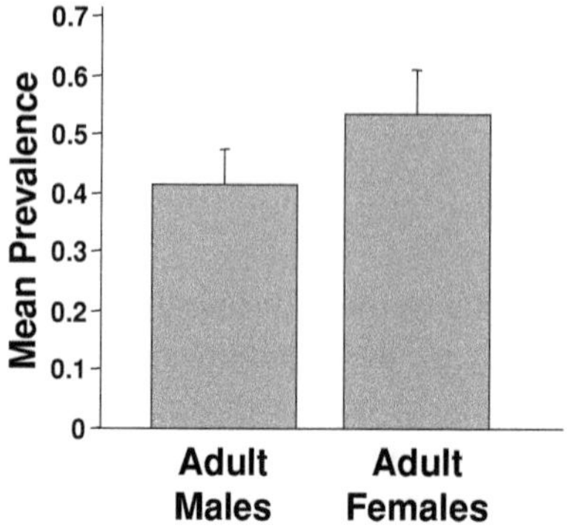

Figure 5.5. Sex differences in prevalence of STDs in wild primates. Bars represent mean prevalence of two STDs in males and females, +1 SE. This difference is significant in a pairs t-test ($t_{18} = 2.49$, $P = 0.011$, one-tailed), with higher prevalence in females. See Nunn & Altizer (2004) for further details on the data set and analysis.

they can easily be understood by others, including through replication and extension of the original model. It is often difficult to explain all the details of an agent-based model, and if such a model is developed, the burden is on the user to make this process as transparent as possible. Grimm & Railsback (2005) and Grimm *et al.* (2006) provide advice on this process, including clear guidelines on how to describe the model.

A number of interesting questions could be addressed with agent-based models in primate disease ecology. A major area involves the coevolution of primate behavior and disease. While many primatologists have speculated about the ways in which primate behavior might be shaped by disease, few have rigorously investigated whether their intuition is correct. It would be relatively straightforward to build models to investigate, for example, how disease impacts the evolution of xenophobia (i.e. resistance to potentially infected immigrants), primate mate choice, and even cultural transmission of behavioral counter-strategies to disease, such as medicinal plants, under variable risk of disease in primate populations. Other potentially fruitful areas of research involve the dynamics of disease in protected populations of primates. As a related issue, we know little about the effects of disease spillover from humans and domesticated animals into primate populations; an agent-based modeling approach could be used to investigate the parameters that influence cross-species transmission of disease, and thus could pinpoint variables to focus on in field research.

In conclusion, this is an exciting time for studying primate disease ecology (Nunn & Altizer, 2006). While many researchers are applying laboratory and field tools to study infectious diseases in wild primates, fewer are taking advantage of the computer revolution to model the spread of disease. We are now at a point where computer software and hardware are less of a restriction on building agent-based models, and the interdisciplinary frameworks developed over the past decade are providing opportunities for linking up researchers with different backgrounds, skills, and interests. Perhaps the most fruitful direction for future research is to take an interdisciplinary approach that combines field, laboratory, and modeling approaches, while also implementing practical applications of disease ecology for primate conservation.

Acknowledgements

I thank the editors for the invitation to participate in this volume, and I wish to acknowledge my collaborators who participated in the projects discussed above, including Sonia Altizer, Peter Thrall, Sandy Harcourt, and Kelly Stewart. Discussions with Volker Grimm provided many useful insights to agent-based modeling. This research was made possible with support from the Max Planck Society, Conservation International, and the NSF (DEB-0211908).

References

Alexander, N. J. (1990). Sexual transmission of Human Immunodeficiency Virus – virus entry into the male and female genital tract. *Fertility and Sterility*, **54**, 1–18.

Altizer, S., Nunn, C. L. & Lindenfors, P. (2007). Do threatened hosts have fewer parasites? A comparative study in primates. *Journal of Animal Ecology*, **76**, 304–314.

Anderson, R. M. & May, R. M. (1991). *Infectious Diseases of Humans: Dynamics and Control*. Oxford: Oxford University Press.

Augustine, D. J. (1998). Modelling Chlamydia-koala interactions: coexistence, population dynamics and conservation implications. *Journal of Applied Ecology*, **35**, 261–272.

Begon, M., Hazel, S. M., Baxby, D. *et al.* (1999). Transmission dynamics of a zoonotic pathogen within and between wildlife host species. *Proceedings of the Royal Society of London Series B-Biological Sciences*, **266**, 1939–1945.

Bermejo, M., Rodriguez-Teijerio, J. D., Illera, G. *et al.* (2006). Ebola outbreak killed 5000 gorillas. *Science*, **314**, 1564.

Blower, S. M. & Dowlatabadi, H. (1994). Sensitivity and uncertainty analysis of complex-models of disease transmission – an HIV model as an example. *International Statistical Review*, **62**, 229–243.

Boots, M. & Sasaki, A. (1999). "Small worlds" and the evolution of virulence: infection occurs locally and at a distance. *Proceedings of the Royal Society of London – Series B-Biological Sciences*, **266**, 1933–1938.

Brown, J. H. (1995). *Macroecology*. Chicago, IL: University of Chicago Press.

Canfield, P. J., Love, D. N., Mearns, G. & Farram, E. (1991). Chlamydial infection in a colony of captive koalas. *Australian Veterinary Journal*, **68**, 167–169.

Carlsson-Graner, U. & Thrall, P. H. (2002). The spatial distribution of plant populations, disease dynamics and evolution of resistance. *Oikos*, **97**, 97–110.

Chapman, C. A., Gillespie, T. R. & Goldberg, T. L. (2005). Primates and the ecology of their infectious diseases: how will anthropogenic change affect host-parasite interactions? *Evolutionary Anthroplogy*, **14**, 134–144.

Cross, P. C., Lloyd-Smith, J. O., Johnson, P. L. & Getz, W. M. (2005). Dueling time scales of host movement and disease recovery determine invasion of disease in structured populations. *Ecology Letters*, **8**, 587–595.

De Gruijter, J. M., Ziem, J., Verweij, J. J., Polderman, A. M. & Gasser, R. B. (2004). Genetic substructuring within *Oesophagostomum bifurcum* (Nematoda) from human and non-human primates from Ghana based on random amplified polymorphic DNA analysis. *American Journal of Tropical Medicine and Hygiene*, **71**, 227–233.

De Gruijter, J. M., Gasser, R. B., Polderman, A. M., Asigri, V. & Dijkshoorn, L. (2005). High resolution DNA fingerprinting by AFLP to study the genetic variation among *Oesophagostomum bifurcum* (Nematoda) from human and non-human primates from Ghana. *Parasitology*, **130**, 229–237.

De'ath, G. & Fabricius, K. E. (2000). Classification and regression trees: a powerful yet simple technique for ecological data analysis. *Ecology*, **81**, 3178–3192.

Dobson, A. P. & Meagher, M. (1996). The population dynamics of brucellosis in the Yellowstone National Park. *Ecology*, **77**, 1026–1036.

Dunbar, R. I. M. (2002). Modelling primate behavioral ecology. *International Journal of Primatology*, **23**, 785–819.

Ezenwa, V. O., Price, S. A., Altizer, S. *et al.* (2006). Host traits and parasite species richness in even and odd-toed hoofed mammals, Artiodactyla and Perissodactyla. *Oikos*, **115**, 526–536.

Freeland, W. J. (1976). Pathogens and the evolution of primate sociality. *Biotropica*, **8**, 12–24.

Galindo, P. & Srihongse, S. (1967). Evidence of recent jungle yellow-fever activity in Eastern Panama. *Bulletin of the World Health Organization*, **36**, 151–161.

Gandon, S., Capowiez, Y., Dubois, Y., Michalakis, Y. & Olivieri, I. (1996). Local adaptation and gene-for-gene coevolution in a metapopulation model. *Proceedings of the Royal Society of London Series B-Biological Sciences*, **263**, 1003–1009.

Graves, B. M. & Duvall, D. (1995). Effects of sexually-transmitted diseases on heritable variation in sexually selected systems. *Animal Behaviour*, **50**, 1129–1131.

Grimm, V. & Railsback, S. F. (2005). *Individual-based Modeling and Ecology*. Princeton, NJ: Princeton University Press.

Grimm, V., Revilla, E., Berger, U., Jeltsch, F. *et al.* (2005). Pattern-oriented modeling of agent-based complex systems: lessons from ecology. *Science*, **310**, 987–991.

Grimm, V., Berger, U., Bastiansen, F. *et al.* (2006). A standard protocol for describing individual-based and agent-based models. *Ecological Modelling*, **198**, 115–126.

Gurnell, J., Rushton, S. P., Lurz, P. W. W. *et al.* (2006). Squirrel poxvirus: landscape scale strategies for managing disease threat. *Biological Conservation*, **131**, 287–295.

Hopkins, M. E. & Nunn, C. L. (2007). A global gap analysis of infectious agents in wild primates. *Diversity and Distributions*, in press.

Hudson, P. J., Rizzoli, A., Grenfell, B. T., Heesterbeek, H. & Dobson, A. P. (2002). *The Ecology of Wildlife Diseases*. Oxford: Oxford University Press.

Kokko, H., Ranta, E., Ruxton, G. & Lundberg, P. (2002). Sexually transmitted disease and the evolution of mating systems. *Evolution*, **56**, 1091–1100.

Leendertz, F. H., Pauli, G., Maetz-Rensing, K. *et al.* (2006). Pathogens as drivers of population declines: the importance of systematic monitoring in great apes and other threatened mammals. *Biological Conservation*, **131**, 325–337.

Leroy, E. M., Rouquet, P., Formenty, P. *et al.* (2004). Multiple ebola virus transmission events and rapid decline of Central African wildlife. *Science*, **303**, 387–390.

Lindenfors, P., Nunn, C. L., Jones, K. E. *et al.* (2007). Parasite species richness in carnivores: effects of host body mass, latitude, geographic range and population density. *Global Ecology and Biogeography*, **16**, 496–509.

Lockhart, A. B., Thrall, P. H. & Antonovics, J. (1996). Sexually transmitted diseases in animals: ecological and evolutionary implications. *Biological Reviews of the Cambridge Philosophical Society*, **71**, 415–471.

Loehle, C. (1995). Social barriers to pathogen transmission in wild animal populations. *Ecology*, **76**, 326–335.

McColl, K. A., Martin, R. W., Gleeson, L. J., Handasyde, K. A. & Lee, A. K. (1984). Chlamydia infection and infertility in the female koala (*Phascolarctos cinereus*). *Veterinary Record*, **115**, 655.

McGrew, W. C., Tutin, C. E. G. & File, S. K. (1989a). Intestinal parasites of two species of free-living monkeys in far Western Africa, *Cercopithecus (aethiops) sabaeus* and *Erythrocebus patas patas*. *African Journal of Ecology*, **27**, 261–262.

McGrew, W. C., Tutin, C. E. G., Collins, D. A. & File, S. K. (1989b). Intestinal parasites of sympatric *Pan troglodytes* and *Papio spp.* at two sites: Gombe (Tanzania) and Mt. Assirik (Senegal). *American Journal of Primatology*, **17**, 147–155.

McKay, M. D., Beckrnan, R. J. & Conover, W. J. (1979). A comparison of three methods for selecting values of input variables in the analysis of output from a computer code. *Technometrics*, **21**, 239–245.

Moore, S. L. & Wilson, K. (2002). Parasites as a viability cost of sexual selection in natural populations of mammals. *Science*, **297**, 2015–2018.

Morand, S. & Harvey, P. H. (2000). Mammalian metabolism, longevity and parasite species richness. *Proceedings of the Royal Society of London Series B-Biological Sciences*, **267**, 1999–2003.

Muller-Graf, C. D. M., Collins, D. A. & Woolhouse, M. E. J. (1996). Intestinal parasite burden in five troops of olive baboons (*Papio cynocephalus anubis*) in Gombe Stream National Park, Tanzania. *Parasitology*, **112**, 489–497.

Muller-Graf, C. D. M., Collins, D. A., Packer, C. & Woolhouse, M. E. J. (1997). *Schistosoma mansoni* infection in a natural population of olive baboons (*Papio cynocephalus anubis*) in Gombe Stream National Park, Tanzania. *Parasitology*, **115**, 621–627.

Nunn, C. L. & Altizer, S. (2004). Sexual selection, behaviour and sexually transmitted diseases. In *Sexual Selection in Primates: New and Comparative Perspectives*, ed. P. M. Kappeler & C. P. van Schaik. Cambridge: Cambridge University Press, pp. 117–130.

Nunn, C. L. & Altizer, S. M. (2006). *Infectious Diseases in Primates: Behavior, Ecology and Evolution*. Oxford: Oxford University Press.

Nunn, C. L., Altizer, S., Jones, K. E. & Sechrest, W. (2003). Comparative tests of parasite species richness in primates. *American Naturalist*, **162**, 597–614.

Nunn, C. L., Altizer, S., Sechrest, W. *et al.* (2004). Parasites and the evolutionary diversification of primate clades. *American Naturalist*, **164**, S90–S103.

Nunn, C. L., Altizer, S. M., Sechrest, W. & Cunningham, A. (2005). Latitudinal gradients of disease risk in primates. *Diversity and Distributions*, **11**, 249–256.

Nunn, C. L., Rothschild, B. M. & Gittleman, J. L. (2007). Why are some species more commonly afflicted by arthritis than others? A comparative study of spondyloarthropathy in primates and carnivores. *Journal of Evolutionary Biology*, **20**, 460–470.

Nunn, C. L., Thrall, P. H., Harcourt, A. H. & Stewart, K. (2008). Emerging infectious diseases and animal social systems. *Evolutionary Ecology*, **22**, 519–543.

O'Keefe, K. J. & Antonovics, J. (2002). Playing by different rules: the evolution of virulence in sterilizing pathogens. *American Naturalist*, **159**, 597–605.

Ostfeld, R. S., Glass, G. E. & Keesing, F. (2005). Spatial epidemiology: an emerging (or reemerging) discipline. *Trends in Ecology and Evolution*, **20**, 328–335.

Padian, N. S., Shiboski, S. C., Glass, S. O. & Vittinghoff, E. (1997). Heterosexual transmission of Human Immunodeficiency Virus (HIV) in Northern California: results from a ten-year study. *American Journal of Epidemiology*, **146**, 350–357.

Pedersen, A. B., Poss, M., Altizer, S., Cunningham, A. & Nunn, C. (2005). Patterns of host specificity and transmission among parasites of wild primates. *International Journal for Parasitology*, **35**, 647–657.

Pope, T. R. (1998). Effects of demographic change on group kin structure and gene dynamics of populations of red howling monkeys. *Journal of Mammalogy*, **79**, 692–712.

Poulin, R. (1995). Phylogeny, ecology, and the richness of parasite communities in vertebrates. *Ecological Monographs*, **65**, 283–302.

Poulin, R. (1997). Species richness of parasite assemblages: evolution and patterns. *Annual Review of Ecology and Systematics*, **28**, 341–358.

Poulin, R. & Morand, S. (2004). *Parasite Biodiversity*. Washington, DC: Smithsonian Institution Press.

Ramos-Fernandez, G., Boyer, D. & Gomez, V. P. (2006). A complex social structure with fission-fusion properties can emerge from a simple foraging model. *Behavioral Ecology and Sociobiology*, **60**, 536–549.

Reynolds, J. C. (1985). Details of the geographic replacement of the red squirrel (*Sciurus vulgaris*) by the grey squirrel (*Sciurus carolinensis*) in Eastern England. *Journal of Animal Ecology*, **54**, 149–162.

Robbins, A. M. & Robbins, M. M. (2005). Fitness consequences of dispersal decisions for male mountain gorillas (*Gorilla beringei beringei*). *Behavioral Ecology and Sociobiology*, **58**, 295–309.

Robbins, M. M. & Robbins, A. M. (2004). Simulation of the population dynamics and social structure of the Virunga mountain gorillas. *American Journal of Primatology*, **63**, 201–223.

Rouquet, P., Froment, J. M., Bermejo, M. *et al.* (2005). Wild animal mortality monitoring and human Ebola outbreaks, Gabon and Republic of Congo, 2001–2003. *Emerging Infectious Diseases*, **11**, 283–290.

Roy, B. A. & Kirchner, J. W. (2000). Evolutionary dynamics of pathogen resistance and tolerance. *Evolution*, **54**, 51–63.

Rushton, S. P., Lurz, P. W. W., Gurnell, J. & Fuller, R. (2000). Modelling the spatial dynamics of parapoxvirus disease in red and grey squirrels: a possible cause of the decline in the red squirrel in the UK? *Journal of Applied Ecology*, **37**, 997–1012.

Rushton, S. P., Lurz, P. W. W., Gurnell, J. *et al.* (2006). Disease threats posed by alien species: the role of a poxvirus in the decline of the native red squirrel in Britain. *Epidemiology and Infection*, **134**, 521–533.

Sainsbury, A. W., Nettleton, P., Gilray, J. & Gurnell, J. (2000). Grey squirrels have high seroprevalence to a parapoxvirus associated with deaths in red squirrels. *Animal Conservation*, **3**, 229–233.

Seaholm, S. K., Ackerman, E. & Wu, S. C. (1988). Latin hypercube sampling and the sensitivity analysis of a Monte Carlo epidemic model. *International Journal of Bio-Medical Computing*, **23**, 97–112.

Smith, D. L., Lucey, B., Waller, L. A., Childs, J. E. & Real, L. A. (2002). Predicting the spatial dynamics of rabies epidemics on heterogeneous landscapes. *Proceedings of the National Academy of Sciences, USA*, **99**, 3668–3672.

Sorci, G., Morand, S. & Hugot, J.-P. (1997). Host-parasite coevolution: comparative evidence for covariation of life history traits in primates and oxyurid parasites. *Proceeding of the Royal Society London B*, **264**, 285–289.

te Boekhorst, I. J. A. & Hogeweg, P. (1994). Self-structuring in artificial chimps offers new hypotheses for male grouping in chimpanzees. *Behaviour*, **130**, 229–252.

Thrall, P. H. & Antonovics, J. (1995). Theoretical and empirical-studies of metapopulations – population and genetic dynamics of the *Silene ustilago* system. *Canadian Journal of Botany*, **73**, S1249–S1258.

Thrall, P. H., Antonovics, J. & Dobson, A. P. (2000). Sexually transmitted diseases in polygynous mating systems: prevalence and impact on reproductive success. *Proceeding of the Royal Society London B*, **267**, 1555–1563.

Tompkins, D. M., White, A. R. & Boots, M. (2003). Ecological replacement of native red squirrels by invasive greys driven by disease. *Ecology Letters*, **6**, 189–196.

Wallis, J. & Lee, D. R. (1999). Primate conservation: the prevention of disease transmission. *International Journal of Primatology*, **20**, 803–826.

Walsh, P. D., Abernethy, K. A., Bermejo, M. *et al.* (2003). Catastrophic ape decline in Western Equatorial Africa. *Nature*, **422**, 611–614.

Walsh, P. D., Biek, R. & Real, L. A. (2005). Wave-like spread of Ebola Zaire. *PLoS Biology*, **3**, e371.
Walsh, P. D., Breuer, T., Sanz, C., Morgan, D. & Doran-Sheehy, D. (2007). Potential for Ebola transmission between gorilla and chimpanzee social groups. *American Naturalist*, **169**, 684–689.
Wilensky, U. (1999). NetLogo. http://ccl.northwestern.edu/netlogo/. Center for Connected Learning and Computer-Based Modeling. Evanston, IL: Northwestern University.

Part II

The natural history of primate–parasite interactions

6 *What does a parasite see when it looks at a chimpanzee?*

MICHAEL V. K. SUKHDEO AND SUZANNE C. SUKHDEO

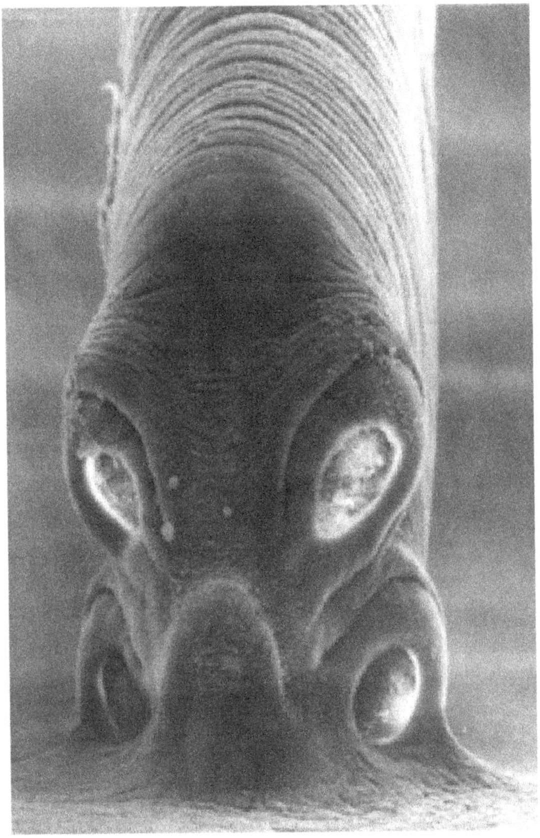

SEM by Thushara Chandrasiri. Scolex (head) of a tapeworm (*Hymenolepis diminuta*) showing four attachment suckers

Primate Parasite Ecology. The Dynamics and Study of Host–Parasite Relationships, ed. Michael A. Huffman and Colin A. Chapman. Published by Cambridge University Press.

Introduction

From the perspective of parasitologists working in other fields, there has been astounding progress in the understanding of the biology of non-human primate parasites. Indeed, the chapters in this book are a testament to the recent explosion of ideas on primate parasites at all levels of ecological organization. At the largest scales, strong inferences exist on the effects of global climate change, emerging diseases, and other worldwide anthropogenic insults on the interactions between primates and their parasites (Wolfe *et al.*, 1998; Chapman *et al.*, 2005a; Nunn & Altizer, 2006). At regional scales, numerous studies detail the effects of seasonality, conservation policy, habitat encroachment, disturbance, logging, fragmentation, and other ecological variables on primate infections (Wallis & Lee, 1999; Ashford *et al.*, 2000; Wallis, 2000; Lilly *et al.*, 2002; Nunn *et al.*, 2003; Nunn & Altizer, 2005, 2006; Altizer *et al.*, 2006; Gillespie & Chapman, 2006). At local scales, there are exciting studies on parasite infracommunities and the effects of primate self-medication, parasite avoidance behaviors, or social rank (Wrangham & Nishida, 1983; Eley *et al.*, 1989; Huffman & Seifu, 1989; Fa *et al.*, 1995; Huffman *et al.*, 1996, 1998; Thoisy *et al.*, 2001; Hahn *et al.*, 2003; Huffman & Hirata, 2004).

In fairness, this paltry list does not do full justice to the diversity of investigations on primate parasites, but these examples suffice to illustrate a most amazing fact about primate parasitology . . . the raw data for almost all of these studies comes from fecal samples! This is impressive at many levels, and it is generally agreed that, for rigor in methodology and analysis, primatologists are the true masters of the art of fecal sampling. However, the real untold story lies in what determines success when sampling feces in the wild. Simply stated, investigators need to have an intimate knowledge of the behaviors of their subjects. Success often requires a deep and comprehensive understanding of the likes, dislikes, and habits of the individual primate, and all the interactions that occur with other members of the group and with the environment. Good stool collectors must be good at reading the behaviors of their subjects, and can often see the world the way their primate subjects see the world.

In primate–parasite interactions, the benefits of understanding host behavior are supported by many studies, e.g. the roles of self-medication or social rank on parasite burdens (Wrangham & Nishida, 1983; Clayton & Wolfe, 1993; Huffman *et al.*, 1996, 1998). Primatologists are keen to get an intimate knowledge of primates, why not their parasites? We argue that exploring the parasite's perspective might also be important to primatologists. For example, consider the way we subscribe to the notion that as anthropogenic habitat changes force humans and primates into closer and more frequent contact, the risks of interspecific disease transmission increases. This seems to hold true at large scales (Daszak *et al.*, 2001; Dobson & Foufopoulus, 2001; Chapman *et al.*, 2005a; Nunn & Altizer, 2006), but at smaller scales the picture is less clear. Studies of *Oesophagostomum bifurcum* infections in humans and primates in Togo and Ghana have found an opposite relationship. In the north, where interspecific contact is limited and infrequent, both humans and primates are heavily infected. In the south where there are much larger human communities and much greater primate/human contact, the primates are heavily infected with this parasite but the humans are not (Polderman *et al.*, 1999; Gasser *et al.*, 2006). Clearly, the parasites are not behaving as expected.

How animals (and parasites) see their worlds

The first thing to acknowledge is that parasites do not actually "see" their world, at least not in the same visual sense that we do. However, parasitic worms have large numbers of receptors and other sophisticated sensory organs that are often more complex than those of their free-living relatives (Sukhdeo & Mettrick, 1987; Rohde, 1989, 1993). In addition, neuroanatomical and neurophysiological studies support the idea that their brains (cerebral ganglia) can be more elaborate than their free-living relatives. For example, the brains of the common liver fluke *Fasciola hepatica* contains specialized structures for faster signal transmission (large diameter axons and specialized trophospongium glial cells) that do not appear again until the evolution of insect brains (Bullock & Horridge, 1965; Sukhdeo *et al.*, 1988; Sukhdeo & Sukhdeo, 1994).

It is therefore not very surprising that parasitic worms are capable of astonishing behavioral feats. Since Thomas (1883) and Leuckart (1884) published the first descriptions of miracidia of *F. hepatica* responding to their lymnaeid snail hosts, parasitologists have been gathering a lot of data on the signals from the host and environment that elicit behavioral responses in parasites (Ulmer, 1971; Lackie, 1975; Sukhdeo & Mettrick, 1987; Sukhdeo & Sukhdeo, 2004). Parasites often make tortuous migrations inside the host before arriving at their final sites. For example, the migrating larvae of the common liver fluke *F. hepatica* penetrate the gut to enter the viscera, and then make a treacherous journey through an abdomen filled with writhing intestines and mesenteries to reach the liver. This was thought to be the equivalent of a man swimming 60 km across rough open water, without a compass! Yet, more than 95% of the worms make this migration in their vertebrate hosts successfully (Dawes, 1963). Outside the host, parasite behaviors are equally incredible. For example, the miracidia of the human blood fluke *Schistosoma mansoni* can locate and infect their specific snail host *Biomphalaria glabrata* at distances of over 70 m (Webbe, 1996). For a human swimmer, this would be ten times back and forth across the English Channel! These stories of complex or inexplicable parasite peregrinations inside and outside the host abound in the parasitology literature. The challenge has been how to translate this large body of information on parasite behaviors into a coherent picture of how parasites see their worlds.

Among the many insights for which Konrad Lorenz, Nikolaas Tinbergen, and Karl von Frisch were awarded their Nobel Prize in 1973, was the idea that all animals live in their own species-specific perceptual worlds. One of their critical proofs was that the way an animal sees the world is linked physiologically to the mechanisms by which the animal responds to its world (von Uexküll, 1934; Tinbergen & Perdeck, 1950; von Frisch, 1950; Lorenz & Tinbergen, 1957). Consequently, it was possible to deduce an animal's perceptual space from its behavioral responses. This is easier said than done. Reconstructing perception is not always a simple task because the linkages may not be direct, and interpretation often requires imagination and an open mind (especially when dealing with perception in invertebrate subjects). For example, our principal senses for locating objects in space are touch and vision, and that is why we speak of tactile and visual space. We do not think of ourselves as having an auditory space because our sense of hearing is not very accurate at localizing sounds in space. On the other hand, bats and owls have excellent acoustic localization, and when we "imagine" their auditory space, it is in the same sense as we think of our visual space. We also understand that different animals may perceive the same objects differently depending on their sensory equipment. Humans (and octopi) possess the most sophisticated eyes on earth, with image quality and resolution that far surpasses the compound eye of the fly or the eyespot in a trematode miracidium. Frogs cannot distinguish between an artificial black dot and a real fly, and respond to both stimuli with the same feeding strike (Wehler, 1981; Inge, 1983). One cannot argue that frogs are poor fly catchers, but they clearly must not be too picky about the "look" of their food.

The term "umwelt," usually translated as "subjective universe," was used to describe these species-specific perceptual worlds (von Uexküll, 1934). Interestingly, the umwelt of parasites is considered to be an impoverished world with little sensory stimulation. This is illustrated by ticks foraging for their hosts; even though a tick's environment is noisy with complex stimuli, ticks recognize only three signals, and have only three associated responses. Butyric acid, emitted by all mammals, induces the tick to drop from its perch; the mechanical stimulation of fur induces crawling; and warmth of host skin induces ticks to bore in for their meal (von Uexküll, 1934).

How parasites see their hosts, looking from the inside

There are at least two stages in a parasite's life cycle when accurate perception of the environment is crucial to its survival. The first stage occurs during host-finding, and the second occurs during their migration inside the host. There have been many more studies on the behavior of parasites *inside* the host when compared with studies on the behavior of parasites *outside* the host. Partly, this can be explained by the difficulties of working with parasites in the field, and especially when working on parasites with life cycles that utilize more than one host. A second reason for the prevalence of "inside-the-host studies" comes from the biomedical perspective which dominates much of parasitology research. It has long been recognized that parasites which become lost during the migration to their site will almost always disintegrate and die without causing much pathology (Fülleborn & Schilling-Torgau, 1911). The classic textbook example of "larva migrans" (which occurs when dog hookworms *Ancylostoma caninum* infect humans) is often used to illustrate this phenomenon; the paths of the larvae under the skin are clearly visible as they get "lost" and wander around aimlessly until destroyed by the immune response. It was reasoned that if worm migration could be blocked in some way, for example by blocking a sensory receptor, it could lead to a novel treatment (Ulmer, 1971; Bailey, 1982). This simple idea fueled a tremendous drive to understand the signals that guided the worms to their sites, and virtually every parasitologist tried their hand at answering this question (see reviews by Smyth, 1966; Ulmer, 1971; Crompton, 1973; Bailey, 1982; Kemp & Devine, 1982; Sukhdeo & Mettrick, 1987).

However, it was all a colossal failure! After hundreds of attempts using several parasite species, no one was ever able to demonstrate that parasites were attracted to their preferred sites, nor could anyone identify any chemical or other signals from these sites that elicited directional movements by the parasites (Sukhdeo & Sukhdeo, 2004). We now know that it was because we had the wrong paradigm about how parasites see their worlds. The problem was that parasitologists all firmly believed that parasites were attracted by signals emanating from their preferred sites. According to this view, for example, liver flukes should be attracted to the liver, lungworms should be attracted to the lungs, and eyeworms should be attracted to the eyes. However, this attraction could not be demonstrated. The desperate search for these attractive signals sometimes produced strange speculations, for example, the idea that human blood flukes (*Schistosoma mansoni*) were guided to their mesenteric sites by responding to a gradient of fecal secretions (Awwad & Bell, 1982). This is not true, but despite much evidence to the contrary, the idea that parasites must migrate up chemical gradients to their sites remained incontestable. One

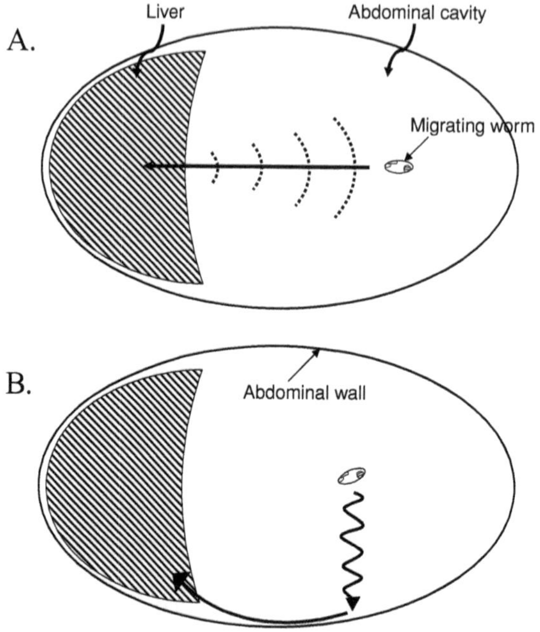

Figure 6.1. Diagrammatic representation of the migration of *Fasciola hepatica* to the liver. A. The migration route in response to chemical gradients from the liver that was thought to occur. B. The actual migration route of the worms, first to the abdominal wall, and then to the liver (data from Sukhdeo, 1990; Sukhdeo & Sukhdeo, 2004).

eminent parasitologist spent as much as 20 years unsuccessfully trying to find the liver chemicals that attracted the migrating larvae of *F. hepatica* (Dawes & Hughes, 1964).

In the free-living world, orientation towards habitat signals is a feature that is common to all animals (Jennings, 1906; Fraenkel & Gunn, 1940), and even simple bacteria can orient away from noxious stimuli (Adler, 1983; Eisenbach, 2004). These responses support a principal point of view when we think of how animals see their worlds, that is, in terms of gradients of signals. Even though we cannot actually see these gradients, we accept that they exist whenever we see vultures circling, butterflies chasing each other, or sea turtles migrating. However, parasites do not see their world in this way. This is because these gradients simply do not exist inside the host. Two factors prevent their existence in the host, turbulence and saturation. For example, consider the larvae of *F. hepatica* migrating through the abdominal cavity to the liver (Figure 6.1A). The turbulence of the constantly writhing intestines and mesenteries will disrupt any gradient of chemicals produced by the liver. In addition, the creation of gradients requires a sink (diffusion to infinity), but the abdominal cavity is a closed system and any potential gradient would quickly become saturated (Figure 6.1A). Finally, if there were liver gradients in the abdominal cavity that attracted the worms, these gradients would have to be so steep to attract these tiny organisms (250 μ) that they would have been easy to detect by investigators (Sukhdeo, 1990).

The evidence now shows that *F. hepatica* does not even migrate directly to the liver. The worms first migrate out of the mesenteries and onto the abdominal wall, where they change their behavior and then migrate along the abdominal wall to the liver (Figure 6.1B). This migration is successful because the inside of the abdominal cavity is a closed ovoid with the liver making intimate contact with a large area of the abdominal wall, and so any direction taken by the worms (up, down, or around) will always bring them to the liver. In addition, these worms do not receive navigation signals from their liquid milieu but from their food, and they feed voraciously and sequentially on different tissues during their migration (Dawes, 1963). Thus, when the worms feed on mesenteric tissue it triggers a peculiar crimping pattern of behavior that causes them to fall out of the mesenteries and onto the abdominal wall. Feeding on abdominal wall tissue triggers a different pattern of behavior that keeps them tightly adhered to the abdominal wall while migrating. During this phase, the worms are difficult to dislodge from the wall and do not easily fall back into the mesenteries. The worms continue this migration along the inside abdominal wall until they reach the liver, which they recognize when they begin feeding on it. At no time during this migration is it necessary for the worms to recognize or respond to chemical gradients (Sukhdeo & Sukhdeo, 2002).

This migration behavior says a lot about how these worms see their world. From the parasite's perspective, the environment in the abdominal cavity must be totally predictable for the migration to work. Indeed, the worms encounter identical chemical and morphological topologies in every host that they infect. In all, the intestines and mesenteries are always surrounded by the abdominal wall, and the abdominal wall is always in contact with the liver. This predictable internal environment of the host is considered to be earth's third environment because it is so distinct from aquatic and terrestrial free-living environments (Allee *et al.*, 1949; Sukhdeo, 1990). Once parasites enter the host, they do not need to prepare for sudden storms or other random environmental catastrophes because there are homeostatic mechanisms to ensure physiological and biochemical constancy. Even the host immune response, designed to repel the parasite, occurs in a standard choreographed sequence that parasites easily avoid. Each host is also topologically identical to every other member of the same host species. A parasite entering the host's mouth will travel through the esophagus, stomach, small intestine, and large intestine, always in that order, and each region will have its own predictable set of physiochemical conditions.

A typical response by animals when evolving under predictable or constant environmental conditions is the development of innate or genetically fixed behaviors that are adaptive (Lorenz & Tinbergen, 1957; Krebs & Davies, 1989; Alcock, 1998). The classic example is the complex egg-retrieval behavior in the graylag goose *Anser anser*. When the goose sees an egg that has rolled outside the nest, it triggers a complicated sequence of distinct behaviors (more than a dozen steps) that bring the egg back into the nest. Once started the cycle goes to completion. If in the middle of the sequence the egg is removed, the entire sequence of behavior continues as if the egg was still there (Pelkwijk & Tinbergen, 1937; Tinbergen & Perdeck, 1950). This is an example of a genetically hard-wired behavior pattern released by a very specific stimulus, and these are common in animals of all phyla (Krebs & Davies, 1989; Alcock, 1998). Natural selection will operate similarly in parasites that encounter identical conditions every time they enter a host; behaviors that are adaptive will become genetically fixed over evolutionary time. For example, the crimping behavior of *F. hepatica* is a fixed or hard-wired pattern that is released when the parasite feeds on mesenteric tissue. The parasite does not require ultimate purpose or directionality to the response, and it only matters that crimping behavior, when triggered, will always take the worms automatically to the abdominal wall.

We will focus on gastrointestinal parasites because this group is of most interest to primate parasitologists. Incoming gastrointestinal parasites typically enter the host as metabolically dormant stages (to conserve energy while waiting to be consumed), and are often contained within an impermeable cyst, egg, or sheath. These infective stages are taken on a predictable one-way trip down the gastrointestinal tube, propelled passively by peristalsis and the inexorable flow of ingesta. The common strategy is for the parasites to recognize signals from the region just anteriad of their preferred sites, which trigger fixed behaviors that lead to emergence and establishment in the intestinal wall. For example, the mouse hookworm *Heligmosomoides polygyrus*, which lives in the anterior small intestine, is triggered by bile secretions which normally enter in the very first section of the intestines (Sukhdeo & Croll, 1981). If bile secretion is surgically re-routed to more posterior (abnormal) sites, the worms will go past their preferred sites and blindly establish in the abnormal sites (Figure 6.2).

In general, the "umwelts" of parasite infective stages inside the host seem to be impoverished, and their perceptual worlds comprise very small subsets of the large number of potential signals in the gastrointestinal milieu. If the parasites fail to recognize their specific trigger signals, they will end up being carried to less optimal sites. At the worst, they could be swept out with the

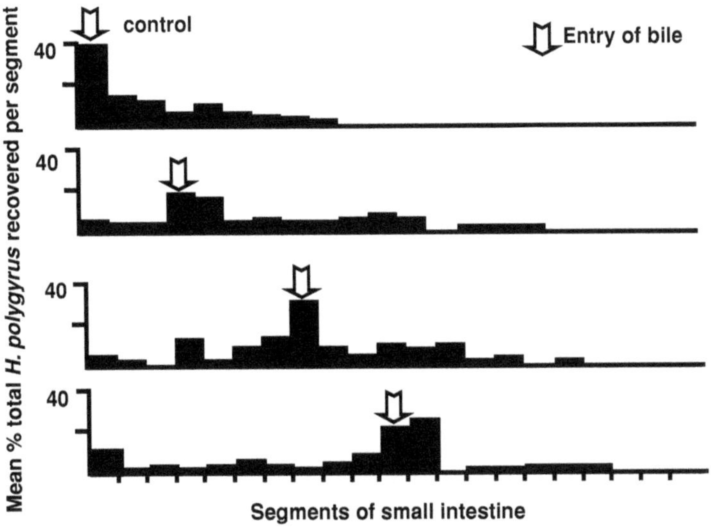

Figure 6.2. The establishment behavior of *Heligmosomoides polygyrus* in the small intestine of their rodent hosts following surgical re-routing of the bile duct to more posterior (abnormal) sites. In the control surgical sham group, the worms established normally in the anterior small intestine. In the surgical treatments, the worms bypass their normal sites to establish in the abnormal regions where bile now enters (data from Sukhdeo & Croll, 1981).

feces and end up back in the free-living world. After recognizing their specific triggers, parasites respond with behaviors which continue until the worms encounter the intestinal epithelium or until they run out of energy. Worms that fail to establish are swept out with the feces, so there is no need to conserve energy or effort. Even under the most abnormal conditions in a laboratory Petri dish, the infective stages always respond to the trigger stimuli, and the activated worms will keep moving until they run out of energy (which takes several hours because they never encounter the intestinal wall in vitro). Thus, even a single parasite egg or larva from a fresh stool sample can yield significant clues about how the worms must see their worlds.

Studying activation signals in parasites is very easy, and that is why parasitologists know more about the triggers of activation in gastrointestinal parasites than any other aspect of their behaviors. As evidence of this, there have been several hefty reviews written on this subject (Erasmus, 1972; Lackie, 1975; Bailey, 1982; Smyth & Halton, 1983; Sommerville & Rogers, 1987; Sukhdeo & Mettrick, 1987; Fried, 1994; Rea & Irwin, 1994). A general consensus is that the optimal stimuli are found in the intestines of the specific host. For example, *Cryptocoyle lingua*, a parasite of gulls, excysts optimally in the intestinal juice of gulls (Stunkard, 1930; McDaniel, 1966). The specific triggers of activation can vary greatly across parasite taxa, but bile is a common component (Lackie, 1975); tapeworms often require proteases and **bile** (Silverman, 1954; Smyth & Haslewood, 1963; Schiller, 1965), roundworms often require pCO_2 and **bile** (Rogers, 1960; Chapman & Undeen, 1968; Mapes, 1972), and trematodes often require pCO_2, proteases, and **bile** (Dawes & Hughes, 1964; Dixon, 1966). Bile is clearly an important trigger signal for gastrointestinal worms. Of more than 50 species from 19 different gastrointestinal parasite families studied in detail, bile was the stimulus or co-stimulus in greater than 65% of the parasites (Lackie, 1975; Sommerville & Rogers, 1987; Fried, 1994).

Bile is not a monovalent solution, but consists of a complex mixture of bile pigments, mucins, proteins, neutral lipids, conjugated and unconjugated bile salts, phospholipids, and inorganic ions (Haslewood, 1978). Each parasite species responds to different constituents of bile that reflect their local host population. For example, taurine-conjugated bile acids are characteristic of carnivore bile, and the trigger stimulus for the pork parasite *Trichinella spiralis* (which enters the host encysted in meat) is taurodeoxycholic acid. On the other hand, glycine-conjugated bile acids are characteristic of herbivore bile. *Fasciola hepatica* is typically a parasite of large herbivores, and it is triggered to emerge from its softened cyst by glycocholic acid (Sukhdeo & Mettrick, 1987). Bile constituents can vary between different populations of the same host based on differences in diet, presence of pathogens, microflora, and other environment variables (Smyth & Haslewood, 1963; Mettrick & Podesta, 1974; Haslewood, 1978).

Thinking outside the box about primate fecal samples

It is probably only wishful thinking to suggest that additional information can be extracted from individual eggs or larvae in primate fecal samples. However, the techniques to study activation signals are simple, and there is a need for non-invasive procedures in primate parasitology (Wolfe *et al.*, 2005). There can be very expensive options based on recent advances in modern tissue culture which permit surgical operations on single mammalian cells, and it is easy to substitute the cell with a parasite egg or larva. However, a far better alternative would be to use a low-tech approach that will yield the same results, and which will be accessible to all stool collectors. All that is required is a microscope and a depression slide to hold the infective stage while a drop of the stimulus is added. Nothing fancy. The right trigger stimulus will elicit a response even under abnormal conditions, because that is how the worms see the world. In addition, potential trigger signals and other treatments are easy to obtain because they are all gut-derived, and all gastric, biliary, pancreatic, and intestinal secretions are cheaply and widely available from any of several biomedical suppliers.

This sort of information on trigger stimuli might be useful in situations where real biological connections need to be made between primates and their parasite eggs. For example, it may help in the situation with *O. bifurcum* in Ghana where it is not clear why some groups of humans are infected with the primate parasite, but other groups are not (Polderman *et al.*, 1991, 1999; see also Gasser *et al.*, Chapter 3, this volume). There have been several large-scale efforts to elucidate these infection patterns using molecular tools to dissect eggs and larvae (from coproculture). Sequences of nuclear and mitochondrial DNA have been subjected to extensive analysis utilizing diverse techniques, including species-specific PCR methods, mutation scanning approaches with SSCP, and AFLP and RAPD fingerprinting (Gasser *et al.*, 1999; Gasser, 2005; de Gruijter *et al.*, 2004, 2005; Gasser *et al.*, Chapter 3, this volume). Overall, these data suggest that there may exist genetically distinct host-species affiliated variants of *O. bifurcum* in Ghana (de Gruijter *et al.*, 2005; Gasser, 2005). In this case, trigger stimuli might offer a way to get biologically relevant responses at the level of the egg that can then be used to test this hypothesis of host-affiliated variants. There may also be other cases where differences in primate groups might shape the infection strategies of their parasites. Parasites tend to adapt to their local hosts (Combes & Théron, 2000; Barger & Esch, 2002; Poulin & Mouillot, 2003; Dobson, 2004; Poulin, 2007), and this seems to be a common feature in parasites of primates. Parasite patterns in wild primate populations are often best explained by group membership and social group size (Freeland, 1979; Vitone *et al.*, 2004), and it is generally thought that parasites treat primate groups as biological islands (Wolfe *et al.*, 1998; Vitone *et al.*, 2004). Parasite activation responses might illuminate infection patterns in primates that are known to vary with arboreal versus ground-dwelling habits, insectivory, water contact patterns, fruiting patterns, or when the same parasite infects several primate species (Müller-Graf *et al.*, 1997; Ashford *et al.*, 2000; Lilly *et al.*, 2002; Chapman *et al.*, 2005b; Pedersen *et al.*, 2005).

"Not all those who wander are lost" (Tolkien 1892–1973)

Inside the host, parasites may see their host world in predictable ways and respond optimally to local conditions, but this does not imply that the paths they take to their final sites are the shortest routes. For example, after emergence in the small intestine, *F. hepatica* does not migrate the easy way from the intestinal lumen up the bile duct to the liver, but it goes on a circuitous route eating its way through the intestines and then migrating through the abdominal cavity (Faust & Khaw, 1927; Wykoff & Lepes, 1957). These migrations are among several parasite behaviors that might be considered "unnecessary" (Anderson, 2000). The best examples come from *Ascaris lumbricoides*, the most common helminth infection of humans worldwide, and *Strongylus vulgaris,* the large strongyle of horses. In these parasites, after the infective stages are ingested, they penetrate out of the intestines and make a long migration through the host's body to end up *right back into the intestines* where they began their journey. Parasitologists have long puzzled over the reason why the worms make these apparently unnecessary migrations. The costs of these migrations can be high for the parasite. In *S. vulgaris* infections, up to 90% of the migrating worms are killed by the host's immune response (Duncan, 1972). The costs are also very high for the hosts, and they suffer extensive pathology from the migrating worms (McGraw & Slocombe, 1976). In humans infected with *Ascaris*, worms breaking out of the lung capillaries can often produce Löffler's pneumonia where large areas of the lung become diseased, and which may result in death (Löffler, 1956).

The answers to these mysterious migrations may lie in the details of the evolutionary transition from free-living soil nematodes to animal-parasitic forms (Anderson, 2000). Free-living nematodes represent one of the most successful animal groups on earth. Four out of every five metazoan animals on the planet are nematodes, and they occupy every conceivable niche in marine, freshwater, and terrestrial environments (Bongers & Ferris, 1999). Yet, it is curious that parasitism in this group did not evolve until animals invaded land (Chitwood & Chitwood, 1950; Chabaud, 1954; Bain & Chabaud, 1979; Anderson, 2000). Nematode parasites are rarely found among marine and freshwater invertebrates, such as molluscs, polychaetes, and crustaceans, but a rich nematode fauna is found in their terrestrial counterparts, such as earthworms, insects, and terrestrial molluscs (Bain & Chabaud, 1979; Anderson, 1988, 2000). On land, it is believed that the first step in vertebrate animal parasitism was the penetration of the hosts' skin (Dougherty, 1951; Anderson, 1984, 1988; Durette-Desset, 1985; Adamson, 1986, 1989; Durette-Desset *et al.*, 1994).

In this scenario, free-living soil nematodes became associated with the moist skin of the early amphibians for paratenic transport, or to survive dry spells in the environment (Figure 6.3). The first penetrations may have been accidental outcomes of free-living burrowing behaviors, but once inside the host, migration through the host tissue was necessary for the worms to get to the bacteria in the host's gut (Durette-Desset, 1985; Adamson, 1986, 1989). This group of skin penetrators is represented by extant hookworms (*Necator* sp. and *Ancylostoma* sp.). Subsequently, with the drier climate of the Permian and the evolution of large tetrapods whose bodies became raised off the ground, skin penetration

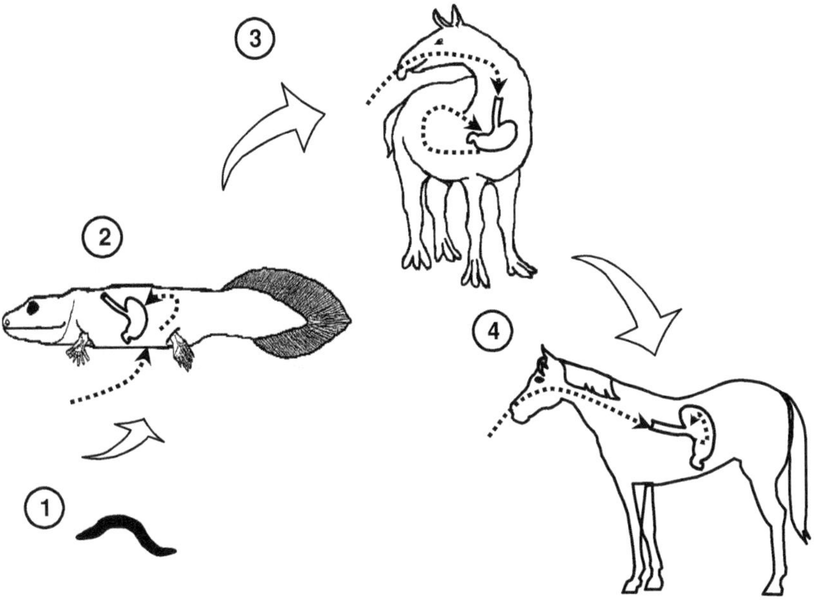

Figure 6.3. Diagrammatic representation of the major stages in the evolution of nematode parasites in terrestrial vertebrate hosts. Ancestral soil-inhabiting nematodes (1) penetrated the skin of early amphibian-like terrestrial vertebrates and migrated to the intestines (2). As tetrapods evolved and raised themselves off the ground, skin-penetrating nematodes adapted to oral infection but retained the tissue migration phase (3). Subsequently, because of the high costs associated with tissue migration, this phase was reduced to a short developmental phase within intestinal tissue (4) (Sukhdeo *et al.*, 1997).

became less efficient as an infection strategy and some species switched to oral infections. The oral route is considered a derived trait that followed evolution of hosts whose feeding habits and/or behavior make fecal contamination likely and who were large volume indiscriminate feeders, i.e. herbivores (Adamson, 1989). Thus, skin-penetrating nematodes adapted to infecting the host via ingestion with food, but the migration to the gut was retained in these worms, probably because of developmental constraints that required a tissue phase. This explains the genesis of the unnecessary tissue migrations in the group that includes *S. vulgaris* which infect orally but penetrate out of the gut to migrate back to the gut. Subsequently, due to the costs associated with tissue migration, some worms abbreviated their migration route significantly. In this group, after penetrating the gut, the worms remain and develop locally within the intestinal tissue, and then emerge into the lumen as adults (Figure 6.3). *Oesophagostomum* spp. belong to this group.

We tested the hypothesis that skin penetration evolved prior to oral infection by reconstructing a phylogeny of the order Strongylida using sequences of the cytochrome c oxidase (subunit I) gene. Strongylida is thought to have evolved from free-living, soil-dwelling ancestors from the order Rhabditida (Dougherty, 1951; Chabaud, 1955; Anderson, 1984; Durette-Desset *et al.*, 1994). Members of this group infect all vertebrates except fishes, but are primarily found in mammals (Anderson, 2000). These parasites have similar life cycles, and the third stage larva (L_3) is responsible for infecting the host either orally or through skin penetration, and it is also the stage responsible for tissue migration in these parasites (Chabaud, 1955; Anderson, 1984; Durette-Desset *et al.*, 1994). A total of 19 taxa from the order Strongylida were used, 11 (including *Oesophagostomum*) were newly sequenced, and eight from a previous study (Sukhdeo *et al.*, 1997).

Our reconstructed phylogenetic tree for the Strongylida is shown in Figure 6.4. Maximium parsimony analysis (MP) provided strong support for monophyly and there is good resolution at base and terminal branches. Maximum likelihood heuristic search produced only a single tree that recovered the same topological arrangement as the MP analyses. These data indicate that skin penetration is an ancestral strategy, and strongly supports the hypothesis of skin penetration as the first step in nematode parasitism. The second step in this process must have been a migration to the lungs. Almost all parasites that skin-penetrate into their vertebrate host "migrate" to the lungs. This is not an active migration, and the worms are simply taken passively to the lungs by venous flow, where they become trapped in the pulmonary capillaries (Crabtree & Wilson, 1980; Kemp & Devine, 1982). Thus, during the course of evolution the lungs should be the first organ colonized, followed by colonization of the intestines. This scenario is supported by this phylogenetic analysis which places the lung-inhabiting nematodes at more basal branches of the tree, before the intestinal infecting species appeared (Figure 6.4).

Overall, morphological, life history, biogeographical, and molecular data support the idea that the original animal parasitic nematodes were skin penetrators, and that oral infection developed secondarily in this group (Anderson, 1984; Sukhdeo *et al.*, 1997). These analyses also support the notion of how these parasites view their worlds. Parasite only "see" small subsets of their total environments, and the fixed nature of their responses probably does not allow for much flexibility in the choice of better routes. Thus, unnecessary migrations can become part of the parasites' behavioral legacy. The route taken by the Strongylida is only one example of how "odd" behaviors might have evolved, and different evolutionary paths have been taken by other groups. In the case of migration in *F. hepatica*, it is thought this trematode evolved from ancestors that were similar to its close relative *Fasciolopsis buski*, a parasite of humans

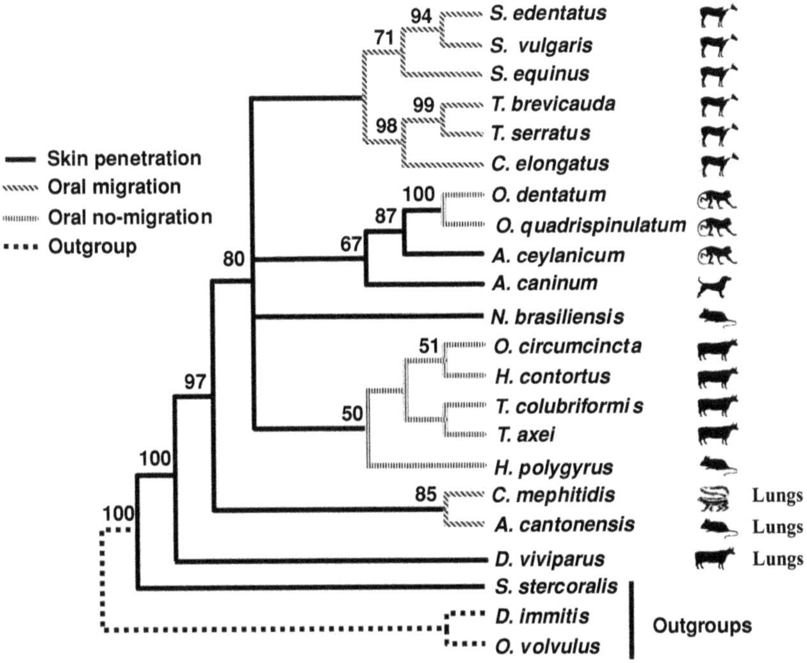

Figure 6.4. Phylogeny of the order Strongylida (Nematoda) reconstructed from cytochrome c oxidase subunit 1 genes; maximum parsimony and maximum likehood analyses recovered the same tree topologies. Three major nematode transmission strategies: skin penetration, oral infection with migration, and oral infection with no migration are mapped onto the tree. This analysis supports skin penetration as the first step in animal parasitism (Sukhdeo *et al.*, 1997; unpublished results).

in Asia, which browses on intestinal tissue in the duodenum. One can imagine these intestinal worms browsing too far down through the mucosa, accidentally finding themselves in the abdominal cavity of their host, and discovering a whole new world (Sukhdeo & Sukhdeo, 2002).

With regards to primate parasites, one must be careful with the limited inferences that can be made from these phylogenetic data. The two important primate-associated genera *Oesophagostomum* and *Ancylostoma* (hookworms) fall out as sister taxa in the same clade, with the oral-infecting larvae in *Oesophagostomum* being the derived strategy (Figure 6.4). Ancestral species (hookworms in this case) tend be generalist parasites on their host's descendants, while more recently evolved species tend to be host-specific. This might help explain the situation in some areas of Ghana with the overlapping populations of primates and humans where both hosts are infected with hookworms, but only the primates are infected with *Oesophagostomum* (Gasser *et al.*,

2006). However, this interpretation is confounded by many other variables that can determine host specificity, including mode of transmission. The skin-penetration strategy itself might contribute towards a generalist habit. In studies of the schistosomes, a trematode group that also includes both skin-penetrating and oral-infecting members, the skin-penetrating species are less host-specific and tend to have many definitive hosts, while the oral-infecting species tended towards high host specificity with > 50% of these species being monospecific (Noble *et al.*, 1989; Gregory *et al.*, 1991; Poulin, 1992). Among primate parasites, host specificity and mode of transmission are clearly associated. A study of 415 parasites (including microparasites) from 119 wild primates species found that parasites which are transmitted by sexual or physical contact were highly host-specific, parasites using intermediate hosts were much less host-specific, and vector transmitted parasites were intermediate (Pedersen *et al.*, 2005). Unfortunately, skin penetration was not included as a distinct category in these analyses. It is also clear that ecological and evolutionary pressures that influence host specificity act on both parasite and hosts, and that both groups must be included in any meaningful analyses. For example, a coevolutionary biogeographic analysis of *Oesophagostomum* in primates indicates a history of host-switching and alternating episodes of migrations between Africa and Asia, and shows high concordance between parasite and primate phylogenies (Brooks & Ferrao, 2005). *Enterobius* sp. (pinworms) also seems to have undergone a very strict coevolution with its primate hosts (Brooks & Glen, 1982). This complex subject is beyond the scope of this discussion, but the future directions in the field are clear. We concur with the many calls for more primate data sets, and for more rigorous studies of primate parasite coevolution (Hasegawa, 1999; Hugot, 1999; Brooks & McLennan, 2003; Brooks & Ferrao, 2005).

How parasites see their hosts, looking from the outside

Modes of parasite transmission are recognized as important variables in primate parasite interactions, and this area has received considerable attention from primatologists (Pedersen *et al.*, 2005). Primatologists understand the nuances of parasite transmission, and patterns of parasite infections in primates have been linked to diverse ecological variables including diet, water contact patterns, position in food chain, and anthropogenic disturbances (Müller-Graf *et al.*, 1997; Ashford *et al.*, 2000; Lilly *et al.*, 2002; Chapman *et al.*, 2005a; Pedersen *et al.*, 2005). Thus, in this section we will only briefly summarize host-finding in one group, the trematodes, to illustrate a recent and significant change in the way parasitologists look at host-finding.

Trematode miracidia make excellent models to study host-finding behavior because they are active ciliated organisms, large enough to be visible under magnifiers, and they infect snails which are easily reared in the laboratory (MacInnis, 1976). These methods are equally available to all parasitologists, so it was odd that the first reports of miracial attraction to snails (Thomas, 1883; Leuckart, 1884) provoked a huge argument among parasitologists, with vehement denunciations from both sides. This battle continued for almost 90 years before it was settled (Ulmer, 1971). One camp said there was attraction, and the other camp said there was no attraction. The "no attraction" camp reported that the miracidia were indifferent to the snail hosts, and infection was the result of accidental encounters. This story, now used as a pedagogical tool, is still amusing to parasitologists because it was all caused by a tiny difference in the methods used by the two camps. The "no attraction" camp used freshly hatched miracidia, and the "attraction" camp used aged miracidia.

There is now a lot of hard evidence that miracidia (and insect parasitoids) employ a three-step strategy during host-finding (Ulmer, 1971; Sukhdeo & Mettrick, 1987; Vet *et al.*, 2002). For miracidia that normally hatch in ponds, the first step is a migration to the snail's habitat using responses to light and gravity. For example, the miracidia of *S. mansoni* are photopositive and geonegative, and this brings them to the shallow edges of the pond which are brighter, and where their snail hosts *B. glabrata* are normally found (Upatham, 1972a, 1972b). The second step is a random swimming pattern through the snail's habitat using a program that is optimal for searching in three-dimensional space (Plorin & Gilbertson, 1981). When the miracidia get near to their hosts, the third step is a specific attraction to mucus secretions that bring them into contact with the snail (Ulmer, 1971). Like egg-retrieval behavior in graylag geese, miracidial host-finding is a fixed behavior pattern released during hatching, and consists of a defined sequence of activities. During step one of the sequence, which lasts 1–3 hours, the miracidia are refractory to snail signals and only respond to light and gravity. This is why the "no attraction" camp did not see attraction; they were using freshly hatched miracidia that had not arrived at the attraction part of the sequence. However, the more important lesson from this story is what it says about how these parasites see their worlds. Host-finding is a fixed sequence of behaviors, and the fixed nature of this response suggests that the parasites are responding to predictable conditions related to the snails' biology. This three-step pattern has been reported so consistently among the many miracidial species tested, and in all parasitoid species, that for a long time, it was thought to be the universal strategy for all active host-finding stages. Thus, it was quite a surprise that the other infective stages of trematodes, cercariae, do not follow this pattern. Cercariae have a two-step pattern because they infect large vertebrates, e.g. chimpanzees!

Cercariae produced in the snail can emerge directly out of the snail's body to infect vertebrate hosts. The first step in host-finding is similar to that of miracidia, and the cercariae use responses to light and gravity to swim to their host's habitat. The second step is also similar and the cercariae engage in repetitive behaviors that keep them in the habitat. In *S. mansoni,* this behavior consists of a characteristic pattern of swimming up the water column, and then slowly parachuting down, with the cycle repeating endlessly until they encounter a host, or until they run out of energy and die (Haas *et al.*, 1990a, 1990b; Haberl *et al.*, 2000). The big problem with cercarial host-finding was that there seemed to be no step three, i.e. there was no attraction to the host.

The war of "attraction" versus "no attraction" started all over again; you would think that we could learn from history, but again, we rejected the evidence for a long time. It is only now after several decades of study, that we have finally accepted that cercariae are not attracted to their vertebrate hosts (Faust & Meleney, 1924; McCoy, 1935; Neuhaus, 1952; Cheng, 1963; Haas, 1994a, 1994b; Haberl *et al.*, 2000; Combes *et al.*, 2002). Attraction to the host is probably adaptive in cases where the host is slow moving like a snail, or like caterpillars being hunted by parasitoids. However, this strategy is probably not efficient for finding large mobile hosts, like primates, that do not remain in one place long enough for gradients to develop. Instead of a host-attraction response, the parasites respond to dark shadows cast, or turbulence caused by large bodies moving through the water. These surrogate signals trigger a frenzied pattern of search behavior in the cercariae, and they attach to anything they encounter during this phase, including the definitive host, twigs, leaves, or abnormal hosts (Haas, 1994a, 1994b). If these were parasites of chimpanzees out foraging for a chimpanzee to infect, they wouldn't even see the chimpanzee until they ran into it!

This has led to a profound change in the way we now think about how parasites see their worlds during host-finding. Before this, our view of this process was host-centric, and we thought that the host was the sole focus of the parasite's point-of-view. Here, the evidence clearly indicates that parasites cannot distinguish between their normal hosts and a twig, and this is somewhat reminiscent of the way that frogs look at moving black dots. However, the more significant realization was that these parasites were not looking for the physical presence of their hosts, instead they were looking for a place and a time where the probability of contact with their host would be maximized (Combes *et al.*, 1994). A series of elegant studies by Combes and his colleagues has demonstrated that cercarial emergence in several schistosome species corresponded closely with the times when their definitive hosts were most likely to visit the aquatic habitat where transmission occurs (Figure 6.5). For example, schistosome species that infected humans tended to emerge during the middle of the

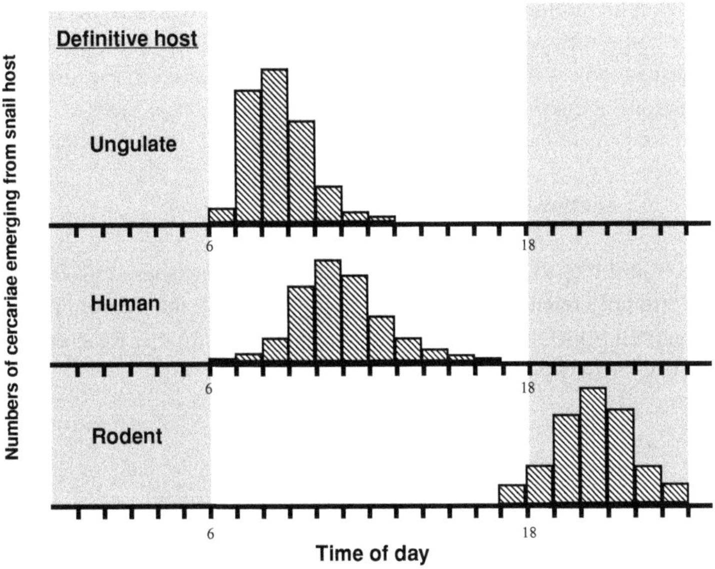

Figure 6.5. Cercarial emergence patterns in some schistosome species. Emergence coincides with the times that their definitive hosts are most likely to be in the water (adapted from Combes *et al.*, 1994).

day when man was most likely to be in the water, schistosome species that infected nocturnal rodents emerged in the evenings when the rodents tended to visit the water, and schistosomes of ungulates tended to emerge in the mornings (Combes *et al.*, 1994; Combes, 2001). These cercarial emergence rhythms were independent of the molluscan host's activities, and also independent of the effective presence of their definitive hosts.

In terms of the relevance of these ideas to primate parasitology, again we can only speculate. Parasitologists are only now beginning to grapple with this new idea that parasites are not always looking directly for their hosts, and the ways this might impact upon models of transmission strategies. As yet, there are not enough data to make any rigorous inferences about primate parasites, or about many other parasites. Nevertheless, one can imagine that if the infective stages of parasites are selected for a time and space where encounters with the host are maximized, then cross infections with other host species will occur when they enter the parasite's time/space intersection zone, especially if the parasite is a generalist as seems to be the case with most primate parasites (Pedersen *et al.*, 2005). It is not clear how wide or narrow these time/space intersection zones may be, or even if they exist. Thus, we can end our discussion with speculations on the potential utility of these changing ideas to the design of novel epidemiological experiments. However, we feel that the more important take-home message from this chapter should be that thinking about the way parasites look at their worlds might offer opportunities to explore the same old questions about primate parasites with completely new eyes.

Acknowledgements

We would like to thank Stacey Lettini, Wayne Rossiter, and Tavis Anderson for critically reading the manuscript and help with the figures. This study was supported from USDA Hatch funds, Rutgers Parasitology Research Grants, and the Center for Research on Animal Parasites.

References

Adamson, M. L. (1986). Modes and transmission and evolution of life histories in zooparasitic nematodes. *Canadian Journal of Zoology*, **64**, 1375–1384.

Adamson, M. L. (1989). Constraints in the evolution of life histories in zooparasitic Nematoda. In *Current Concepts in Parasitology*, ed. R. C. Ko. Hong Kong: University Press, pp. 221–253.

Adler, J. (1983). Bacterial chemotaxis and molecular neurobiology. *Cold Spring Harbor Symposium of Quantitative Biology*, **48**, 803–804.

Alcock, J. (1998). *Animal Behavior: An Evolutionary Approach.* Sunderland, MA: Sinauer Associates Inc.

Allee, W. C., Emerson, A. E., Park, O., Park, T., & Schmidt, K. P. (1949). *Principles of Animal Ecology*. Philadelphia, PA: W. B. Saunders.

Altizer, S., Dobson, A., Hosseini, P. *et al.* (2006). Seasonality and the dynamics of infectious diseases. *Ecology Letters*, **B**, 467–484.

Anderson, R. C. (1984). The origins of zooparasitic nematodes. *Canadian Journal of Zoology*, **62**, 317–328.

Anderson, R. C. (1988). Nematode transmission patterns. *Journal of Parasitology*, **74**, 30–45.

Anderson, R. C. (2000). *Nematode Parasites of Vertebrates. Their Development and Transmission*, 2nd edn. Wallingford: CABI Publishing.

Ashford, R. W., Reid, G. D. F. & Wrangham, R. W. (2000). Intestinal parasites of the chimpanzee *Pan troglodytes* in Kibale Forest, Uganda. *Annals of Tropical Medicine and Parasitology*, **94**, 173–179.

Awwad, M. & Bell, D. R. (1982). Faecal extract attracts copulating schistosomes. *Annals of Tropical Medicine and Parasitology*, **72**, 389–390.

Bailey, W. S. (ed.) (1982). *Cues that Influence Behavior of Internal Parasites.* New Orleans, LA: United States Department of Agriculture.

Bain, O. & Chabaud, A. G. (1979). Sur les Muspiceidae (Nematoda-Dorylaimina). *Annales de Parasitologie Humaine et Comparée*, **54**, 207–225.

Barger, M. A. & Esch, G. W. (2002). Host specificity and the distribution abundance relationship in a community of parasites infecting fishes in streams of North Carolina. *Journal of Parasitology*, **88**, 446–453.

Bongers, T. & Ferris, H. (1999). Nematode community structure as a bioindicator in environmental monitoring. *Trends in Evolution and Ecology*, **14**, 224–228.

Brooks, D. R. & Ferrao, A. L. (2005). The historical biogeography of co-evolution: emerging infectious diseases are evolutionary accidents waiting to happen. *Journal of Biogeography*, **32**, 1291–1299.

Brooks, D. R. & Glen, D. R. (1982). Pinworms and primates: a case study in coevolution. *Proceedings of the Helminthological Society of Washington*, **49**, 76–85.

Brooks, D. R. & McLennan, D. A. (2003). Extending phylogenetic studies of coevolution: secondary Brooks parsimony analysis, parasites and the Great Apes. *Cladistics*, **19**, 104–119.

Bullock, T. H. & Horridge, G. A. (1965). *Structure and Function in the Nervous Systems of Invertebrates*. Vol. I., San Francisco, CA: W. H. Freeman and Co.

Chabaud, A. G. (1954). Sur le cycle évolutif des spirurides et nematodes ayant une biologie comparable. Valeur systematique des caractères biologique. *Annales de Parasitologie Humaine et Comparée*, **29**, 42–88.

Chabaud, A. G. (1955). Essai d'interprétation phylétique des cycles évolutifs chez les Nématodes, parasites de Vertébrés. Conclusion taxonomiques. *Annales de Parasitologie Humaine et Comparée*, **30**, 83–126.

Chapman, C. A., Gillespie, T. R. & Goldberg, T. L. (2005a). Primates and the ecology of their infectious diseases: how will anthropogenic change affect host-parasite interactions. *Evolutionary Anthropology*, **14**, 134–144.

Chapman, C. A., Gillespie, T. R. & Speirs, M. L. (2005b). Dynamics of gastrointestinal parasites in two colobus monkeys following a dramatic increase in host density: contrasting density-dependent effects. *American Journal of Primatology*, **67**, 259–266.

Chapman, W. H. & Undeen, A. H. (1968). *In vivo* and *in vitro* hatching of eggs of *Trichosomoides crassicauda*. *Experimental Parasitology*, **22**, 213–218.

Cheng, T. C. (1963). Activation of *Gorgodera amplicava* cercariae by molluscan sera. *Experimental Parasitology*, **13**, 342–347.

Chitwood, B. G. & Chitwood, M. B. (1950). *An Introduction to Nematology*. Baltimore, MD: University Park Press.

Clayton, D. H. & Wolfe, N. D. (1993). The adaptive significance of self-medication. *Trends in Ecology and Evolution*, **8**, 60–63.

Combes, C. (2001). *The Art of Being a Parasite*. Chicago, IL: University of Chicago Press.

Combes, C. & Théron, A. (2000). Metazoan parasites and resource heterogeneity: constraints and benefits. *International Journal for Parasitology*, **30**, 299–304.

Combes, C., Fournier, A., Mone, H. & Théron, A. (1994). Behaviours in trematode cercariae that enhance parasite transmission. *Parasitology*, **109**, S3–S13.

Combes, C., Bartoli, P. & Théron, A. (2002). Trematode transmission strategies. In *The Behavioural Ecology of Parasites*, ed. E. E. Lewis, J. F. Campbell & M. V. K. Sukhdeo. Wallingford: CABI Publishing, pp. 1–12.

Crabtree, J. E. & Wilson, R. A. (1980). *Schistosoma mansoni*: a scanning electron microscope study of the developing schistosomulum. *Parasitology*, **81**, 553–564.
Crompton, D. W. T. (1973). The sites occupied by some parasitic helminths in the alimentary tract of vertebrates. *Biological Reviews*, **48**, 27–83.
Daszak, P., Cunningham, A. A. & Hyatt, A. D. (2001). Anthropogenic environmental change and the emergence of infectious diseases in wildlife. *Acta Tropica*, **78**, 103–116.
Dawes, B. (1963). The migration of juvenile forms of *Fasciola hepatica* L. through the wall of the intestines in the mouse, with some observations on food and feeding. *Parasitology*, **53**, 109–122.
Dawes, B. & Hughes, D. L. (1964). *Fasciola* and fascioliasis. *Advances in Parasitology*, **2**, 97–168.
de Gruijter, J. M., Ziem, J., Verweij, J. J., Polderman, A. M. & Gasser, R. B. (2004). Genetic substructuring within *Oesophagostomum bifurcum* (Nematoda) from human and non-human primates from Ghana based on random amplification of polymorphic DNA analysis. *American Journal of Tropical Medicine and Hygiene*, **71**, 227–233.
de Gruijter, J. M., Gasser, R. B., Polderman, A. M., Asigri, V. & Dijkshoorn, L. (2005). High resolution DNA fingerprinting by AFLP to study the genetic variation among *Oesophagostomum bifurcum* (Nematoda) from human and non-human primates in Ghana. *Parasitology*, **130**, 229–237.
Dixon, K. E. (1966). The physiology of excystment of the metacercariae of *Fasciola hepatica*. *Parasitology*, **56**, 431–436.
Dobson, A. P. (2004). Population dynamics of pathogens with multiple host species. *American Naturalist*, **164**, S64–S78.
Dobson, A. P. & Foufopoulus, J. (2001). Emerging infectious pathogens on wildlife. *Philosophical Transactions of the Royal Society of London Series B*, **356**, 1001–1012.
Dougherty, E. C. (1951). Evolution of zooparasitic groups in the phylum Nematoda, with special reference to host-distribution. *Journal of Parasitology*, **37**, 353–378.
Duncan, J. L. (1972). The life cycle, pathogenesis and epidemiology of *S. vulgaris* in the horse. *Equine Veterinary Journal*, **5**, 20–25.
Durette-Desset, M. C. (1985). Trichostrongyloid nematodes and their vertebrate hosts: reconstruction of the phylogeny of a parasitic group. *Advances in Parasitology*, **24**, 239–306.
Durette-Desset, M. C., Beveridge, I. & Spratt, D. M. (1994). The origins and evolutionary expansion of the Stronglyida (Nematoda). *International Journal for Parasitology*, **24**, 1139–1165.
Eisenbach, M. (2004). *Chemotaxis*. London: Imperial College Press.
Eley, R. M., Strum, S. C., Muchemi, G. & Reid, G. D. F. (1989). Nutrition, body condition, activity patterns, and parasitism of free-ranging troops of olive baboons (*Papio anubis*) in Kenya. *American Journal of Primatology*, **18**, 209–219.
Erasmus, D. A. (1972). *The Biology of Trematodes*. London: Edward Arnold Limited.
Fa, J., Juste, J., Del Val, J. & Castroviejo, J. (1995). Impact of market hunting on mammal species in Equatorial Guinea. *Conservation Biology*, **9**, 1107–1115.

Faust, E. C. & Khaw, O. K. (1927). Studies on *Clonorchis sinensis* (Cobbold). *American Journal of Hygiene Monogram Series No. 8.*
Faust, E. C. & Meleney, H. E. (1924). Studies on *Schistosoma japonica. American Journal of Hygiene Monogram Series No. 3.*
Fraenkel, G. S. & Gunn, D. L. (1940). *The Orientation of Animals.* Oxford, UK: Oxford University Press.
Freeland, W. J. (1979). Social organization and population density in relation to food use and availability. *Folia Primatology*, **32**, 108–124.
Fried, B. (1994). Metacercarial excystment of trematodes. *Advances in Parasitology*, **33**, 91–138.
Fülleborn, F. & Schilling-Torgau, V. (1911). Untersuschungen über den Infektionsweg bei Strongyloides und Ancyclostomum Vorläufige Mitteilung. *Archiv für Schiffs und Tropenhygiene*, **15**, 569–571.
Gasser, R. B. (2005). Molecular tools – advances, opportunities and prospects. *Veterinary Parasitology*, **136**, 69–89.
Gasser, R. B., Woods, W. G., Blotkamp, J. *et al.* (1999). Screening for nucleotide in ribosomal DNA arrays of *Oesophagostomum bifurcum* by polymerase chain reaction-coupled single-strand conformation polymorphism. *Electrophoresis*, **20**, 1486–1491.
Gasser, R. B., de Gruijter, J. M. & Polderman, A. M. (2006). Insights into the epidemiology and genetic make-up of *Oesophagostomum bifurcum* from human and non-human primates using molecular tools. *Parasitology*, **132**, 453–460.
Gillespie, T. R. & Chapman, C. A. (2006). Forest fragment attributes predict parasite infection dynamics in primate metapopulations. *Conservation Biology*, **20**, 441–448.
Gregory, R. D., Keymer, I. E. & Harvey, P. H. (1991). Life history, ecology and parasite community structure in Soviet birds. *Biological Journal of the Linnean Society*, **43**, 249–262.
Haas, W. (1994a). Physiological analyses of host-finding behaviour in trematode cercariae: adaptations for transmission success. *Parasitology*, **109** (Suppl.), S15–S29.
Haas, W. (1994b). *Schistosoma haematobium* cercarial host-finding and host-recognition differs from that of *S. mansoni. Journal of Parasitology*, **80**, 345–353.
Haas, W., Granzer, M. & Brockelman, C. (1990a). *Opisthorchis viverrini*: finding and recognition of the fish host by the cercariae. *Experimental Parasitology*, **71**, 422–431.
Haas, W., Granzer, M. & Brockelman, C. (1990b). Finding and recognition of the bovine host by the cercariae of *Schistosoma spindale. Parasitology Research*, **76**, 343–350.
Haberl, B., Kroner, M., Spengler, Y. *et al.* (2000). Host-finding in *Echinostoma caproni*: miracidia and cercariae use different signals to identify the same snail species. *Parasitology*, **120**, 479–486.
Hahn, N. E., Proulx, D., Muruthi, P. M., Alberts, S. & Altman, J. (2003). Gastrointestinal parasites in free-ranging Kenyan baboons (*Papio cynocephalus* and *P. anubis*). *International Journal of Primatology*, **24**, 271–279.

Hasegawa, H. (1999). Phylogeny, host-parasite relationship and zoogeography. *Korean Journal of Parasitology*, **37**, 197–213.

Haslewood, G. A. D. (1978). *The Biological Importance of Bile Salts*. Amsterdam: North-Holland Publishing Co.

Huffman, M. A. & Hirata, S. (2004). An experimental study of leaf swallowing in captive chimpanzees – insights into the origin of a self-medicative behavior and the role of social learning. *Primates*, **45**, 113–118.

Huffman, M. A. & Seifu, M. (1989). Observations on the illness and consumption of a possibly medicinal plant *Vernonia amygdalina* by a wild chimpanzee in the Mahale Mountains, Tanzania. *Primates*, **30**, 51–63.

Huffman, M. A., Page, J. E., Sukhdeo, M. V. K. *et al.* (1996). Leaf-swallowing by chimpanzees, a behavioral adaptation for the control of strongyle nematode infections. *International Journal of Parasitology*, **17**, 475–503.

Huffman, M. A., Ohigashi, H., Kawanaka, M. *et al.* (1998). African great ape self-medication: a new paradigm for treating parasite disease with natural medicines? In *Towards Natural Medicine Research in the 21st Century*, ed. Y. Ebizuka. Amsterdam: Elsevier Science B.V., pp. 113–123.

Hugot, J. P. (1999). Primates and their pinworm parasite: the Cameron hypothesis revisited. *Systematic Biology*, **48**, 523–546.

Inge, D. (1983). Brain mechanisms of visual localisation in frogs and toads. In *Advances in Vertebrate Neuroethology*, ed. J.-P. Ewert, R. R. Capranica & D. Ingle. New York, NY: Plenum Press, pp. 177–226.

Irwin, S. W. B. (1983). In vitro excystment of the metacercariae of *Maritrema arenaria* (Digenea: Microphallidae). *International Journal for Parasitology*, **13**, 191–196.

Jennings, H. S. 1906. *Behavior of the Lower Organisms*. Bloomington, IN: Indiana University Press.

Kemp, W. M. & Devine, D. R. (1982). Behavioral cues in trematode life cycles. In *Cues that Influence Behavior of Internal Parasites*, ed. W. S. Bailey. New Orleans, LA: United States Department of Agriculture, pp. 67–80.

Krebs, J. R. & Davies, N. B. (1989). *Behavioral Ecology. An Evolutionary Approach*. Oxford, UK: Blackwell.

Lackie, A. M. (1975). The activation of infective stages on endoparasites of vertebrates. *Biological Reviews*, **50**, 285–323.

Leuckart, K. G. F. (1884). *Die Parasiten des Menschen und die von ihnen Lerruhreden Krankheiten*. Part 5., Vol. 1. Leipzig. Germany: Verlagshandlung.

Liang, J. (1937). Host finding by insect parasites. I. Observations on the finding of hosts by *Alysia manducator*, *Mormoniella vitropennis* and *Trichogramma evanescens*. *Journal of Animal Ecology*, **6**, 298–317.

Lilly, A. A., Mehlman, P. T. & Doran, D. (2002). Intestinal parasites in gorillas, chimpanzees, and humans at Mondika Research Site, Dzanga–Ndoki National Park, Central African Republic. *International Journal of Primatology*, **23**, 555–573.

Lorenz, K. Z. & Tinbergen, N. (1957). Taxis and instinct. In *Instinctive Behavior; The Development of a Modern Concept*, ed. C. H. Schiller. New York, NY: International Universities Press, pp. 176–208.

Löffler, W. (1956). Transient lung infiltrations with blood eosinophilia. *International Archives of Allergy*, **8**, 54–59.

MacInnis, A. J. (1976). How parasites find hosts: some thoughts on the conception of host parasite integration. In *Ecological Aspects of Parasitology*, ed. C. R. Kennedy. Amsterdam: North-Holland Publishing Co, pp. 3–20.

Mapes, C. J. (1972). Bile and bile salts and exsheathment of the intestinal nematodes *Trichostrongylus colubriformis* and *Nematodirus battus*. *International Journal for Parasitology*, **2**, 433–438.

McCoy, O. R. (1935). The physiology of helminth parasites. *Physiology Reviews*, **15**, 221–240.

McDaniel, J. S. (1966). Excystment of *Crytocotyle lingua* metacercariae. *Biological Bulletin (Woods Hole)*, **130**, 369–377.

McGraw, B. M. & Slocombe, J. O. D. (1976). *Strongylus equinus*: development and pathological effects in the equine host. *Canadian Journal of Comparative Medicine*, **49**, 372–383.

Mettrick, D. F. & Podesta, R. B. (1974). Ecological and physiological aspects of helminth-host interactions in the mammalian gastrointestinal canal. *Advances in Parasitology*, **12**, 183–278.

Müller-Graf, C. D., Collins, D. A., Packer, C. & Woolhouse, M. E. (1997). *Schistosoma mansoni* infection in a natural population of olive baboons (*Papio cynocephalus anubis*) in Gombe Stream National Park, Tanzania. *Parasitology*, **14**, 621–627.

Neuhaus, W. (1952). Biologie und Entwicklung von *Trichobilharzia szidati* n. sp. (Trematoda, Schistosomatidae), einem Erreger von Dermatitis beim Menschen. *Zeitschrift fur Parasitenkunde*, **15**, 203–266.

Noble, E. R., Noble, G. A., Schad, G. A. & MacInnes, A. J. (1989). *Parasitology: The Biology of Animal Parasites*, 6th edn. Philadelphia, PA: Lea & Febiger.

Nunn, C. L. & Altizer, S. (2005). The global mammal parasite database: an online resource for infectious disease records in wild primates. *Evolutionary Anthropology*, **14**, 1–2.

Nunn, C. L. & Altizer, S. (2006). *Infectious Diseases in Primates: Behavior, Ecology and Evolution*. New York, NY: Oxford University Press.

Nunn, C. L., Altizer, S., Jones, K. E. & Sechrest, W. (2003). Comparative tests of parasite species richness in primates. *American Naturalist*, **162**, 597–614.

Pedersen, A. B., Poss, M., Altizer, S., Cunningham, A. & Nunn, C. L. (2005). Patterns of host specificity and transmission among parasites of wild primates. *International Journal for Parasitology*, **35**, 647–657.

Pelkwijk, J. J. ter & Tinbergen, N. (1937). Eine reizbiologisch Analyse einiger Verhaltensweisen von *Gasterosteus aculeatus* L. *Zeitschrift fur Tierpsychologie*, **1**, 193–200.

Plorin, G. G. & Gilbertson, D. E. (1981). Behavior of *Schistosoma mansoni* miracidia upon contacting solid surfaces. *Journal of Parasitology*, **67**, 727–728.

Polderman, A. M., Anemana, S. D. & Asigri, V. (1999). Human oesophagostomiasis: a regional public health problem in Africa. *Parasitology Today*, **15**, 129–130.

Polderman, A. M., Krepel, H. P., Baeta, S., Blotkamp, J. & Gigase, P. (1991). Oesophagostomiasis, a common infection of man in northern Togo and Ghana. *American Journal of Tropical Medicine and Hygiene*, **44**, 336–344.

Poulin, R. (1992). Determinants of host-specificity in parasites of freshwater fishes. *International Journal for Parasitology*, **22**, 753–758.

Poulin, R. (2007). *Evolutionary Ecology of Parasites*, 2nd edn. Princeton, NJ: Princeton University Press.

Poulin, R. & Mouillot, D. (2003). Parasite specialization from a phylogenetic perspective: a new index of host specificity. *Parasitology*, **126**, 473–480.

Rea, J. G. & Irwin, S. W. B. (1994). The ecology of host-finding behaviour and parasite transmission: past and future perspectives. *Parasitology*, **109** (Suppl.), S31–S39.

Rogers, W. P. (1960). The physiology of infective processes of nematode parasites. The stimulus from the animal host. *Proceedings of the Royal Society London B*, **152**, 367–386.

Rohde, K. (1989). At least eight types of sense receptors in an endoparasitic flatworm: a counter trend to sacculinization. *Naturwissenschaften*, **76**, 383–385.

Rohde, K. (1993). *Ecology of Marine Parasites*, 2nd edn. Wallingford, UK: CAB International.

Schiller, E. L. (1965). A simplified method for the *in vitro* cultivation of the rat tapeworm *Hymenolepis diminuta*. *Journal of Parasitology*, **51**, 516–518.

Silverman, P. H. (1954). Studies on the biology of some tapeworms of the genus *Taenia*. I. Factors affecting hatching and activation of taenid ova, and some criteria of their viability. *Annals of Tropical Medicine and Parasitology*, **48**, 207–215.

Smyth, J. D. (1966). *The Physiology of Trematodes*. Edinburgh, UK: Oliver and Boyd.

Smyth, J. D. & Halton, D. W. (1983). *The Physiology of Trematodes*. Cambridge: Cambridge University Press.

Smyth, J. D. & Haslewood, G. A. D. (1963). The biochemistry of bile as a factor in determining host specificity in intestinal parasites, with particular reference to *Echinococcus granulosus*. *Annals of the New York Academy of Science*, **113**, 234–260.

Sommerville, R. I. & Rogers, W. P. (1987). The nature and action of host signals. *Advances in Parasitology*, **26**, 239–293.

Stunkard, H. W. (1930). The life history of *Cryptocotyle lingua* (Creplin), with notes of the physiology of the metacercariae. *Journal of Morphology and Physiology*, **50**, 143–191.

Sukhdeo, M. V. K. (1990). Habitat selection by helminths, a hypothesis. *Parasitology Today*, **6**, 234–237.

Sukhdeo, M. V. K. (1994). Parasites and behaviour. In *Parasites and Behaviour*, ed. M. V. K. Sukhdeo. Cambridge: Cambridge University Press, pp. 1–151.

Sukhdeo, M. V. K. & Croll, N. A. (1981). The location of parasites in their hosts. Bile and site selection behaviour in *Nematospiroides dubius*. *International Journal for Parasitology*, **11**, 157–162.

Sukhdeo, M. V. K. & Mettrick, D. F. (1987). Parasite behaviour: understanding Platyhelminth responses. *Advances in Parasitology*, **26**, 73–144.

Sukhdeo, M. V. K. & Sukhdeo, S. C. (2002). Fixed behaviours and migration in parasitic flatworms. *International Journal for Parasitology*, **32**, 329–342.

Sukhdeo, M. V. K. & Sukhdeo, S. C. (2004). Trematode behaviours and the perceptual worlds of parasites. *Canadian Journal of Zoology*, **82**, 292–315.

Sukhdeo, S. C. & Sukhdeo, M. V. K. (1994). Mesenchyme cell of *Fasciola hepatica* (Platyhelminthes): primitive gala? *Tissue and Cell*, **26**, 123–131.

Sukhdeo, S. C., Sukhdeo, M. V. K., Black, M. B. & Vrijenhoek, R. C. (1997). The evolution of tissue migration in parasitic nematodes (Nematoda: Strongylida) inferred from a protein-coding mitochondrial gene. *Biological Journal of the Linnean Society*, **61**, 281–298.

Sukhdeo, S. C., Sukhdeo, M. V. K. & Mettrick, D. F. (1988). Neurocytology of the cerebral ganglia of *Fasciola hepatica* (Platyhelminthes). *Journal of Comparative Neurology*, **278**, 337–343.

Thoisy, B., Vogel, I., Reynes, J.-M. *et al.* (2001). Health evaluation of translocated free-ranging primates in French Guiana. *American Journal of Primatology*, **54**, 1–16.

Thomas, A. P. (1883). The life history of the liver fluke (*Fasciola hepatica*). *Quarterly Journal of Microscopic Science*, **223**, 99–121.

Tinbergen, N. & Perdeck, A. C. (1950). On the stimulus situation releasing the begging response in the newly hatched Herring Gull chick (*Larus a. argentatus* Pontopp.). *Behaviour*, **3**, 1–38.

Ulmer, M. J. (1971). Site-finding behaviour in helminths in intermediate and definitive hosts. In *Ecology and Physiology of Parasites*, ed. A. M. Fallis. Toronto: University of Toronto Press, p. 267.

Upatham, E. S. (1972a). Effect of water depth on the infection of *Biomphalaria glabrata* by miracidia of St. Lucian *Schistosoma mansoni* under laboratory and field conditions. *Journal of Helminthology*, **46**, 317–325.

Upatham, E. S. (1972b). Exposure of caged *Biomphalaria glabrata* (Say) to investigate dispersion of miracidia of *Schistosoma mansoni* Sambon in outdoor habitats in St. *Lucia. Journal of Helminthology*, **46**, 297–306.

Vet, L. E. M., Hemerik, L., Visser, M. E. & Wäckers, F. L. (2002). Flexibility in host-search and patch-use strategies of insect parasitoids. In *The Behavioural Ecology of Parasites*, ed. E. E. Lewis, J. F. Campbell & M. V. K. Sukhdeo. Wallingford, UK: CABI Publishing, pp. 39–64.

Vitone, N. D., Altizer, S. & Nunn, C. L. (2004). Body size, diet and sociality influence the species richness of parasitic worms in anthropoid primates. *Evolutionary Ecology Research*, **6**, 183–199.

von Frisch, K. (1950). *Bees – Their Vision, Chemical Senses and Language*. New York, NY: Cornell University Press.

von Uexküll, J. (1934). Streifzge durch die Umwelten von Tieren und Menschen (A stroll through the worlds of animals and men). In *Instinctive Behavior*, ed. C. H. Schiller. New York, NY: International Universities Press, pp. 5–80.

Wallis, J. (2000). Prevention of disease transmission in primate conservation. *Annals of the New York Academy of Sciences*, **916**, 691–693.

Wallis, J. & Lee, D. R. (1999). Primate conservation: the prevention of disease transmission. *International Journal of Primatology*, **20**, 803–826.

Webbe, G. (1996). The effect of water velocities on the infection of *Biomphalaria sudanica tanganycencis* exposed to different numbers of *Schistosoma mansoni* miracidia. *Annals of Tropical Medicine and Parasitology*, **75**, 49–62.

Wehler, R. (1981). Spatial vision in arthropods. In *Handbook of Sensory Physiology*, ed. H. Autrum. Berlin: Springer-Verlag, pp. 287–316.

Wolfe, N. C., Escalante, A. A., Karesh, W. B. *et al.* (1998). Wild primate populations in emerging infectious disease research: the missing link. *Emerging Infectious Diseases*, **4**, 149–158.

Wolfe, N. D., Daszak, P., Kilpatrick, A. M. & Burke, D. S. (2005). Bushmeat hunting, deforestation, and predicting a zoonotic emergence. *Emerging Infectious Diseases*, **11**, 1822–1827.

Wrangham, R. & Nishida, T. (1983). *Aspilia* spp. leaves: a puzzle in the feeding behavior of wild chimpanzees. *Primates*, **24**, 276–282.

Wykoff, D. E. & Lepes, T. J. (1957). Studies on *Clonorchis sinensis*. I. Observation on the route of migration in the definitive host. *American Journal of Medicine and Hygiene*, **6**, 1061–1065.

7 *Primate malarias: evolution, adaptation, and species jumping*

ANTHONY DI FIORE, TODD DISOTELL, PASCAL GAGNEUX, AND FRANCISCO J. AYALA

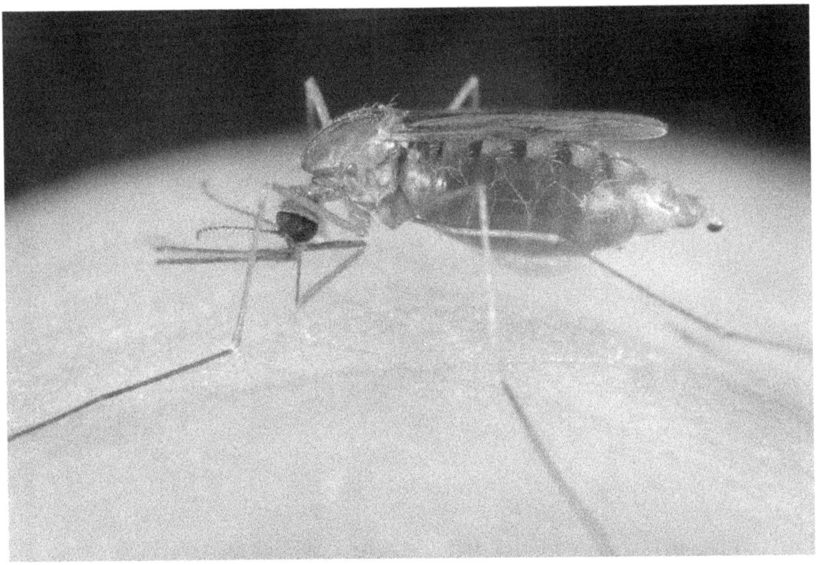

Photograph by James Gathany. Female *Anopheles freeborni* taking a blood meal from a human host. CDC http://phil.cdc.gov/phil/home.asp

Introduction

Malaria is one of the most widespread infectious diseases of modern vertebrates, with an endemic distribution that spans the globe's tropic, subtropic, and some temperate regions (Figure 7.1). The disease is caused by

Primate Parasite Ecology. The Dynamics and Study of Host–Parasite Relationships, ed. Michael A. Huffman and Colin A. Chapman. Published by Cambridge University Press.

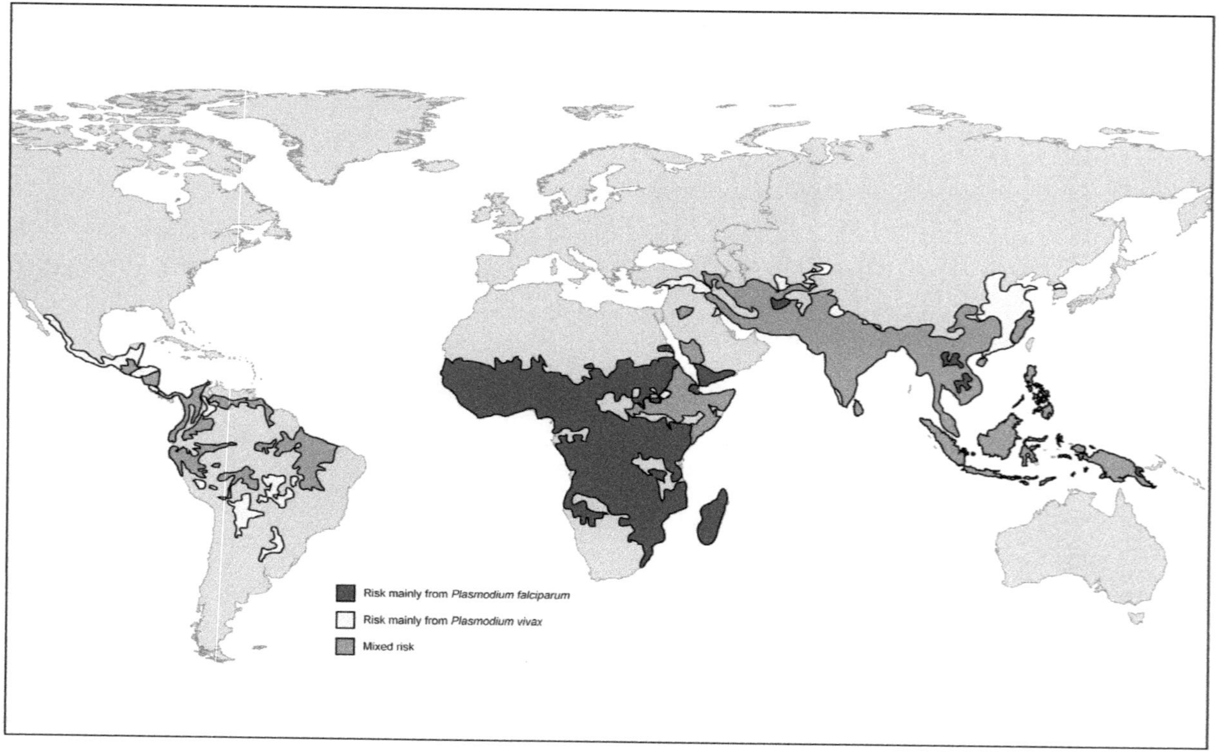

Figure 7.1. Distribution of human malarial risk in 2005. Redrawn from Guerra *et al.* (2006).

protozoan parasites belonging to the phylum Apicomplexa (Table 7.1). The most common agents of malaria are members of the genus *Plasmodium*, but species from several other genera can also cause disease in birds, squamates (lizards and snakes), and some mammals. Currently close to 200 species of the genus *Plasmodium* are recognized, based primarily on their life-history traits, morphology at different life cycle stages, and host species they infect. More than 50 species infect mammals (mainly primates and rodents) (Collins & Aikawa, 1993), over 30 infect birds (van Riper *et al.*, 1994), and close to 90 infect squamates (Telford, 1994). Genetic studies suggest that there also exist numerous "cryptic" species of *Plasmodium* and other genera of malarial parasites in lizards (Perkins, 2000) and birds (Bensch *et al.*, 2000, 2004; Hellgren, 2005). Thus, current estimates of the diversity of malaria-causing protists are likely to be low.

The global biological significance of malarial parasites is staggering. The World Health Organization estimated that 396 million cases of human malaria (and 1.1 million deaths) occurred in 2001 due to infection by *Plasmodium falciparum*, the most malignant of the four parasite species responsible for the disease in humans (WHO, 2003). For 2002, the estimated number of human cases of malaria due to *P. falciparum* infection was 515 million, with over 70% of those occurring in Africa (Snow *et al.*, 2005). Another 71 to 80 million cases are estimated to be caused annually by *P. vivax*, which is responsible for more than half of the cases of human malaria outside of Africa (Mendis *et al.*, 2001). More than 3.2 billion people – roughly half of the world's human population – were estimated to be at risk for malarial infection in 2005 (Guerra *et al.*, 2006) (Figure 7.1), and projections suggest that over 80% of the global population, or more than 8.8 billion people, will be at risk for infection in 2080 (Arnell *et al.*, 2002).

In the future, the worldwide morbidity (i.e. rate of incidence) of human malaria is also expected to increase, in some areas, dramatically, because of population growth and urbanization in regions of high malarial risk, changing patterns of land use and land cover in the tropics, increased population mobility, and global climate change (Hay *et al.*, 2004; Sutherst, 2004). Climate changes – particularly altered rainfall patterns and increased global temperatures associated with higher levels of atmospheric CO_2 – are likely to impact the distribution of areas suitable for parasite persistence and influence the length of the potential transmission season (Kovats *et al.*, 2001; McMichael *et al.*, 2003, 2006; Patz *et al.*, 2005). Changes in temperature and rainfall can also influence the geographic distributions of the insect vectors most responsible for transmission of malarial parasites, not just among humans but among other animal species as well (Sutherst, 1998; Harvell *et al.*, 2002). Although not without controversy (Rogers & Randolph, 2000; Hay *et al.*, 2002a, 2002b; Thomas *et al.*, 2004), a number of recent modeling studies suggest that over the next

Table 7.1. *Common genera and species of protozoal parasites from the Phylum Apicomplexa*

Genus		Species	Common definitive host(s)	Common intermediate host(s)	Associated disease in host or zoonosis in humans
Conoidasina	Coccidiasina	*Cryptosporidium*	Rodents	None	Cryptosporidiosis
		Eimera	Waterfowl, marine birds	None	Coccidiosis
		Neospora	Canids	None, mammals	Neosporis
		Toxoplasma	Felids	Mammals	Toxoplasmosis
		Sarcocystis	Canids	Oviids, rodents, waterfowl	Sarcocystosis
Aconoidasina	Piroplasmorida	*Theilera*	Ticks	Bovids	East Coast Fever, theileriosis
		Babesia	Ticks	Ungulates	Babeosis
		Cytauxzoon	Ticks	Felids	Cytauxzoonosis
	Haemospororida	*Haemoproteus*	Midges, louse-flies	Birds, squamates	Malaria
		Saurocytozoon	*Culex* mosquitoes	Squamates	Malaria
		Leucocytozoon	Simuliid flies	Birds	Malaria
		Hepatocystis	Midges	Squirrels, Old World monkeys, bats	Malaria
		Plasmodium falciparum	*Anopheles* mosquitoes	Humans	Malaria
		vivax	*Anopheles* mosquitoes	Humans	Malaria
		malariae	*Anopheles* mosquitoes	Humans	Malaria
		ovale	*Anopheles* mosquitoes	Humans	Malaria
		reichenowi	*Anopheles* mosquitoes	Chimpanzees	Malaria
		gonderi	*Anopheles* mosquitoes	Old World monkeys	Malaria
		simiovale	*Anopheles* mosquitoes	Old World monkeys	Malaria
		knowlesi	*Anopheles* mosquitoes	Old World monkeys	Malaria
		cynomolgi	*Anopheles* mosquitoes	Old World monkeys	Malaria
		simium	*Anopheles* mosquitoes	New World monkeys	Malaria
		brasilianum	*Anopheles* mosquitoes	New World monkeys	Malaria
		berghei	*Anopheles* mosquitoes	Rodents	Malaria
		elongatum	*Culex* mosquitoes	Birds	Malaria
		gallinaceum	*Aedes* and *Culex* mosquitoes	Birds	Malaria
		wenyoni	*Culex* mosquitoes	Snakes	Malaria
		mexicanum	Sand flies, mites	Lizards	Malaria

50–75 years warming global climates could lead to dramatic increases in the number of people at risk for malaria and in the transmission potential of the disease, particularly in temperate latitudes, as well as result in modest increases in its likely latitudinal and altitudinal distribution (Martens *et al.* 1995, 1999; Tol & Dowlatabadi, 2001; Hartman *et al.*, 2002; Tanser *et al.*, 2003; van Lieshout *et al.*, 2004; Ebi *et al.*, 2005; Pascual *et al.*, 2006). As just one example, Martens *et al.* (1999) have projected that by 2080 more than 450 million *additional* people may be at risk for infection by *P. falciparum* and *P. vivax* as a result of global climate change than would be if current climate conditions remain stable.

Overview of the biology of malarial parasites

Plasmodium and other related genera of malaria-inducing protists are digenetic or "two-host" parasites – their life cycles involve both sexual reproduction (in the parasites' "definitive," invertebrate hosts) and asexual, clonal multiplication (in a vertebrate, "intermediate" host species) (Figure 7.2). Malarial parasites are transmitted between vertebrate hosts by hematophagous ("blood-eating") insects and, less commonly, arachnid vectors. For mammals, these vectors are typically mosquitoes of the genus *Anopheles*, but other mosquito genera (e.g. *Aedes*, *Culex*) and other hematophagous arthropods (e.g. sand flies, midges, louse-flies, mites) may also serve as either common or occasional vectors for transmission of some primate, avian, and squamate malarial parasites.

A typical *Plasmodium* infection cycle in a human or non-human primate host is shown in Figure 7.2. The life cycles of *Plasmodium* in other hosts and of other genera of malarial parasites in their vertebrate hosts are fundamentally similar, although differences can be found in specific aspects of the cycle. For example, when squamates and birds are infected with *Plasmodium*, the early rounds of asexual multiplication tend to take place in epithelial cells. The daughter merozoites produced are then released to the bloodstream both to invade circulating blood cells and to colonize solid tissues (e.g. liver, spleen) (Paul *et al.*, 2003). In other genera of malarial parasites (e.g. *Haemoproteus*, *Hepatocystis*, and *Leucocytozoon*), asexual multiplication takes place solely within solid tissues in the body and not within circulating red blood cells in the peripheral blood system (Paul *et al.*, 2003).

Evolutionary history of malarial parasites

Origins and host-transfer in vertebrates

Over the last decade, as genetic data have accumulated on the malarial parasites and other Apicomplexa, a preliminary picture of the evolutionary history of the

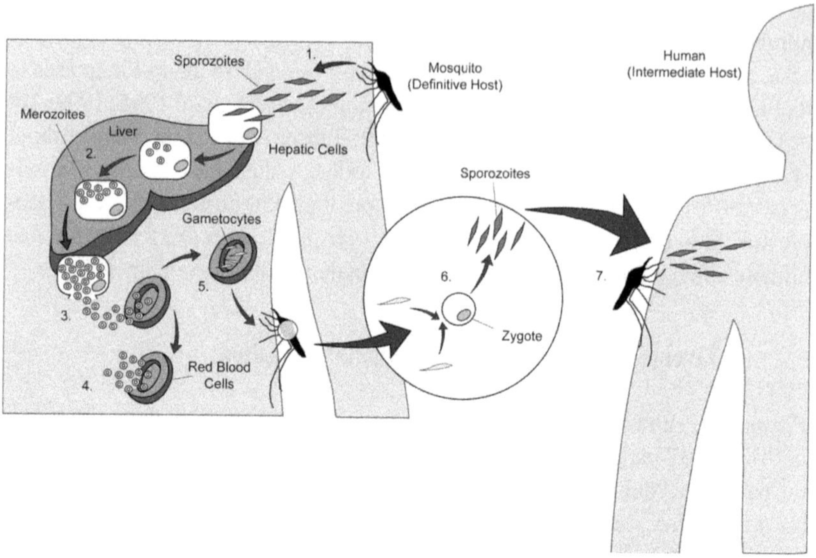

Figure 7.2. Basic life cycle of primate malarial parasites. Redrawn from www.encarta.msn.com. (1) Sporozoites are inoculated into a primate host through the bite of an infected insect vector and move quickly via the bloodstream to the liver, where they penetrate hepatic parenchymal cells. (2) In the liver, the parasite multiplies asexually to form haploid merozoites. Some merozoites in the liver can remain dormant and become reactive years later. (3) After several replication cycles, these hepatic cells burst, releasing their merozoites into the bloodstream where they invade red blood cells and continue reproducing asexually. The merozoites can infect either circulating immature or mature red blood cells, or, for some species of malarial parasites, white blood cells. (4) Once inside red blood cells, the merozoites multiply further, breaking down the constituent hemoglobin in those cells for nutrients and causing anemia in the host. On a ~24, ~48, or ~72 hour cycle, depending on the infecting species of *Plasmodium*, blood cells burst open synchronously, releasing large numbers of merozoites, which infect additional blood cells for further rounds of asexual multiplication. Successive rounds of parasite reproduction and bursting of infected red blood cells triggers an immune system response characterized by the periodic high fevers typically preceded by chills that are the classic clinical symptom of malaria in humans. (5) Within some infected blood cells, the merozoites develop further into haploid gametocytes, which can be ingested by another vector individual. (6) Within the vector's gut, the gametocytes are released from the blood cells, mature, and fuse to form diploid parasite zygotes. Zygotes develop into oocysts on the stomach wall of the vector, which then produce new sporozoites. (7) These sporozoites migrate to the salivary glands of the vector where they can be passed into a new host through subsequent blood feeding. See also Coatney *et al.* (2003) and Bannister & Mitchell (2003).

malarial parasites and their relatives has emerged. Escalante & Ayala (1995) were the first to examine the evolutionary history of malarial parasites within the context of the phylum Apicomplexa using molecular data. Based on a phylogenetic analysis of ~1550 base pairs of nuclear DNA sequence data from the slowly evolving small subunit ribosomal RNA (18S SSU rRNA) genes of *Plasmodium* – plus seven other genera of apicomplexans (including *Babesia*, *Toxoplasma*, and *Cryptosporidium*, all of which are potentially zoonotic for modern humans; Polley, 2005) and nine outgroup taxa – they concluded that the origins of the phylum Apicomplexa may date to as early as ~825 million years ago. This date precedes by several hundred million years the emergence of the land vertebrates that are the contemporary intermediate hosts for many apicomplexans. It similarly predates emergence of the dipteran insects that are the definitive hosts for most malarial parasites (Benton & Donaghue, 2007). Apicomplexans most likely evolved originally as monogenetic parasites of marine invertebrates, with digenesis arising independently – and much more recently – in several of the major Apicomplexa lineages, as the parasites adapted to hematophagy on emerging terrestrial vertebrate hosts (Barta, 1989) (Figure 7.3).

Within the phylum Apicomplexa, the radiation of the genus *Plasmodium* likely dates to sometime during the Middle to Late Mesozoic. Based on applying a crude molecular clock to their 18S SSU rRNA sequence data, for example, Escalante & Ayala (1994, 1995) estimated the age of the last common ancestor of species of *Plasmodium* from birds, rodents, and humans as ~130 to 150 million years ago. Since this date is more recent than the divergence of mammals from birds and squamates (Benton & Donaghue, 2007) – the three groups of modern vertebrate intermediate hosts for malarial parasites – the present distribution of *Plasmodium* species among vertebrates requires, at minimum, several instances of lateral transfer across the vertebrate classes. Moreover, most analyses of 18S SSU rRNA data suggest that malarial parasites infecting mammals do not form a monophyletic group. Rather, avian and squamate malarias appear to nest within that group, again implying multiple cases of lateral transfer among vertebrates, although different analyses provide contradictory assessments of the position of rodent malarial parasites as either within (Escalante & Ayala, 1994, 1995; Qari *et al.*, 1996; Hagner *et al.*, 2007) or basal to (Escalante *et al.*, 1997; Leclerc *et al.*, 2004b) the remainder of the *Plasmodium* clade. Notably, in two early studies based on 18S SSU rRNA genes, the human parasite *P. faliciparum* was found to be most closely related to certain avian species of *Plasmodium*, prompting the suggestion that the causative agent of the most virulent human malarial was acquired via recent lateral transfer from birds (Waters *et al.*, 1991, 1993), a position that several subsequent phylogenetic studies have disputed.

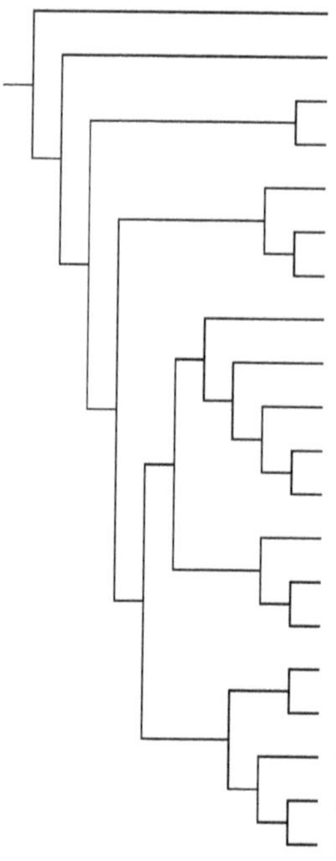

	Definitive Host	Intermediate Host	Blood Parasite	Life Cycle
Ciliates (Ciliophora)	none	none	no	monogenetic
Dinoflagellates (Dinoflagellata)	none	none	no	monogenetic
Cryptosporidium parvum	rodents	none	no	monogenetic
Cryptosporidium wrairi	rodents	none	no	monogenetic
Sarcocystis gigantea	canids	waterfowl, rodents, bovids	no	digenetic
Toxoplasma gondii	felids	mammals	no	digenetic
Neospora caninum	canids	none, mammals	no	monogenetic, digenetic
Cytauxzoon felis	ticks	felids	yes	digenetic
Babesia equi	ticks	equids	yes	digenetic
Theileria bufelli	ticks	bovids	yes	digenetic
Theileria parva	ticks	bovids	yes	digenetic
Theileria taurotragi	ticks	bovids	yes	digenetic
Babesia canis	ticks	canids	yes	digenetic
Babesia bovis	ticks	bovids	yes	digenetic
Babesia caballi	ticks	equids	yes	digenetic
Plasmodium falciparum	dipterans	primates	yes	digenetic
Plasmodium gallinaceum	dipterans	primates	yes	digenetic
Plasmodium malariae	dipterans	primates	yes	digenetic
Plasmodium vivax	dipterans	primates	yes	digenetic
Plasmodium berghei	dipterans	rodents	yes	digenetic

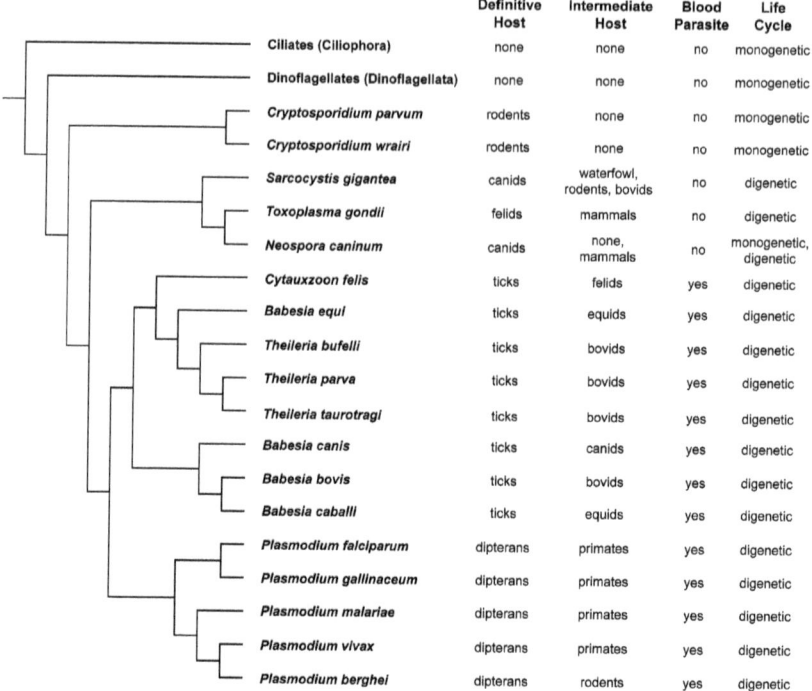

Figure 7.3. 18S SSU rRNA phylogeny of apicomplexan species and associated life-history characteristics based on Escalante & Ayala (1995) and Barta (1989).

A number of issues concerning the use of SSU rRNA sequence data require that caution be applied when interpreting some of these conclusions about the evolutionary history of *Plasmodium* and other apicomplexans. First, multiple copies of SSU rRNA genes are present in the apicomplexan genome, making it difficult to ensure that the sequence alignments used for phylogenetic analysis are based on comparing genes that are orthologous (identical by descent). In recognition of this issue, some analyses based on SSU rRNA data have used information on the secondary structure of the molecule to help guide alignments (Escalante *et al.*, 1997; Hagner *et al.*, 2007). Second, gene conversion (a poorly understood process of intra-chromosomal recombination) among different SSU rRNA genes can potentially confound inferences of phylogenetic relationships based on these loci. Finally, given the immense time depths under consideration and the lack of solid calibration points for determining the evolutionary rate of SSU rRNA loci, any divergence times assigned under a simple molecular clock model can only be regarded as tentative.

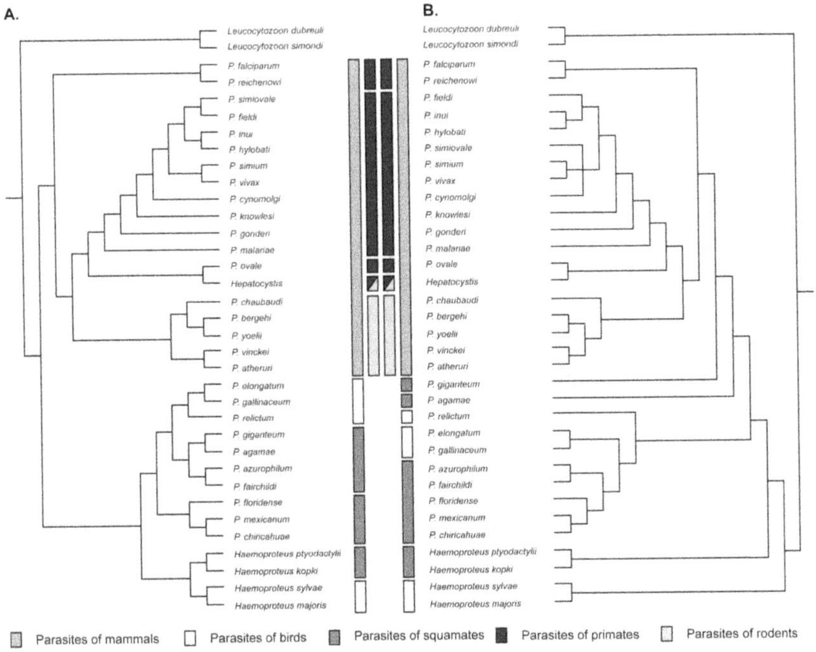

Figure 7.4. Simplified phylogenies of vertebrate malarial parasites based on cytochrome b mitochondrial DNA sequences from (A) Perkins & Schall (2002) and (B) Yotoko & Elisei (2006). The major vertebrate host of each parasite clade is indicated in the boxes adjacent to the clade at the center of the figure (i.e. bird vs. squamate vs. mammal and, within mammals, primate vs. rodent).

Other molecular studies of the evolutionary relationships among malarial parasites have focused on only a handful of additional loci. These include several, single-copy nuclear genes – circumsporozoite surface protein (*Csp*, ~1050 bases) (Escalante *et al.*, 1995; McCutchan *et al.*, 1996; Vargas-Serrato *et al.*, 2003), merozoite surface proteins 1 (*Msp-1*, ~6600 bases) (Polley *et al.*, 2005; Tanabe *et al.*, 2007) and 9 (*Msp-9*, ~2300 bases) (Vargas-Serrato *et al.*, 2003), and adenylosuccinate lyase (*ASL*, ~1400 bases) (Kedzierski *et al.*, 2002) – as well as one plastid gene, caseinolytic protease C (*Clp-C*, ~640 bases) (Rathore *et al.*, 2001; Hagner *et al.*, 2007), and the mitochondrial gene cytochrome b, (*cyt b*, ~1100 bases) (Escalante *et al.*, 1998a; Perkins & Schall, 2002; Ricklefs *et al.*, 2004; Yotoko & Elisei, 2006). With the exception of cytochrome b, none of these have been looked at in a broad range of parasite taxa.

Figure 7.4 shows two simplified evolutionary trees resulting from recent phylogenetic analyses of cytochrome b sequence data from more than 50 malarial parasites comprising a large number of species of *Plasmodium* that infect mammals, birds, lizards, and snakes, plus representatives of additional genera of malaria-inducing protists (*Hepatocystis*, *Haemoproteus*, *Leucocytozoon*). While there are some inconsistencies between the trees, several important conclusions concerning the evolutionary history of vertebrate malarial parasites can be drawn (Escalante *et al.*, 1998a; Perkins & Schall, 2002; Yotoko & Elisei, 2006). First, all of the malarial parasites of mammals fall into a single clade that includes species attributed to the genera *Plasmodium* and *Hepatocystis*. Interestingly, this result stands in contrast to most of the phylogenies inferred from 18S SSU rRNA sequence data, and to the results of subsequent studies of *Csp*, *Clp-C*, *MSP-1*, and *ASL*, in which at least some avian species of *Plasmodium* fall inside the grouping of mammalian malarial parasites (Rathore *et al.*, 2001; Kedzierski *et al.*, 2002; Vargas-Serrato *et al.*, 2003; Polley *et al.*, 2005; Tanabe *et al.*, 2007).

Second, the primary causative agents of malaria in humans (*P. falciparum*, *P. vivax*, *P. malariae*, and *P. ovale*) do not form a monophyletic group but rather have multiple, independent evolutionary origins that date to different times, a robust result found in all molecular phylogenies for *Plasmodium*. Nonetheless, according to the cytochrome b data, all human malarial agents do fall within a clade of mammalian parasites. With respect to *P. falciparum*, this observation runs counter to the hypothesis that the parasite entered the human population recently via lateral transmission from birds (Waters *et al.*, 1991, 1993; McCutchan *et al.*, 1996).

Third, the cytochrome b phylogeny corroborates a sister-taxon relationship of human *P. falciparum* and *P. reichenowi*, a malarial parasite of wild chimpanzees – an observation that was previously suggested by phylogenetic analysis of both 18S SSU rRNA and *Csp* sequence data (Escalante & Ayala, 1994; Escalante *et al.*, 1995, 1996; Qari *et al.*, 1996; Escalante *et al.*, 1997). Notably, in the cytochrome b phylogeny, as in most of the 18S SSU rRNA phylogenies, the *P. falciparum/reichenowi* clade appears basal with the mammalian malarial parasites, and these two species are quite divergent from other mammalian *Plasmodium*. Rodent *Plasmodium* then diverges subsequent to the origins of the *P. falciparum/reichenowi* clade, thus the group of malarial species that infect primates is paraphyletic. Given the estimated rate of nucleotide substitution at the 18S SSU rRNA and *Csp* loci, the human and chimpanzee malarial parasites are estimated to have diverged roughly 8–11 million years ago (Escalante & Ayala, 1994; Escalante *et al.*, 1995), which pre-dates the time of divergence of their hosts. This suggests that the last common ancestor of those parasites transferred into a hominoid host prior to the human–chimpanzee split (Rich & Ayala, 2003).

Using a cytochrome b based phylogenetic tree and information on the class of vertebrate host infected by each extant parasite taxon, Yotoko & Elisei

(2006) reconstructed the most likely ancestral hosts at each internal node of the *Plasmodium* phylogeny, repeating the procedure for several different tree topologies. Their analyses yielded several key findings. First, under all topologies, a squamate host was reconstructed as most likely for the last common ancestor of all extant *Plasmodium*, *Haemoproteus*, and *Hepatocystis* species. Thus, a minimum of four host shifts across vertebrate classes would be required to arrive at the host distribution of modern malarial parasites – one from squamates to mammals, one from squamates to birds, plus various numbers of additional squamates–bird or bird–squamates switches, depending on the specific tree topology used. Second, for all topologies, only a single host-switch into mammals is required, although within the mammals, at least two additional ordinal host shifts (between primates and rodents and between primates and bats) are implied.

Despite the progress made thus far in understanding the deep evolutionary history of *Plasmodium*, further work is needed, particularly the accumulation of additional sequence data from multiple, independent loci in the parasite genomes. It is worth noting that all of the studies mentioned above were based on alignments of very small segments of DNA – only one was more than 2500 bases in length out of a genome more than 10 000 times that size – and it is not surprising that analyses of different, single loci yield incongruous, poorly resolved gene trees. In fact, in a recent reanalysis of the SSU rRNA, *Clp-C*, and cytochrome b data sets, among the largest and most comprehensive available, Hagner *et al.* (2007) concluded that the phylogenetic signal provided by the first two of these loci is insufficient for resolving the question of whether mammalian malarial parasites indeed form a monophyletic clade, or for addressing the related issue of whether *P. falciparum* might represent an avian zoonosis. Even for the cytochrome b data set, the putative monophyly of mammalian parasites was only weakly supported. Fortunately, as genomic data become available for more species of *Plasmodium*, it should prove easier to accumulate the comparative data needed to develop a more complete picture of the evolutionary history of malarial parasites.

Host-shifts and parasite–host coevolution in primates

Primates are by far the most common mammalian intermediate hosts for malarial parasites, and the extent to which the relationships among primate malarial agents and their host species have been shaped by either cospeciation or by the lateral transfer of parasites among different primate lineages is of great interest, particularly given the global human burden of *Plasmodium* infection. To investigate the coevolution of malarial parasites and their primate hosts, well-resolved phylogenetic trees for each set of taxa must be available. To date,

Table 7.2. *Species of* Plasmodium *infecting primates and their natural primate hosts*

Plasmodium species	Genera of Natural Primate Host(s)
P. brasilianum	*Alouatta, Aotus, Brachyteles, Ateles, Lagothrix, Cacajao, Chiropotes, Callicebus, Cebus, Pithecia, Saguinus, Saimiri*
P. bucki	*Eulemur*
P. coatneyi	*Macaca*
P. coulangesi	*Eulemur*
P. cynomolgi	*Macaca, Presbytis*
P. eylesi	*Hylobates*
P. falciparum	*Homo*
P. fieldi	*Macaca*
P. foleyi	*Eulemur*
P. fragile	*Macaca*
P. georgesi	*Cercocebus*
P. girardi	*Eulemur*
P. gonderi	*Cercocebus, Mandrillus*
P. hylobati	*Hylobates*
P. inui	*Macaca, Presbytis*
P. jefferyi	*Hylobates*
P. knowlesi	*Macaca, Presbytis,* occasionally *Homo*
P. lemuris	*Eulemur*
P. malariae	*Homo,* perhaps *Pan*
P. ovale	*Homo*
P. percygarnhami	*Eulemur*
P. petersi	*Cercocebus*
P. pitheci	*Pongo*
P. reichenowi	*Pan, Gorilla*
P. rodhaini	*Pan, Gorilla*
P. schwetzi	*Pan, Gorilla*
P. shortii	*Macaca*
P. silvaticum	*Pongo*
P. simiovale	*Macaca*
P. simium	*Alouatta, Brachyteles, Ateles*
P. uilenbergi	*Eulemur*
P. vivax	*Homo*
P. "vivax-like"	*Homo*
P. youngi	*Hylobates*
P. species (undescribed)	*Eulemur*
P. species (undescribed)	*Mandrillus*

however, the most complete phylogenetic trees for primate malarial parasites only include around half of the more than 30 species of *Plasmodium* that infect various primates (Gysin, 1998; Coatney *et al.*, 2003) (Table 7.2). Nonetheless, it is very clear that the coevolutionary history of *Plasmodium* and its primate hosts is a complex one, involving multiple host switches within the primates.

Recently, Mu *et al.* (2005) examined the coevolution of 14 species of primate *Plasmodium* and their hosts, using a parasite phylogeny inferred from sequence data for the complete mitochondrial genome of eight species plus cytochrome b data from the remaining parasite taxa. Figure 7.5 summarizes the results of their study, incorporating one additional malarial parasite of New World monkeys, *P. brasilianum*, at the appropriate position in the parasite phylogeny (Ayala *et al.*, 1998). Two key points are clear from the figure. First, humans are host to several very divergent strains of malarial parasites, whose most recent common ancestor likely predated the split between primates and other orders of mammals, roughly 95 million years ago (Hedges *et al.*, 1996; Arnason *et al.*, 1998; Kumar & Hedges, 1998). *Macaca* is also host to a number of species of *Plasmodium*, but these most likely arose via cospeciation within the macaque radiation. Second, some distantly related primates are host to closely related parasite species (e.g. *Hylobates* with *P. hylobati* and *Macaca* with *P. inui*). To account for the current distribution of *Plasmodium* species among modern primates, a minimum of five cases of lateral transfer are required – one from macaques to gibbons (involving the common ancestor of *P. inui* and *P. hylobati*), one from macaques to humans (involving *P. knowlesi* as a zoonosis) (Jongwutiwes *et al.*, 2004; Singh *et al.*, 2004), one from macaques to either New World monkeys or humans (involving the common ancestor of *P. simium* and *P. vivax*) followed by more recent transfer into the other of those primate taxa, and a second case of transfer between humans and New World monkeys (involving *P. brasilianum* and *P. malariae*) where the direction of transfer is also controversial. As additional primate malarial parasites are sequenced and added to this phylogeny, it is possible that the number of identifiable cases of host-switching will increase. It is noteworthy that within birds, host-switching by *Plasmodium* and *Haemoproteus* malarial parasites also appears to be common (Bensch *et al.*, 2000; Ricklefs & Fallon, 2002; Ricklefs *et al.*, 2004).

Timing of *Plasmodium* transfers into humans

As alluded to above, there remains some controversy over the specifics of when and where various species of *Plasmodium* came to adopt humans as their intermediate hosts. At the source of the debate are the close phylogenetic relationships seen among several pairs of malarial parasites, one of which infects human and the other a non-human primate host. In one pair, the closest phylogenetic relative of the human parasite, *P. falciparum*, is *P. reichenowi*, a parasite of chimpanzees. Though sister taxa, these parasite species shared a last common ancestor ~8 million years ago, prompting the conclusion that an ancestral hominoid was likely to have been a host to the common ancestor

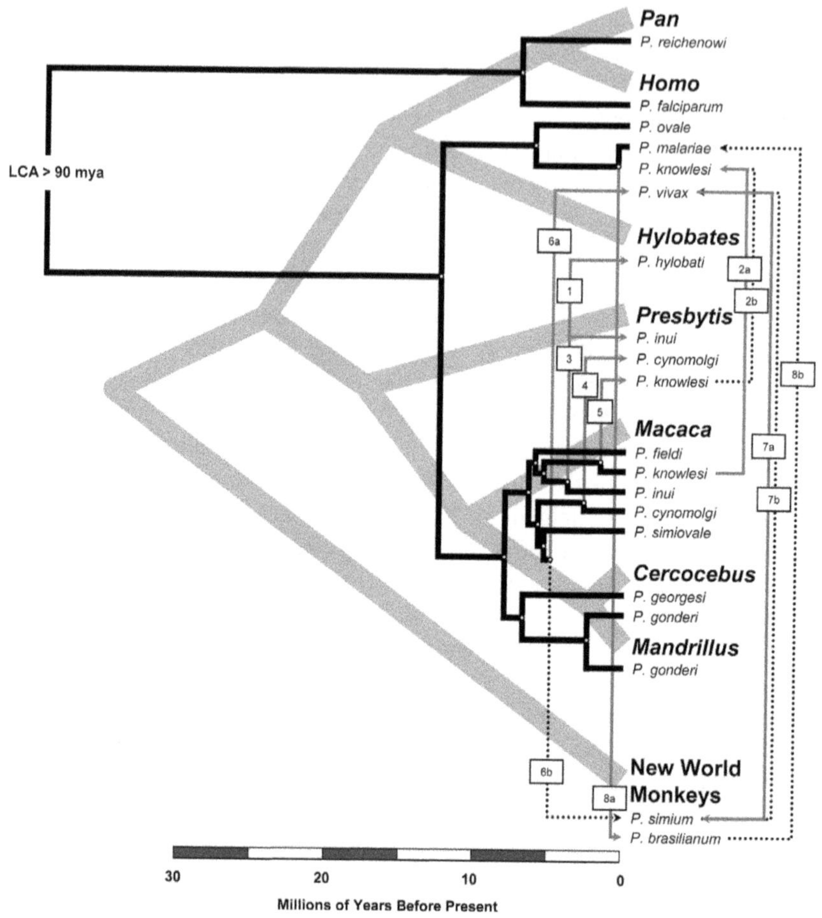

Figure 7.5. Comparison of consensus primate (thick gray) and primate malarial parasite (black) phylogenies showing eight putative host-switching events (thin numbered lines, with the direction of the switch indicated by the arrowhead) based on Ayala *et al.* (1998) and Mu *et al.* (2005). Alternative scenarios for several host switches are indicated by both solid and dotted numbered lines, identified, respectively, by the "a" or "b" following the host-switch number. (1) *Macaca* (Cercopithecine) to *Homo*, (2) *Macaca* or *Presbytis* (Old World monkey) to *Homo*, (3) to (5) *Macaca* to *Presbytis*, (6) and (7) *Macaca* (Cercopithecine) to *Homo* followed by *Homo* to New World monkeys or *Macaca* (Cercopithecine) to New World monkeys followed by New World monkeys to *Homo*, (8) *Homo* to New World monkeys or New World monkeys to *Homo*.

of *P. falciparum* and *P. reichenowi*, and that the ancestral parasite cospeciated along with the divergence of its hominoid host into chimpanzees and humans (Escalante *et al.*, 1995; Rich & Ayala, 2003).

Two other pairs of closely related *Plasmodium* species show a similar pattern, with one member being found in humans and the other in a non-human primate taxon. However, in these two cases the non-human primate host is only very distantly related to *Homo sapiens*. Thus, *P. vivax*, found in humans, is very closely related to (and perhaps genetically indistinguishable from) *P. simium*, a parasite of three genera of New World primates, and *P. malariae* in humans is a closely related sister taxon to *P. brasilianum*, a generalist parasite that infects 12 genera of New World primates (Escalante *et al.*, 1995; Ayala *et al.*, 1998) (Table 7.2). For these two cases, it seems most likely that host-switches between humans and non-human primate species have occurred in the very recent past, within the last several hundred years.

Emergence and expansion of **Plasmodium falciparum**

There is little doubt that the *P. falciparum* lineage has deep evolutionary roots. Phylogenetic analyses based on multiple regions of the genome all indicate, first, that *P. falciparum* from humans and *P. reichenowi* from chimpanzees are each other's closest relatives and, second, that these taxa are very distantly related to other human malarial parasites, having diverged from other lineages of *Plasmodium* prior to the origins of primates roughly 90 million years ago. Moreover, historical records strongly suggest that highly virulent *falciparum*-like malaria has been impacting human populations for several thousands of years, making *P. falciparum* the malarial agent with the longest known history of association with humans (Sherman, 1998, 2006).

At issue, then, is precisely how and when the modern population of *P. falciparum* came to assume its current global distribution and population size. Assuming that the parasite has been present in the *Homo* lineage since our divergence from chimpanzees, has it enjoyed a widespread distribution and large effective population size for hundreds of thousands to millions of years ago, pre-dating or coinciding with the initial spread of humans around the globe? Or, has the parasite population only recently expanded from a small effective population size – e.g. within the last several thousand years – a scenario referred to as the "Malaria's Eve" hypothesis (Rich *et al.*, 1998)?

One way to address this issue is with data on the amount and structuring of genetic variation found within modern-day *P. falciparum*. If the parasite has enjoyed a large effective population size throughout a long history of association with humans, we would expect to see high levels of polymorphism in the modern *P. falciparum* population, particularly at selectively neutral or nearly-neutral sites within the genome. Limited neutral polymorphism, by contrast, would reflect more recent expansion of the parasite from a small ancestral population. With data on the amount of genetic variation seen among modern-day *P. falciparum* and information on the rate of sequence evolution in the parasite, it should be possible to estimate the timing of the parasite's first significant emergence in humans – the so-called coalescence time for the last common ancestor of the modern parasite population.

Substantial genetic variation does exist in the global population of *P. falciparum* in some parasite genes. Rather than being neutral, however, much of the variation seen involves non-synonymous substitutions – i.e. single base pair mutations in the DNA that result in a different protein product being made – in loci coding for a variety of cell membrane molecules that contribute to antigenic variation in the parasite, an adaptation for evading the host's immune system (Hughes, 1991, 1992; Hughes & Hughes, 1995; Escalante *et al.*, 1998b; Polley *et al.*, 2005). Coalescence times have been estimated at tens of millions of years for alleles at some of these loci. These non-neutral polymorphisms might have been maintained by positive selection over a time scale much greater than the parasite's association with humans, and indeed, much greater than the divergence between *P. falciparum* and *P. reichenowi* (Hughes, 1992; Hughes & Hughes, 1995; Hughes & Verra, 1998). However, since alleles subject to strong positive selection (such as those involved in the immune response or in resistance to antimalarial drugs) can greatly increase in frequency or become fixed in a population within a few generations, conclusions drawn from this type of analysis are speculative and should be interpreted with caution (Rich & Ayala, 1998, 2003; Rich *et al.*, 1998).

Genetic studies that have focused on neutral or nearly-neutral variation – e.g. synonymous or "silent" substitutions in protein-coding regions and sequence polymorphisms in introns – may be more suitable for reconstructing the history of *P. falciparum*'s association with humans. These have provided widely divergent coalescence date estimates ranging from less than 10 000 years ago to more than 400 000 years (Conway & Baum, 2002; Hartl, 2004). Thus, at least some estimates based on coalescent analyses of neutral polymorphisms are compatible with alternative scenarios concerning the parasite's effective population size throughout its association with humans.

Rich *et al.* (1998) examined sequence data published in GenBank for 10 nuclear protein-coding genes in *P. falciparum* strains isolated from around the globe and found absolutely no synonymous (i.e. neutral) variation in more than 16 000 base pairs of compared sequence. Using estimates of the rate of silent mutation at these loci derived from comparing *P. falciparum* with other species of *Plasmodium*, Rich *et al.* (1998) calculated that the common ancestor ("Malaria's Eve") of all modern-day populations of *P. falciparum* dates to 24 500 to 57 500 years ago at the earliest. Recent analysis of more than 22 000 base pairs of aligned data from 20 additional nuclear protein-coding loci likewise revealed very little synonymous variation within the global population of *P. falciparum* and lend support to Rich *et al.*'s (1998) hypothesis of a recent expansion of the parasite in humans (Hartl, 2004).

Additional support for the "Malaria's Eve" hypothesis comes from studies of neutral single nucleotide polymorphisms (SNPs) found in introns. On the basis of the very limited sequence diversity seen in 25 intronic regions from genes on two parasite chromosomes, Volkman *et al.* (2001) inferred the age of the most recent common ancestor of all modern *P. falciparum* to be at most 9500 to 23 000 years. Similarly, mitochondrial DNA sequence data also support the conclusion of a relatively recent though slightly older date for the origins and expansion of the parasite in humans. For example, Conway *et al.* (2000) found very little synonymous sequence diversity in the complete mitochondrial genomes of a worldwide sample of *P. falciparum* when compared to the divergence between *P. falciparum* and *P. reichenowi*, and they inferred an age for the last common ancestor of modern *P. falciparum* of less than 50 000 years. Similarly, based on the number of synonymous substitutions seen in protein-coding regions in a sample of 100 complete mitochondrial genomes from around the globe, Joy *et al.* (2003) concluded that the last common ancestor of worldwide populations of *P. falciparum* existed 70 000 to 98 000 years ago. Interestingly, though, Joy *et al.*'s (2003) analysis also suggests a dramatic increase in the parasite population in Africa within the last 10 000 years, coinciding roughly with the origins and spread of agriculture during the Neolithic and with the origin of the mosquito *Anopheles gambiae*, the main African vector of *P. falciparum* (Coluzzi, 1999).

Not all genetic data, however, support the conclusion of a recent expansion of *P. falciparum* in humans. For example, based on coalescent analysis of the sequence diversity found in a set of 23 nuclear protein-coding loci that show no evidence of having been under positive selection, Hughes & Verra (2001) have argued that the age of the last common ancestor of modern *P. falciparum* existed 290–390 thousand years ago. Similarly, in a large-scale survey of SNP variation on chromosome 3 of the parasite genome, Mu *et al.* (2002) used the number of neutral SNPs (i.e. synonymous substitutions and polymorphisms found non-coding regions) to estimate a date of 102 000 to 177 000 years ago for the common ancestor of all modern *P. falciparum*.

At present, based on the low level of genetic polymorphism seen in a broad range of neutral markers and loci, the genetic evidence seems to come down in favor of a recent and precipitous increase in the population size of *P. falciparum* from a relatively small ancestral population. Some portions of the *P. falciparum* genome do have much older coalescence times, but that presumably reflects balancing selection to maintain diversity in the parasite's antigenic proteins, which help the parasite evade its hosts' immune defenses. Recent work has also indicated that some of the polymorphisms recognized in those studies estimating an older common ancestor for modern *P. falciparum* likely resulted from sequencing errors and undue reliance on unverified sequence data from GenBank and other databases (Barry *et al.*, 2003; Rich & Ayala, 2003; Hartl, 2004). Moreover, some putative polymorphisms are likely to result from gene conversion between paralogous copies of the gene in question that have arisen from a past gene duplication event (Nielsen *et al.*, 2003) and should not be used in estimating coalescence times under a model of neutral sequence evolution.

Origins and expansion of **Plasmodium vivax**

Plasmodium vivax is second only to *P. falciparum* in the number of cases of human malaria it is responsible for each year (Figure 7.1) (Mendis *et al.*, 2001), although the mortality rate is far lower. As noted above, it is genetically very similar to (and perhaps indistinguishable from) *P. simium*, a parasite of three genera of New World monkeys. As with *P. falciparum*, however, questions remain as to the timing of its significant expansion in humans and over the geographic region where the parasite originated, with various researchers suggesting Africa (Carter, 2003), South-East Asia (Escalante *et al.*, 2005; Jongwutiwes *et al.*, 2005; Cornejo & Escalante, 2006), and even the Americas (Ayala *et al.*, 1999; Rich & Ayala, 2003; Lim *et al.*, 2005).

As in the case of *P. falciparum*, there is a debate whether the association of *P. vivax* with the human lineage is ancient or recent. Part of the controversy results from the fact that different studies paint contrasting pictures of the level of genetic diversity present within modern-day populations of *P. vivax*, which then influences the conclusions drawn about the timing of emergence of the parasite as a significant agent of human disease. For example, two independent studies using sequence data from the complete mitochondrial genomes of a worldwide sample of *P. vivax* – one involving 106 isolates (Jongwutiwes *et al.*, 2005) and one 176 isolates (Mu *et al.*, 2005) – have yielded coalescence time estimates for the most recent common ancestor of the modern parasite population ranging from 53 000 to more than 300 000 years ago, depending on the combination of nucleotide substitution rate and population demographic history assumed. Cornejo & Escalante (2006) recently combined and reanalyzed the data sets used in these two studies. While they caution that any estimate of the date of the expansion of *P. vivax* is sensitive to assumptions made about the neutral mutation rate and the parasite population's demographic history and geographic structuring, all of the different permutations of these variables in their analyses nonetheless yield estimates in the range of 162 000 to 465 000 years, or well before the spread of modern humans around the globe (Cornejo & Escalante, 2006). Sequence data from *MSP-1* – a locus which shows evidence of having undergone positive and diversifying selection – also suggest a Middle Pleistocene coalescence date of ~594 000 years ago for modern *P. vivax* (Tanabe *et al.*, 2007). Finally, in a large-scale sequencing study covering roughly 100 kilobases from the genome of five *P. vivax* isolates, Feng *et al.* (2003) found a greater number of non-coding and synonymous SNPs than are present in the homologous region of the *P. falciparum* genome. They conclude that the *P. vivax* genome is highly diverse and, by implication, its radiation could not have been very recent. Thus, all of these studies support the idea that *P. vivax* underwent a relatively ancient population expansion prior to the emergence of modern *Homo sapiens*.

By contrast, Leclerc *et al.* (2004a) examined the variation seen at rapidly evolving microsatellite and other tandem-repeat loci – including the most polymorphic of those found in Feng *et al.*'s (2003) study – in a set of ~100 isolates of *P. vivax* from a worldwide sample and found far less diversity than is seen in *P. falciparum*. They conclude that the modern-day population of *P. vivax* is genetically depauperate, at least at neutral sites, which suggests that the population has either undergone an implausible series of recent selective sweeps or has rapidly expanded from a small effective population size in the very recent past, likely within the last 10 000 years (Leclerc *et al.*, 2004a). This position finds support from a more recent study of variation in *Csp* gene sequences, which found very few synonymous polymorphisms within a global sample of *P. vivax* isolates and likewise concludes that *P. vivax* became a significant human parasite with a global distribution only since the Holocene (Lim *et al.*, 2005). Interestingly, it appears that there may have been two independent host transfers of *P. vivax/simium* between humans and New World monkeys during this time, based on the fact that the same two strain types are found in both populations (Lim *et al.*, 2005).

Multiple lines of evidence implicate South-East Asia for the origin of *P. vivax*, particularly if the association of the parasite with humans is an ancient one. First, molecular phylogenies based on sequence data from multiple nuclear loci as well as complete mitochondrial genomes place both *P. vivax* and *P. simium* squarely within a monophyletic clade whose other members comprise only malarias of Asian primates – macaques, leaf-monkeys and gibbons – with the African primate malarias more basal (Escalante *et al.*, 1998a, 2005; Perkins & Schall, 2002; Mu *et al.*, 2005; Yotoko & Elisei, 2006). Within this clade, *P. vivax* and *P. simium* are also more similar to the malarial parasites of macaques than to parasites of gibbons, suggesting that if the parasite indeed was introduced into humans in South-East Asia then the source was a cercopithecoid rather than a hominoid primate. Results from cross-species screening of microsatellite and other simple sequence repeat loci echo these points: *P. vivax*-derived tandem-repeat loci amplify more reliably in several macaque parasites than in *P. hylobati*, a malarial agent in gibbons, or *P. gonderi*, a parasite of African cercopithecoids (Leclerc *et al.*, 2004a). Finally, when different geographical subsamples of complete *P. vivax* mitochondrial genomes are analyzed separately, the Asian sample contains greater haplotype diversity and yields a much older estimate of the age of the most recent common ancestor than do the samples from either Africa or the Americas, lending additional support to the conclusion of an Asian origin for the parasite (Cornejo & Escalante, 2006).

The inclusion of *P. simium*, a parasite of New World monkeys, in a clade of principally South-East Asian parasites is something of a paradox. Under the Asian origins scenario, the close phylogenetic relationship between *P. vivax* and *P. simium* is interpreted as an anthroponosis – a case of host switching from humans to monkeys. This scenario would involve at least two host switches because two different strain types of *P. vivax/simium* infect both humans and platyrrhines (Lim *et al.*, 2005). The number of host transfers between humans and monkeys may have been greater than two because *P. simium* is a parasite of multiple platyrrhine genera. Alternatively, several transfers between platyrrhine species would have to have occurred after the original two transfers from humans. The original direction of transfer, however, could have been the reverse – i.e. *P. vivax* might be a zoonosis introduced very recently into humans via lateral transfer from New World monkeys, rather than vice versa (Escalante & Ayala, 1995; Ayala *et al.*, 1998; Rich & Ayala, 2003).

Several points, in fact, make a New World monkey-to-human host-switch more plausible than a human-to-monkey transfer. First, modern humans shared a common ancestor much more recently with Old World than New World monkeys and, likewise, have a much longer history of geographic contact with Old World primates – on the order of millions rather than thousands of years. Thus, if parasite transfer from humans to monkeys were have occurred, it seems far more likely that it would have taken place in the Old World. Moreover, given that *P. simium* infects several different genera of New World monkeys that shared a common ancestor millions of years before the very recent emergence of the parasite, multiple host shifts between humans and monkeys (or at least two, followed by additional transfers among monkey species) would be required to explain the parasite's current distribution. As others have noted, evolutionary parsimony thus favors a monkey-to-human host switch (Escalante & Ayala, 1995; Ayala *et al.*, 1998; Rich & Ayala, 2003).

Origins of Plasmodium malariae

Another agent of human malaria, *Plasmodium malariae*, is also very closely related to a malarial parasite of New World monkeys, *P. brasilianum*. To date, intraspecific variation in *P. malariae* has not been well studied. Based on a very small number of samples (n = 2), Ayala *et al.* (1998) found the level of polymorphism at the *Csp* locus to be comparable to that seen in *P. vivax*, perhaps implying a similar time frame for the expansion of *P. malariae* in humans. Clearly, however, additional loci and a much larger set of samples need to be studied before any robust conclusions might be drawn.

As is the case for *P. vivax/simium*, the question of whether humans acquired *P. malariae/brasilianum* from or transferred the parasite to New World monkeys remains unresolved, but given the fact that *P. brasilianum* is known to infect at least 12 of the roughly 16 currently recognized genera of platyrrhines (Table 7.2), a strong, parsimonious case can be made for a single platyrrhine-to-human transfer. By contrast, any scenario involving host-switching from humans to monkeys would have to involve either numerous transfers from humans to different platyrrhines or multiple transfers among platyrrhine genera following an introduction from humans.

As others have noted (Ayala *et al.*, 1998; Rich & Ayala, 2003; Lim *et al.*, 2005), the critical data needed for resolving outstanding questions about the direction of transfer for both *P. malariae/brasilianum* and *P. vivax/simium* are estimates of the amount of neutral polymorphism found in natural populations of the New World parasites. If New World monkeys are in fact the source of the two human malarial agents, then the expected neutral genetic polymorphisms in *P. brasilianum* and *P. simium* should be much greater and should coalesce further back in time than those in *P. malariae* and *P. vivax*.

Parasite evolutionary dynamics and host adaptations

Virulence differences within and between species

A staple topic of study in host–parasite interactions is the evolutionary dynamics of parasite virulence. Conventional wisdom suggests that parasites that have a long relationship with a particular host taxon should, over time, evolve to become less virulent in their hosts, since overly virulent parasites can cause host death, thereby curtailing the parasite's ability to transfer into a new host and its eventual reproductive success. However, other features of host demography (e.g. host population density), as well as features of parasite population ecology (e.g. the timing of transmission relative to host mortality, the parasite's dependence upon or independence from one or more vectors, or the prevalence of host coinfection by multiple parasite strains, which engenders within-host competition among parasites) are also expected to influence parasite virulence (Bull, 1994; Day, 2001, 2003). Some researchers have further suggested a tradeoff between host specialization and virulence, with lower virulence characterizing more "generalist" parasites (parasites that infect multiple host species) and higher virulence characterizing more "specialized" parasites (Woolhouse *et al.*, 2001; Gandon, 2004). All of these factors may contribute to the wide variation in virulence seen among and within parasite species.

Within primates, different species of *Plasmodium* are known to infect different numbers of host species, from one, in many cases (e.g. *P. falciparum*, which is specific to humans) to more than two dozen (*P. brasilianum* in New World monkeys) (Table 7.2). Recently, Garamszegi (2006) used the primate malaria host–parasite complex to test the hypothesis that parasite virulence covaries with host specialization. He found that average peak parasitemias – i.e. counts of infected cells per volume of blood in experimentally inoculated animals averaged across the set of host species – were negatively associated with the degree of host specialization of the parasite, lending support to the hypothesis that more generalist parasite species are less virulent.

With respect to intraspecific variation in virulence, it is known that in humans the severity of malaria caused by different strains of *P. falciparum* can differ markedly, from "mild" (associated with low host mortality) to "severe" (associated with host mortalities >10%, even with treatment; Gupta *et al.*, 1994). The most severe cases of *P. falciparum* malaria result when infected blood cells adhering to the walls of blood vessels – one of the parasite's strategies for circumventing the host's immune system by allowing it to remain sequestered in the peripheral circulatory system (Craig & Scherf, 2001; Beeson & Brown, 2002) – cause those vessels in the brain and other vital organs to become blocked and rupture. Malarial parasites accomplish this adhesion by causing several types of ligand proteins to be expressed on the surface of infected blood cells of their host – proteins which are coded for by several families of genes (e.g. *var*, *rif*, and *stevor* in *P. falciparum*; Crabb & Cowman, 2002). *Plasmodium falciparum* is unique among human malarias in that its merozoites are able to infect mature red blood cells.

Recent experimental studies have demonstrated that genetic variation among parasite strains within some species of *Plasmodium* is associated with differences in virulence, both in the intermediate host (Mackinnon & Read, 1999; Chotivanich *et al.*, 2000) and in their mosquito vectors (Ferguson & Read, 2002). In fact, in at least one model system involving the rodent parasite, *P. chaubaudi*, Ferguson & Read (2002) found a gene-by-environment effect on virulence in the parasite's definitive host, the mosquito *Anopheles stephensi*. Thus, a clear mechanism exists whereby phenotypic variation in virulence may be maintained in a population. The extent to which the combination of environmental and genetic variation account for variability in pathogenicity associated with *P. falciparum* infection in humans remains to be investigated.

Human and non-human primate adaptations to malaria

Over the course of modern human evolutionary history, malaria is possibly responsible for the deaths of more than half of all people who have ever lived (Sherman, 2006), and the disease is likely to be the single most important selective force to which modern humans have had to adapt (Kwiatkowski, 2005). In turn, the evolutionary dynamics of *Plasmodium* infecting humans and other non-human primates has also been shaped by selection pressures associated with the evolved defenses of their hosts. Not surprisingly, then, a wide range of genetic polymorphisms seen in modern humans have been linked to their role in resistance to infection by malarial parasites (Flint *et al.*, 1998; Evans & Wellems, 2002; Fortin *et al.*, 2002; Kwiatkowski, 2005; Williams, 2006). Many of these polymorphisms involve changes to the cell surface proteins and/or internal structure of red blood cells that either render them less susceptible to invasion by circulating *Plasmodium* merozoites, greatly reduce or destroy the ability of the parasites to grow in red blood cells, or enhance the process by which hosts develop natural immunity to *Plasmodium* infection (Friedman, 1978; Williams *et al.*, 2005a).

The classic example of one such polymorphism involves the ß-hemoglobin (HBB) gene and is responsible for "sickle-cell anemia" in humans. The HBB gene codes for one of the peptides that makes up hemoglobin, the molecule in red blood cells responsible for binding and transporting oxygen. One form of the HBB gene, the hemoglobin A (HbA) allele, is by far the most common variant in human populations worldwide. However, in some populations where the prevalence of malaria is high – particularly in sub-Saharan Africa – an alternative allele known as HbS, achieves a high frequency. The HbS allele arises from a single base pair missense mutation, which changes one amino acid in the ß subunits of the hemoglobin molecule. This change causes red blood cells to assume a reversible, sickled shape under hypoxic conditions; hence, HbS is often referred to as the "sickle-cell" allele. While individuals homozygous for the HbS allele suffer from debilitating anemia and painful vascular infarctions caused by sickled red blood cells blocking and sometime bursting vessels of the circulatory system, heterozygotes possessing one HbA allele and one HbS allele experience very few of the symptoms of the condition, except when oxygen deprived. Additionally, heterozygous individuals show resistance to *Plasmodium* infection, seemingly either because the normal metabolism of the parasite in erythrocytes is disrupted (Friedman, 1978) or because infected blood cells are more effectively recognized and removed from the blood and destroyed in the spleen (Kwiatkowski, 2005). Thus, the HbA–HbS polymorphism is apparently maintained due to an increased fitness of heterozygotes in environments where malarial risk is high.

Two other single nucleotide polymorphisms in the HBB gene are responsible for yet other hemoglobin alleles that also convey substantial protection against *Plasmodium* infection (Williams, 2006): HbC, which like HbS is common in sub-Saharan and western Africa, and HbE, which is most common in South-East Asia (Hutagalung *et al.*, 1999; Agarwal *et al.*, 2000; Modiano *et al.*, 2001; Chotivanich *et al.*, 2002; Ohashi *et al.*, 2004). Linkage disequilibrium studies of the various HBB polymorphisms suggest that the HbS allele has arisen independently in several different geographical regions within Africa (Mears *et al.*, 1981; Antonarakis *et al.*, 1984; Pagnier *et al.*, 1984; Chebloune *et al.*, 1988; Flint *et al.*, 1998), and the same may be true for HbE. Moreover, these alleles seemingly arose relatively recently (<5 000 years ago) in human populations (Flint *et al.*, 1998; Currat *et al.*, 2002; Ohashi *et al.*, 2004), providing testament to the powerful selective role that malaria has played in recent human evolution and lending additional support to the idea that *P. falciparum* has only emerged as a significant pathogen of modern humans within the last several thousand years.

A large number of additional red blood cell polymorphisms have also been maintained in humans presumably as a result of the strong selective pressures imposed by malaria. For example, many different polymorphisms in the genes coding for either the α or ß subunits of the hemoglobin molecule (and in the regulatory regions influencing transcription of these two genes) are responsible for thalassemias, a family of blood disorders in which red blood cells underproduce hemoglobin, which may result in mild to severe anemia. As in the case of heterozygous carriers of the HbS allele, individuals heterozygous for certain α- and ß-thalassemias suffer from mild anemia but show markedly increased resistance to severe malaria (Flint *et al.*, 1986; Allen *et al.*, 1997; Williams *et al.*, 2005b).

Polymorphism at the X-linked glucose-6-phosphate dehydrogenase (G6PD) locus has also long been known to be associated with the occurrence of malaria in humans (Allison & Clyde, 1961; Gilles *et al.*, 1967; Beutler, 1994). The G6PD enzyme is ubiquitous in animal cells where it plays a major role in glucose metabolism and in the production of nicotinamide adenine dinucleotide phosphate (NADPH), which is critical for cells – particularly red blood cells – to be able to cope with oxidative stress (Greene, 1993; Ruwende & Hill, 1998). A variety of mutations in the gene result in deficiencies in G6PD production, and the geographic distribution of G6PD-deficient variants corresponds well with areas of high malaria risk, suggestive of a selection-driven link between the condition and malaria resistance. Some in vitro studies have demonstrated that G6PD deficiency inhibits the growth of *P. falciparum*, at least in the early stages of infection (Roth *et al.*, 1983; Ruwende & Hill, 1998), and field studies have revealed that for both heterozygous female and hemizygous male children, one form of G6PD deficiency (G6PD A-) was associated with a 46–58% reduction in the risk of severe malaria in two African populations (Ruwende *et al.*, 1995). Different G6PD polymorphisms have arisen and been selected for in different parts of the world, and recent haplotype analysis of two of these resistance-conferring variants (G6PD A- and G6PD Med) suggest that they arose and spread rapidly within African and circum-Mediterranean populations within the last 1500 to 12 000 years (Tishkoff *et al.*, 2001; Saunders *et al.*, 2002). Another common G6PD deficiency allele variant in modern humans, G6PD A, also appears to be maintained through selection, but the age of that allele predates the recent emergence of severe malaria, suggesting a different adaptive function than malaria resistance (Verrelli *et al.*, 2002). Interestingly, a recent parallel study of variation at the G6PD locus in a large set of chimpanzees plus exemplar individuals from several other non-human primate taxa concluded that the evolution of the enzyme has been strongly constrained over the 30- to 40-million year history of anthropoid primates (Verrelli *et al.*, 2006). Thus, in contrast to the situation for humans, there is no evidence to support the idea of positive selection for malaria resistance at the G6PD locus in chimpanzees (Verrelli *et al.*, 2006).

Additionally, variation in the structure of several types of red blood cell membrane proteins also plays an important role in human susceptibility and resistance to *Plasmodium* infection. For example, to gain access to human erythrocytes, both *P. vivax* and *P. knowlesi* merozoites must recognize and bind to a specific chemokine receptor protein, the Duffy antigen, which is expressed on the surface of red blood cells (Miller *et al.*, 1975, 1976; Barnwell *et al.*, 1989). The Duffy antigen is coded for by a gene known as FY, which has three main allele types in humans, FY^{A}, FY^{B}, and FY^{null}. Both *P. vivax* and *P. knowlesi* merozoites express a ligand protein that contains a Duffy-binding-like (DBL) domain, which allows the parasite to bind to and enter red blood cells that bear either FY^{A} or FY^{B} coded Duffy antigens. Most native sub-Saharan Africans are homozygous for the Duffy-negative allele (FY^{null}/FY^{null}) and thus fail to produce the Duffy antigen, rendering them effectively immune to infection by either *P. vivax* or *P. knowlesi*. Studies of both sequence polymorphism and microsatellite variation in the FY-gene region suggest that the locus has been under positive, directional selection in African populations, with the near fixation of the FY^{null} allele in Africa estimated to have arisen within the last 33 000 years, i.e. after the dispersal of anatomically modern humans out of

Africa (Hamblin & Rienzo 2000; Hamblin *et al.*, 2002), and perhaps much more recently (Seixas *et al.*, 2002).

In contrast to *P. vivax*, which can only infect human red blood cells that bear the Duffy antigen, *P. falciparum* merozoites express several different erythrocyte binding ligands that contain Duffy-binding-like domains and can use multiple, redundant invasion pathways to gain entry into mature red blood cells (Dolan *et al.*, 1994; Okoyeh *et al.*, 1999; Adams *et al.*, 2001; Chitnis, 2001; Gaur *et al.*, 2004). Not all erythrocyte surface proteins recognized by these ligands are known, but the glycophorins – membrane proteins that are highly glycosylated, bearing O-linked and N-linked glycans (oligosaccharide chains) rich in the terminal sugar sialic acid – are among the set of targets that are most commonly utilized by the parasite (Pasvol *et al.*, 1982a, 1982b, 1993; Friedman *et al.*, 1984; Dolan *et al.*, 1994; Lobo *et al.*, 2003; Mayer *et al.*, 2006). Glycophorin loci are among the fastest evolving genes in humans, especially at sites where glycans are attached (Baum *et al.*, 2002; Wang *et al.*, 2003), strongly suggesting that these glycoproteins have been the targets of positive selection at least in part due to the risk of *P. falciparum* infection. Indeed, variation in some human genes coding for the protein components of several glycophorins (e.g. GYPA, GYPB, GYPC) influences how readily those membrane proteins are bound by *P. falciparum* ligands (Gaur *et al.*, 2004; Mayer *et al.*, 2006), which, in turn, influences how susceptible the red blood cells bearing those proteins are to malarial infection.

The sialic acid component of red blood cell glycophorins is particularly important in the recognition of erythrocyte receptors by certain *P. falciparum* ligands. Thus, genes involved in sialic acid biochemistry and modification are also likely to be associated with human susceptibility/resistance to malaria and other pathogens that target sialic acids to invade animals cells (Varki, 2001). For example, the dominant invasion pathway for *P. falciparum* involves recognition of the sialic acid residue of glycophorin A (N-acetylneuraminic acid, or Neu5Ac) on human red blood cells by the erthyrocyte-binding antigen (EBA) 175 of the parasite. Humans are almost unique among mammals in having Neu5Ac as the principal sialic acid associated with erythrocyte cell membrane proteins, while the red blood cells of chimpanzees (and most other mammals) instead carry a mixture of Neu5Ac and Neu5Gc, a related sialic acid synthesized from Neu5Ac (Varki, 2001). Since the human–chimpanzee divergence, the gene encoding the enzyme CMP-N-acetylneuraminic acid hydroxylase (CMAH), which is centrally involved in synthesis of Neu5Gc from Neu5Ac, has become deactivated in the human lineage (Chou *et al.*, 1998, 2002). Martin *et al.* (2005) have demonstrated that this difference in the form of sialic acid associated with glycophorin A is likely to be responsible for the remarkable host-specificity of *P. falciparum* and *P. reichenowi* for humans and chimpanzees, respectively, where inoculation of one of these hominoid species with the parasite of the other fails to produce sustained infection or significant parasitemia. Interestingly, the same study found that red blood cells of *Aotus*, a New World monkey genus commonly used in *P. falciparum* research, resembles human red blood cells by only carrying Neu5Ac.

With respect to non-human primates, a potential case of the impact of *Plasmodium* on the evolution of orangutans (*Pongo pygmaeus*) has recently come to light. A duplication of the α-globin gene has been discovered in orangutans that Steiper *et al.* (2006) suggest may be of adaptive significance. It is hypothesized that the activity of this locus can result in thalassemia-like conditions, which, as in humans, may provide some resistance to *Plasmodium* infection. Orangutans can be naturally infected by two malaria species, *P. pitheci* and *P. silvaticum* (Table 7.2), and evidence now indicates that some orangutans can be infected with the human parasite, *P. vivax*, and with the macaque malarial species, *P. cynomolgi* and *P. inui* (Wolfe *et al.*, 2002; Reid *et al.*, 2006). Given that orangutans once ranged all the way from China to the Celebes Islands, they have likely been under considerable pressure from malaria throughout much of their evolutionary history (Peters *et al.*, 1976).

Other α-globin duplications have been found in gorillas, chimpanzees, and crab-eating macaques (Takenaka *et al.*, 1993), raising the possibility that malaria has also been a selective force driving α-globin evolution in other tropical primates. It has even been hypothesized that the driving force behind the speciation of *Macaca mulatta* and *M. fascicularis* may have been malarial pressure (Wheatley, 1980). Whereas rhesus macaques (*M. mulatta*) show little to no variation in the constituent chains of the hemoglobin molecule, crab-eating macaques (*M. fascicularis*) show several variants (Barnicot *et al.*, 1966). This variation was hypothesized to correlate with the different selective pressures of malaria on these two species. *Macaca mulatta* is widely used as a model organism in malaria research because they show a strong, usually fatal, response to infection with *P. knowlesi* (a *falciparum*-like species). *Macaca fascicularis*, on the other hand, only exhibit minor, chronic infection with low level parasitemias when similarly infected (Schmidt *et al.*, 1977). However, peninsular Malaysian *M. fascicularis* gets as sick as rhesus macaques. This is interesting because molecular phylogenetic studies reveal that mainland *M. fascicularis* have hybridized with *M. mulatta* (Tosi *et al.*, 2002) and therefore may have similar genetic background in key anti-inflammatory and malaria resistance loci (Praba-Egge *et al.*, 2002; Ylostalo *et al.*, 2005).

From the parasite's point of view, it is clear that human and non-human primate adaptations to *Plasmodium* infection have also significantly influenced the course of the parasite's evolution, by pressuring the parasite to find ways to circumvent its hosts' defenses. One example is the development of multiple and widespread drug-resistant strains of *P. falciparum* and *P. vivax* within the last half century, following concerted efforts by the World Health Organization and world governments to eradicate the disease. Another example is the rapid evolution of diversity in erythrocyte membrane molecules, which seems to have been matched by a corresponding evolution of diversity in parasite ligands. For example, recent studies have found a high level of non-synonymous polymorphism in the gene coding for EBA 175, the principal parasite ligand allowing *P. falciparum* merozoites to bind and invade mature erythrocytes (Baum *et al.*, 2003; Wang *et al.*, 2003), and additional variation has been seen in other ligands involved in the infection of immature red blood cells (reticulocytes) (Taylor *et al.*, 2002). Coupled with the high rate of evolution seen at the human glycophorin A locus, this strongly suggests an ongoing evolutionary arm's race between *P. falciparum* and its human hosts (Wang *et al.*, 2003). Indeed, the evolution and maintenance of diversity in the glycans associated with animal cell membrane glycoproteins and glycolipids may, in general, be driven by the coevolutionary arms struggles between microbial pathogens and their hosts (Gagneux & Varki, 1999; Bishop & Gagneux, 2007). Consistent with this idea of rapid adaptive evolution between parasite and host, comparative genomic data for *P. falciparum* have shown that genes presumably coding for antigenic cell surface molecules – which influence the parasite's ability to invade host cells and evade its host's immune defenses – are characterized by much greater diversity than genes associated with more basic metabolic functions (Volkman *et al.*, 2007). Moreover, a recent genomic comparison between *P. falciparum* and the chimpanzee parasite, *P. reichenowi*, found that the key functional differences between these parasite genomes are also primarily found in those genes involved in mediating parasite–host interactions. For example, loci coding for membrane proteins have evolved at a much faster rate since the *P. falciparum/reichenowi* split than have genes coding for proteins active primarily within the cell (Jeffares *et al.*, 2007).

Given their much larger population sizes and much shorter generation times, the evolutionary dynamics of a parasite taxon theoretically should often outpace that of its host(s), thus the fact that malaria and other parasite-induced infectious diseases today remain such a challenge for long-lived species such as humans and other non-human primates is perhaps not surprising. Although the situation is obviously complex in a two-host system like malaria – where the life histories and biochemical milieu of the invertebrate "definitive" and vertebrate "intermediate" hosts differ dramatically from one another, as well as from that of the parasite – the relative speed of adaptation is still likely to be far more rapid for the parasite than for either host. Still, the complexity of a two-host system may make various vector control strategies (e.g. larviciding of vector hatching sites, use of insecticide-treated bed nets to reduce vector-human contact) a viable option for reducing the human toll of the parasite, rather than focusing public health efforts solely on the development of vaccines or more effective antimalarial drugs.

Future directions

Over the last decade, substantial progress has been made in understanding the evolutionary history of the malarial parasites, particularly as they relate to the various primate genera that are their most common mammalian intermediate hosts. Still, much work needs to be done. First, to date, fewer than half of the *Plasmodium* species known to infect non-human primates have been included in phylogenetic studies (Table 7.2), thus a complete picture of the evolutionary history of the malarial parasites and their primate hosts is lacking. Importantly, none of the parasite species infecting strepsirhine primates (e.g. *P. lemuris*) have been included in phylogenetic analyses, nor have most of the hominoid parasites (e.g. *P. rhodani, P. schwetzi, P. pitheci, P. youngi*). Inclusion of these parasite species might reveal evidence of additional host transfers among non-human primates, influence the debate over the geographic origins of *P. vivax*, and provide insight into the evolutionary history of malaria among the Malagasy primates. For example, have the *Eulemur* malarias coevolved with their hosts over a long period of time or do they represent anthroponoses acquired in the last several thousand years since humans first colonized the island of Madagascar?

Second, further efforts are needed to characterize the natural variation found within populations of additional species of malarial parasites. To date, large-scale studies of intraspecific variation have only been carried out for two of the human parasites, *P. falciparum* and *P. vivax*, and not for any of the parasites targeting other genera of primates. Data on intraspecific variation in additional species of *Plasmodium* would be desirable not only for evaluating hypotheses about the source and timing of emergence of other human malarial agents (e.g. *P. ovale* and *P. malariae*), but also of widespread non-human primate malarial parasites like *P. knowlesi* and *P. brasilianum*. The debate over whether host transfer by *P. vivax/simium* and *P. malariae/brasilianum* occurred from humans to New World monkeys or vice versa could be answered by comparing the relative amount of neutral genetic variation found in the parasites (Rich & Ayala, 2003; Lim *et al.*, 2005). Given that studies of intraspecific variation in *P. falciparum* have invigorated the search for ways to reduce the human toll of malaria (e.g. by suggesting novel vaccination strategies and therapies), it is likely that further appreciation of the variation within other species of *Plasmodium* may do the same.

Finally, apart from G6PD in chimpanzees (Verrelli *et al.*, 2006) and a few known hemoglobin variants in hominoids and macaques (Barnicot *et al.*, 1966; Takenaka *et al.*, 1993; Steiper *et al.*, 2006), the extent to which any non-human primate taxon exhibits variation at any of the loci that have been implicated in malaria resistance in humans is unknown, as are the functional reasons why some species of primates are resistant to infection by species of *Plasmodium* that infect their close phylogenetic relatives. Presumably, if malaria has asserted a significant evolutionary selective pressure on primates other than humans, then these species, too, should show genetic and functional signatures of their adaptations to infection by *Plasmodium*, but such signatures have not yet been widely looked for outside of humans.

Conclusions

The late 20th and early 21st centuries have been marked by substantial progress in genetic research on malaria. Phylogenetic data clearly demonstrate that the primate malarias have a complex evolutionary history vis-à-vis their hosts – a history characterized by both coevolution and cases of host-switching between sometimes very distantly related primate taxa. Among the human parasites, while some of the genetic variation present in *P. falciparum* and *P. vivax* appears to be ancient, most of the genetic data suggests that modern-day populations of these two species are descended relatively recently from a very small number of founders. It appears that malaria only became a significant health burden for humans recently in our evolutionary past, most likely within the last 6000 to 30 000 years, and certainly well after the origins of modern *Homo sapiens*. As others have suggested, the presumed timing of emergence of *P. falciparum*, in particular, seems to coincide well with the Neolithic transition to agriculture in Africa and the origins of anthropophilic species of mosquitoes such as *Anopheles gambiae*, the primary vector for *P. falciparum* transmission in sub-Saharan African populations (Coluzzi, 1999; Hume *et al.*, 2003; Rich & Ayala, 2003; Ayala & Coluzzi, 2005).

In addition to the results reviewed above, a wealth of comparative genomic data on *Plasmodium* is either currently available or forthcoming. For example, the complete genome of *P. falciparum* has recently been sequenced, assembled, and published (Gardner *et al.*, 2002), and for a number of additional *Plasmodium* species, whole genome data in various stages of assembly and annotation are also available (Carlton *et al.*, 2002; Hall *et al.*, 2005). These include three rodent malarial agents (*P. berghei*, *P chaubaudi*, and *P. yoelii*), one Old World monkey parasite (*P. knowlesi*), one avian parasite (*P. gallinaceum*), and one additional human parasite (*P. vivax*) that have been sequenced by either the

Wellcome Trust's Sanger Institute or by The Institute for Genome Research (TIGR). Additionally, the Wellcome Trust is part-way through the process of sequencing the *P. reichenowi* genome. Thus, complete genomes will soon be available for a suite of *Plasmodium* species that infect a wide range of vertebrate hosts, thereby facilitating comparative analyses of gene evolutionary history, structure, and function. PlasmoDB (www.plasmodb.org) is a comprehensive, searchable, web-based database for comparative *Plasmodium* genomics that makes much of this data publicly available (Kissinger *et al.*, 2002; Bahl *et al.*, 2003; Stoeckert *et al.*, 2006).

Compared to the progress made in understanding the evolutionary history of the malarial parasites, progress on reducing the human toll of malaria has been far less impressive. While concerted efforts to combat the disease succeeded in reducing the global burden of malaria for a portion of the mid-20th century, malaria is once again on the rise and is a considered a re-emerging infectious disease that, in many places, has evolved resistance to some of the most effective treatments previously used (Carter & Mendis, 2002). The Roll Back Malaria Partnership – launched in 1998 by the World Health Organization in collaboration with the United National Development Program, UNICEF, and the World Bank – was aimed at reversing the disappointing increase in the global malarial burden that followed the interruption of eradication programs in the 1970s. However, the Partnership's goal of reducing the annual number of worldwide deaths due to malaria by 50% by the year 2010 seems unreachable. Given that close to half the world's human population is at risk for malarial infection (Guerra *et al.*, 2006), that resistance to standard antimalarial drugs is increasing among parasite populations, and that global climate change is likely to dramatically increase the world regions facing malaria risk (Hay *et al.*, 2004; Sutherst, 2004), the contemporary significance of malaria for humans cannot be underestimated.

References

Adams, J. H., Blair, P. L., Kaneko, O. & Peterson, D. S. (2001). An expanding *ebl* family of *Plasmodium falciparum*. *Trends in Parasitology*, **17**, 297–299.

Agarwal, A., Guindo, A., Cissoko, Y. *et al.* (2000). Hemoglobin C associated with protection from severe malaria in the Dogon of Mali, a West African population with a low prevalence of hemoglobin S. *Blood*, **96**, 2358–2363.

Allen, S. J., O'Donnell, A., Alexander, N. D. *et al.* (1997). a^{+}-thalassemia protects children against disease caused by other infections as well as malaria. *Proceedings of the National Academy of Sciences*, USA, **94**, 14736–14741.

Allison, A. C. & Clyde, D. F. (1961). Malaria in African children with deficient erythrocyte glucose-6-phosphate dehydrogenase. *British Medical Journal*, **1**, 1346–1349.

Antonarakis, S. E., Boehm, C. D., Serjeant, G. R. *et al.* (1984). Origin of the b^S-globin gene in Blacks: The contribution of recurrent mutation or gene conversion or both. *Proceedings of the National Academy of Sciences*, USA, **81**, 853–856.

Arnason, U., Gullberg, A. & Janke, A. (1998). Molecular timing of primate divergences as estimated by two nonprimate calibration points. *Journal of Molecular Evolution*, **47**, 718–727.

Arnell, N. W., Cannell, M. G. R., Hulme, M. *et al.* (2002). The consequences of CO_2 stabilisation for the impacts of climate change. *Climatic Change*, **53**, 413–446.

Ayala, F. J. & Coluzzi, M. (2005). Chromosome speciation: humans, *Drosophila*, and mosquitoes. *Proceedings of the National Academy of Sciences, USA*, **102** (Suppl. 1), 6535–6542.

Ayala, F., Escalante, A., Lal, A. & Rich, S. (1998). Evolutionary relationships of human malarias. In *Malaria: Parasite Biology, Pathogenesis, and Protection*, ed. I. W. Sherman. Washington, DC: American Society of Microbiology, pp. 285–300.

Ayala, F., Escalante, A. A. & Rich, S. M. (1999). Evolution of *Plasmodium* and the recent origin of the world populations of *Plasmodium falciparum*. *Parassitologia*, **41**, 55–68.

Bahl, A., Brunk, B., Crabtree, J. *et al.* (2003). PlasmoDB: the *Plasmodium* genome resource. A database integrating experimental and computational data. *Nucleic Acids Research*, **31**, 212–215.

Bannister, L. & Mitchell, G. (2003). The ins, outs and roundabouts of malaria. *Trends in Parasitoogy*, **19**, 209–213.

Barnicot, N. A., Huehns, E. R. & Jolly, C. J. (1966). Biochemical studies on haemoglobin variants of the irus macaque. *Proceedings of the Royal Society of London, B*, **165**, 224–244.

Barnwell, J. W., Nichols, M. E. & Rubinstein, P. (1989). *In vitro* evaluation of the role of the Duffy blood group in erythrocyte invasion by *Plasmodium vivax*. *Journal of Experimental Medicine*, **169**, 1795–1802.

Barry, A. E., Leliwa, A., Choi, M. *et al.* (2003). DNA sequence artifacts and the estimation of time to the most recent common ancestor (TMRCA) of *Plasmodium falciparum*. *Molecular and Biochemical Parasitology*, **130**, 143–147.

Barta, J. R. (1989). Phylogenetic analysis of the class Sporozoea (Phylum Apicomplexa Levine, 1970): evidence for the independent evolution of heteroxenous life cycles. *Journal of Parasitology*, **75**, 195–206.

Baum, J., Ward, R. H. & Conway, D. J. (2002). Natural selection on the erythrocyte surface. *Molecular Biology and Evolution*, **19**, 223–229.

Baum, J., Thomas, A. W. & Conway, D. J. (2003). Evidence for diversifying selection on erythrocyte-binding antigens of *Plasmodium falciparum* and *P. vivax*. *Genetics*, **163**, 1327–1336.

Beeson, J. G. & Brown, G. V. (2002). Pathogenesis of *Plasmodium falciparum* malaria: the roles of parasite adhesion and antigenic variation. *Cellular and Molecular Life Sciences*, **59**, 258–271.

Bensch, S., Stjernman, M., Hasselquist, D. *et al.* (2000). Host specificity in avian blood parasites: a study of *Plasmodium* and *Haemoproteus* mitochondrial DNA amplified from birds. *Proceedings of the Royal Society of London B, Biological Sciences*, **267**, 583–1589.

Bensch, S., Pearez-Tris, J., Waldenstroum, J. & Hellgren, O. (2004). Linkage between nuclear and mitochondrial DNA sequences in avian malaria parasites: multiple cases of cryptic speciation? *Evolution*, **58**, 1617–1621.

Benton, M. J. & Donaghue, P. C. J. (2007). Paleontological evidence to date the tree of life. *Molecular Biology and Evolution*, **24**, 26–53.

Beutler, E. (1994). G6PD deficiency. *Blood*, **11**, 3613–3636.

Bishop, J. & Gagneux, P. (2007). Evolution of carbohydrate antigens: microbial forces shaping host glycomes? *Glycobiology*, **7**, 23R–34R.

Bull, J. J. (1994). Virulence. *Evolution*, **48**, 1423–1437.

Carlton, J. M., Angiuoli, S. V., Suh, B. B. *et al.* (2002). Genome sequence and comparative analysis of the model rodent malaria parasite *Plasmodium yoelii yoelii*. *Nature*, **419**, 512–519.

Carter, R. (2003). Speculations on the origins of *Plasmodium vivax* malaria. *Trends in Parasitology*, **19**, 214–219.

Carter, R. & Mendis, K. N. (2002). Evolutionary and historical aspects of the burden of malaria. *Clinical Microbiology Reviews*, **15**, 564–594.

Chebloune, Y., Pagnier, J., Trabuchet, G. *et al.* (1988). Structural analysis of the 5' flanking region of the beta-globin gene in African sickle cell anemia patients: further evidence for three origins of the sickle cell mutation in Africa. *Proceedings of the National Academy of Sciences*, USA, **85**, 4431–4435.

Chitnis, C. E. (2001). Molecular insights into receptors used by malaria parasites for erythrocyte invasion. *Current Opinion in Hematology*, **8**, 85–91.

Chotivanich, K., Udomsangpetch, R., Simpson, J. A. *et al.* (2000). Parasite multiplication potential and the severity of *falciparum* malaria. *Journal of Infectious Disease*, **181**, 1206–1209.

Chotivanich, K., Udomsangpetch, R., Pattanapanyasat, K. *et al.* (2002). Hemoglobin E: a balanced polymorphism protective against high parasitemias and thus severe *P. falciparum malaria*. *Blood*, **100**, 1172–1176.

Chou, H.-H., Takematsu, H., Diaz, S. *et al.* (1998). A mutation in human CMP-sialic acid hydroxylase occurred after the *Homo-Pan* divergence. *Proceedings of the National Academy of Sciences, USA*, **95**, 11751–11756.

Chou, H.-H., Hayakawa, T., Diaz, S. *et al.* (2002). Inactivation of CMP-N-acetylneuraminic acid hydroxylase occurred prior to brain expansion during human evolution. *Proceedings of the National Academy of Sciences*, USA, **99**, 11736–11741.

Coatney, G. R., Collins, W. E., Warren, M. & Contacos, P. G. (2003). *The Primate Malarias* (originally published in 1971), CD-ROM version 1.0 edn. Atlanta, FL: CDC Division of Parasitic Disease.

Collins, W. H. & Aikawa, M. (1993). Plasmodia of non-human primates. In *Parasitic Protozoa* 5, 2nd edn, ed., J. Kreier. London: Academic Press, pp. 105–133.

Coluzzi, M. (1999). The clay feet of the malaria giant and its African roots: hypotheses and inferences about origin, spread and control of *Plasmodium falciparum*. *Parassitologia*, **41**, 277–283.

Conway, D. J. & Baum, J. (2002). In the blood – the remarkable ancestry of *Plasmodium falciparum*. *Trends in Parasitology*, **18**, 351–355.

Conway, D. J., Fanello, C., Lloyd, J. M. *et al.* (2000). Origin of *Plasmodium falciparum* malaria is traced by mitochondrial DNA. *Evolution*, **111**, 163–171.

Cornejo, O. E. & Escalante, A. A. (2006). The origin and age of *Plasmodium vivax*. *Trends in Parasitology*, **22**, 557–563.

Crabb, B. S. & Cowman, A. F. (2002). *Plasmodium falciparum* virulence determinants unveiled. *Genome Biology*, **3**, 103.1–103.4.

Craig, A. & Scherf, A. (2001). Molecules on the surface of the *Plasmodium falciparum* infected erythrocyte and their role in malaria pathogenesis and immune evasion. *Molecular and Biochemical Parasitology*, **115**, 129–143.

Currat, M., Trabuchet, G., Rees, D. *et al.* (2002). Molecular analysis of the ß-globin gene cluster in the Niokholo Mandenka population reveals a recent origin of the $ß^S$ Senegal mutation. *American Journal of Human Genetics*, **70**, 207–223.

Day, T. (2001). Parasite transmission modes and the evolution of virulence. *Evolution*, **55**, 2389–2400.

Day, T. (2003). Virulence evolution and the timing of disease life-history events. *Trends in Ecology and Evolution*, **18**, 113–118.

Dolan, S. A., Proctor, J. L., Alling, D. W. *et al.* (1994). Glycophorin B as an EBA-175 independent *Plasmodium falciparum* receptor of human erythrocytes. *Molecular and Biochemical Parasitology*, **64**, 55–63.

Ebi, K., Hartman, J., Chan, N. *et al.* (2005). Climate suitability for stable malaria transmission in Zimbabwe under different climate change scenarios. *Climatic Change*, **73**, 375–393.

Escalante, A. A. & Ayala, F. J. (1994). Phylogeny of the malarial genus *Plasmodium* derived from rRNA gene sequences. *Proceedings of the National Academy of Sciences*, USA, **91**, 11373–11377.

Escalante, A. A. & Ayala, F. J. (1995). Evolutionary origin of *Plasmodium* and other Apicomplexa based on rRNA genes. *Proceedings of the National Academy of Science, USA*, **92**, 5793–5797.

Escalante, A., Barrio, E. & Ayala, F. (1995). Evolutionary origin of human and primate malarias: evidence from the circumsporozoite protein gene. *Molecular Biology and Evolution*, **12**, 616–626.

Escalante, A. A., Goldman, I. F., de Rijk, P. *et al.* (1997). Phylogenetic study of the genus *Plasmodium* based on the secondary structure-based alignment of the small subunit ribosomal RNA. *Molecular and Biochemical Parasitology*, **90**, 317–321.

Escalante, A. A., Freeland, D. E., Collins, W. E. & Lal, A. A. (1998a). The evolution of primate malaria parasites based on the gene encoding cytochrome b from the linear mitochrondrial genome. *Proceedings of the National Academy of Sciences, USA*, **95**, 8124–8129.

Escalante, A. A., Lal, A. A. & Ayala, F. J. (1998b). Genetic polymorphism and natural selection in the malaria parasite *Plasmodium falciparum*. *Genetics*, **149**, 189–202.

Escalante, A. A., Cornejo, O. E., Freeland, D. E. *et al.* (2005). A monkey's tale: the origin of *Plasmodium vivax* as a human malaria parasite. *Proceedings of the National Academy of Sciences*, USA, **102**, 1980–1985.

Evans, A. G. & Wellems, T. E. (2002). Coevolutionary genetics of *Plasmodium* malaria parasites and their human hosts. *Integrative and Comparative Biology*, **42**, 401–407.

Feng, X., Carlton, J. M., Joy, D. A. *et al.* (2003). Single-nucleotide polymorphisms and genome diversity in *Plasmodium vivax*. *Proceedings of the National Academy of Sciences, USA*, **100**, 8502–8507.

Ferguson, H. M. & Read, A. F. (2002). Genetic and environmental determinants of malaria parasite virulence in mosquitoes. *Proceedings of the Royal Society of London B, Biological Sciences*, **269**, 1217–1224.

Flint, J., Hill, A. V. S., Bowden, D. K. *et al.* (1986). High frequencies of a-thalassaemia are the result of natural selection by malaria. *Nature*, **321**, 744–750.

Flint, J., Harding, R. M., Boyce, A. J. & Clegg, J. B. (1998). The population genetics of the haemoglobinopathies. *Baillière's Clinical Hematology*, **11**, 1–51.

Fortin, A., Stevenson, M. M. & Gros, P. (2002). Susceptibility to malaria as a complex trait: big pressure from a tiny creature. *Human Molecular Genetics*, **11**, 2469–2478.

Friedman, M. J. (1978). Erythrocytic mechanism of sickle cell resistance to malaria. *Proceedings of the National Academy of Sciences, USA*, **75**, 1994–1997.

Friedman, M. J., Blankenburt, T., Sensabaugh, G. & Tenforde, T. F. (1984). Recognition and invasion of human erythrocytes by malarial parasites: contribution of sialoglycoproteins to attachment and host specificity. *Journal of Cell Biology*, **98**, 1682–1687.

Gagneux, P. & Varki, A. (1999). Evolutionary considerations in relating oligosaccharide diversity to biological function. *Glycobiology*, **9**, 747–755.

Gandon, S. (2004). Evolution of multihost parasites. *Evolution*, **58**, 455–469.

Garamszegi, L. Z. (2006). The evolution of virulence and host specialization in malaria parasites of primates. *Ecology Letters*, **9**, 933–940.

Gardner, M. J., Hall, N., Fung, E. *et al.* (2002). Genome sequence of the human malaria parasite *Plasmodium falciparum*. *Nature*, **419**, 498–511.

Gaur, D., Mayer, D. C. G. & Miller, L. H. (2004). Parasite ligand–host receptor interactions during invasion of erythrocytes by *Plasmodium* merozoites. *International Journal for Parasitology*, **34**, 1413–1429.

Gilles, N. H., Hendrickse, R. G., Linder, R., Reddy, S. & Allan, N. (1967). Glucose-6-phosphate dehydrogenase deficiency, sickling, and malaria in African children in southwestern Nigeria. *Lancet*, **1**, 138–140.

Greene, L. S. (1993). G6PD deficiency as protection against *falciparum* malaria: an epidemiologic critique of population and experimental studies. *Yearbook of Physical Anthropology*, **36**, 153–178.

Guerra, C. A., Snow, R. W. & Hay, S. I. (2006). Defining the global spatial limits of malaria transmission in 2005. *Advances in Parasitology*, **62**, 157–179.

Gupta, S., Hill, A. V. S., Kwiatkowski, D. *et al.* (1994). Parasite virulence and disease patterns in *Plasmodium falciparum* malaria. *Proceedings of the National Academy of Science, USA*, **91**, 3715–3719.

Gysin, J. (1998). Animal models: primates. In *Malaria: Parasite Biology, Pathogenesis, and Protection*, ed. I. W. Sherman. Washington, DC: ASM Press, pp. 419–441.

Hagner, S. C., Misof, B., Maier, W. A. & Kampen, H. (2007). Bayesian analysis of new and old malaria parasite DNA sequence data demonstrates the need for more phylogenetic signal to clarify the descent of *Plasmodium falciparum*. *Parasitology Research*, **101**, 493–503.

Hall, N., Karras, M., Raine, J. D. *et al.* (2005). A comprehensive survey of the *Plasmodium* life cycle by genomic, transcriptomic, and proteomic analyses. *Science*, **307**, 82–86.

Hamblin, M. T. & Rienzo, A. D. (2000). Detection of the signature of natural selection in humans: evidence from the Duffy blood group locus. *American Journal of Human Genetics*, **66**, 1669–1679.

Hamblin, M. T., Thompson, E. E. & Rienzo, A. D. (2002). Complex signatures of natural selection at the Duffy blood group locus. *American Journal of Human Genetics*, **70**, 369–383.

Hartl, D. L. (2004). The origin of malaria: mixed messages from genetic diversity. *Nature Reviews: Microbiology*, **2**, 15–22.

Hartman, J., Ebi, K., McConnell, K. J., Chan, N. & Weyant, J. (2002). Climate suitability for stable malaria transmission in Zimbabwe under different climate change scenarios. *Global Change and Human Health*, **3**, 42–54.

Harvell, C. D., Mitchell, C. E., Ward, J. R. *et al.* (2002). Climate warming and disease risks for terrestrial and marine biota. *Science*, **296**, 2158–2162.

Hay, S. I., Cox, J., Rogers, D. J. *et al.* (2002a). Regional warming and malaria resurgence. *Nature*, **420**, 628.

Hay, S. I., Rogers, D. J., Randolph, S. E. *et al.* (2002b). Hot topic or hot air? Climate change and malaria resurgence in East African highlands. *Trends in Parasitology*, **18**, 530–534.

Hay, S. I., Guerra, C. A., Tatem, A. J., Noor, A. M. & Snow, R. W. (2004). The global distribution and population at risk of malaria: past, present, and future. *The Lancet Infectious Diseases*, **4**, 327–336.

Hedges, S. B., Parker, P. H., Sibley, C. G. & Kumar, S. (1996). Continental breakup and the ordinal diversification of birds and mammals. *Nature*, **381**, 226–229.

Hellgren, O. (2005). The occurrence of haemosporidian parasites in the Fennoscandian bluethroat (*Luscinia svecica*) population. *Journal of Ornithology*, **146**, 55–60.

Hughes, A. L. (1991). Circumsporozoite protein genes of malaria *Plasmodium falciparum*. *Journal of Molecular Biology*, **195**, 273–287.

Hughes, A. L. (1992). Positive selection and interallelic recombination at the merozoite surface antigen-1 (MSA-1) locus of *Plasmodium falciparum*. *Molecular Biology and Evolution*, **9**, 381–393.

Hughes, A. L. & Verra, F. (1998). Ancient polymorphism and the hypothesis of a recent bottleneck in the malaria parasite *Plasmodium falciparum*. *Genetics*, **150**, 511–513.

Hughes, A. L. & Verra, F. (2001). Very large long-term effective population size in the virulent human malaria parasite *Plasmodium falciparum*. *Proceedings of the Royal Society of London B: Biological Sciences*, **268**, 1855–1860.

Hughes, M. K. & Hughes, A. L. (1995). Natural selection on *Plasmodium* surface proteins. *Molecular and Biochemical Parasitology*, **71**, 99–103.

Hume, J. C. C., Lyons, E. J. & Day, K. P. (2003). Human migration, mosquitoes and the evolution of *Plasmodium falciparum*. *Trends in Parasitology*, **19**, 144–149.

Hutagalung, R., Wilairatana, P., Looareesuwan, S. *et al.* (1999). Influence of hemoglobin E trait on the severity of *Falciparum* malaria. *Journal of Infectious Diseases*, **179**, 283–286.

Jeffares, D. C., Pain, A., Berry, A. *et al.* (2007). Genome variation and evolution of the malaria parasite *Plasmodium falciparum*. *Nature Genetics*, **39**, 120–125.

Jongwutiwes, S., Putaporntip, C., Iwasaki, T., Sata, T. & Kanbara, H. (2004). Naturally acquired *Plasmodium knowlesi* malaria in human, Thailand. *Emerging Infectious Diseases*, **10**, 2211–2213.

Jongwutiwes, S., Putaporntip, C., Iwasaki, T. *et al.* (2005). Mitochondrial genome sequences support ancient population expansion in *Plasmodium vivax*. *Molecular Biology and Evolution*, **22**, 1733–1739.

Joy, D. A., Feng, X., Mu, J. *et al.* (2003). Early origin and recent expansion of *Plasmodium falciparum*. *Science*, **300**, 318–321.

Kedzierski, L., Escalante, A. A., Isea, R. *et al.* (2002). Phylogenetic analysis of the genus *Plasmodium* based on the gene encoding adenylosuccinate lyase. *Infection, Genetics and Evolution*, **1**, 297–301.

Kissinger, J. C., Brunk, B. P., Crabtree, J. *et al.* (2002). PlasmoDB: the *Plasmodium* genome database. Designing and mining a eukaryotic genomics resource. *Nature*, **419**, 490–492.

Kovats, R. S., Campbell-Lendrum, D. H., McMichael, A. J., Woodward, A. & St. H. Cox, J. (2001). Early effects of climate change: do they include vector-borne disease? *Philosophical Transactions of the Royal Society of London, B*, **356**, 1057–1068.

Kumar, S. & Hedges, S. B. (1998). A molecular timescale for vertebrate evolution. *Nature*, **392**, 917–920.

Kwiatkowski, D. P. (2005). How malaria has affected the human genome and what human genetics can teach us about malaria. *American Journal of Human Genetics*, **77**, 171–190.

Leclerc, M. C., Durand, P., Gauthier, C. *et al.* (2004a). Meager genetic variability of the human malaria agent *Plasmodium vivax*. *Proceedings of the National Academy of Sciences, USA*, **101**, 14455–14460.

Leclerc, M. C., Hugot, J. P., Durand, P. & Renaud, F. (2004b). Evolutionary relationships between 15 *Plasmodium* species from New and Old World primates (including humans): an 18S rDNA cladistic analysis. *Parasitology*, **129**, 677–684.

Lim, C. S., Tazi, L. & Ayala, F. J. (2005). *Plasmodium vivax*: recent world expansion and genetic identity to *Plasmodium simium*. *Proceedings of the National Academy of Sciences, USA*, **102**, 15523–15528.

Lobo, C. A., Rodriguez, M., Reid, M. & Lustigman, S. (2003). Glycophorin C is the receptor for the *Plasmodium falciparum* erythrocyte binding ligand PfEBP-2 (baebl). *Blood*, **101**, 4628–4631.

Mackinnon, M. J. & Read, A. F. (1999). Genetic relationships between parasite virulence and transmission in the rodent malaria *Plasmodium chabaudi*. *Evolution*, **53**, 689–703.

Martens, P., Kovats, R. S., Nijhof, S. *et al.* (1999). Climate change and future populations at risk of malaria. *Global Environmental Change*, **9**, S89–S107.

Martens, W. J. M., Niessen, L. W., Rotmans, J., Jetten, T. H. & McMichael, A. J. (1995). Potential impact of global climate change on malaria risk. *Environmental Health Perspectives*, **103**, 458–464.

Martin, M. J., Rayner, J. C., Gagneux, P., Barnwell, J. W. & Varki, A. (2005). Evolution of human–chimpanzee differences in malaria susceptibility: relationship to human genetic loss of N-glycolylneuraminic acid. *Proceedings of the National Academy of Sciences, USA*, **102**, 12819–12824.

Mayer, D. C. G., Jiang, L., Achur, R. N. *et al.* (2006). The glycophorin C N-linked glycan is a critical component of the ligand for the *Plasmodium falciparum* erythrocyte receptor BAEBL. *Proceedings of the National Academy of Sciences, USA*, **103**, 2358–2362.

McCutchan, T. F., Kissinger, J. C., Touray, M. G. *et al.* (1996). Comparison of circumsporozoite proteins from avian and mammalian malarias: biological and phylogenetic implications. *Proceedings of the National Academy of Sciences, USA*, **93**, 11889–11894.

McMichael, A. J., Campbell-Lendrum, D. H., Corvalán, C. F. *et al.* (2003). *Climate Change and Human Health: Risks and Responses*. Geneva: WHO.

McMichael, A. J., Woodruff, R. E. & Hales, S. (2006). Climate change and human health: present and future risks. *Lancet*, **367**, 859–869.

Mears, J. G., Lachman, H. M., Cabannes, R. *et al.* (1981). Sickle gene: its origin and diffusion from West Africa. *Journal of Clinical Investigation*, **68**, 606–610.

Mendis, K., Sina, B. J., Marchesini, P. & Carter, R. (2001). The neglected burden of *Plasmodium vivax* malaria. *American Journal of Tropical Medicine and Hygiene*, **64**, 97–106.

Miller, L. H., Mason, S. J., Dvorak, J. A., McGinniss, M. H. & Rothman, I. K. (1975). Erythrocyte receptors for (*Plasmodium knowlesi*) malaria: Duffy blood group determinants. *Science*, **189**, 561–563.

Miller, L. H., Mason, S. J., Clyde, D. F. & McGinniss, M. H. (1976). The resistance factor to *Plasmodium vivax* in blacks: the Duffy blood-group genotype, FyFy. *New England Journal of Medicine*, **295**, 302–304.

Modiano, D., Luoni, G., Sirima, B. S. *et al.* (2001). Haemoglobin C protects against clinical *Plasmodium falciparum* malaria. *Nature*, **414**, 305–308.

Mu, J., Duan, J., Makova, K. D. *et al.* (2002). Chromosome-wide SNPs reveal an ancient origin for *Plasmodium falciparum*. *Nature*, **418**, 322–326.

Mu, J., Joy, D. A., Duan, J. *et al.* (2005). Host switch leads to emergence of *Plasmodium vivax* malaria in humans. *Molecular Biology and Evolution*, **22**, 1686–1693.

Nielsen, K. M., Kasper, J., Choi, M. *et al.* (2003). Gene conversion as a source of nucleotide diversity in *Plasmodium falciparum*. *Molecular Biology and Evolution*, **20**, 726–734.

Ohashi, J., Naka, I., Patarapotikul, J. *et al.* (2004). Extended linkage disequilibrium surrounding the hemoglobin E variant due to malarial selection. *American Journal of Human Genetics*, **74**, 1198–1208.

Okoyeh, J. N., Pillai, C. R. & Chitnis, C. E. (1999). *Plasmodium falciparum* field isolates commonly use erythrocyte invasion pathways that are independent of sialic acid residues of glycophorin A. *Infection and Immunity*, **67**, 5784–5791.

Pagnier, J., Mears, J. G., Dunda-Belkhodja, O. *et al.* (1984). Evidence for the multicentric origin of the sickle cell hemoglobin gene in Africa. *Proceedings of the National Academy of Sciences, USA*, **81**, 1771–1773.

Pascual, M., Ahumada, J. A., Chaves, L. F., Rodó, X. & Bouma, M. (2006). Malaria resurgence in the East African highlands: temperature trends revisited. *Proceedings of the National Academy of Sciences, USA*, **13**, 5829–5834.

Pasvol, G., Jungery, M., Weatherall, D. J. *et al.* (1982a). Glycophorin as a possible receptor for *Plasmodium falciparum*. *Lancet*, **2**, 947–950.

Pasvol, G., Wainscoat, J. S. & Weatherall, D. J. (1982b). Erythrocytes deficient in glycophorin resist invasion by the malarial parasite *Plasmodium falciparum*. *Nature*, **297**, 64–66.

Pasvol, G., Carlsson, J. & Clough, B. (1993). The red cell membrane and invasion by malarial parasites. *Baillière's Clinical Hematology*, **6**, 513–534.

Patz, J. A., Campbell-Lendrum, D., Holloway, T. & Foley, J. A. (2005). Impact of regional climate change on human health. *Nature*, **438**, 310–317.

Paul, R. E. L., Ariey, F. & Robert, V. (2003). The evolutionary ecology of *Plasmodium*. *Ecology Letters*, **6**, 866–880.

Perkins, S. L. (2000). Species concepts and malaria parasites: detecting a cryptic species of *Plasmodium*. *Proceedings of the Royal Society of London B, Biological Sciences*, **267**, 2345–2350.

Perkins, S. L. & Schall, J. J. (2002). A molecular phylogeny of malarial parasites recovered from cytochrome b gene sequences. *Journal of Parasitology*, **88**, 972–978.

Peters, W., Garnham, P. C. C., Killick-Kendrick, R. *et al.* (1976). Malaria of the orang-utan (*Pongo pygmaeus*) in Borneo. *Philosophical Transactions of the Royal Society of London, B*, **275**, 439–482.

Polley, L. (2005). Navigating parasite webs and parasite flow: emerging and re-emerging parasitic zoonoses of wildlife origin. *International Journal for Parasitology*, **35**, 1279–1294.

Polley, S. D., Weedall, G. D., Thomas, A. W., Golightly, L. M. & Conway, D. J. (2005). Orthologous gene sequences of merozoite surface protein 1 (MSP1) from *Plasmodium reichenowi* and *P. gallinaceum* confirm an ancient divergence of *P. falciparum* alleles. *Molecular and Biochemical Parasitology*, **142**, 25–31.

Praba-Egge, A. D., Cogswell, F. B., Montenegro-James, S., Kohli, T. & James, M. A. (2002). Cytokine responses during acute simian malaria infection in the rhesus monkey. *American Journal of Tropical Medicine and Hygiene*, **67**, 586–596.

Qari, S. H., Shi, Y. A., Pieniazek, N. J., Collins, W. E. & Lal, A. A. (1996). Phylogenetic relationship among the malaria parasites based on small subunit rRNA gene sequences: monophyletic nature of the human malaria parasite, *Plasmodium falciparum*. *Molecular Phylogenetics and Evolution*, **6**, 157–165.

Rathore, D., Wahl, A. M., Sullivan, M. & McCutchan, T. F. (2001). A phylogenetic comparison of gene trees constructed from plastid, mitochondrial and genomic

DNA of *Plasmodium* species. *Molecular and Biochemical Parasitology*, **114**, 89–94.

Reid, M. J. C., Ursic, R., Cooper, D. *et al.* (2006). Transmission of human and macaque *Plasmodium* spp. to ex-captive orangutans in Kalimantan, Indonesia. *Emerging Infectious Diseases*, **12**, 1902–1908.

Rich, S. M. & Ayala, F. J. (1998). The recent origin of allelic variation in antigenic determinants of *Plasmodium falciparum*. *Genetics*, **150**, 515–517.

Rich, S. M. & Ayala, F. J. (2003). Progress in malaria research: the case for phylogenetics. *Advances in Parasitology*, **54**, 255–280.

Rich, S. M., Licht, M. C., Hudson, R. R. & Ayala, F. J. (1998). Malaria's Eve: evidence of a recent bottleneck in the global *Plasmodium falciparum* population. *Proceedings of the National Academy of Science, USA*, **95**, 4425–4430.

Ricklefs, R. E. & Fallon, S. M. (2002). Diversification and host switching in avian malaria parasites. *Proceedings of the Royal Society of London B, Biological Sciences*, **269**, 885–892.

Ricklefs, R. E., Fallon, S. M. & Bermingham, E. (2004). Evolutionary relationships, cospeciation, and host switching in avian malaria parasites. *Systematic Biology*, **53**, 111–119.

Rogers, D. J. & Randolph, S. E. (2000). The global spread of malaria in a future, warmer world. *Science*, **289**, 1763–1766.

Roth, E. F., Raventos-Suarez, C., Rinaldi, A. & Nagel, R. L. (1983). Glucose-6-phosphate dehydrogenase deficiency inhibits *in vitro* growth of *Plasmodium falciparum*. *Proceedings of the National Academy of Sciences, USA*, **80**, 298–299.

Ruwende, C. & Hill, A. (1998). Glucose-6-phosphate dehydrogenase deficiency and malaria. *Journal of Molecular Medicine*, **76**, 581–588.

Ruwende, C., Khoo, S. C., Snow, R. W. *et al.* (1995). Natural selection of hemi- and heterozygotes for G6PD deficiency in Africa by resistance to severe malaria. *Nature*, **376**, 246–249.

Saunders, M. A., Hammer, M. F. & Nachman, M. W. (2002). Nucleotide variability at *G6pd* and the signature of malarial selection in humans. *Genetics*, **162**, 1849–1861.

Schmidt, L. H., Fradkin, R., Harrison, J. & Rossan, R. (1977). Differences in the virulence of *Plasmodium knowlesi* for *Macaca iris* (*fascicularis*) of Philippine and Malayan origins. *American Journal of Tropical Medicine and Hygiene*, **26**, 612–622.

Seixas, S., Ferrand, N. & Rocha, J. (2002). Microsatellite variation and evolution of the human Duffy blood group polymorphism. *Molecular Biology and Evolution*, **19**, 1802–1806.

Sherman, I. W. (1998). A brief history of malaria and discovery of the parasite's life cycle. In *Malaria: Parasite Biology, Pathogenesis, and Protection*, ed. I. W. Sherman. Washington, DC: ASM Press, pp. 3–10.

Sherman, I. W. (2006). *The Power of Plagues*. Washington, DC: ASM Press.

Singh, B., Sung, L. K., Matusop, A. *et al.* (2004). A large focus of naturally acquired *Plasmodium knowlesi* infections in human beings. *Lancet*, **363**, 1017–1024.

Snow, R. W., Guerra, C. A., Noor, A. M., Myint, H. Y. & Hay, S. I. (2005). The global distribution of clinical episodes of *Plasmodium falciparum* malaria. *Nature*, **434**, 214–217.

Steiper, M. E., Wolfe, N. D., Karesh, W. B. *et al.* (2006). The phylogenetic and evolutionary history of a novel alpha-globin-type gene in orangutans (*Pongo pygmaeus*). *Infection, Genetics, and Evolution*, **6**, 277–286.

Stoeckert, Jr., C. J., Fischer, S., Kissinger, J. C. *et al.* (2006). PlasmoDB v5: new looks, new genomes. *Trends in Parasitology*, **22**, 543–546.

Sutherst, R. W. (1998). Implications of global change and climate variability for vector-borne diseases: generic approaches to impact assessments. *International Journal of Parasitology*, **28**, 935–947.

Sutherst, R. W. (2004). Global change and human vulnerability to vector-borne diseases. *Clinical Microbiology Reviews*, **17**, 136–173.

Takenaka, A., Udono, T., Miwa, N., Varavudhi, P. & Takenaka, O. (1993). High frequency of triplicated a-globin genes in tropical primates, crab-eating macaques (*Macaca fascicularis*), chimpanzees (*Pan troglodytes*), and orang-utans (*Pongo pygmaeus*). *Primates*, **34**, 55–60.

Tanabe, K., Escalante, A., Sakihama, N. *et al.* (2007). Recent independent evolution of msp1 polymorphism in *Plasmodium vivax* and related simian malaria parasites. *Molecular and Biochemical Parasitology*, **156**, 74–79.

Tanser, F. C., Sharp, B. & le Sueur, D. (2003). Potential effect of climate change on malaria transmission in Africa. *Lancet*, **362**, 1792–1798.

Taylor, H. M., Grainger, M. & Holder, A. A. (2002). Variation in the expression of a *Plasmodium falciparum* protein family implicated in erythrocyte invasion. *Infection and Immunity*, **70**, 5779–5789.

Telford, S. R. (1994). Plasmodia of reptiles. In *Parasitic Protozoa 7*, 2nd edn, ed. J. Kreier. London: Academic Press, pp. 1–71.

Thomas, C. J., Davies, G. & Dunn, C. E. (2004). Mixed picture for changes in stable malaria distribution with future climate in Africa. *Trends in Parasitology*, **20**, 216–220.

Tishkoff, S. A., Varkonyi, R., Cahinhinan, N. *et al.* (2001). Haplotype diversity and linkage disequilibrium at human G6PD: recent origin of alleles that confer malarial resistance. *Science*, **293**, 455–462.

Tol, R. S. J. & Dowlatabadi, H. (2001). Vector-borne diseases, development and climate change. *Integrated Assessment*, **2**, 173–181.

Tosi, A., Morales, J. & Melnick, D. (2002). Y-chromosome and mitochondrial markers in *Macaca fascicularis* indicate introgression with Indochinese *M. mulatta* and a biogeographic barrier in the Isthmus of Kra. *International Journal of Primatology*, **23**, 161–178.

van Lieshout, M., Kovats, R. S., Livermore, M. T. J. & Martens, P. (2004). Climate change and malaria: analysis of the SRES climate and socio-economic scenarios. *Global Environmental Change*, **14**, 87–99.

van Riper, C., Atkinson, C. T. & Seed, T. M. (1994). Plasmodia of birds. In *Parasitic Protozoa 7*, 2nd edn, ed. J. Kreier. London: Academic Press, pp. 73–140.

Vargas-Serrato, E., Corredor, V. & Galinski, M. R. (2003). Phylogenetic analysis of CSP and MSP-9 gene sequences demonstrates the close relationship of

Plasmodium coatneyi to *Plasmodium knowlesi*. *Infection, Genetics and Evolution*, **3**, 67–73.

Varki, A. (2001). Loss of N-glycolylneuraminic acid in humans: mechanisms, consequences, and implications for hominid evolution. *Yearbook of Physical Anthropology*, **44**, 54–69.

Verrelli, B. C., McDonald, J. H., Argyropoulos, G. *et al.* (2002). Evidence for balancing selection from nucleotide sequence analyses of human G6PD. *American Journal of Human Genetics*, **71**, 1112–1128.

Verrelli, B. C., Tishkoff, S. A., Stone, A. C. & Touchman, J. W. (2006). Contrasting histories of G6PD molecular evolution and malarial resistance in humans and chimpanzees. *Molecular Biology and Evolution*, **23**, 1592–1601.

Volkman, S. K., Barry, A. E., Lyons, E. J. *et al.* (2001). Recent origin of *Plasmodium falciparum* from a single progenitor. *Science*, **293**, 482–484.

Volkman, S. K., Sabeti, P. C., DeCaprio, D. *et al.* (2007). A genome-wide map of diversity in *Plasmodium falciparum*. *Nature Genetics*, **39**, 113–119.

Wang, H.-Y., Tang, H., Shen, C.-K. J. & Wu, C.-I. (2003). Rapidly evolving genes in human. I. The glycophorins and their possible role in evading malaria parasites. *Molecular Biology and Evolution*, **20**, 1795–1804.

Waters, A. P., Higgins, D. G. & McCutchan, T. F. (1991). *Plasmodium falciparum* appears to have arisen as a result of lateral transfer between avian and human hosts. *Proceedings of the National Academy of Science, USA*, **88**, 3140–3144.

Waters, A. P., Higgins, D. G. & McCutchan, T. F. (1993). Evolutionary relatedness of some primate models of *Plasmodium*. *Molecular Biology and Evolution*, **10**, 914–923.

Wheatley, B. P. (1980). Malaria as a possible selective factor in the speciation of macaques. *Journal of Mammalogy*, **61**, 307–311.

Williams, T. N. (2006). Red blood cell defects and malaria. *Molecular and Biochemical Parasitology*, **149**, 121–127.

Williams, T. N., Mwangi, T. W., Roberts, D. J. *et al.* (2005a). An immune basis for malaria protection by the sickle cell trait. *PLoS Medicine*, **2**, 441–445.

Williams, T. N., Wambua, S., Uyoga, S. *et al.* (2005b). Both heterozygous and homozygous a^+ thalassemias protect against severe and fatal *Plasmodium falciparum* malaria on the coast of Kenya. *Blood*, **106**, 368–371.

Wolfe, N. D., Karesh, W. B., Kilbourn, A. M. *et al.* (2002). The impact of ecological conditions on the prevalence of malaria among orangutans. *Vector-Borne and Zoonotic Disease*, **2**, 97–103.

Woolhouse, M. E. J., Taylor, L. H. & Haydon, D. T. (2001). Population biology of multihost pathogens. *Science*, **292**, 1109–1112.

World Health Organization (2003). Worldwide malaria distribution in 2002. Geneva: Public Health Mapping Group, WHO.

Ylostalo, J., Randall, A. C., Myers, T. A. *et al.* (2005). Transcriptome profiles of host gene expression in a monkey model of human malaria. *Journal of Infectious Disease*, **191**, 400–409.

Yotoko, K. S. C. & Elisei, C. (2006). Malaria parasites (Apicomplexa, Haematozoea) and their relationships with their hosts: is there an evolutionary cost for the specialization? *Journal of Zoological Systematics*, **44**, 265–273.

8 *Disease avoidance and the evolution of primate social connectivity: Ebola, bats, gorillas, and chimpanzees*

PETER D. WALSH, MAGDALENA BERMEJO, AND
JOSÉ DOMINGO RODRÍGUEZ-TEIJEIRO

Photograph by Angelique Todd (*Gorilla gorilla*)

Introduction

During the 1970s and 1980s ecology experienced a vigorous, one might even say raucous, debate over whether biological communities were structured by competition or predation (Connell, 1975; Menge & Sutherland, 1976). The answer that emerged may have been predictable . . . "both, of course" . . . but the

Primate Parasite Ecology. The Dynamics and Study of Host–Parasite Relationships, ed. Michael A. Huffman and Colin A. Chapman. Published by Cambridge University Press.

debate energized the field. Spurred on by an almost apostolic vigor, ecologists cranked out a raft of exciting new theories and boatloads of great empirical work to test these theories (reviewed in Sih *et al.*, 1985; Gurevitch *et al.* 2000).

At about the same time, there were signs that a similar debate might take hold in primatology. The melding of ethology and ecology brought a wave of new ideas on how behavior might interact with ecological factors and, in particular, on the way in which competition for resources might influence social organization. Factors such as territoriality, social group size, and rates of association within groups were all viewed in terms of the way resource distribution influenced competition (e.g. Eisenberg *et al.*, 1972; Mitani & Rodman, 1979; van Schaik & van Hooff, 1983; Terborgh & Janson, 1986). At about the same time, William J. Freeland wrote a seminal paper discussing the influence of disease on the evolution of primate social structure (Freeland, 1976, see discussion in Chapman *et al.*, Chapter 21, this volume). Freeland's central insight was that the risk of disease transmission was proportional to the number of individuals one interacted with and, therefore, that natural selection should promote behavioral mechanisms that limited social contact. He looked at the same phenomena as the competitionistas . . . territoriality, group size, and rates of association within groups . . . and came to very different conclusions about their evolutionary origin.

Unfortunately, the debate never materialized. Since the 1980s primatology has danced to the beat of one hand clapping, developing an almost religious devotion to the idea that resource distribution and abundance is the (only) important evolutionary driver of social structure. Territoriality, social group size, rates of association within groups, and other attributes of social organization are now almost universally viewed as adaptive consequences of competition for food and/or mates. A few devoted acolytes of the disease cult have labored diligently to keep Freeland's candle burning (Nunn & Alizer, 2006). But even they have tended to treat disease transmission as an emergent consequence of primate social structure rather than a selection pressure driving its evolution (e.g., Nunn & Dokey 2006).

This chapter is an attempt to stir the pot: to rekindle wider interest in Freeland's ideas by presenting empirical data that illustrate just how strong a selective pressure disease can be and how much primate social structure can influence disease impact. To that end we present data on western gorilla (*Gorilla gorilla*) and chimpanzee (*Pan troglodytes*) mortality from Ebola virus at our study site in and around Lossi Sanctuary in northwest Republic of Congo. Over the last 15 years Ebola has caused massive gorilla and chimpanzee population declines in Congo and neighboring Gabon (Huijbregts *et al.*, 2003; Walsh *et al.*, 2003; Bermejo *et al.*, 2006; Caillaud *et al.*, 2006), killing about one-third of the world's protected area gorilla population (IUCN, 2007). Ebola makes a nice case study in that gorillas and chimpanzees differ both in their social structure and in the extent of their contact with the putative reservoir hosts for Ebola, bats (Leroy *et al.*, 2005). Conveniently, the bat and social contact patterns make opposing predictions about which species, gorillas or chimpanzees, should be more susceptible to Ebola infection.

Here we first outline the differences between gorillas and chimpanzees in social structure and bat contact rates and make predictions about how these differences should translate into different rates of Ebola infection. We then present survey data on patterns of Ebola survivorship collected in and around Lossi after Ebola outbreaks in 2002–2003 and evaluate different scenarios for the Ebola transmission dynamics underlying both the survey data and auxiliary data on bats. Finally, we discuss the implications of our results for hypotheses on the evolution of social organization in primates, with a particular focus on how body size affects disease exposure risk.

Social contact structures and spillover rates

Western gorilla and chimpanzee social organization differ in three ways that have direct implications for disease transmission. First, neighboring western gorilla social groups show substantial ranging overlap. They have direct social encounters about every 2 weeks (Bermejo, 2004; Doran-Sheehy *et al.*, 2004) and more often visit fruiting trees and clearings visited by other groups on the same day (Walsh *et al.*, 2007). Interactions between groups are often tolerant if not affiliative. Juveniles from different groups occasionally play together and neighboring groups have even been observed to nest together (Bermejo, 2004). In stark contrast, chimpanzee communities vigorously defend territories to the point of killing intruders. Second, western gorillas live in much smaller social groups than chimpanzees. Fourteen gorilla groups at Lossi averaged 14 individuals (Bermejo *et al.*, 2006; see also Robbins *et al.*, 2004), while chimpanzee communities typically contain about 50–75 individuals (Wrangham, 2000). Third, gorillas live in cohesive groups which forage together each day and sleep together every night. Chimpanzees spend much of the year in smaller parties, with subgroup composition turning over on a daily basis (Chapman *et al.*, 1995).

Exposure to disease spillover from bats also differs between the two ape species as a function of three major differences in their diets. The first involves the rate at which they consume the fruit of trees from the genus *Ficus*. Rates of *Ficus* consumption are probably good indicators of spillover rates because the three bat species implicated as reservoirs for Ebola are obligate frugivores which tend to specialize on small seeded fruit, particularly *Ficus*. For example, the

Table 8.1. *Minimum convex polygon estimates of Apollo group home range size*

Year	MCP area	Obs days
1996	6.05	64
1997	5.99	68
1998	5.31	128
2000	5.84	37
2002	6.17	86
Mean	5.87	

MCP = minimum convex polygon.
Observation days = observation days per year.

only systematic study of *Hypsygnathus monstrosus*, the largest of the putative Ebola reservoirs, was conducted about 200 km west of our study site at Lossi and showed that *Ficus* seeds were present in 85% of *H. monstrosus* dung (Bradbury, 1977).

Ficus fruit are a prominent component of chimpanzee diet, with *Ficus* seeds appearing in 50% or more of chimpanzee dung piles (Morgan & Sanz, 2007). In fact, feeding at a *Ficus* tree was one of the risk factors observed during an Ebola outbreak amongst habituated chimpanzees in Cote d'Ivoire (Formenty *et al.*, 1999). In contrast, during 440 days of all-day follows at Lossi, gorillas fed on *Ficus* trees on only 38 days or one out of every 11.6 days. If we interpret the 50% of chimpanzee dung piles containing *Ficus* seeds to mean that chimpanzees eat *Ficus* every other day, these results imply a nearly six-fold greater rate of chimpanzee feeding overlap with bats at *Ficus* trees than experienced by Lossi gorillas. This difference may stem from a specific preference of chimps for *Ficus*, or it may, in part, reflect the fact that chimpanzee community territories typically cover a larger area than gorilla home ranges and, therefore, contain more *Ficus* trees. For instance, chimpanzee communities typically cover about 20 km^2 (Herbinger *et al.*, 2001), while annual minimum convex polygon estimates of home range size for the Apollo gorilla group averaged only 5.87 km^2 (Table 8.1). The ratio of home range sizes ($\frac{20\,km^2}{5.87\,km^2} = 3.45$) is only about half the ratio of *Ficus* feeding rates ($\frac{1\ event/2\ days}{1\ event/11.6\ days} = 5.8$), suggesting that home range size may explain some, but not all, of the difference in *Ficus* feeding rates.

The second major dietary difference is that gorillas consume substantially more plant vegetative material than chimpanzees. This does not result in a substantive decrease in the diversity of fruit species eaten by gorillas, as sympatric populations of chimpanzees and gorillas show a very high overlap in which species they consume (Tutin & Fernandez, 1993). However, it does result in a reduction in the number of different trees visited per day. All-day follows showed that the Apollo gorilla group fed on fruit from an average of 1.92 trees/day ($n = 199$ days, $SE = 0.06$) and had a mean day path length of only 1.31 km, an estimate comparable to other western gorilla studies (Cipolletta, 2004). This compares to a typical day path length of more than 2 km for chimpanzees (Pontzer & Wrangham, 2006). Thus, even if spillover from bats occurs at fruit trees other than *Ficus*, chimpanzees should have a substantially higher rate of exposure than gorillas.

The third dietary difference between gorillas and chimpanzees is in their propensity to hunt other vertebrates. Chimpanzees do. Gorillas do not. The tendency for chimpanzees to hunt other primates is pertinent to Ebola transmission because, in addition to being reservoir hosts in their own right (Courgnaud *et al.*, 2003), other frugivorous primates may be intermediate hosts for bat viruses. In fact, consumption of monkey meat was a second risk factor during the Ebola outbreak amongst chimpanzees in the Cote d'Ivoire outbreak (Formenty *et al.*, 1999).

These differences between gorillas and chimpanzees in social contact structure and spillover risk provide one very clear prediction about the expected impact of Ebola. If Ebola outbreaks are caused by "massive spillover" in which each ape is infected directly from the reservoir host (Leroy *et al.*, 2004, 2005), then chimpanzees should suffer higher rates of mortality than gorillas as a consequence of their greater dietary overlap with bats. On the other hand, if spillover is rare and most apes are infected through subsequent ape-to-ape transmission, then gorillas should suffer higher mortality rates as a consequence of their higher rates of social contact.

Ape mortality at Lossi

One pattern that is immediately obvious in the nest survey data from Lossi is that gorillas were almost totally extirpated from 2700 km^2 of survey zone lying west of 14.55 degrees. Comparisons of nest encounter rates east and west of 14.55 suggest gorilla mortality rates of about 96% (Bermejo *et al.*, 2006). This estimate is consistent with the mortality rate estimated from 14 known social groups (243 individuals) in our primary study area in the Lossi Sanctuary (Bermejo *et al.*, 2006) and the rate estimated for more than 350 known gorillas at Lokoue clearing in nearby Odzala National Park (Caillaud *et al.*, 2006). Although chimpanzees also suffered very high mortality rates over much of the western part of the survey zone, there were regions of chimpanzee survival along the northern and southern borders. Consequently, chimpanzee mortality rates averaged about 86% in the western zone.

Given the greater exposure of chimpanzees to bats, higher rates of chimpanzee survival are not consistent with a "massive spillover" scenario in which most apes are infected directly from the reservoir host. Rather, higher rates of survival in chimpanzees seem more likely a consequence of lower rates of social contact amongst chimpanzees than gorillas. High rates of transmission within gorilla social groups is suggested by observations both at Lokoue (Caillaud *et al.*, 2006), where adult males (silverbacks) suffered higher mortality rates than solitary males and at Lossi, where deaths within groups were not confined to a single event, but spread out over 6 weeks or more (Bermejo *et al.*, 2006). At both sites, all affected groups lost a large proportion of group members. Patterns of chimpanzee mortality at Taï forest, Cote d'Ivoire were also suggestive of chains of transmission involving multiple individuals, but mortality rates were lower than for gorillas: only 28% (12 of 43) of habituated chimpanzees died (Formenty *et al.*, 1999).

The difference between ape species in population impact is not well explained in terms of differences in the virulence of Ebola infection to individual animals, as over large swathes of the western survey zone chimpanzees suffered population declines just as extreme as gorillas. Furthermore, the surviving habituated chimpanzees at Taï showed no symptoms of Ebola infection. Thus, the population impact differences between chimpanzees and gorillas appear to be due to differences in exposure rate rather than differences in infection virulence.

Based only on mean rates, it is difficult to discern whether transmission between social groups contributed to the observed mortality rate difference between Lossi gorillas and chimpanzees. For instance, one can imagine two contrasting scenarios that might produce higher survival in chimpanzees. In the first, spillover rates are high enough to ensure that all social groups are infected, but not high enough to kill all members within each social group. In this case, higher contact rates within gorilla groups would lead to higher gorilla mortality rates. An alternative scenario is that spillover rates are so low that only a small proportion of ape social groups are infected directly from the reservoir. Ebola then spreads laterally amongst ape groups, with gorillas infected at a higher rate because of their greater ranging overlap. Cross-species transmission from gorillas to chimpanzees is even plausible in that chimpanzees have very high rates of ranging and feeding overlap with sympatric gorillas. For instance, a study at Nouabale-Ndoki National Park in northeast Congo found that gorillas and chimpanzees fed simultaneously in the same tree at least once every 15 days (Walsh *et al.*, 2007).

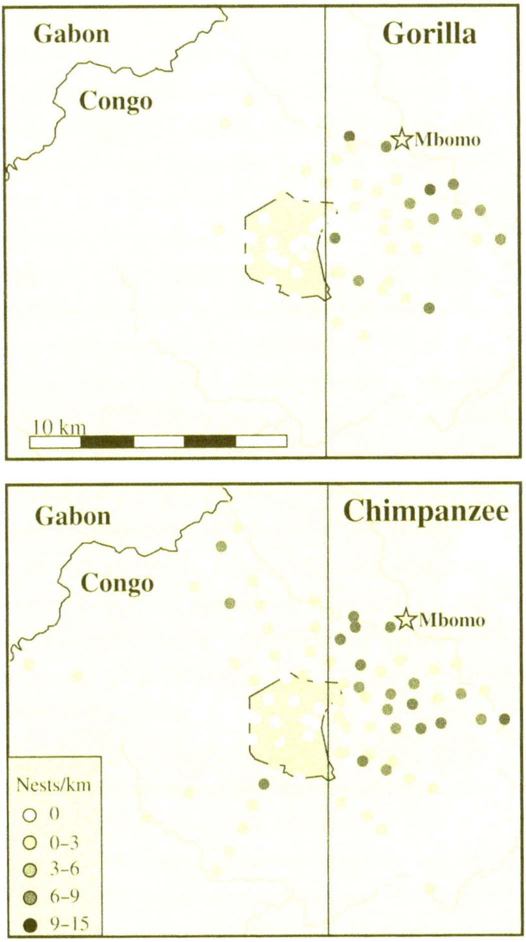

Figure 8.1. Gorilla and chimpanzee nest encounter rates in Lossi survey zone. Each hash mark represents ape nest encounter rate for 5 km segment of reconnaissance survey. Lossi Sanctuary at center and roads (light gray lines) on periphery of survey zone. Region is fully forested except for savanna patches (light areas). Vertical line at 14.55 degrees east longitude.

Although mean mortality rates do not provide information for discriminating between these hypotheses, the spatial pattern of mortality does. In particular, post-Ebola densities of both chimpanzees and gorillas at Lossi showed a strong negative correlation with distance from roads (Figure 8.1, 8.2A). This pattern was likely a consequence of a gradient in hunting intensity, which is evident in an increase in the density of elephants and duiker with increasing distance

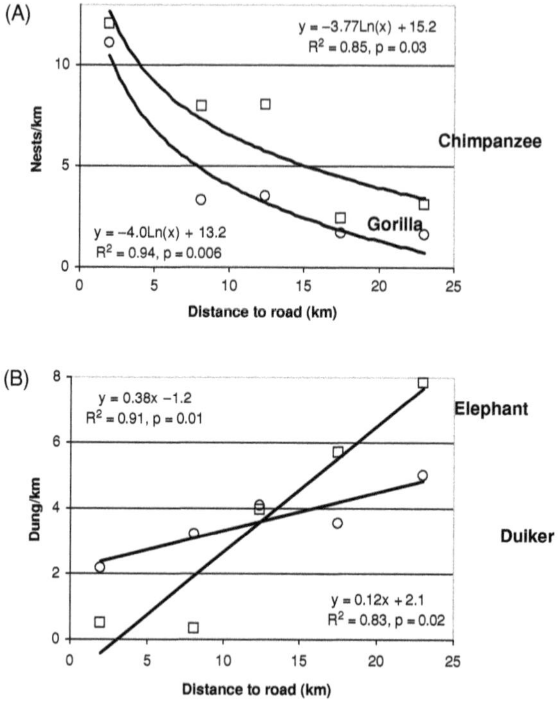

Figure 8.2. Ebola and hunting induced gradients in large mammal density. (A) Decreasing gorilla and chimpanzee nest encounter rates decrease with increasing distance to road suggest density dependent Ebola transmission. (B) Increasing elephant and duiker dung encounter rates increase illustrate impact of hunting. Data are for 137 5-km survey segments pooled into 5-km distance-to-road classes. Linear regression analyses without pooling into distance classes give similar results.

from roads (Figure 8.2B). More apes likely survived near roads because the lower ape densities reduced rates of Ebola transmission.

In principle, the observed gradient in ape mortality might be explained in terms of a gradient in bat density causing a gradient in spillover rates. In practice, this hypothesis is not tenable because larger game have yet to be depleted in the Lossi region (Walsh *et al.*, in review). Although bats are taken opportunistically, hunting rates are not high enough to deplete bat densities over the spatial scale on which the ape mortality gradient was observed, particularly given the resilience to offtake afforded by high bat reproductive rates (Langevin & Barclay, 1990). Consequently, there is no gradient in bat density on a comparable scale as the gradients in gorilla and chimpanzee density.

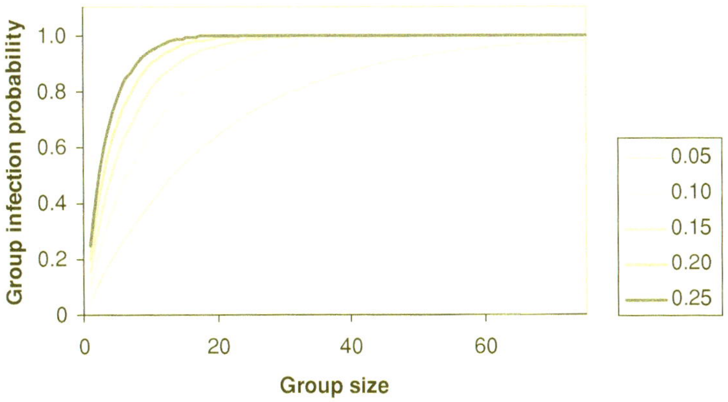

Figure 8.3. Effect of group size on probability that at least one individual in a group is infected by reservoir spillover. Model assumes each individual in a group suffers independent risk of spillover from reservoir host.

Another alternative form of the spillover hypothesis is that hunting induced a pre-Ebola gradient not just in ape densities, but also in ape group sizes. Because the probability that at least one group member is infected by reservoir spillover increases with group size (assuming independent, random infection of individual apes), the proportion of groups in which at least one individual is infected by reservoir spillover increases along the gradient in group size. There are two problems with this hypothesis. The first is that the effect of group size on the probability that at least one individual in a group is infected by reservoir spillover should be highly non-linear (Figure 8.3). This non-linearity makes it hard to maintain between-group heterogeneity in infection probability at both the group sizes typical of gorillas and the community sizes typical of chimpanzees. For example, an individual spillover risk that is high enough to ensure that at least one individual is infected in most larger, but not smaller, gorilla groups will result in infection of virtually all chimpanzee communities, regardless of their sizes. However, lowering the spillover rate to allow heterogeneity amongst chimpanzee communities in the probability that at least one individual is infected will drop the infection probability for large gorilla groups far below the level observed in the high impact zone at Lossi. Keeping spillover probability high but assuming that individuals are infected in clusters (rather than independently) stretches heterogeneity in chimpanzee community infection probability over a wider range of community sizes. However, increasing spillover cluster size makes it more likely that a substantial proportion of larger gorilla groups (i.e. groups far from the road) would escape infection entirely: a result not observed at Lossi. Thus, this version of the spillover hypothesis cannot explain the observed gradients in both gorilla and chimpanzee density.

The second problem with this form of the spillover hypothesis is that our survey data showed no correlation between distance from roads and the size of gorilla social groups. Combined with the observation of strong correlation between gorilla nest density and distance from roads, this lack of a correlation between group size and distance from roads implies that the mechanism underlying the mortality gradient was not a group size dependent gradient in the probability that each group was infected, but rather a gradient in the rate of transmission amongst groups, presumably caused by a gradient in group density.

A further problem with the massive spillover hypothesis is that genetic testing of bats captured at Mbomo, on the edge of our study area, revealed that bats already showed a 22.6% prevalence of Ebola infection in March 2003 (Leroy *et al.*, 2005). This is a problem for the spillover hypothesis because ape densities in the zone immediately surrounding Mbomo showed no sign of Ebola impact despite the high prevalence of infection in sympatric bats. This might be explained away in terms of the lack of permissive spillover conditions (e.g. weather or fruit availability) in March 2003 except for the fact that gorillas were dying 15 km to the southwest at Lossi, both 1 month earlier and 6 months later (Bermejo *et al.*, 2006). There were also ape deaths (and human spillover) 3 months later near Mbandza, 15 km to the north (Leroy *et al.*, 2004). It seems more than slightly implausible that in this extremely flat and floristically homogenous region, weather or fruiting phenology would be unsuitable for spillover in an area covering hundreds of square kilometers, but suitable for massive spillover events in areas covering thousands of square kilometers only 15 km to the north and south.

A more likely explanation for the observed spatial gap in ape mortality is that Ebola spillover from bats is not common enough to infect every single ape social group. Rather, spillover is a relatively rare event, with the great majority of ape deaths resulting from transmission within and between ape social groups. This conclusion is supported by observations on known gorilla groups during the 2003 outbreak at Lossi, where a series of groups showed a sequence of mortality clearly suggestive of transmission between groups, with lags in mortality onset between neighboring groups comparable to the infection cycle length for Ebola (Bermejo *et al.*, 2006). The distance separating neighboring home range centers, about 2 km, was also much smaller than the typical night range distance of *H. monstrosus*: about 8 km (Bradbury, 1977). It seems unlikely that spillover events could follow such a tight spatial progression if the reservoir host was visiting transmission foci (e.g. fig trees) distributed over a much larger scale.

Furthermore, pulses of spillover from bats have been proposed to result from the high viral titers that develop as a consequence of maternal immunosuppression around the time of parturition (Leroy *et al.*, 2005). However, the two annual bat birthing seasons are in July/August and January/February (Bradbury, 1977), while gorilla deaths at Lossi were detected from September through January (Bermejo *et al.*, 2006). Thus, although maternal immunosuppression might be a reasonable explanation for initial spillover into apes, it should not result in sustained spillover for several following months. In this context, it is worth mentioning that viral titers in the body fluids of symptomatic primates are always extremely high as a consequence of the great virulence of Ebola infection in primates (Geisbert *et al.*, 2003). In fact, contact with carcasses found in the forest has been the primary mode of human outbreak initiation in Gabon and Congo, while touching of corpses at funerals is a major mechanism of human outbreak amplification (Roels *et al.*, 1999). Thus, contact with carcasses covered in or surrounded by infective body fluids is one plausible mechanism of transmission within or between gorillas and chimpanzees. Both ape species are curious about conspecific carcasses and will closely inspect and even groom them (Walsh *et al.*, 2007).

Evolutionary implications

Much of the preceding discussion has involved the role of reservoir spillover in Ebola transmission. However, rather than focusing on the minutiae of feeding overlap between chimpanzees and bats it may ultimately be more instructive to cast the more general issue of disease transmission in primates as a body size scaling problem. As a consequence of body size effects on factors such as thermal efficiency, home range size within a given taxonomic group tends to scale roughly with the 3/4; power of body size (Peters, 1983). Thus, the number of fruit trees falling within a given primate's home range should also scale approximately with the 3/4; power of body size. For instance, the home range of a 45 kg chimpanzee should contain roughly seven times more fruit trees than the home range of a 3.5 kg cephus monkey (*Cercopithecus cephus*) ($45^{3/4}/3.5^{3/4} = 6.8$). Now, each fruit tree is a potential focal point for contact with infective body fluids, deposited by conspecifics but also by other primates and by other frugivorous reservoir hosts such as bats, rodents, or birds. Consequently, to the extent that infection risk is linearly proportional to the number of focal points visited, each chimpanzee should suffer an infection risk that is roughly seven times higher than each cephus monkey.

Of course infection risk will tend to vary with factors other than the number of focal points in a given primate's home range. But the point is that ranging over larger scales inherently increases infection risk. Thus, the strength of selection on behavioral traits that limit potential for disease transmission from conspecifics should be stronger for large-bodied chimpanzees than for smaller monkeys. In principle, larger body size should make such selection even stronger for gorillas. However, in practice, the greater rate at which gorillas exploit more uniformly distributed, non-fruit resources (another body size effect) appears to have resulted in smaller day range and home range sizes for gorillas. Thus, the selection pressure for behavioral mechanisms to avoid disease transmission from conspecifics may have been weaker than would be predicted by their body size.

Although our observations at Lossi illustrate how massive a selection pressure disease can be and how social structure can influence disease transmission dynamics, they are clearly very far from "proving" that the territoriality and fission-fusion structure of chimpanzees evolved specifically to minimize disease transmission. They do, however, merit a more critical attitude towards the assumption that primate social structure is driven entirely by feeding and mate competition. The challenge ahead will be to devise empirical tests that discriminate between competition and disease avoidance as evolutionary drivers of social structure, as both hypotheses make similar predictions. For instance, under the "Resource Defensibility" hypothesis a correlation between territoriality and the ratio of day range length to home range size is interpreted as evidence that resource defense is the primary modulator of territoriality (Mitani & Rodman, 1979). However, such a correlation is equally consistent with the hypothesis of territoriality as a means of disease avoidance in that species that "trapline" between high yield fruit trees should have longer day ranges and higher disease exposure risk than species that exploit more uniformly distributed resources.

Finally, we suspect that the answer to the question "competition or disease?" will ultimately prove to be "both." We also suspect that resolving the relative importance of competition and disease avoidance in structuring primate social systems will require an expansion beyond the traditional primatological focus on the behavior of individuals and their response to ecological conditions. The lesson of apes and Ebola is that interactions between disease and social structure can only be fully understood by integrating behavior at the individual and social group levels with constraints and pressures imposed by processes at the population and community levels.

Acknowledgements

We thank our field assistants for their personal investment; the European Union (EU) ECOFAC program, the Congolese Ministry of the Environment, and

T. Smith for facilitation; C. Aveling, J. Nadal, and D. Vinyoles for advice, support, and encouragement; and Energy Africa Oil Company, the EU Espèces-Phares program, and the University of Barcelona (Socrates Project) for funding.

References

Bermejo, M. (2004). Home-range use and intergroup encounters in western gorillas (*Gorilla gorilla gorilla*) at Lossi Forest, North Congo. *American Journal of Primatology*, **64**, 223–232.

Bermejo, M., Rodríguez-Teijeiro, J. D., Illera, G. *et al.* (2006). Ebola outbreak kills 5000 gorillas. *Science*, **314**, 1564.

Bradbury, J. W. (1977). Lek mating behavior in the hammer-headed bat. *Zeitschrift für Tierpsychologie*, **45**, 225–255.

Caillaud, D., Levréro, F., Cristescu, R. *et al.* (2006). Gorilla susceptibility to Ebola virus: the cost of sociality. *Current Biology*, **16**, 489–491.

Chapman, C. A., Wrangham, R. & Chapman, L. J. (1995). Ecological constraints on group size: an analysis of spider monkey and chimpanzee subgroups. *Behavioural Ecology and Sociobiology*, **36**, 59–70.

Cipolletta, C. (2004). Effects of group dynamics and diet on the ranging patterns of a western gorilla group (*Gorilla gorilla gorilla*) at Bai Hokou, Central African Republic. *American Journal of Primatology*, **64**, 193–205.

Connell, J. H. (1975). Some mechanisms producing structure in natural communities: a model and evidence from field experiments. In *Ecology and Evolution of Communities*, ed. M. L. Cody& J. Diamond. Cambridge, MA: Belknap Press, pp. 460–490.

Courgnaud, C., Abela, B. & Pourrut, X. (2003). Identification of a new simian immunodeficiency virus lineage with a vpu gene present among different *Cercopithecus* monkeys (*C. mona, C. cephus*, and *C. nictitans*) from Cameroon. *Journal of Virology*, **77**, 12523–12534.

Doran-Sheehy, D. M., Greer, D., Mongo, P. & Schwindt, D. (2004). Impact of ecological and social factors on ranging in western gorillas. *American Journal of Primatology*, **64**, 207–222.

Eisenberg, J. F., Muckenhirn, N. A. & Rudran, R. (1972). The relation between ecology and social structure in Primates. *Science*, **176**, 863–874.

Formenty, P., Boesch, C., Wyers, M. *et al.* (1999). Ebola virus outbreak among wild chimpanzees living in a rain forest of Côte d'Ivoire. *Journal of Infectious Diseases*, **179** (Suppl. 1), S120–S126.

Freeland, W. J. (1976). Pathogens and the evolution of primate sociality. *Biotropica*, **8**, 12–24.

Geisbert, T. W., Hensley, L. E., Larsen, T. *et al.* (2003). Pathogenesis of Ebola hemorrhagic fever in cynomolgus macaques – evidence that dendritic cells are early and sustained targets of infection. *American Journal of Pathology*, **163**, 2347–2370.

Gurevitch, J., Morrison, J. A. & Hedges, L. V. (2000). The interaction between competition and predation: a meta-analysis of field experiments. *American Naturalist*, **155**, 435–453.

Herbinger, I., Boesch, C. & Rothe, H. (2001). Territory characteristics among three neighboring chimpanzee communities in the Taï National Park, Cote d'Ivoire. *International Journal of Primatology*, **22**, 143–167.

Huijbregts, B., De Wachter, P., Obiang, L. S. N. & Akou, M. E. (2003). Ebola and the decline of gorilla *Gorilla gorilla* and chimpanzee *Pan troglodytes* populations in Minkebe Forest, north-eastern Gabon. *Oryx*, **37**, 437–443.

IUCN (2007). *IUCN Red List of Threatened Species*. http://www.iucnredlist.org/.

Langevin, P. & Barclay, M. R. (1990). *Hypsignathus monstrosus*. *Mammalian Species*, **357**, 1–4.

Leroy, E. M., Rouquet, P., Formenty, P. *et al.* (2004). Multiple Ebola virus transmission events and rapid decline of central African wildlife. *Science*, **303**, 387–390.

Leroy, E. M., Kumulungui, B., Pourrut, X. *et al.* (2005). Fruit bats as reservoirs of Ebola virus. *Nature*, **438**, 575–576.

Menge, B. A. & Sutherland, J. P. (1976). Species diversity gradients: synthesis of the roles of predation, competition and temporal heterogeneity. *American Naturalist*, **110**, 351–369.

Mitani, J. C. & Rodman, P. S. (1979). Territoriality: the relation of ranging pattern and home range size to defendability, with an analysis of territoriality among primate species. *Behavioural Ecology and Sociobiology*, **5**, 241–251.

Morgan, D. & Sanz, C. (2007). Chimpanzee feeding ecology and comparisons with sympatric gorillas in the Goualougo Triangle, Republic of Congo. In *Feeding Ecology of Apes and Other Primates*, ed. G. Hohmann, M. M. Robbins & C. Boesch. Cambridge: Cambridge University Press, pp. 95–120.

Nunn, C. L. & Dokey, A. T.-W. (2006). Ranging patterns and parasitism in primates. *Biology Letters*, **2**, 351–354.

Robbins, M. M., Bermejo, M., Cipolletta, C. *et al.* (2004). Social structure and life-history patterns in western gorillas (*Gorilla gorilla gorilla*). *American Journal of Primatology*, **64**, 145–159.

Peters, R. H. (1983). *The Ecological Implications of Body Size*. Cambridge: Cambridge University Press.

Pontzer, H. & Wrangham, R. W. (2006). Ontogeny of ranging in wild chimpanzees. *International Journal of Primatology*, **27**, 295–309.

Roels, T. H., Bloom, A. S., Buffington, J. *et al.* (1999). Ebola hemorrhagic fever, Kikwit, Democratic Republic of the Congo, 1995: risk factors for patients without a reported exposure. *Journal of Infectious Diseases*, **179**, S92–S97.

Sih, A., Crowley, P., McPeek, M., Petranka, J. & Strohmeier, K. (1985). Predation, competition, and prey communities: a review of field experiments. *Annual Review Ecology and Systematics*, **16**, 269–311.

Terborgh, J. & Janson, C. H. (1986). The socioecology of primate groups. *Annual Review Ecology and Systematics*, **17**, 111–135.

Tutin, C. E. G. & Fernandez, M. (1993). Composition of the diet of chimpanzees and comparisons with that of sympatric lowland gorillas in the Lopé Reserve, Gabon. *American Journal of Primatology*, **30**, 195–211.

van Hooff, J.A.R.A.M. (1983). On the ultimate causes of primate social systems. *Behaviour*, **85**, 91–117.

Walsh, P. D., Abernethy, K. A., Bermejo, M. *et al.* (2003). Catastrophic ape decline in western equatorial Africa. *Nature*, **422**, 611–622.

Walsh, P. D., Breuer, T., Sanz, C., Morgan, D. & Doran-Sheehy, D. (2007). Potential for Ebola transmission between gorilla and chimpanzee social groups. *American Naturalist*, **169**, 684–689.

Wrangham, R. W. (2000). Why are male chimpanzees more gregarious than mothers? A scramble competition hypothesis. In *The Socioecology of Primate Males*, ed., P. K. Kappeler. Cambridge: Cambridge University Press, pp. 248–258.

9 *Primate–parasitic zoonoses and anthropozoonoses: a literature review*

TARANJIT KAUR AND JATINDER SINGH

Photograph by Jatinder Singh (*Pan troglodytes schweinfurthii*)

Introduction

Parasites by definition are organisms that obtain some advantage from living within another organism or by living upon another organism, and in general, exist at the expense of their host. They inhabit four general anatomic locations in their hosts: (1) within body tissues (i.e. lumen of the lungs, blood vessels, and intestines or in the subcutis), (2) within body cells (i.e. blood, skeletal and cardiac muscle, brain, kidney, spleen, or liver cells), (3) within body cavities

Primate Parasite Ecology. The Dynamics and Study of Host–Parasite Relationships, ed. Michael A. Huffman and Colin A. Chapman. Published by Cambridge University Press.

(i.e. thoracic, abdominal, or peritoneal), and (4) on the body surface (Dorlands Illustrated Medical Dictionary, 2000). Parasite infections (internal) or infestations (external) may cause a variety of disease manifestations in the host. For example, in primates, physiological disturbances, nutritional loss, pathologic lesions leading to serious debilitation, secondary infections in already compromised hosts, and sometimes death have been reported (Ratcliffe, 1931; Vickers, 1968; Burrows, 1972; Benirschke & Adams, 1980; Wilson *et al.*, 1984). The type and degree of disturbance is dependent on the type and degree of parasite burden, as well as a number of host factors, such as age, immune status, and underlying physiological condition (Colford *et al.*, 1996; Guerrant, 1997; Schuster & Visvesvara, 2004; Chapman *et al.*, 2006; Park *et al.*, 2007).

Parasites can be transmitted from infected vertebrates to humans (zoonoses), or from infected humans to other vertebrates (anthropozoonoses). Some parasite transmissions can occur in either direction naturally, with infections being maintained in both humans and non-human primates (amphixenoses; non-human primates hereafter called primates). Distinction or determination as to the direction of transmission is typically based on different parameters, such as a parasite's life cycle and infective stages, hosts' health status, history and exposure potential, population densities and dynamics, and associated risk factors. Prevailing environmental conditions, including prevalence of the disease occurrence in the human versus animal populations (e.g. endemic, epidemic, or pandemic vs. enzootic, epizootic, or panzootic) also needs to be taken into consideration.

The risk of parasite transmission is linked to the parasite's life cycle (direct or indirect) and the potential for hosts' exposure under local environmental conditions. Parasites that have a direct life cycle do not require another animal host. However, in some cases it may require development to the infective stage outside of a host in the environment. Transmission can occur when a susceptible host encounters secretions, excretions, exudates, tissues, or aerosols from an infected host (i.e. feces, saliva, urine, blood, milk, skin, or hair). Also, transmission can occur through contact with mechanical vectors (i.e. insects) or abiota (i.e. inanimate surfaces), or even through auto-inoculation via skin sticks from contaminated needles or scalpel blades.

Transmission of parasites that have an indirect life cycle requires biological vectors and intermediate hosts for parasite multiplication and/or development. The parasite's life cycle may require more than one vertebrate host (cyclozoonoses), an invertebrate host prior to transmission to a vertebrate host (metazoonoses), or a non-animal organic host such as food plants or soil (saprozoonoses).

We have organized this chapter into direct and indirect transmission of diseases based on the life cycle of the parasite. We include reports of parasitic transmission in the order Primates, and families Platyrrhini (New World monkeys), Cercopithecoidea (Old World monkeys), Hylobatidae (gibbons and siamangs), and Pongidae (orangutans, gorillas, chimpanzees, and bonobos) in the sub-order Anthropoidea. For the most part, we have excluded those for the sub-order Prosimii, and family Tupaiidae (e.g. lorises, lemurs, and tarsiers). We omitted parasitic diseases where primates have, to date, not been reported as having a significant role in the epidemiology of the human disease, reports of only rare or sporadic infections in primates and/or humans, and parasites which are not known to be pathogenic to humans and thus, to date, have little to no reported public health significance.

Direct transmission

Amebiasis

Amebiasis (synonyms amebic colitis, amebic dysentery) is a serious protozoal disease of the large intestine and is caused by the ameba *Entamoeba histolytica*. *Entamoeba histolytica* has worldwide distribution. Primates and swine are reservoirs for the organism (Schuster & Visvesvara, 2004). The disease can be severe in both humans and primates (Ravdin, 1995). Transmission occurs directly when parasite cysts passed in the feces of an infected host are ingested. This may occur by ingestion of sewage-contaminated water and food, or through direct ingestion of cysts after inadvertently coming in direct contact with infected fecal matter (Schuster & Visvesvara, 2004). Cysts can be passed in the stool of asymptomatic carriers, or those with mild disease. They can remain viable in moist, cool conditions for over 12 days and in water for up to 30 days. Flies and other insects can serve as mechanical vectors, transferring cysts from primate feces to other surfaces with subsequent ingestion by unsuspecting hosts (Lehner, 1984). Person-to-person transmission does occur, and infected humans may transmit the organism to humans and primates, particularly if they are preparing or handling food (Krauss *et al.*, 2003; Schuster & Visvesvara, 2004).

Entamoeba histolytica is reported to be non-pathogenic while living in the intestinal lumen, however, upon invasion of the mucosa, can cause mild or severe disease (Levin, 1970). Reports indicate that pathogenicity can vary depending on a number of different factors such as nutritional status and bacterial co-inhabitants in the host's gastrointestinal tract (McCarrison, 1920; Levin, 1970; Burrows, 1972; Flynn, 1973; Shadduck & Pakes, 1978). Amebic abscesses can form in other organs of the body, particularly in the liver, lungs, or central nervous system, sometimes with fatal consequences (Shadduck &

Pakes, 1978; Lehner, 1984). Fatalities have been reported in a chimpanzee (Fremming *et al.*, 1955), an orangutan (Patten, 1939), a group of spider monkeys (Amyx *et al.*, 1978), douc langurs (Frank, 1982), and several colobus monkeys (Frank, 1982; Loomis & Britt, 1983; Loomis *et al.*, 1983).

Other non-pathogenic amebic organisms (i.e. *E. dispar*, *E. coli*, and *E. hartmanni*) can inhabit the lumen of the gastrointestinal tract and their cysts may be found in stool samples. Microscopic analysis alone may not always be sufficient to differentiate them from *E. histolytica*, unless ingested red blood cells are actually observed, confirming invasion of the mucosa (Schuster & Visvesvara, 2004).

Strict sanitation and personal hygiene practices, including thorough hand washing, prevention of fecal contamination of drinking water, and washing of raw fruits and vegetables with uninfected water prior to ingestion are important measures for preventing transmission (Chin, 2000). Boiling of water to 50 °C (Fox *et al.*, 2002), or chlorine levels of 10 ppm will kill the cysts. Treatment for amebiasis is available and should be administered to those with disease following definitive diagnosis. Because of the serious nature of the disease, asymptomatic carriers should also be treated.

Giardiasis

Giardiasis is a gastrointestinal disease caused by the flagellated protozoa *Giardia lamblia* (synonyms *G. intestinalis, G. duodenalis*). This organism is found throughout the world, and there are reportedly avian, mammalian, and amphibian species. Cysts have been reported in the feces of New World monkeys, Old World monkeys, apes, and humans (Sleeman *et al.*, 2000; Graczyk *et al.*, 2002a; Hope *et al.*, 2004; Vitazkova & Wade, 2006; Naumova *et al.*, 2007; Salzer *et al.*, 2007). Historical classifications were based on host origin; however, the lack of morphologic differences did not allow accurate distinctions to be made between different genetic variants (Appelbee *et al.*, 2005). Molecular-genotyping tools now allow characterization of isolates directly from feces so different *Giardia* spp. are being distinguished and their host occurrence and zoonotic potential can be determined (Thompson, 2000; Thompson *et al.*, 2000; Graczyk *et al.*, 2002a; Thompson & Monis, 2004; Appelbee *et al.*, 2005; Vitazkova & Wade, 2006). Most *Giardia* spp. are host-adapted, however, *G. duodenalis* has six recognized variants (or assemblages) that have different host specificities. Assemblages A and B infect a broad range of mammals, including humans and primates, and are considered zoonotic. Assemblage A has been reported in fecal samples of free-ranging human-habituated mountain gorillas and people sharing gorilla habitats in Uganda (Graczyk *et al.*, 2002a, b).

Although commonly observed inhabitants of the small intestine of several species of primates, they are considered as pathogens (Armstrong & Hertzog, 1979). The presence of clinical signs and pathology varies. The finding of *Giardia* cysts in the stool when diarrhea is present indicates that giardiasis should be considered in the differential diagnosis. However, *Giardia* cannot be the causative agent of the illness without a critical assessment and consideration of other possible causes (Armstrong & Hertzog, 1979).

Transmission occurs when infective cysts shed in the feces are ingested by susceptible hosts. Primates not showing any signs of disease may be infected and shed cysts that can serve as a source of infection to other hosts (Overturf, 1994; Faubert, 1996). Sewage contamination of water can lead to water-borne transmission. Use of human or animal feces for fertilizers of fruits and vegetables can lead to food-borne transmission (Davies & Hibler, 1979; Juranek, 1979; Fox *et al.*, 2002; Krauss *et al.*, 2003). Strict sanitation and personal hygiene practices, including thorough hand washing, prevention of contamination of drinking water, and washing of fruits and vegetables with uninfected water are important measures for preventing transmission. Treatment is available.

Cryptosporidiosis

Cryptosporidiosis is an intestinal disease caused by various *Cryptosporidium* species, a coccidian protozoan parasite that can infect mammals, reptiles, fish, and birds. *Cryptosporidium* species have been reported in the digestive tracts of various primate species, New World monkeys, Old World monkeys, and apes (Cockrell *et al.*, 1974; Wilson *et al.*, 1984; Miller *et al.*, 1990; Nizeyi *et al.*, 1999; Graczyk *et al.*, 2001a; Ekanayake *et al.*, 2006; Salzer *et al.*, 2007). Of the 15 named species of *Cryptosporidium*, *C. parvum* has been reported to be the most widely distributed, having the broadest host range and being the most commonly associated with infections in mammals, with *C. hominis* being most commonly reported in humans and monkeys (Morgan-Ryan *et al.*, 2002; Fayer, 2004; Xiao *et al.*, 2004; Appelbee *et al.*, 2005; Fayer *et al.*, 2006). Severe intestinal disease may occur in infant and juvenile primates with adults being relatively resistant to the disease (Cockrell *et al.*, 1974; Wilson *et al.*, 1984). Although once reported to be an infrequent and non-pathogenic finding, it now is considered an important pathogen in humans, with both children and adults being susceptible to disease (Appelbee *et al.*, 2005; Tumwine *et al.*, 2005). Of particular concern are those with a compromised immune system, such as patients suffering from Human Immuno-deficiency Virus (HIV) and Acquired Immune Deficiency Syndrome (AIDS) (Colford *et al.*, 1996; Agarwal *et al.*, 1998).

Fecal contamination of water with *Cryptosporidium* by infected hosts and subsequent ingestion of infectious oocysts is a major route of transmission. Because infective sporulated oocysts are passed in the stool there is a high potential for immediate infection of susceptible hosts. Even if not immediately ingested, the oocysts are resistant to drying, and remain viable under cool, moist conditions for many months, especially in cool water in lakes, rivers, and ponds; they are also highly resistant to conventional water treatment by chlorination and chemical disinfection (Navin & Juranek, 1984; Miller *et al.*, 1990; Carey *et al.*, 2004; Fayer, 2004). The risk of waterborne transmission of *Cryptosporidium* is a serious global issue in drinking water safety. No antimicrobial agents have proved to be effective for treating the disease. Infection persists until cleared by the immune system of the host. Immune-compromised hosts remain persistently infected (Tumwine *et al.*, 2005; Hughes & Kelly, 2006).

Additional research into improving water treatment and sewage treatment practices, and a timely and efficient means to concentrate, purify, and detect *C. parvum* and *C. hominis* oocysts in environmental samples, while limiting the presence of extraneous materials, is needed (Carey *et al.*, 2004). Molecular-based techniques, particularly a multiplex PCR for the simultaneous detection of *C. parvum*, *C. hominis*, and other waterborne pathogens such as *Giardia lamblia* would greatly benefit the water industry and protect human health (Carey *et al.*, 2004; see Chapman *et al.*, Chapter 21, this volume).

Cyclosporiasis

Cyclosporiasis is an intestinal illness of humans caused by the coccidia, *Cyclospora cayetenensis*. It frequently results in severe watery diarrhea, weight loss, and fatigue. The range of potential primate hosts may possibly be extensive (Auerbach, 1953; Smith *et al.*, 1996; Fox *et al.*, 2002). An identical or similar organism has been reported in chimpanzees from Uganda, and it is considered common in baboons in Tanzania (Smith *et al.*, 1996). The route of transmission is fecal-oral, but the oocysts are unsporulated when passed in the feces of infected hosts. Under appropriate conditions, sporulation will occur within about 10 days, at which time infection can occur if ingested by susceptible hosts. Treatment with antimicrobial agents is effective (Fox *et al.*, 2002).

Balantidiasis

Balantidiasis is a severe, ulcerative disease of the gastrointestinal tract caused by *Balantidium coli*, a large ciliated protozoan. The organism has worldwide distribution and is commonly found in the gastrointestinal tract of New World monkeys, Old World monkeys, and apes showing no symptoms of disease (Nakauchi, 1999; Drevon-Gaillot *et al.*, 2006; Weyher *et al.*, 2006). Although *B. coli* is usually non-pathogenic, diarrhea or dysentery may occur, and in severe cases, blood and mucus may be found in the stool (Chin, 2000). It is the only ciliate known to cause infections in humans and disease is uncommon (Schuster & Visvesvara, 2004; Karanis *et al.*, 2007). Chin (2000) has reported a high natural incidence in humans without symptoms; however, in cases of acute onset of diarrhea, disease from *B. coli* should be considered.

Primates and swine are reservoirs for *B. coli*. Ingestion of fecal matter from poor hygiene, or poor sanitation, contaminated food or sewage-contaminated water leads to infection (Schuster & Visvesvara, 2004). Control measures are similar to those described above for amebiasis. Treatment is available and has been successful in eliminating *B. coli* infections (Teare & Loomis, 1982).

Balamuthiasis

Balamuthiasis is a disease of the central nervous system caused by the organism *Balamuthia mandrillaris*, a free-living soil-dwelling ameba. The disease was first described after the organism was isolated from inflamed brain tissue of a mandrill that died at the San Diego Wild Animal Park (Visvesvara *et al.*, 1990, 1993). Infections have since been identified in a white-cheeked gibbon, two western lowland gorillas, and a kibuya colobus monkey (Rideout *et al.*, 1997). At least 60 human cases of amebic meningoencephalitis (inflammation of brain tissues and the membranes surrounding tissues of the central nervous system) from the same organism have been reported (Martínez *et al.*, 1994). Although much remains unknown about the life cycle of *B. mandrillaris*, the potential for transmission from infected animals to humans exists when performing post-mortem examinations on infected animals. The disease is often fatal and no effective treatment has yet been identified (Krauss *et al.*, 2003).

Microsporidiosis

Microsporidiosis is caused by organisms that are obligate intracellular parasites. *Encephalitozoon cuniculi*, *E. bienusi*, and other microsporidia have been reported as causes of the disease. Microsporidia species may infect the brain, kidneys, heart, lungs, adrenals, intestine, liver, gall bladder, or other tissues. Although reported in a wide variety of species, natural infections in primates have been reported only in squirrel monkeys, and there is one reported case in a dusky titi monkey (Anver *et al.*, 1972; Brown *et al.*, 1972; Seibold & Fussell,

1973; Canning, 1977; Shadduck & Pakes, 1978; Zeman & Baskin, 1985). If clinical signs occur prior to death in monkeys, they are usually non-specific (Zeman & Baskin, 1985).

Infection with *E. cuniculi* can occur when spores contaminating the environment or food are ingested. Transmission also occurs when handling primates and excrement, particularly urine (Flynn, 1973; Zeman & Baskin, 1985). A single genotype of *E. intestinalis* has been reported to infect free-ranging gorillas and people sharing their habitats in Uganda (Graczyk *et al.*, 2002b). Microsporidiosis has been reported as an emerging and opportunistic infection in immune-suppressed or immune-competent individuals (Chacin-Bornilla, 2006). No treatment is known to be effective (Lehner, 1984).

Ascariasis

Ascariasis is a disease caused by ascarid nematodes, also often called roundworms. Ascarid eggs in feces and adult roundworms in the intestinal tract are commonly reported in primates, New World monkeys, Old World monkeys, and apes. Occasionally larvae have been reported in body tissues. In general, ascarids are considered an incidental finding, although fatalities have been reported in monkeys and apes (Pillers, 1924; Stam, 1960; Hayama & Nigi, 1963; McClure & Guilloud, 1971; Orihel & Seibold, 1972). Ascarids found in primates have been reported as being indistinguishable from the human roundworm, *Ascaris lumbricoides* (Thornton, 1924; Augustine, 1939; Dunn & Greer, 1962; Yamashita, 1963; Orihel & Seibold, 1972; although see Gasser *et al.*, Chapter 3, this volume for the difficulty of positively identifying transmission). Although cross-infection has not yet been documented, ascariasis is considered a potentially transmissible infection between primate species.

Trichuriasis

Trichuriasis is a disease caused by whipworms, which are common nematode parasites of the cecum and large intestine of primates (Muriuki *et al.*, 1998; Murray *et al.*, 2000; Sleeman *et al.*, 2000; Michaud *et al.*, 2003; Mengistu & Erko, 2004). Infection is based on finding characteristic eggs in the feces or adult whipworms in the cecum. Infections have been reported in New World monkeys, Old World monkeys, and apes. Usually infections are of no significant clinical consequence, although heavy infections may cause disease and sometimes even death (Ruch, 1959; Graham, 1960; Thienpont *et al.*, 1962; Flynn, 1973). Whipworms found in primates are reported to be indistinguishable from whipworms infecting humans, and thus cross-infection from animals to humans is possible and experimental transmission from monkeys to humans has been reported (Ruch, 1959; McClure & Guilloud, 1971; Flynn, 1973; Horii & Usui, 1985).

Strongyloidiasis

Strongyloidiasis is a disease that results from infection with species of the nematode genus *Strongyloides*. They are most prevalent in tropical and subtropical regions of the world. Infections have been reported in New World monkeys (*S. cebus)*, Old World monkeys, apes, and humans (*S. fulleborni* and *S. stercoralis).* Although there are a wide variety of clinical signs, the most common clinical sign is diarrhea, which may be hemorrhagic or mucoid (De Paoli & Johnsen, 1978). The most common lesions reported consist of catarrhal to hemorrhagic or necrotizing enterocolitis with or without secondary peritonitis. Outside the digestive tract, pulmonary hemorrhage is the most common lesion. Reports of fatal cases exist in chimpanzees, gibbons, orangutans, patas monkeys, and wooly monkeys (Pillers & Southwell, 1929; McClure *et al.*, 1973; De Paoli & Johnsen, 1978; Benirschke & Adams, 1980; Penner, 1981; Harper *et al.*, 1982).

Strongyloides has a unique and complex dual life cycle, with distinctive filariform infective larvae being produced in two different ways. *Strongyloides* first stage rhabditiform larvae are passed in the feces, and within 48 hours, rapidly develop into filariform third stage infective larvae without passing through a free-living adult generation in the soil. Through an indirect life cycle, first stage rhabditiform larvae undergo at least one cycle or a succession of free-living generations in the soil before adults produce third stage infective filariform larvae (Flynn, 1973). In both cases, the third stage infective filariform larvae enter the body of the host by penetrating the skin or oral mucosa. They find their way to the bloodstream, and migrate via the blood to the heart and lungs, molt, and pass through the airways into the mouth. They are swallowed and molt finally into a parthenogenetic adult female worm in hosts' small intestine.

Particularly noteworthy is *S. stercoralis*, which is an important parasite in humans. Fatal strongyloidosis has occurred in humans undergoing corticosteroid therapy with an unknown and therefore untreated *S. stercoralis* infection (Suvajdzic *et al.*, 1999; Mora *et al.*, 2006; Guyomard *et al.*, 2007). Although considered a transitory infection or one that will die out early in the process, *S. stercoralis* can occur in chimpanzees and other apes (Murata *et al.*, 2002; Gillespie *et al.*, 2005).

Strongyloides fulleborni is also zoonotic and has been reported in humans in sub-Saharan Africa; an *S. fulleborni*-like parasite has been found in humans in Papua New Guinea even though primates have apparently never inhabited New Guinea (Kelly *et al.*, 1976; Hira & Patel, 1977, 1980; Muriuki *et al.*, 1998). Generally, finding *Strongyloides* larvae in fresh fecal samples suggests infection with *S. stercoralis*, whereas finding eggs in fresh feces suggests infection with *S. fulleborni*. Since the rhabditiform first-stage larvae are morphologically indistinguishable from free-living nematodes and hookworm larvae, stool culture to the free-living adult stage should be used to confirm diagnosis (Mori *et al.*, 1998; Gillespie *et al.*, 2004, 2005).

Strongyloides stercoralis is capable of re-invading its host by autoinfection, whereby adult worms in the hosts' intestinal tract produce first-stage larvae which rapidly molt therein into third-stage infective larvae and directly penetrate the wall of the intestine from its lumen or pass through the intestinal tract, out the anus, and then re-invade the host by penetrating through the perineal or perianal skin. In both cases, the third-stage larvae access the bloodstream of the host and migrate to the liver, heart, and lung, and move through the airways to the mouth, and mature in the host's intestine. Thus, infections can be sustained, manifesting in clinical disease with severe damage to bodily organs and subsequent death. Hyperinfection can occur when many larvae re-invade the host, as may be the case with immunosuppressed individuals. Strongyloidiasis is another important reason why special care must be taken when handling excrement from primates. Treatment is available, but sanitation is important to prevent re-infection (Renquist & Whitney, 1987).

Oxyuriasis

Oxyuriasis is caused by pinworms, parasitic nematodes that inhabit the colon and cecum. These nematodes are geographically widely distributed. Pinworms are the most common helminth infection of humans in the USA and Western Europe (Burkhart & Burkhart, 2005). The parasite has a direct life cycle and eggs are deposited by the adult female in the perianal or perineal area. They contain fully developed larvae in 6 hours or less. Infections in New World monkeys (*Trypanoxyuris* and *Oxyuronema* species), and Old World monkeys, and apes (*Enterobius vermicularis, Enterobius anthropopitheci*, and other *Enterobius* spp.) have been reported. Clinical signs of anal pruritis and irritation are reported in cases of oxyuriasis which may lead to self-mutilation, restlessness, and increased aggressiveness (Ruch, 1959; Christensen, 1964; Das, 1965; McClure & Guilloud, 1971; Orihel & Seibold, 1972; Flynn, 1973; King, 1976; Schmidt, 1978; Shadduck & Pakes, 1978). With coprophagy in chimpanzees, re-infection is a common occurrence. Transmission between humans and primates occurs with transmission being reported in both directions. Enterobiasis has been reported as fatal in chimpanzees with reports of extensive enterocolitis, peritonitis, and necrogranulomatous lymphadenitis of the mesenteric lymph nodes (Schmidt & Prine, 1970; Keeling & McClure, 1974; Schmidt, 1978; Holmes *et al.*, 1980; Murata *et al.*, 2002). Treatment is available and effective (Nakano *et al.*, 2003).

Oesophagostomiasis

Oesophagostomiasis is a disease caused by the nodular worm, *Oesophasostomum* spp. This is the most common nematode infection reported in Old World monkeys and apes (Visvesvara *et al.*, 1990; Huffman *et al.*, 1997; Ashford *et al.*, 2000; Sleeman *et al.*, 2000; Krief *et al.*, 2005; van Lieshout *et al.*, 2005; Huffman *et al.*, Chapter 16, this volume). It is rarely found in New World monkeys (Yamashita, 1963; Dunn, 1968; Orihel & Seibold, 1972). They are fairly universal in geographic distribution, and have a direct life cycle with a minimum of approximately 1 week for eggs passed in the feces to become infective. Eggs of *Oesophagostomum* spp. are difficult to differentiate from one another, but they can be differentiated from the eggs of *Ternidens* and hookworm species by experienced individuals based on size, under the microscope. Definitive species identification and diagnosis should always be based on stool culture and positive identification of larvae. If adults are passed in the feces, they can be identified to species (Flynn, 1973; Shadduck & Pakes, 1978; Huffman *et al.*, 1996). Although asymptomatic and light infections are usually reported, heavy infections led to weight loss and diarrhea with debilitation and mortality (Shadduck & Pakes, 1978; Kalter, 1989). Lesions are reported in various other organs (Orihel & Seibold, 1972; Shadduck & Pakes, 1978). Typically nodular lesions are firm, elevated, and about 2–4 mm in diameter, and are found on the serosal surface of the large intestine and cecum and their mesentery (Ruch, 1959; Flynn, 1973; Shadduck & Pakes, 1978). In young nodules worms may be alive, however, more typically the parasite is dead and surrounded by caseous debris, and in older nodules, foci of mineralization (Habermann & Williams, 1957; Flynn, 1973; Shadduck & Pakes, 1978). *Oesophagostomum* spp. infect humans (Ziem *et al.*, 2004). Treatment is effective (Ziem *et al.*, 2004).

Ternideniasis

Ternideniasis is caused by *Ternidens deminutus*, a parasite that inhabits the cecum and colon. It is a strongyle related to oesophagostomes and hookworms, and has been reported in Old World monkeys and apes (Habermann & Williams, 1957; Ruch, 1959; Sasa *et al.*, 1962; Tanaka *et al.*, 1962; Nelson, 1965; Reardon & Rininger, 1968; Orihel & Seibold, 1972; Flynn, 1973; Kalter, 1989). Adult worms and eggs are morphologically similar to *Oesophagostomum*, and have a similar life cycle (Ruch, 1959; Tanaka *et al.*, 1962; Flynn, 1973). Blood loss, even anemia, and cystic nodules in the intestinal wall can occur from extensive mucosal damage when worms move around in the intestinal lumen. *Ternidens deminutus* does infect humans, with treatment being the same as that for *Oesophagostomum* spp.

Ancyclostomiasis and necatoriasis

Ancyclostomiasis and necatoriasis are diseases caused by hookworms usually found in humans, *Ancylostoma duodenale* and *Necator americanus.* They are occasionally reported in Old World monkeys and apes and rarely reported in New World monkeys (Murray *et al.*, 2000; Michaud *et al.*, 2003). They have a direct life cycle and infective larvae develop after about 1 week. Hookworms attach to the intestinal mucosa and heavy infections produce the classic clinical profile of anemia, eosinophilia, pot-bellied appearance, dyspnea on exertion, and general debilitation (Hamerton, 1933, 1942; Ruch, 1959; Flynn, 1973; Pawlowski *et al.*, 1991). Since eggs cannot be morphologically differentiated from other strongyle species, definitive diagnosis should always be based on stool culture and positive identification of larvae or using PCR with specific primers (Shadduck & Pakes, 1978; Pawlowski *et al.*, 1991; Verweij *et al.*, 2001; de Gruijter *et al.*, 2005). Treatment is effective.

Flea infestation

Flea infestation by the stick-tight, jigger or chigoe flea, *Tunga penetrans*, has been reported in Old World monkeys and apes. Humans are often infested in endemic areas. *Tunga penetras* invade the hard skin over the ischial tuberosities. Female fleas become firmly attached and penetrate the epidermis, which then proliferates around the flea. This can cause severe irritation and pruritis with the potential for secondary bacterial infection after flea removal (Renquist & Whitney, 1987). Appropriate precautions should be taken under natural conditions in endemic areas; and if infested, new arrivals into a captive environment should be treated with surgery and sterilization of the wound (Fiennes, 1972; Flynn, 1973).

Dermal myiasis

Dermal myiasis is a disease caused by larvae (bots) of several species of flies. New World monkeys are reportedly a natural host for *Cuterebra* spp. Infestation primarily occurs in the cervical region and parasites produce dermal cysts or swellings, with inflammation and sometimes exudate from the lesion. After the larvae emerge lesions heal. Secondary bacterial infections can occur and may be severe. Fly larvae do affect humans living in endemic areas, but insect control for primates living under captive conditions will prevent them from being a direct public health hazard (Flynn, 1973).

Tick infestation

Tick infestation is a public health concern since ticks can serve as vectors of zoonotic diseases (Williams *et al.*, 2002). Thus, primates with tick infestation should be handled with caution. Manual removal of ticks from primates should be done carefully to prevent contamination of open wounds or mucous membranes with any blood from crushed ticks (Flynn, 1973).

Cutaneous acariasis

Cutaneous acariasis is caused by *Sarcoptes scabiei*, the human itch mite. The disease has been reported in Old World monkeys and apes. A closely related species has been reported in Old World and New World monkeys in captivity (Fain, 1963; Flynn, 1973). Signs associated with primate infestation include intense pruritis, anorexia, weakness, weight loss, tremors, and emaciation. Thickening and scaling of the skin and severe alopecia have been reported in juvenile human-habituated mountain gorillas (Graczyk *et al.*, 2001b). Itching can lead to self-mutilation with secondary hemorrhage and suppurative bacterial dermatitis. Death of a chimpanzee and an infant mountain gorilla has been reported from severe *S. scabiei* infestion (Pillers, 1921; Kalema-Zikusoka *et al.*, 2002). *Sarcoptes scabiei* are transmissible by direct contact so precautions should be taken to prevent infestations between primates.

Dinobdellaiasis

Dinobdellaiasis is caused by the leech, *Dinobdella ferox*, which is found throughout southern Asia. It is a frequent finding in the nasal cavities of macaques. *Dinobdella ferox* lays eggs in cocoons near ponds. When the eggs hatch, immature leeches go to the surface of the water and infection typically occurs while the host is drinking water. The leech enters the body through oral and nasal cavities, sucks blood, matures, and detaches, dropping off through the nostrils. This leech does infect humans and presents some public health concern so proper precautions should be taken when removing leeches from primates (Flynn, 1973).

Indirect transmission

Toxoplasmosis

Toxoplasmosis is a systemic disease caused by the coccidian protozoa, *Toxoplasma gondii*. It occurs in humans and most other mammals and even birds that serve as the intermediate hosts, with extra-intestinal invasion causing systemic disease. Domestic and wild cats are the definitive hosts and the primary reservoir with shedding of oocysts in their feces. Nearly one-third of all adult humans in the USA and Europe have been exposed to *T. gondii*. It is transmitted to susceptible hosts through ingestion (or possibly inhalation) and swallowing of oocysts (Teutsch *et al.*, 1979; Krauss *et al.*, 2003). Transmission can also occur through the consumption of tissue cysts in infected undercooked or uncooked meat (Herwaldt, 2001). Human infections have also been reported after ingestion of oocyst-contaminated water, transplacental transfer, and blood transfusions (Benenson *et al.*, 1982; Shulman, 1994; Wong & Remington, 1994; Neto *et al.*, 2004). Insects can serve as mechanical vectors and spread the organism (Wallace, 1972; Flynn, 1973; Frenkel, 1973, 1974; Duszynski & File, 1974; Shadduck & Pakes, 1978).

Infections with *T. gondii* are very common in people, although disease occurs only sporadically. Under certain circumstances transmission can have severe consequences. For example, when infection occurs during pregnancy, a transplacental transmission can lead to severe neuropathological changes and disease in the nervous system of the developing fetus (Hohlfeld *et al.*, 1989). Also, it is a major concern whenever an immune deficient individual becomes infected with *T. gondii* (Antinori *et al.*, 2004). Reports in the USA show that between 10% and 40% of patients infected with the HIV have antibodies against *T. gondi*, and those with CD4 T-cell counts $< 100/\mu l$ can develop toxoplasmic encephalitis, an inflammation of the brain tissue (Grant *et al.*, 1990; Luft & Remington, 1992). In HIV-infected patients with AIDS, toxoplasmic encephalitis is almost always caused by re-activation of a chronic infection with the incidence of this disease correlating directly with the prevalence of anti-*T. gondii* antibodies. Early studies reported that between 24 and 47% of *T. gondii*-seropositive AIDS patients ultimately developed toxoplasmic encephalitis (Grant *et al.*, 1990; Zangerle *et al.*, 1991; Luft & Remington, 1992). After introduction of primary prophylaxis and effective anti-retroviral therapy, the risk of disease from *T. gondii* infection decreased, and the incidence of toxoplasmic encephalitis declined from 2.1/100 person-years in 1992 to 0.7/100 person-years in 1997 among AIDS patients in the USA (Jones *et al.*, 1999).

Naturally occurring infections have been reported in Old World monkeys, New World monkeys, and apes, as well as Prosimians. New World monkeys are reportedly more susceptible to disease, particularly marmosets that may die within 5–6 days. Marmosets and owl monkeys both uniformly develop acute fatal infections after inoculation by a wide variety of routes. The effectiveness of treatment for primates is not known, and the combined treatment regimes used for human toxoplasmosis have been suggested with pediatric doses being proposed for New World monkeys (Fiorello *et al.*, 2006).

American trypanosomiasis

American trypanosomiasis (synonyms South American trypanosomiasis, Chagas disease) is a disease caused by the flagellated protozoan, *Trypanosoma cruzi*, a pathogen that lives in the bloodstream of the host. Infections occur under natural conditions in humans and in a wide variety of New World primate species, with geographic distribution of the disease being similar to that of the biological vector (reduviid bugs) and the sylvatic host reservoir (rodents). It is endemic in Latin America and a few cases have been reported to occur in the southern USA and California (Krauss *et al.*, 2003). High infection rates are found among poor people in rural areas, slums under poor sanitary conditions, and wild animal refuges in countries in South America (Krauss *et al.*, 2003). The disease has been reported to occur in other primate species under captive conditions in endemic areas in the western hemisphere including the USA. Fetal infection via transplacental transmission and infection of infants via breast milk have also been reported (Krauss *et al.*, 2003).

Trypanosoma cruzi has an indirect life cycle with kissing bugs and other members belonging to the family Reduviidae (cone-nose bugs and assassin bugs) becoming infected after taking a blood meal from an infected host. Following a developmental stage in the gut of the bug, the parasite reproduces and develops into a stage infectious to humans and animals. Primate hosts become infected after a reduviid bug bites, and then defecates before leaving. The infectious trypanosomes are contained in the bug's fecal droplet and enter the primate's blood vessels through the feeding lesion, skin abrasions, or mucous membranes (Krauss *et al.*, 2003). After a short stay in the peripheral blood, the organism travels and invades cells of the internal organs (such as skeletal and cardiac muscle, spleen, liver, and other tissues) where it differentiates into its non-flagellated reproductive form, propagating and forming parasite-filled pseudocysts. Cells rupture and the organisms transform, develop flagella and travel, invading nearby tissue cells and the bloodstream, which carries them to distant tissue cells to initiate further cycles of reproduction (Krauss *et al.*, 2003).

In primates, a wide range of non-specific clinical signs have been reported (i.e. generalized swelling, anemia, and weight loss), with inflammation of heart muscle being the most often noted lesion in naturally occurring cases (Bernacky *et al.*, 2002). In humans, Chagas disease has two phases, the acute phase and the chronic phase. During the acute phase, about 7 days after the bite from an infected bug, a lesion appears at the cutaneous site of entry and within 10–30 days generalized disease occurs. Clinical signs include fever, malaise, enlargement of liver, spleen, and lymph nodes, and swelling (edema) of the face and lower extremities. In some people, especially young children, there is severe inflammation of the heart muscle with congestive heart failure or inflammation of the brain and membranes surrounding the tissue of the central nervous system. During the chronic phase, the infection may remain silent for decades or even for life, or in some cases, serious and even life-threatening cardiac or intestinal disease develops.

There is no effective treatment. People working with New World monkeys under natural conditions or any captive primate in endemic areas in the USA or South America must prevent exposure to the biological vector. Also, extreme caution must be exercised to prevent accidental autoinoculation or exposure of mucous membranes or skin to contaminated primate tissues including blood from infected animals (Baker, 1972; Flynn, 1973). Interventions, such as insect vector control and improvement of housing and living conditions in rural areas where the disease is endemic, would help reduce transmission (Faust *et al.*, 1968; Flynn, 1973; Lehner, 1984; Krauss *et al.*, 2003).

African trypanosomiasis

African trypanosomiasis (synonyms sleeping sickness, *maladie du sommeil*) is caused by two different organisms that are morphologically identical, *Trypanosoma brucei gambiense* and *T. brucei rhodesiense*, flagellated protozoa that live in the bloodstream of the host. Disease occurrence coincides with geographic distribution of different species of the tsetse fly vector (family Glossinidae). For instance, flies of the group *G. palpalis* need high humidity and live in the forest zones of West and Central Africa, and mainly transmit *T. brucei gambiense* infecting primarily humans, but also wild animals (including monkeys). Flies of the *G. morsitans* group inhabit drier savanna areas in East Africa and transmit *T. brucei rhodesiense*, which more commonly affects wild animals (including monkeys), and may infect humans. The parasite develops and multiplies in the gut of the tsetse fly and then migrate to the salivary gland. Transmission occurs when the infected fly bites the host. The parasite multiplies in the blood, tissue fluid, and cerebral spinal fluid of the host.

In humans, an inflamed, tender nodule develops at the site of inoculation days to 2 weeks after the infected tsetse fly bites. Often the lesion has a centrally located pustule extending up to several centimeters in diameter, and heals after 2–3 weeks. The disease progresses in two stages, with Stage 1 being the hemolymphatic stage (infection of the blood and lymphatics) and Stage 2 being the meningoencephalitic stage (infections of the brain and membranes surrounding the tissue of the central nervous system). In animals, no clinical signs are usually observed (Krauss *et al.*, 2003). The drug of choice for treatment depends on the stage of infection and species of *Trypanosoma* involved. Because of toxicity treatment should be administered under the supervision of a physician (Krauss *et al.*, 2003).

Malaria

Malaria is one of the most important human parasitic diseases and occurs in tropical and semi-tropical regions of the world (Di Fiore *et al.*, Chapter 7, this volume). It is caused by various *Plasmodium* species, protozoa which invade red blood cells of the host. Over 172 different species have been named and are reported to infect a wide range of mammals, birds, reptiles, and amphibians (Krauss *et al.*, 2003). Malaria in most primates is not fatal, and in the natural hosts, the organisms do not produce a very severe disease. However, when a *Plasmodium* sp. infects an aberrant host, infection produces severe disease, often causing death.

The parasite's life cycle is indirect with *Anopheles* mosquitoes serving as biological vectors. Four different *Plasmodium* species commonly infect humans: *P. falciparum*, *P. vivax*, *P. ovale*, and *P. malariae*. Humans are the sole natural host for the first three species, and chimpanzees may also be the natural host for *P. malariae* (Krauss *et al.*, 2003). Natural infection with *Plasmodium* spp. has been reported in all primates, except rhesus monkeys, tamarins and marmosets, and owl monkeys (Marinkelle & Grose, 1968; Lehner, 1984). Malaria in Old World monkeys is caused by eight *Plasmodium* spp. with humans being susceptible to at least three of them; malaria in New World monkeys is caused by two *Plasmodium* spp. with humans being susceptible to both of them (Fonseca, 1951; Garnham, 1963, 1967, 1980; Deane *et al.*, 1966; Coatney *et al.*, 1971; Voller, 1972; Flynn, 1973; Loeb *et al.*, 1978); and malaria of Anthropoid apes is caused by five *Plasmodium* spp. with three of them being identical (homologous and morphologically indistinguishable) or very similar to *Plasmodium* spp. that cause malaria in humans.

Although malaria preventative and therapeutic agents are available, insect vector control is the most affordable and practical means of preventing disease. Mosquitoes feed at night, biting primarily at dawn and dusk. Wearing bright clothes with long sleeves and trousers from evening through daybreak are measures recommended to prevent infection.

Filariasis

Filariasis is a commonly reported disease occurring in primates (Old World and New World monkeys, as well as apes). It is caused by many different species of filarial nematodes, which are long, slender worms that inhabit tissues outside the gastrointestinal tract (Webber & Hawking, 1955; Ruch, 1959; Orihel & Seibold, 1972; Flynn, 1973; King, 1976; Levin, 1976; Shadduck & Pakes, 1978). Adult worms can vary in length (depending on species) from a few cm up to 30 cm, with females being much larger. Females produce small primitive larvae that circulate in the blood or live in the skin of the definitive hosts (Webber & Hawking, 1955; Ruch, 1959; Deinhardt *et al.*, 1967; Orihel & Seibold, 1972; Flynn, 1973; King, 1976). The life cycle is indirect, with biting, or blood- or lymph-sucking insects serving as the required intermediate host (King, 1976; Shadduck & Pakes, 1978). *Brugia malayi* has been reported in Asian monkeys, particularly macaques, and in humans (Buckley, 1960; Laing *et al.*, 1960; Orihel & Seibold, 1972). *Meningonema peruzzii* has been reported to infect African Old World monkeys; it has also been reported in humans with marked symptoms of the central nervous system. *Onchocerca volvulus* and *Loa loa* are human parasites reported to infect gorillas, and chimpanzees and gorillas, respectively (Orihel & Seibold, 1972). *Mansonella vanhoffi*, a filariid parasite of the chimpanzee, is very similar if not identical to *M. perstans* found in humans (Toft & Eberhard, 1998).

Schistosomiasis

Schistosomiasis (bilharziasis) is caused by several species of flatworms (also known as trematodes or flukes): *S. mansoni* in New World monkeys, Old World monkeys, and apes; *S. haematobium* in Old World monkeys and apes;

and *S. mattheei* in Old World monkeys (Flynn, 1973; Damian *et al.*, 1976; Sturrock *et al.*, 1976). Of these, most human disease is caused by *S. mansoni* (blood fluke) and *S. hematobium* (bladder fluke). In terms of socioeconomic and public health impact, schistosomiasis is second only to malaria as the most devastating parasitic disease in tropical countries. An estimated 200 million people in 74 countries are infected with the disease – 100 million in Africa alone (King, 2001; Farah *et al.*, 2003).

Trematodes have a complex and unique indirect life cycle that requires a mollusk (e.g. freshwater snail) as an intermediate host; hence, the disease has also been referred to as snail fever. Infected snails release free-swimming larvae which can survive in fresh water up to 48 hours. During this time, attachment to the skin of a susceptible host must occur or they die. Following attachment, larvae migrate through intact skin into dermal veins, travel to the pulmonary vasculature and metamorphose to schistosomula that are highly resistant to the host's immune response. They travel through the systemic circulation to the portal veins where they mature and mate. Adult worms parasitize mesenteric (*S. mansoni*) or vesicular (*S. haematobium*) blood vessels. Eggs are passed through urine or feces to fresh water, where larval stages can infect a new host by penetrating the skin (Flynn, 1973; Damian *et al.*, 1976).

Acute schistosomiasis may occur, however, chronic schistosomiasis is far more common and is due to granuloma formation. The most frequently reported lesion is thickening of the intestinal and urinary bladder walls due to chronic inflammation. Lesions have also been reported in the liver, brain, spleen, and other parts of the urogenital tract (Strong *et al.*, 1961; Kuntz *et al.*, 1979). To prevent the transmission of schistosomiasis avoid bathing and drinking from contaminated streams or irrigation ditches. Also, precautions should be taken when handling or eating crustaceans (i.e. crabs and crayfish) that may serve as second intermediate hosts for some flukes. Also, because schistosomiasis is an important and serious disease in humans, to minimize public health concerns excreta from captive primates or field samples brought to the laboratory should be decontaminated before disposal even though there is a requirement for a mollusk intermediate host (El-Ansary & Al-Daihan, 2006). Although schistosomiasis is not eradicable, the disease can be prevented and transmission controlled with the drug praziquantel (Stelma *et al.*, 1995).

Summary

Cross-species transmission of parasites occurs between all families (New World monkeys, Old World monkeys, apes, and humans) within the order Primates, sub-order Anthropoidea. A variety of parasite, host, and environmental factors contribute to transmission, infection, and disease occurrence. Infections

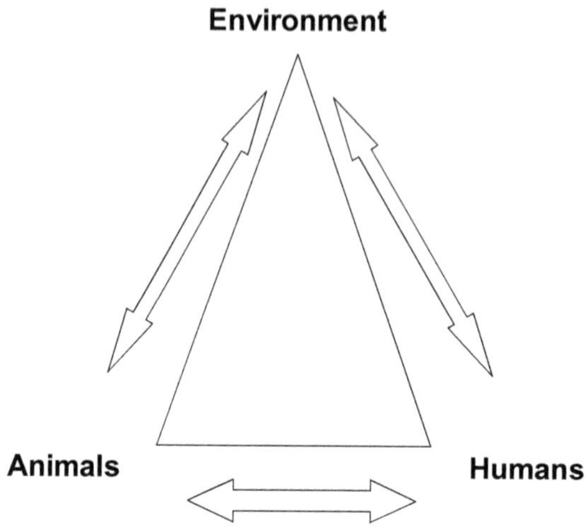

Figure 9.1. The Public Health Triad illustrates the dynamic interplay between humans, animals, and the environment.

may cause physiological disturbances, nutritional loss, and pathologic lesions, including serious debilitation, opportunistic infections, organ dysfunction, and sometimes death. In some cases, primates may be reservoirs and serve as carriers, shedding infective stages of parasites while showing absolutely no clinical signs of infection. Typically humans unknowingly participate in parasite–host cycles and become a link in the transmission cycle. In cases where a primate species is an aberrant host not adapted to a pathogenic organism, infection could lead to high levels of morbidity and mortality. Primates under significant physiological stress or with a compromised immune system are at high risk of severe consequences and even death from infectious agents, even those with low virulence. This is of particular importance during this era of the AIDS pandemic.

The Public Health Triad is an organizing principle used to study dynamic interrelationships between humans, animals, and the environment (Figure 9.1). It provides a useful broad, holistic, and systematic framework to study host–parasite interactions. Change in any one of these may affect another, potentially influencing infectious organisms, altering parasite transmission and distribution, as well as host specificities and degree of pathogenicity. Human activities over the past couple of hundred years have begun to dramatically change the natural order and global distribution of animal and human populations, as well as alter local and global environmental conditions. This is leading to shifts in the global distribution of susceptible hosts, vectors, and infectious agents. The human population explosion, human displacement and migration due to wars and famine with encroachment into remote or previously unpopulated areas, human-habituation of wild primates for ecotourism purposes, deforestation for logging and cultivation with wildlife habitat destruction have all brought humans into wilderness areas, once inhabited only by wild primates. Unknown, previously unreported and new disease transmissions can occur as wildlife and humans compete for precious limited natural resources.

Wildlife have been transported from remote environments into other very different non-native environments because of international travel and trade either by legal or illegal importation of exotic animals for exhibition at zoos, research purposes, or private ownership. Often inspection and quarantine practices are not sufficient for detection and control of infectious agents. Animal organs, fecal samples, and even cultures of animal cells may contain infectious agents. Moreover, infected hosts may not show any clinical signs while harboring an infectious agent, and yet be capable of transmitting them.

Widespread industrialization, the clearing of woodlands, and the burning of fossil fuels are leading to global climate change and wildlife habitat destruction. Humans, animals, and animal products are being transported over long distances and in many cases into vastly differing environments inhabited by new and sometimes even aberrant susceptible hosts. Direct and indirect contact between previously isolated populations of animals and humans is increasing dramatically and under changing environmental conditions. Parasites may be transmitted to new areas of the world, new susceptible hosts, totally naïve hosts, and even aberrant hosts, altering historical patterns of parasite distribution and hence, disease occurrence. Understanding the dynamics and complexities of primate parasite interaction, and recognizing, reporting, preventing, and treating morbidity and mortality requires multidisciplinary teams of professionals, including environmental biologists, epidemiologists, primatologists, wildlife, zoo, and laboratory animal veterinarians, physicians (e.g. private practitioners, as well as occupational health and safety specialists), public health professionals, and infectious disease specialists. Cross-species transmission within the Primate sub-order, Anthropoidea, as well as from other animal species and biological vectors to anthropoid primates is a major public health concern. Ongoing surveillance and reporting of known, uncommon, and new infectious/parasitic agents in aberrant hosts, as well as known hosts, is invaluable for the future of public health. New, sensitive, and advanced methods of detection of microorganisms need to be developed and would undoubtedly contribute toward the recognition of unidentified and newly emerging infectious agents, before becoming a public health hazard.

References

Agarwal, A., Ninqthouja, S., Sharma, D., Mohen, Y. & Singh, N. B. (1998). *Cryptosporidium* and HIV. *Journal of the Indian Medical Association*, **96**, 276–277.

Appelbee, A. J., Thompson, R. C. A. & Olson, M. E. (2005). Giardia and *Cryptosporidium* in mammalian wildlife – current status and future needs. *Trends in Parasitology*, **21**, 370–376.

Amyx, H. L., Asher, D. M., Nash, T. E., Gibbs, C. J. Jr. & Gajdusek, D. C. (1978). Hepatic amebiasis in spider monkeys. *American Journal of Tropical Medicine and Hygiene*, **27**, 888–891.

Antinori, A., Larussa, D., Cingolani, A. *et al.* (2004). Prevalence, associated factors, and prognostic determinants of AIDS-related toxoplasmic encephalitis in the era of advanced highly active antiretroviral therapy. *Clinical Infectious Diseases*, **39**, 1681–1691.

Anver, M. R., King, N. W. & Hunt, R. D. (1972). Congenital encephalitozoonosis in a squirrel monkey (*Saimiri sciureus*). *Veterinary Pathology*, **9**, 475–480.

Armstrong, J. & Hertzog, R. E. (1979). Giardiasis in apes and zoo attendants, Kansas City, Missouri. *Veterinary Public Health Notes*, **1**, 7–8.

Ashford, R. W., Reid, G. D. & Wrangham. R. W. (2000). Intestinal parasites of the chimpanzee *Pan troglodytes* in Kibale Forest, Uganda. *Annals of Tropical Medicine and Parasitology*, **94**, 173–179.

Auerbach, E. (1953). A study of *Balantidium coli* Stein, 1863 in relation to cytology and behavior in culture. *Journal of Morphology*, **93**, 404–445.

Augustine, D. L. (1939). Some observations on some ascarids from a chimpanzee (*Pan troglodytes*) with experimental studies on the susceptibility of monkeys (*Macaca mulatta*) to infection with human pig ascaris. *American Journal of Hygiene*, **30**, 29–33.

Baker, J. R. (1972). Protozoa of tissues and blood (other than the Haemosporina). In *Pathology of Simian Primates*, Part 2, ed. R. N. Fiennes. Basel: Karger, pp. 29–56.

Benenson, M. W., Takafuji, E. T., Lemon, S. M., Greenup, R. L. & Sulzar, A. J. (1982). Oocyst-transmitted toxoplasmosis associated with ingestion of contaminated water. *New England Journal of Medicine*, **307**, 666–669.

Benirschke, K. & Adams, F. D. (1980). Gorilla diseases and causes of death. *Journal of Reproduction and Fertility*, **28** (Suppl.), 139–148.

Bernacky, B. J., Gibson, S. V., Keeling, M. & Abee, C. R. (2002). Nonhuman primates. In *Laboratory Animal Medicine*, 2nd edn, ed. J. G. Fox, L. C. Anderson, F. M. Loew & F. W. Quimby. San Diego, CA: Academic Press, pp. 675–791.

Brown, R. J., Hinkle, D. K., Trevethan, S. P., Kupper, J. L. & McKee, A. E. (1972). Nosematosis in a squirrel monkey (*Saimiri sciureus*). *Journal of Medical Primatology*, **2**, 114–123.

Buckley, J. J. (1960). On Brugia gen. nov. for *Wuchereria* spp. Of the malayi group i.e. *Wuchereria malayi* (Brug, 1927) *Wuchereria pahangi* Buckley and Edeson 1956 and *Wuchereria patei* Buckley, Nelson, Heisch 1958. *Annals of Tropical Medicine and Parasitology*, **54**, 75–77.

Burkhart, C. N. & Burkhart, C. G. (2005). Assessment of frequency, transmission, and genitourinary complications of enterobiasis (pinworms). *International Journal of Dermatology*, **44**, 837–840.

Burrows, R. B. (1972). Protozoa of the intestinal tract. In *Pathology of Simian Primates*, ed. R. N. Fiennes. Basel: Karger, pp. 2–28.

Canning, E. U. (1977). Microsporidea. In *Parasitic Protozoa*, Vol. 4, ed. J. P. Kreier. New York, NY: Academic Press, pp. 155–196.

Carey, C. M., Lee, H. & Trevors, J. T. (2004). Biology, persistence and detection of *Cryptosporidium parvum* and *Cryptosporidium hominis* oocyst. *Water Research*, **38**, 818–862.

Chacin-Bornilla, L. (2006). Microsporidiosis: an emerging and opportunistic infection. *Investigación Clínica*, **47**, 105–107.

Chapman, C. A., Wasserman, M. D., Gillespie, T. R. *et al.* (2006). Do food availability, parasitism, and stress have synergistic effects on Red Colobus populations living in forest fragments? *American Journal of Physical Anthropology*, **131**, 525–534.

Chin, J. (ed.) (2000). *Control of Communicable Diseases Manual*. Washington, DC: American Public Health Association.

Christensen, L. T. (1964). Chimp and owners share worm infestation. *Veterinary Medicine*, **59**, 801–803.

Coatney, G. R., Collins, W. E., Warren M. & Contacos, P. G. (1971). *The Primate Malarias*. Washington, DC: United States Government Printing Office.

Cockrell, B. Y., Valerio, M. G. & Garner, F. M. (1974). Cryptosporidiosis in the intestines of rhesus monkeys (*Macaca mulatta*). *Laboratory Animal Science*, **24**, 881–887.

Colford, J. M., Tager, I. B., Hirozawa, A. W. *et al.* (1996). Cryptosporidiosis among patients infected with human immunodeficiency virus. *American Journal of Epidemiology*, **144**, 807–816.

Damian, R. T., Greene, N. D., Meyer, K. F. *et al.* (1976). *Schistosoma mansoni* in baboons. III. The course and characteristics of infection, with additional observations on immunity. *American Journal of Tropical Medicine and Hygiene*, **25**, 299–306.

Das, K. M. (1965). Discussion. In *Pathology of Laboratory Animals*, ed. W. Ribelin & J. McCoy. Springfield, IL: Charles C. Thomas Publishers, pp. 363–364.

Davies, R. B. & Hibler, C. P. (1979). Animal reservoirs and cross-species transmission of Giardia. In *Waterborne Transmission of Giardiasis*, ed. W. Jakubowski & J. C. Hoff. Springfield, VA: U.S. Environmental Protection Agency, Office of Research and Development, Environmental Research Center, pp. 104–126.

Deane, L. M., Deane, M. P. & Neto, J. F. (1966). Studies on transmission of simian malaria and on a natural infection of man with *Plasmodium simium* in Brazil. *Bulletin of the World Health Organization*, **35**, 805–808.

de Gruijter, J. M., van Lieshout, L., Gasser, R. B. *et al.* (2005). Polymerase chain reaction-based differential diagnosis of *Ancylostoma duodenale* and *Necator americanus* infections in humans in northern Ghana. *Tropical Medicine and International Health*, **10**, 574–580.

Deinhardt, F., Holmes, A. W., Devine, J. & Deinhardt, J. (1967). Marmosets as laboratory animals. IV. The microbiology of laboratory kept marmosets. *Laboratory Animal Care*, **17**, 48–70.

De Paoli, A. & Johnsen, D. O. (1978). Fatal strongyloidiasis in gibbons (*Hylobates lar*). *Veterinary Pathology*, **15**, 31–39.

Dorland's Illustrated Medical Dictionary (2000). Philadelphia, PA: W. B. Saunders Company.

Drevon-Gaillot, E., Perron-Lepage, M., Clément, C. & Burnett, R. (2006). A review of background findings in cynomolgus monkeys (*Macaca fascicularis*) from three different geographical origins. *Experimental and Toxicologic Pathology*, **58**, 77–88.

Dunn, F. L. (1968). The parasites of Saimiri in the context of Platyrrhine parasitism. In *The Squirrel Monkey*, ed. L. A. Rosenblum & R. W. Cooper. New York, NY: Academic Press, pp. 31–68.

Dunn, F. L. & Greer, W. E. (1962). Nematodes resembling *Ascaris lumbricoides* L., 1758, from a Malayan gibbon, *Hylobates agilis* F. Cuvier, 1821. *Journal of Parasitology*, **48**, 150.

Duszynski, D. W. & File, S. K. (1974). Structure of oocyst and excystation of sporozoites of *Isospora endocallimici* n. sp. from marmoset *Callimico goeldii. Transactions of the American Microscopical Society*, **93**, 403–408.

Ekanayake, D. K., Arulkanthan, A., Horadagoda, N. U. *et al.* (2006). Prevalence of *Cryptosporidium* and other enteric parasites among wild non-human primates in Polonnaruwa, Sri Lanka. *American Journal of Tropical Medicine and Hygiene*, **74**, 322–329.

El-Ansary, A. & Al-Daihan, S. (2006). Important aspects of *Biomphalaria* snail schistosome interactions as targets for antischistosome drug. *Medical Science Monitor*, **12**, 282–292.

Fain, A. (1963). Les acarïens producteurs de gale chez les lémuriens et les singes avec une étude des Psoroptidae (Sarcoptiformes). *Bulletin de l'Institut Royal des Sciences Naturelles de Belgique*, **39**, 1–125.

Farah, I., Börjesson, A., Kariuki, T. *et al.* (2003). Morbidity and immune response to natural schisitosomiasis in baboons (*Papio anubis*). *Parasitology Research*, **91**, 344–348.

Faubert, G. M. (1996). The immune response to *Giardia*. *Parasitology Today*, **12**, 140–150.

Faust, E. C., Beaver, P. C. & Jung, R. C. (1968). *Animal Agents and Vectors of Human Disease*, 3rd edn. Philadelphia, PA: Lea & Febiger.

Fayer, R. (2004). *Cryptosporidium*: a water-borne zoonotic parasite. *Veterinary Parasitology*, **126**, 37–56.

Fayer, R., Santín, M., Trout, J. M. *et al.* (2006). Prevalence of Microsporidia, *Cryptosporidium* spp., and *Giardia* spp. in beavers (*Castor canadensis*) in Massachusetts. *Journal of Zoo and Wildlife Medicine*, **37**, 492–497.

Fiennes, R. N. (1972). *Zoonoses of Primates. The Epidemiology and Ecology of Simian Diseases in Relation to Man.* London: Weidenfeld & Nicolson.

Fiorello, C. V., Heard, D. J., Barnes, H. L. & Russell, K. (2006). Medical management of *Toxoplasma* meningitis in a white-throated capuchin (*Cebus capucinus*). *Journal of Zoo and Wildlife Medicine*, **37**, 409–412.

Flynn, R. J. (1973). *Parasites of Laboratory Animals*. Ames, IA: Iowa State University Press.

Fonseca, F. (1951). Plasmódio de primate do Brasil. *Memórias do Instituto Oswaldo Cruz*, **49**, 543–553.

Fox, J. G., Newcomer, C. E. & Rozmiarek, H. (2002). Selected zoonoses. In *Laboratory Animal Medicine*, 2nd edn, ed. J. G. Fox, L. C. Anderson, F. M. Loew & F. W. Quimby. San Diego, CA: Academic Press, pp. 1060–1105.

Frank, H. (1982). Pathology of amebiasis in leaf monkeys (*Colobidae*). *Proceedings of the International Symposium of Diseases of Zoo Animals*, **24**, 321–326.

Fremming, B. D., Vogel, F. S., Benson, R. E. & Young, R. J. (1955). A fatal case of amebiasis with liver abscesses and colitis in a chimpanzee. *Journal of the American Veterinary Medical Association*, **126**, 406–407.

Frenkel, J. K. (1973). Toxoplasmosis: parasite life cycle, pathology and immunology. In *The Coccidia: Eimeria, Toxoplasma, Isospora, and Related Genera*, ed. D. M. Hammond & P. Long. Baltimore, MD: University Park Press, pp. 343–410.

Frenkel, J. K. (1974). Advances in the biology of sporozoa. *Zeitschrift fur Parasitenkunde*, **45**, 125–162.

Garnham, P. C. (1963). Distribution of simian malaria parasites in various hosts. *Journal of Parasitology*, **49**, 905–911.

Garnham, P. C. (1967). Malaria parasites and other haemosporidia. *Transactions of the American Microscopical Society*, **86**, 514.

Garnham, P. C. (1980). Malaria in its various vertebrate hosts. In *Malaria, Epidemiology, Chemotherapy, Morphology, and Metabolism*, Vol. 1, ed. J. P. Kreier. New York, NY: Academic Press, pp. 95–114.

Gillespie, T. R., Greiner, E. C. & Chapman, C. A. (2004). Gastrointestinal parasites of the guenons of western Uganda. *Journal of Parasitology*, **90**, 1356–1360.

Gillespie, T. R., Greiner, E. C. & Chapman, C. A. (2005). Gastrointestinal parasites of the colobus monkeys of Uganda. *Journal of Parasitology*, **91**, 569–573.

Graczyk, T. K., DaSilva, A. J., Cranfield, M. R. *et al.* (2001a). *Cryptosporidium parvum* Genotype 2 infections in free-ranging mountain gorillas (*Gorilla gorilla beringei*) of the Bwindi Impenetrable National Park, Uganda. *Parasitology Research*, **87**, 368–370.

Graczyk, T. K., Mudakikwa, A. B., Cranfield, M. R. & Eilenberger, U. (2001b). Hyperkeratotic mange caused by *Sarcoptes scabiei* (Acariformes: Sarcoptidae) in juvenile human-habituated mountain gorillas (*Gorilla gorilla beringei*). *Parasitology Research*, **87**, 1024–1028.

Graczyk, T. K., Bosco-Nizeyi, J., Ssebide, B. *et al.* (2002a). Anthropozoonotic *Giardia duodenalis* Genotype (assemblage) A infections in habitats of free-ranging human-habituated gorillas, Uganda. *Journal of Parasitology*, **88**, 905–909.

Graczyk, T. K., Bosco-Nizeyi, J., da Silva, A. J. *et al.* (2002b). A single genotype of *Encephalitozoon intestinalis* infects free-ranging gorillas and people sharing their habitats in Uganda. *Parasitology Research*, **88**, 926–931.

Graham, G. L. (1960). Parasitism in monkeys. *Annals of the New York Academy of Sciences*, **85**, 842–860.

Grant, I. H., Gold, J. W., Rosenblum, M., Niedzwiecki, D. & Armstrong, D. (1990). *Toxoplasma gondii* serology in HIV-infected patients: the development of central nervous system toxoplasmosis in AIDS. *AIDS*, **4**, 519–521.

Guerrant, R. L. (1997). Cryptosporidiosis: an emerging, highly infectious threat. *Emerging Infectious Diseases*, **3**, 51–57.

Guyomard, J. L., Chevrier, S., Betholom, J. L., Guigen, C. & Charlin, J. F. (2007). Finding of *Strongyloides stercoralis* infection, 25 years after leaving the endemic area, upon corticotherapy for ocular trauma. *Journal francais d'opthalmologie*, **30**, 4.

Habermann, R. T. & Williams Jr., F. P. (1957). Diseases seen at necropsy of 708 *Macaca mulatto* (rhesus monkey) and *Macaca philippinensis* (cynomolgus monkey). *American Journal of Veterinary Research*, **18**, 419–426.

Hamerton, A. E. (1933). Report on deaths occurring in the Society's gardens during the year 1932. *Proceedings of the Zoological Society of London*, **2**, 451–482.

Hamerton, A. E. (1942). Report on the deaths occurring in the Society's gardens during 1939–1940. *Proceedings of the Zoological Society of London*, **111**, 151–184.

Harper, J. S., Rice, J. M., London, W. T., Sly, D. L. & Middleton, C. (1982). Disseminated strongyloidiasis in *Erthrocebus patas*. *American Journal of Primatology*, **3**, 89–98.

Hayama, S. & Nigi, H. (1963). Investigation on the helminth parasites in the Japan Monkey Centre during 1959–61. *Primates*, **4**, 97–112.

Herwaldt, B. L. (2001). Laboratory-acquired parasitic infections from accidental exposures. *Clinical Microbiology Reviews*, **14**, 659–688.

Hira, P. R. & Patel, B. G. (1977). *Strongyloides fülleborni* infections in man in Zambia. *American Journal of. Tropical Medicine Hygiene*, **26**, 640–643.

Hira, P. R. & Patel, B. G. (1980). Human strongyloidiasis due to the primate species *Strongyloides fülleborni*. *Tropical and Geographical Medicine*, **32**, 23–29.

Hohlfeld, P., Daffos, F., Thulliez, P. *et al.* (1989). *Journal of Pediatrics*, **115**, 765–769.

Holmes, D. D., Kosanke, S. D. & White, G. L. (1980). Fatal enterobiasis in a chimpanzee. *Journal of the American Veterinary Medical Association*, **177**, 911–913.

Hope, K., Goldsmith, M. L. & Graczyk, T. (2004). Parasitic health of olive baboons in Bwindi Impenetrable National Park, Uganda. *Veterinary Parasitology*, **122**, 165–170.

Horii, Y. & Usui, M. (1985). Experimental transmission of *Trichuris ova* from monkeys to man. *Transactions of the Royal Society of Tropical Medicine and Hygiene*, **79**, 423.

Hughes, S. & Kelly, P. (2006). Interactions of malnutrition and immune impairment, with specific reference to immunity against parasites. *Parasite Immunology*, **28**, 577–588.

Huffman, M. A., Page, J. E., Sukhdeo, M. V. K. *et al.* (1996). Leaf-swallowing by chimpanzees, a behavioral adaptation for the control of strongyle nematode infections. *International Journal of Primatolology*, **17**, 475–503.

Huffman, M. A., Gotoh, S., Turner, L. A., Hamai, M. & Yoshida, K. (1997). Seasonal trends in intestinal nematode infection and medicinal plant use among chimpanzees in the Mahale Mountains, Tanzania. *Primates*, **38**, 111–125.

Jones, J. L., Hanson, D. L., Dworkin, M. S. *et al.* (1999). Survellance for AIDS-defining opportunistic illnesses. *Morbidity and Mortality Weekly Report CDC Surveillance Summary*, **48**, 1–22.

Juranek, D. (1979). Waterborne giardiasis. In *Waterborne Transmission of Giardiasis*, ed. W. Jakubowski & J. C. Hoff. Springfield, VA: U.S. Environmental Protection Agency, Office of Research and Development, Environmental Research Center, pp. 150–163.

Kalema-Zikusoka, G., Kock, R. A. & MacFie, E. J. (2002). Scabies in free-ranging mountain gorillas (*Gorilla beringei beringei)* in Bwindi Impenetrable National Park, Uganda. *Veterinary Record*, **150**, 12–15.

Kalter, S. S. (1989). Infectious diseases of nonhuman primates in a zoo setting. *Zoo Biology Supplement*, **1**, 61–76.

Karanis, P., Kourenti, C. & Smith, H. (2007). Waterborne transmission of protozoan parasites: a worldwide review of outbreaks and lessons learnt. *Journal of Water and Health*, **5**, 1.

Keeling, M. E. & McClure, H. M. (1974). Pneumoncoccal meningitis and fatal enterobiasis in a chimpanzee. *Laboratory Animal Science*, **24**, 92–95.

Kelly, A., Little, M. D. & Voge, M. (1976). *Strongyloides fulleborni*-like infections in man in Papua New Guinea. *American. Journal of Tropical Medicine and Hygiene*, **25**, 694–699.

King, C. L. (2001). Initiation and regulation of disease in schistosomiasis. In *Schistosomiasis*, Vol. 3, ed. A. A. F. Mohamud & S. James. London: Imperial College, pp. 213–264.

King Jr., N. W. (1976). Synopsis of the pathology of new world monkeys. *Scientific Publication of the Pan American Health Organization*, **317**, 169–198.

Krauss, H., Weber, A., Appel, M. *et al.* (2003). *Zoonoses. Infectious Diseases Transmissible from Animals to Humans*, 3rd edn. Washington, DC: ASM Press.

Krief, S., Huffman, M. A., Sévenet, T. *et al.* (2005). Noninvasive monitoring of the health of *Pan troglodytes schweinfurthii* in the Kibale National Park, Uganda. *International Journal of Primatology*, **26**, 467–489.

Kuntz, R. E., Moore, J. A. & Huang, T. C. (1979). Distribution of egg deposits and gross lesions in nonhuman primates infected with *Schistosoma haematobium* (Iran). *Journal of Medical Primatology*, **8**, 167–178.

Laing, A. B., Edeson, J. F. & Wharton, R. H. (1960). Studies on filariasis in Malaya; the vertebrate hosts of *Brugia malayi* and *B. pahangi*. *Annals of Tropical Medicine and Parasitology*, **54**, 92–99.

Lehner, N. D. M. (1984). Biology and diseases of Cebidae. In *Laboratory Animal Medicine*, ed. B. Fox, B. Cohen & F. Loew. Orlando, FL: Academic Press, pp. 321–353.

Levin, N. D. (1970). Protozoan parasites of nonhuman primates as zoonotic agents. *Laboratory Animal Care*, **20**, 377–382.

Levin, N. D. (1976). *Nematode Parasites of Domestic Animals and of Man.* Minneapolis, MN: Burgess Publishing Company.

Loeb, W. F., Bannerman, R. M., Rininger, B. F. & Johnson, A. J. (1978). Hematologic disorders. In *Pathology of Laboratory Animals*, Vol. 1, ed. K. Benirschke, F. M. Garner & T. C. Jones. New York, NY: Springer-Verlag, pp. 1000–1021, 1032–1050.

Loomis, M. R. & Britt, J. O. (1983). An epizootic of *Entamoeba histolytica* in colobus monkeys. *Annual Proceedings of the American Association of Zoo Veterinarians*, **1983**, 10.

Loomis, M. R., Britt, J. O., Gendron, A. P., Holshuh, H. J. & Howard, E. B. (1983). Hepatic and gastric amebiasis in black and white colobus monkeys. *Journal of the American Medical Association*, **183**, 1188–1191.

Luft, B. J. & Remington, J. S. (1992). Toxoplasmic encephalitis in AIDS. *Clinical Infectious Diseases*, **15**, 211–222.

Marinkelle, C. J. & Grose, E. (1968). *Plasmodium brasilianum* in Colombian monkeys. *Tropical and Geographical Medicine*, **20**, 276–280.

Martínez, A. J., Guerra, A. E., Garcia-Tamayo, J. *et al.* (1994). Granulomatous amebic encephalitis: a review and report of a spontaneous case from Venezuela. *Acta Neuropathology*, **87**, 430–434.

McCarrison, R. (1920). The effects of deficient dietaries on monkeys. *British Medical Journal*, Feb. 21, 249–253.

McClure, H. M. & Guilloud, N. B. (1971). Comparative pathology of the chimpanzee. In *The Chimpanzee*, Vol. 4, ed. G. H. Bourne. Baltimore, MD: University Park Press, pp. 103–272.

McClure, H. M., Strozier, L. M., Keeling, M. E. & Healy, G. R. (1973). Strongyloidosis in two infant orangutans. *Journal of the American Veterinary Medical Association*, **163**, 629–632.

Mengistu, L. & Erko, B. (2004). Zoonotic intestinal parasites in *Papio anubis* (baboon) and *Cercopithecus aethiops* (vervet) from four localities in Ethiopia. *Acta Tropica*, **90**, 231–236.

Michaud, C., Tantalean, M., Eque, C., Montoya, E. & Gozalo, A. (2003). A survey for helminth parasites in feral New World non-human primate populations and its comparison with parasitological data from man in the region. *Journal of Medical Primatology*, **32**, 341–345.

Miller, R. A., Bronsdon, M. A., Kuller, L. & Morton, W. R. (1990). Clinical and parasitologic aspects of cryptosporidiosis in nonhuman primates. *Laboratory Animal Science*, **40**, 42–46.

Mora, C. S., Segami, M. I. & Hidalgo, J. A. (2006). *Strongyloides stercoralis* hyperinfection in systemic lupus erythematosus and the antiphospholipid syndrome. *Seminars in Arthritis and Rheumatism*, **36**, 135–143.

Morgan-Ryan, U. M., Fall, A., Ward, L. A. *et al.* (2002). *Cryptosporidium hominis* n. sp. (Apicomplexa: Cryptosporidiidae) from *Homo sapiens*. *Journal of Eukaryotic Microbiology*, **49**, 433–440.

Mori, S., Konishi, T., Matsuoka, K. *et al.* (1998). Strongyloidiasis associated with nephrotic syndrome. *Internal Medicine*, **37**, 606–610.

Murata, K., Hasegawa, H., Nakano, T., Noda, A. & Yanai, T. (2002) Fatal infection with human pinworm, *Enterobius vermicularis*, in a captive chimpanzee. *Journal of Medical Primatology*, **31**, 104–108.

Muriuki, S. M., Murugu, R. K., Muene, E., Karere, G. M. & Chai, D. C. (1998). Some gastro-intestinal parasites of zoonotic (public health) importance commonly observed in old world non-human primates in Kenya. *Acta Tropica*, **71**, 73–82.

Murray, S., Stem, C., Boudreau, B. & Goodall, J. (2000). Intestinal parasites of baboons (*Papio cynocephalus anubis*) and chimpanzees (*Pan troglodytes*) in Gombe National Park. *Journal of Zoo and Wildlife Medicine*, **31**, 176–178.

Nakano, T., Murata, K., Ikeda, Y. & Hasegawa, H. (2003). Growth of *Enterobius vermicularis* in a chimpanzee after anthelmintic treatment. *Journal of Parasitology*, **83**, 439–443.

Nakauchi, K. (1999). The prevalence of *Balantidium coli* infection in fifty-six mammalian species. *Journal of Veterinary Medical Science, Japanese Society of Veterinary Science*, **61**, 63–65.

Naumova, E. N., Jagai, J. S., Matyas, B. *et al.* (2007). Seasonality in six enterically transmitted diseases and ambient temperature. *Epidemiology and Infection*, **135**, 281–292.

Navin, T. R. & Juranek, D. D. (1984). Cryptosporidiosis: clinical, epidemiologic and parasitologic review. *Review of Infectious Diseases*, **6**, 313–327.

Nelson, G. S. (1965). The parasitic helminths of baboons with particular reference to species transmissible to man. In *The Baboon in Medical Research*, ed. H. Vagtborg. Austin, TX: University of Texas Press, pp. 441–470.

Neto, E. C., Rubin, R., Schulte, J. & Giugliani, R. (2004). Newborn screening for congenital infectious diseases. *Emerging Infectious Diseases*, **10**, 1069–1073.

Nizeyi, J. B., Mwebe, R., Nanteza, A. *et al.* (1999). *Cryptosporidium* sp. and *Giardia* sp. infections in mountain gorillas (*Gorilla gorilla beringei*) of the Bwindi Impenetrable National Park, Uganda. *Journal of Parasitology*, **85**, 1084–1088.

Orihel, T. C. & Seibold, H. R. (1972). Nematodes of the bowel and tissues. In *Pathology of Simian Primates*, Part 2, ed. R. N. Fiennes. Basel: Karger, pp. 76–103.

Overturf, G. D. (1994). Endemic giardiasis in the United States – role of the day-care centre: editorial response. *Clinical Infectious Diseases*, **18**, 764–765.

Park, W. B., Choe, P. G., Jo, J. H. *et al.* (2007). Amebic liver abscess in HIV-infected patients, Republic of Korea. *Emerging Infectious Diseases*, **13**, 516–517.

Patten, R. A. (1939). Amoebic dysentery in orang-utans (*Simia satyrus*). *Australian Veterinary Journal*, **15**, 68–71.

Pawlowski, Z. S., Schad, G. A. & Stott, G. J. (1991). Hookworm infection an anaemia: approaches to prevention and control. Geneva, Switzerland: World Health Organization.

Penner, L. R. (1981). Concerning threadworm (*Strongyloides stercoralis*) in apes – lowland gorillas (*Gorilla gorilla*) and chimpanzees (*Pan troglodytes*). *Journal of Zoo Animal Medicine*, **12**, 128–131.

Pillers, A. W. (1921). Sarcoptic scabies (or itch) in the chimpanzee. *British Veterinary Journal*, **77**, 329–333.

Pillers, A. W. (1924). *Ascaris lumbricoides* causing fatal lesions in a chimpanzee. *Annals of Tropical Medicine and Parasitology*, **18**, 101–102.

Pillers, A. W. & Southwell, T. (1929). Strongyloidosis of the wooly monkey (*Lagothrix humboldti*). *Annals of Tropical Medicine and Parasitology*, **23**, 129.

Ratcliffe, H. L. (1931). A comparative study of amoebiasis in man, monkeys and cats, with special reference to the formation of the early lesions. *American Journal of Hygiene*, **14**, 337–352.

Ravdin, J. I. (1995). Amebiasis. *Clinical Infectious Diseases*, **20**, 1453–1466.

Reardon, L. V. & Rininger, B. F. (1968). A survey of parasites in laboratory animals. *Laboratory Animal Care*, **18**, 577–580.

Renquist, D. M. & Whitney, R. A. Jr. (1987). Zoonoses acquired from pet primates. *Veterinary Clinics of North America: Small Animal Practice, 17*, **1**, 219–240.

Rideout, B. A., Gardiner, C. H., Stalis, I. H. *et al.* (1997). Fatal infections with *Balamuthia mandrillaris* (a free-living amoebae) in gorillas and other old world primates. *Veterinary Pathology*, **34**, 15–22.

Ruch, T. C. (1959). *Diseases of Laboratory Primates*. Philadelphia, PA: Saunders Publishing.

Salzer, J. S., Rwego, I. B., Goldberg, T. L., Kuhlenschmidt, M. S. & Gillespie, T. R. (2007). *Giardia* sp. and *Cryptosporidium* sp. infections in primates in fragmented and undisturbed forest in western Uganda. *Journal of Parasitology*, **93**, 439–440.

Sasa, M., Tanaka, H., Fukui, M. & Takata, A. (1962). Internal parasites of laboratory animals. In *The Problems of Laboratory Animal Disease*, ed. R. J. Harris. New York, NY: Academic Press, pp. 195–214.

Schmidt, R. E. (1978). Systemic pathology of chimpanzees. *Journal of Medical Primatology*, **7**, 274–318.

Schmidt, R. E. & Prine, J. R. (1970). Severe enterobiasis in a chimpanzee. *Pathologia Veterinaria*, **7**, 56–59.

Schuster, F. L. & Visvesvara, G. S. (2004). Amebae and ciliated protozoa as causal agents of waterborne zoonotic disease. *Veterinary Parasitology*, **126**, 91–120.

Seibold, H. R. & Fussell, E. N. (1973). Intestinal microsporidiosis in *Callicebus moloch*. *Laboratory Animal Science*, **23**, 115–118.

Shadduck, J. A. & Pakes, S. P. (1978). Protozoal and metazoal diseases. In *Pathology of Laboratory Animals*, ed. K. Benirschke, F. M. Garner & T. C. Jones. New York, NY: Springer-Verlag, pp. 1587–1696.

Shulman, I. A. (1994). Parasitic infections and their impact on blood donor selection and testing. *Archives of Pathology and Laboratory Medicine*, **118**, 366–370.

Sleeman, J. M., Meader, L. L., Mudakikwa, A. B., Foster, J. W. & Patton, S. (2000). Gastrointestinal parasites of mountain gorillas (*Gorilla gorilla beringei*) in the Parc National Des Volcans, Rwanda. *Journal of Zoo and Wildlife Medicine*, **31**, 322–328.

Smith, H. V., Paton, C. A., Girdwood, R. W. & Mtambo, M. M. (1996). *Cyclospora* in non-human primates in Gombe, Tanzania. *Veterinary Record*, **138**, 528.

Stam, A. B. (1960). Fatal ascaridosis in a dwarf chimpanzee. *Annales de Parasitologie Humaine et Comparée*, **35**, 675.

Stelma, F. F., Talla, I., Sow, S. *et al.* (1995). Efficacy and side effects of praziquantel in an epidemic focus of *Schistosoma mansoni*. *American Journal of Tropical Medicine and Hygiene*, **53**, 167–170.

Strong, J. P., McGill, Jr., H. C. & Miller, J. H. (1961). *Schistosomiasis mansoni* in the Kenya baboon. *American Journal of Tropical Medicine and Hygiene*, **10**, 25–32.

Sturrock, R. F., Butterworth, A. E. & Houba, V. (1976). *Schistosoma mansoni* in the baboon (*Papio anubis*): parasitological responses of Kenyan baboons to different exposures of a local parasite strain. *Parasitology*, **73**, 239–252.

Suvajdzic, N., Kranjcić-Zec, I., Jovanović, V., Popović, D. & Colović, M. (1999). Fatal strongyloidosis following corticosteroid therapy in a patient with chronic idiopathic thrombocytopenia. *Haematologia*, **29**, 323–326.

Tanaka, H., Fukui, M., Yamamoto, H., Hayama, S. & Kodera, S. (1962). Studies on the identification of common intestinal parasites of primates. *Bulletin of Experimental Animals*, **11**, 111–116.

Teare, J. A. & Loomis, M. R. (1982). Epizootic of balantidiasis in lowland gorillas. *Journal of the American Veterinary Medical Association*, **181**, 1345–1347.

Teutsch, S. M., Juranek, D. D., Sulzer, A., Dubey, J. P. & Sikes, R. K. (1979). Epidemic toxoplasmosis associated with infected cats. *New England Journal of Medicine*, **300**, 695–699.

Thienpont, D., Mortelmans, J. & Vercruysse, J. (1962). Contribution à l'étude de la Trihuriose du chimpanzé et de son traitement avec la methyridine. *Annales de la Societe belge de medecine tropicale*, **2**, 211–218.

Thompson, R. C. (2000). Giardiasis as a re-emerging infectious disease and its zoonotic potential. *International Journal for Parasitology*, **30**, 1259–1267.

Thompson, R. C. & Monis, P. T. (2004). Variation in *Giardia*: implications for taxonomy and epidemiology. *Advanced Parasitology*, **58**, 69–137.

Thompson, R. C., Hopkins, R. M. & Homan, W. L. (2000). Nomenclature and genetic groupings of *Giardia* infecting mammals. *Parasitology Today*, **16**, 210–213.

Thornton, H. (1924). The relationship between the ascarids of man, pig and chimpanzee. *Annals of Tropical Medicine and Parasitology*, **18**, 99–100.

Toft, II, J. D. & Eberhard, M. L. (1998). Parasitic diseases. In *Nonhuman Primates in Biomedical Research: Diseases*, ed. B. T. Bennett, C. R. Abee & R. Henrickson. San Diego, CA: Academic Press, pp. 111–205.

Tumwine, J. K., Kekitiinwa, A., Bakeera-Kitaka, S. *et al.* (2005). Cryptosporidiosis and microsporidiosis in Ugandan children with persistent diarrhea with and without concurrent infection with the human immunodeficiency virus. *American Journal of Tropical Medicine and Hygiene*, **73**, 921–925.

Van Lieshout, L., de Gruijter, J. M., Adu-Nsiah, M. *et al.* (2005). *Oesophagostomum bifurcum* in non-human primates is not a potential reservoir for human infection in Ghana. *Tropical Medicine and International Health*, **10**, 1315–1320.

Verweij, J. J., Pit, D. S., van Lieshout, L. *et al.* (2001). Determining the prevalence of *Oesophagostomum bifurcum* and *Necator americanus* infections using specific PCR amplification of DNA from faecal samples. *Tropical Medicine and International Health*, **6**, 726–731.

Vickers, J. H. (1968). Gastrointestinal diseases of primates. *Current Veterinary Therapy*, **3**, 393–396.

Visvesvara, G. S., Martinez, A. J., Schuster, F. L. *et al.* (1990). *Leptomyxid ameba*, a new agent of amebic meningoencephalitis in humans and animals. *Journal of Clinical Microbiology*, **28**, 2750–2756.

Visvesvara, G. S., Schuster, F. L. & Martinez, A. J. (1993). *Balamuthia mandrillaris*, n. g., n. sp., agent of amebic meningoencephalitis in humans and other animals. *Journal of Eukaryotic Microbiology*, **40**, 504–514.

Vitazkova, S. K. & Wade, S. E. (2006). Parasites of free-ranging black howler monkeys (*Alouatta pigra*) from Belize and Mexico. *American Journal of Primatology*, **68**, 1089–1097.

Voller, A. (1972). *Plasmodium* and *Hepatocystis*. In *Pathology of Simian Primates*, Part 2, ed. R. N. Fiennes. New York, NY: Karger, pp. 57–73.

Wallace, G. D. (1972). Experimental transmission of *Toxoplasma gondii* by cockroaches. *Journal of Infectious Diseases*, **126**, 545–547.

Webber, W. A. & Hawking, F. (1955). The filarial worms *Dipetalonema digitatum* and *D. gracile* in monkeys. *Parasitology*, **45**, 401–408.

Weyher, A. H., Ross, C. & Semple, S. (2006). Gastrointestinal parasites in crop raiding and wild foraging *Papio anubis* in Nigeria. *International Journal of Primatology*, **27**, 1519–1534.

Williams, C. V., Van Steenhouse, J. L., Bradley, J. M. *et al.* (2002). Naturally occurring *Ehrlichia chaffeenis* infection in two prosimian primate species: ring-tailed lemurs (*Lemur catta*) and ruffed lemurs (*Varecia variegata*). *Emerging Infectious Diseases*, **8**, 1497–1500.

Wilson, D. A., Day, P. A. & Brummer, E. G. (1984). Diarrhea associated with *Cryptosporidium* sp. in juvenile macaques. *Veterinary Pathology*, **21**, 447–450.

Wong, S. Y. & Remington, J. S. (1994). Toxoplasmosis in pregnancy. *Clinical Infectious Diseases*, **18**, 853–861.

Xiao, L., Fayer, R., Ryan, U. & Upton, S. J. (2004) *Cryptosporidium* taxonomy: recent advances and implications for public health. *Clinical Microbiology Reviews*, **17**, 72–97.

Yamashita, J. (1963). Ecological relationships between parasites and primates. I. Helminth parasites and primates. *Primates*, **4**, 1–96.

Zangerle, R., Allerberger, F., Pohl, P., Fritsch, P. & Dierich, M. P. (1991). High risk of developing toxoplasmic encephalitis in AIDs patients seropositive to *Toxoplasma gondii*. *Medical Microbiology and Immunology*, **180**, 59–66.

Zeman, D. H. & Baskin, G. B. (1985). Encephalitozoonosis in squirrel monkeys (*Saimiri sciureus*). *Veterinary Pathology*, **22**, 24–31.

Ziem, J. B., Kettenis, I. M., Bayita, A. *et al.* (2004). The short-term impact of albendazole treatment on *Oesophagostomum bifurcum* and hookworm infections in northern Ghana. *Annals of Tropical Medicine and Parasitology*, **98**, 385–390.

10 *Lice and other parasites as markers of primate evolutionary history*

DAVID L. REED, MELISSA A. TOUPS, JESSICA E. LIGHT, JULIE M. ALLEN, AND SHELLY FLANNIGAN

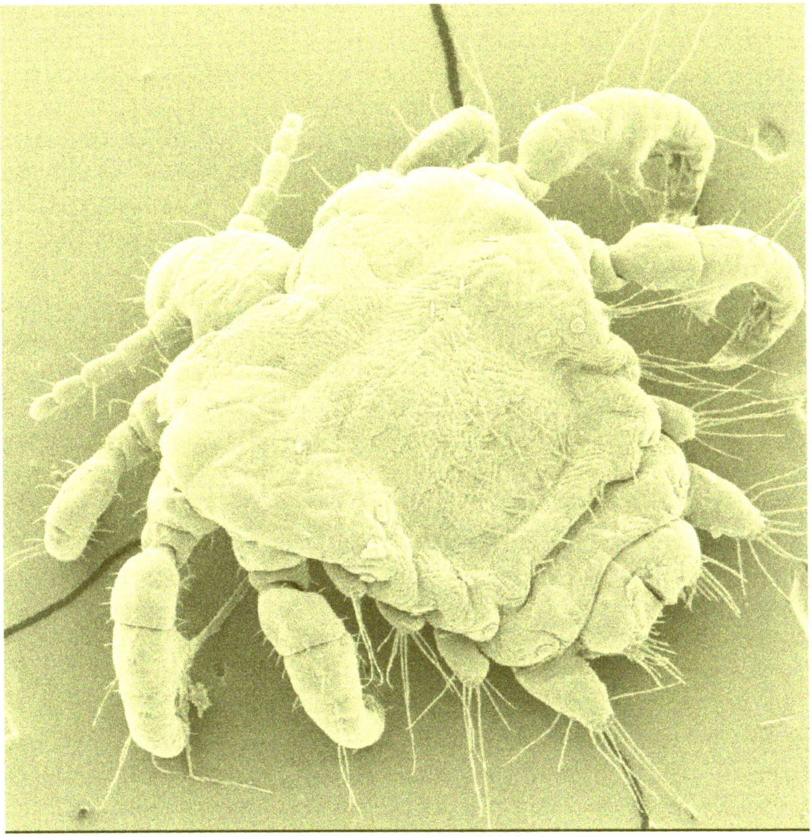

SEM, by Vincent S. Smith (*Pthirus pubis* human pubic louse) http://sid.zoology.gla.ac.uk/upload/user_uploads/1239.tif

Primate Parasite Ecology. The Dynamics and Study of Host–Parasite Relationships, ed. Michael A. Huffman and Colin A. Chapman. Published by Cambridge University Press.

Introduction

"Coevolution" is a term that was first used by Ehrlich & Raven (1964) to describe the intimate association between phytophagous insects and their host plants. Currently, coevolution is a broadly used term meant to describe the relationships between taxa that have been intimately associated over long periods. By definition, coevolution is a combination of both microevolutionary and macroevolutionary processes, specifically an ecological relationship (coaccommodation) and a historical relationship (cospeciation) between taxa (Brooks, 1979). Coaccommodation (later called coadaptation or reciprocal adaptation) is the mutual modification of traits over time and includes parameters such as pathogenicity, host specificity, and synchrony of life cycle stages. Cospeciation, on the other hand, is concurrent speciation (cladogenesis) in two associated lineages of organisms (Brooks, 1979). The interactions between the two associated lineages of organisms vary from mutualistic to antagonistic to parasitic in nature. Examples of cospeciating organisms can be found among plant/herbivore, predator/prey, insect/endosymbiont, host/pathogen, and host/parasite assemblages. For simplicity we will refer to these assemblages as host/parasite assemblages unless we are referring to a specific system.

The process of cospeciation can be identified by comparing the phylogenies of hosts and their parasites to determine if they are significantly congruent. The process of "strict" cospeciation results in a pattern of perfectly congruent phylogenies. Cospeciation does not lead to perfectly congruent phylogenies in all cases however, because other historical events such as host switching, sorting events (e.g. extinction and lineage sorting), parasite speciation, and failure to speciate can result in deviations from perfect congruence. In addition, congruent phylogenies can result from processes other than cospeciation, such as the sequential colonization of closely related hosts (Percy *et al.*, 2004). A variety of methods are currently available to compare host and parasite phylogenies to test for congruence, which is usually the first step in determining whether cospeciation has occurred in an assemblage (Page, 2003).

In assemblages of cospeciating hosts and parasites one has the ability to uncover a shared evolutionary history that is common to two associated organisms. This permits researchers to compare host and parasite characteristics such as the relative and absolute rates of nucleotide substitutions between distantly related hosts and parasites (e.g. Hafner *et al.*, 1994; Moran *et al.*, 1995). Having a documented history of cospeciation also permits the researcher to examine the parasite as a marker of host evolutionary history. The inference of host evolutionary history from parasites can occur over long timescales where entire assemblages of hosts and parasites are being examined (e.g. studies of cospeciation), or shorter evolutionary timescales where populations within one host and parasite species are being examined (e.g. studies of codemography).

Of course, the degree to which parasites can inform us about host evolutionary history varies greatly among parasites. The strength in using parasites as markers of host evolutionary history is the potential for triangulation in biology – the use of multiple, independent data sets to simultaneously infer the evolutionary history of an organism. The use of multiple parasites to study host evolutionary history can be likened to using multiple cameras to record a sporting event. When a play on the field is in dispute, multiple camera angles can help to construct a better-resolved picture of the disputed play, and when all camera angles are considered, a better understanding of the event can be achieved. Similarly, when data from multiple parasites provide consistent information about a particular evolutionary event (e.g. the divergence date between humans and chimpanzees) these data constitute strong evidence in support of that event. In our research program we are interested in events that have happened recently (e.g. the Peopling of the Americas), as well as events that happened during a brief period of time (e.g. emergence of modern humans from Africa). Both recent evolutionary events and those that were brief in duration often suffer from a lack of resolution because neither recent nor rapid events are accompanied by a large amount of evolutionary change. In the case of molecular evidence, there is no great accumulation of nucleotide substitutions in the molecular data that researchers collect, which can lead to poor resolution of the event in question. Because hosts sometimes evolve too slowly to resolve these rapid events, we look to parasites and pathogens that evolve faster than their hosts (i.e. they accumulate more nucleotide substitutions per unit time) to provide more data points with which to resolve the event in question. To return to our analogy, we are searching for a super slow-motion camera that records many frames per second. This parasite or pathogen may only be useful for a narrow window of time, but it may resolve an event that is not well recorded in the data of the host.

Parasites and pathogens can yield valuable information regarding host evolutionary history. In this chapter, we first discuss the characteristics of parasites that make them useful for inferring the evolutionary history of their hosts. We then review parasitic lice as a marker of primate evolutionary history, and finally we look to other parasites and pathogens for insight into the degree to which we can infer primate evolutionary history (in general) and human evolutionary history (in particular). It is important to point out that these are nascent studies, and that the aforementioned concept of triangulation is not yet fully developed.

Choosing an effective pathogen or parasite

If the goal of the study is to use coevolving parasites to *infer* the evolutionary history of their hosts, then there are certain characteristics that make some parasites more appropriate for this task than others. First, parasites with high host specificity are ideal under most circumstances. Generalist parasites – ones that successfully parasitize many host species – are less than ideal because it may be difficult to determine the degree to which other hosts have influenced the evolutionary history of the parasite (Hasegawa, 1999; Holmes, 2004). Second, parasites that are both obligate (cannot be free-living) and permanent (spend their entire life on a single host species) are ideal because they are completely dependent on the host for mobility, and therefore are only transmitted through direct contact between hosts. Third, parasites that are both prevalent and abundant are easier to collect and may also be more likely to mirror the host's true evolutionary history than a rare parasite. Lastly, the parasite must have been associated with the host during the time frame of interest.

When examining large-scale evolutionary processes, parasites that have cospeciated with their primate hosts make excellent markers. Ectoparasitic lice, for example, are appropriate for studying the last 25 million years of primate evolutionary history because they have cospeciated with their hosts during this time frame, show high host specificity, are obligate and permanent parasites, and are often abundant on the host and prevalent on many members of a given species (Reed *et al.*, 2004, 2007).

If, by contrast, the goal of the study is to look at host population dynamics, then faster evolving parasites/pathogens such as bacteria, viruses, or fungi might be more appropriate. Sexually transmitted or vertically transmitted parasites or pathogens (those transmitted from parent to offspring) are better for tracing population dynamics of the host (Holmes, 2004). Vertically transmitted pathogens can be used to track populations in a genealogical manner making direct comparisons to host molecular data possible. In contrast, sexually transmitted pathogens have the ability to more easily cross close genealogical boundaries making them more appropriate for studying very recent population movements or migrations. In general, parasites that have long-term infections or long latency periods are more likely to reflect host population dynamics than a pathogen that spreads epidemically across population boundaries (Holmes, 2004). JC virus, for example, is an excellent marker for examining population dynamics in humans because it is effectively transmitted vertically (i.e. passed from parents to offspring during cohabitation), causes persistent infections in the human host, and is both prevalent and abundant in the human population (Pavesi, 2003).

Lice as markers of host evolution

The insect order Phthiraptera (lice) fit many of the criteria mentioned above as good potential markers for inferring host evolutionary history. Lice are both obligate and permanent parasites of mammals and birds and, in general, are highly host specific. This is partially the result of the very low vagility of the lice; they are wingless parasites with limited dispersal capabilities that are effectively stranded on their hosts. Several vertebrate groups have been studied for patterns of cospeciation with lice, such as gophers (Hafner *et al.*, 1994), heteromyid rodents (Light, 2005), wallabies (Barker, 1991), doves (Johnson & Clayton, 2003), toucans (Weckstein, 2004), and primates (Reed *et al.*, 2004, 2007). Many of these vertebrate hosts have shown convincing evidence of a long shared coevolutionary history with their lice. Furthermore, lice can have very high prevalence and abundance making them easy to collect and study. Lastly, several studies have shown that lice tend to evolve slightly faster than their hosts, which is useful for inferring host evolutionary history (Hafner *et al.*, 1994; Page, 1996; Page *et al.*, 1998; Reed *et al.*, 2004; Light, 2005).

Primate lice (specifically the sucking louse genera *Pediculus*, *Pthirus*, and *Pedicinus*) have been coevolving with their primate hosts for at least the last 25 million years and have been useful as markers of primate evolutionary history (Figure 10.1; Reed *et al.*, 2004). *Pediculus* occurs on chimpanzees and humans, *Pthirus* occurs on humans and gorillas, and *Pedicinus* occurs only on the Cercopithecoid primates. Because host and parasite clade divergences are highly correlated in cospeciating assemblages (Hafner *et al.*, 1994), one benefit of studying cospeciating assemblages is that fossil evidence relating to the host tree can often be used to calibrate the parasite tree. In the absence of a parasite fossil record, the ability to calibrate with host information permits the estimation of other nodes in the parasite tree. However, it should be noted that there is error associated with divergence date estimation, which may be compounded when researchers rely upon host fossil evidence to calibrate parasite phylogenies. Reed *et al.* (2004) used the date of 20–25 million years ago for the Cercopithecoid/Hominoid divergence to calibrate the corresponding node in the phylogenetic tree of primate lice and found the divergence between chimpanzee and human lice (*Pediculus schaeffi* and *P. humanus*, respectively) to be 5.6 million years based on two mitochondrial genes (Figure 10.1). This estimate is similar to the host divergence based on primate mtDNA (5–7 million years ago; Stauffer *et al.*, 2001). The divergence dates among primate lice have recently been re-examined including the additional taxon, *Pthirus gorillae*, parasitic on gorillas (Reed *et al.*, 2007). Divergence dates based on the mitochondrial cytochrome c oxidase subunit I (Cox1) and the nuclear elongation factor 1 alpha (EF1α) genes were estimated using approaches that relax the constraints of a molecular clock. Mean divergence date estimates for the split between the chimpanzee and human head/clothing lice averaged 6.39 MYA (again, roughly contemporaneous with the host split 5–7 million years ago;

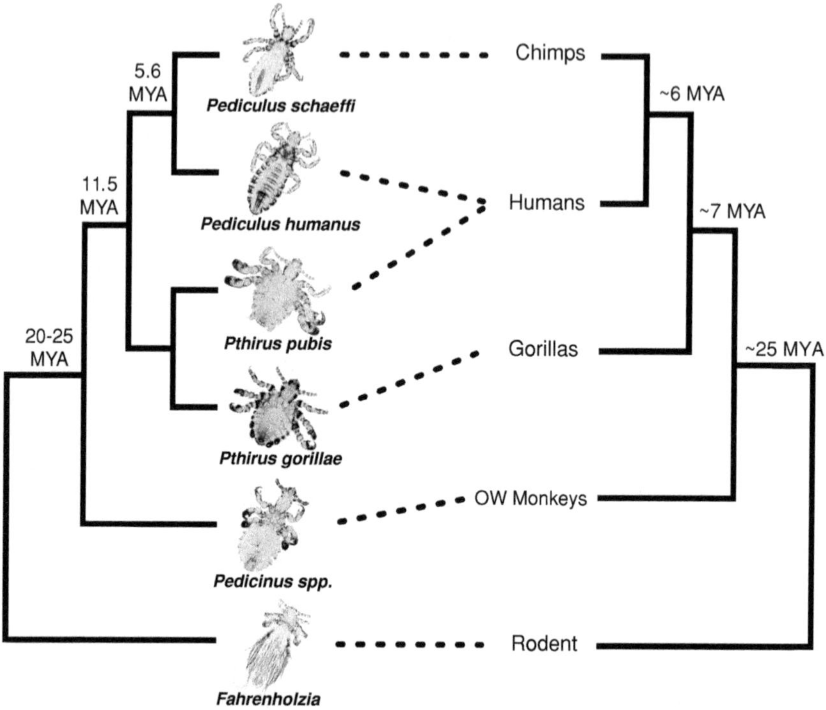

Figure 10.1. Primate and primate louse phylogenies redrawn from Reed *et al.* (2004). The primate tree is based on the Cox1 gene and the louse tree is based on the Cox1 and Cytb genes. Divergence dates are listed above the nodes in millions of years and divergences in lice are those presented in Reed *et al.* (2004).

Figure 10.2). The divergence date estimates for the gorilla and human pubic lice (*Pthirus gorillae* and *Pthirus pubis*, respectively) averaged 3.32 MYA and are noticeably more recent than the split between the two *Pediculus* species (Figure 10.2). Furthermore, the divergence between *Pthirus pubis* and *Pthirus gorillae* is considerably younger than the estimated divergence of their hosts (7–8 million years ago), which Reed *et al.* (2007) determined was the result of a host switch roughly 3.3 million years ago between gorillas and humans. This host switch could have resulted from early human contact with gorilla ancestors through interactions such as sexual contact, scavenging, sharing nesting sites, or acts of aggression. Regardless, this places early humans and gorillas in roughly the same geographic location and in roughly the same habitat 3.3 million years ago. It is important to note here that occasional departures from cospeciation (host switches, parasite duplications, parasite extinctions, etc.) are historical events that potentially add to the information content of the parasite rather than detract from its utility for inferring host evolutionary history.

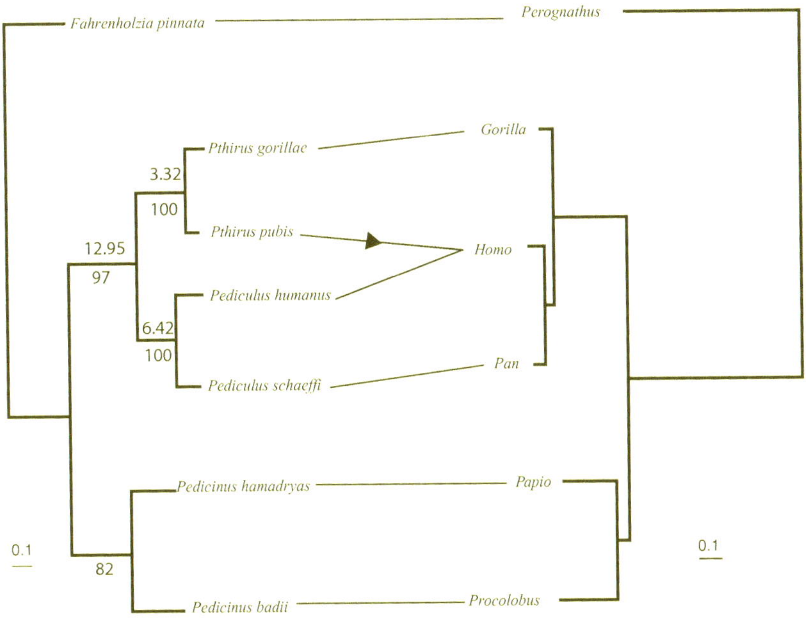

Figure 10.2. Primate and primate louse phylogenies redrawn from Reed *et al.* (2007). Louse phylogeny is based on a maximum likelihood analysis of the Cox1 and EF1α genes. Divergence dates are given above the nodes (in millions of years). Bootstrap values are given below the nodes. The arrow represents the proposed host switch from gorillas to humans *c.* 3.3 million years ago.

Lice appear to be capable of providing insight into primate evolutionary history, but there are many more louse taxa to be examined. For example, Old World Monkeys are parasitized exclusively by a single genus of sucking louse, *Pedicinus*. There are at least 14 recognized species of *Pedicinus*, and it is not known whether these lice have cospeciated with their hosts. Similarly, there are many species of chewing and sucking lice described from the Malagasy primates that also remain untested for patterns of cospeciation. Patterns of primate phylogeography in Madagascar (Yoder *et al.*, 2005) could be tested with the lice, which would add to our understanding of the biogeographical history of this region of high endemism.

Lice as population-level markers

To look at more recent evolutionary events, such as those within a single species, fast-evolving molecular markers are necessary. The mitochondrial markers, Cox1 and cytochrome b (Cytb), have shown the distinct signature of a population expansion in one clade of the human louse *Pediculus humanus* (Reed *et al.*, 2004). Estimates of the timing of the expansion (100 000 years; Reed *et al.*, 2004) were concordant with the estimated emergence of modern humans out of Africa 70 000–100 000 years ago (di Rienzo & Wilson, 1991; Rogers & Harpending, 1992; Harpending *et al.*, 1993). In another study, Kittler *et al.* (2003, 2004) were able to deduce the origin of clothing use in humans since clothing lice (*P. humanus humanus*) could not have existed without this highly specialized niche. By dating the oldest node containing clothing lice Kittler *et al.* (2003, 2004) were able to determine that clothing was in use by modern humans by at least 107 000 years ago.

The Cox1 and Cytb genes have shown that the age of the human louse, *Pediculus humanus*, is an order of magnitude older than its modern human host, despite a long history of coevolution that would predict similar clade ages for hosts and parasites. The deep divergence seen within *P. humanus* was much older than could have been maintained in a large, panmictic population, which indicates that either lice retained much more mtDNA diversity than their hosts (modern human mtDNA coalesce to a single lineage within about 150 000–200 000 years) or that one living lineage of human lice had its evolutionary origins on a now-extinct species of archaic hominid (Reed *et al.*, 2004). If the latter were true this would suggest that modern humans were in direct physical contact with archaic humans during periods of contemporeity.

Although the Cox1 and Cytb markers were suitable for estimating events over 100 000 years, they probably do not evolve fast enough to record more recent events in human evolutionary history such as the Peopling of the Americas, thought to have happened approximately 30 000 years ago. The absolute rate of mutation in Cox1 for these lice was estimated to be approximately 9.0×10^{-9} substitutions per site per generation, which is five to six times faster than human mtDNA excluding the D loop. However, it is considerably slower than the D loop itself (Reed *et al.*, 2004). Faster markers in the lice would be needed to address recent population-level questions. Candidate markers include additional mtDNA sequences, microsatellite DNA (Leo & Barker, 2005), single nucleotide polymorphisms (SNP) data, or even sequence data from fast-evolving endosymbiotic bacteria that live within the lice (see below).

Within the primates, there are numerous population-level questions for which lice might be explored as a marker of host evolutionary history. Currently, the two recognized species of chimpanzees (*Pan troglodytes* and *Pan paniscus*) are hosts to a single species of louse, *Pediculus schaeffi*. It is not known whether *P. schaeffi* contains divergences within it reflecting the two species of host. Because lice evolve at roughly three times the rate of their primate hosts (Reed *et al.*, 2004), it can be assumed that population structure in the chimpanzees might be ascertained from the louse DNA as well. The question of whether the chimp lice show the same age of divergence as their hosts is an intriguing one, and whether the lice of either species can be used to explore within-species evolution remains to be seen. These same types of questions can be addressed with the lice of almost any primate given the degree to which they seem to be tied to their hosts in both ecological and evolutionary time.

Endosymbionts in lice

Many insects feeding on nutrient-poor diets, such as sucking lice feeding on blood, have primary endosymbiotic bacteria which provide the unavailable nutrients (Buchner, 1965). These bacteria are usually vertically transmitted (Douglas, 1989), which leads to a long coevolutionary history between the two taxa. For example, all aphids have a primary endosymbiont in the genus *Buchnera* that has coevolved and cospeciated with aphids (Munson *et al.*, 1991). Furthermore, it has been shown that many primary endosymbionts evolve faster than their aphid hosts (Moran *et al.*, 1995) suggesting that they may be useful fast-evolving markers for host evolutionary history.

The primary endosymbiont of the human head louse (*Candidatus* Riesia pediculicola) was first seen over 300 years ago (Hooke, 1664), yet it was only formally described in 2006 based on molecular data (Sasaki-Fukatsu *et al.*, 2006). *Candidatus* Riesia has cospeciated with primate lice and is evolving faster than its louse hosts (Allen *et al.*, 2007). This suggests that the endosymbiont may be used to infer more recent events in human evolutionary history, such as the Peopling of the Americas, which cannot be resolved fully using human or louse mtDNA markers.

Other parasites as markers of non-human primate evolution

Pinworms

Cameron (1929) hypothesized that the pinworm parasites of primates (Oxyuridae, Nematoda) were highly host specific and that their evolutionary history mirrored that of their hosts. Subsequent work by Brooks & Glen (1982) and by Hugot (1999) have confirmed this assertion based on cladistic analyses of morphological data. The association between pinworms and their parasite hosts is considerably older than that between lice and their primate hosts. Pinworms parasitize members of the Strepsirrhini, Platyrrhini, and Catarrhini, and each primate group is parasitized by a different genus of pinworm. Because of this long-term association, pinworms have been used to evaluate hypotheses relating to the major radiations within living primates. The pinworm genus

Lemuricola accompanied the Strepsirrhini migration from the African mainland to Madagascar, and subsequently radiated in tandem with its host (Hugot, 1999). Species in the genus *Trypanoxyuris* accompanied the spread of the Platyrrhini into South America, and the pinworm genus *Enterobius* accompanied the radiation of the Catarrhini in Africa and the subsequent migrations of some taxa into Asia (Hugot, 1999). Although most pinworms show evidence of cospeciation, there are cases of host switches and sorting events (such as parasite extinction or lineage sorting) within primate orders (Hugot, 1999).

Pinworm species may be useful for finer scale questions because they are host specific (Hugot, 1999). To date, however, no DNA sequence data have been examined to estimate divergences within or among species. DNA sequence data exist for the species that occur in humans, and have even been sequenced from ancient material collected from coprolites of humans demonstrating that even degraded DNA can be sequenced for these taxa (Iniguez *et al.*, 2003) potentially providing a source of information about early humans or other primates.

Pneumocystis fungi

Another intriguing parasite of primates, as well as other mammals, is the fungus assigned to a single taxon, *Pneumocystis carinii*, which has coevolved with primates for presumably tens of millions of years (Demanche *et al.*, 2001; Guillot *et al.*, 2001; Hugot *et al.*, 2003). This fungus appears to be quite host specific as strains from a particular host cluster together phylogenetically even when some of the individuals were wild-caught while others were captive animals. Hugot *et al.* (2003) demonstrated that the primate and *Pneumocystis* tree topologies are much more similar than expected by chance alone, and are therefore likely to be the result of long-term coevolutionary history. Similar to the pinworms mentioned above, *Pneumocystis* fungi are found in Strepsirrhini, Platyrrhini, and Catarrhini. Molecular data have been evaluated for these species, however divergence date estimates have yet to be generated to investigate whether the timing of *Pneumocystis* divergences match those of their hosts.

Plasmodium

There are currently over 30 species in the genus *Plasmodium* that parasitize primates (Gysin, 1998; Di Fiore *et al.*, Chapter 7 this volume). Host switches are common, suggesting a complex evolutionary history between parasite and host species. The virulent human parasite, *P. falciparum*, and its sister taxon, the chimpanzee parasite *P. reichenowi*, consistently form a well-supported clade

(Escalante *et al.*, 1998; Leclerc *et al.*, 2004). *Plasmodium vivax*, which also parasitizes humans, is sister to and genetically similar to *Plasmodium simium*, which parasitizes three genera of New World primates. Both of the human parasites have the potential to provide great insight into human evolution and human migration patterns.

Plasmodium might be useful also in elucidating the evolutionary history of Asian primates. Recently, Mu *et al.* (2005) used cophylogenetic mapping to determine that a member of the genus *Macaca* most likely transferred *Plasmodium* from Africa to Asia where both cospeciation (within the genus *Macaca*) and host switching led to a rapid diversification of *Plasmodium* parasites. *Plasmodium vivax, P. simium,* and *P. hylobati,* which parasitizes gibbons, are nested within the *Plasmodium* species that parasitize macaques, further indicating a macaque malarial parasite ancestor (Escalante *et al.*, 1998). Escalante *et al.* (2005) later identified *P. fieldi* as the most closely related macaque parasite to the *P. vivax*/*P. simium* clade and *P. gonderi*, which parasitizes African mangabeys and mandrills, as sister to all Asian *Plasmodium* parasites. Furthermore, fossil evidence (Delson, 1980) and mtDNA (Hayasaka *et al.*, 1996) place the radiation of Asian macaques to be 2.1–2.5 million years ago, which coincides with estimates for their *Plasmodium* parasites (Escalante *et al.*, 1998).

Although the evolutionary history of *Plasmodium* as it relates to host evolution is obscured by host-switching, the parasite still holds considerable promise as a marker of host evolutionary history in primates. More complete taxon sampling may be the key to a better understanding of the utility of *Plasmodium*. Additionally, it may be effective to look within a few closely related taxa (*P. falciparum* and *P. reichenowi* in humans and chimpanzees) and to examine clades of species that have cospeciated, such as the *Plasmodium* species that infect Asian primates.

Simian foamy virus

Simian foamy viruses (SFV) are persistent, non-pathogenic retroviruses that infect a wide diversity of mammals, including primates. Recent work by Switzer *et al.* (2005) has shown that this virus has been coevolving with primates for at least 30 million years (see also Wolfe & Switzer, Chapter 17, this volume). In comparison with the cytochrome c oxidase subunit II (Cox2) gene of their primate hosts, the integrase gene of the virus seems to be evolving at roughly the same rate as their host's mitochondrial genome, which makes it one of the slowest evolving RNA viruses yet examined (substitution rates $= 1.7 \pm 0.45 \times 10^{-8}$ substitutions/site/year). As such, it can be used to estimate older divergences among primates than typical RNA viruses. For example, Switzer *et al.* (2005) were able to estimate the divergence between chimpanzees and gorillas at 9.8–10.8 million years ago using the Cercopithecoid/Hominoid divergence of 25–30 million years as a calibration. Calattini *et al.* (2006) were able to resolve the relationships of four subspecies of *Pan troglodytes* as well as the two clades of *Gorilla* with the integrase gene from SFV. However, Calattini *et al.* (2006) did not estimate divergence dates within the SFV-based phylogenetic tree.

We estimated SFV divergence dates using data from Calattini *et al.* (2006) to determine if these dates match those known from host DNA sequence data or data from Switzer *et al.* (2005). Data from Calattini *et al.* (2006) were downloaded from GenBank and aligned with Clustal X (Thompson *et al.*, 1997) and corrected by eye. MODELTEST (Posada & Crandall, 1998) was used to find a best-fit model of nucleotide evolution, which was then used in a maximum likelihood phylogeny constructed using a best-fit model of nucleotide evolution in Paup* v. b4.10 (Swofford, 2006). This tree topology (Figure 10.3) was almost identical to the one reported by Calattini *et al.* (2006). We then used non-parametric rate smoothing in the software package r8s (Sanderson, 2003) to estimate divergences within the tree using the Cercopithecoid/Hominoid divergence of 25–30 million years. This resulted in an estimated divergence between chimpanzees and gorillas of 7.46 million years, and an age of the genus *Pan* at 4.65 million years (Figure 10.3). The age of the most recent common ancestor (MRCA) of chimpanzees and gorillas is younger than the age estimated by Switzer *et al.* (2005), but is similar to dates estimated elsewhere (Goodman *et al.*, 1998; Page & Goodman, 2001; Stauffer *et al.*, 2001). Like other parasites/pathogens described in this chapter, SFV looks especially promising as a marker of primate evolutionary history, especially among closely related species with relatively young divergence times.

With SFV, and many of the other parasites and pathogens mentioned above, we are only just beginning to explore host evolutionary history. For example, several parasites are now confirming splits between chimpanzees and humans of roughly 5–7 million years ago. These dates are well established from host data, but the real advancement will come as we assemble parasite data sets for host taxa of which we know considerably less.

Other parasites as markers of human evolution

Most of the work that demonstrates the full potential of parasites and pathogens for inferring host evolutionary history comes from the human evolution literature, which is reviewed elsewhere (Leal & Zanotto, 2000; Disotell, 2003; Pavesi, 2005). Rather than provide yet another review of this literature, we feel

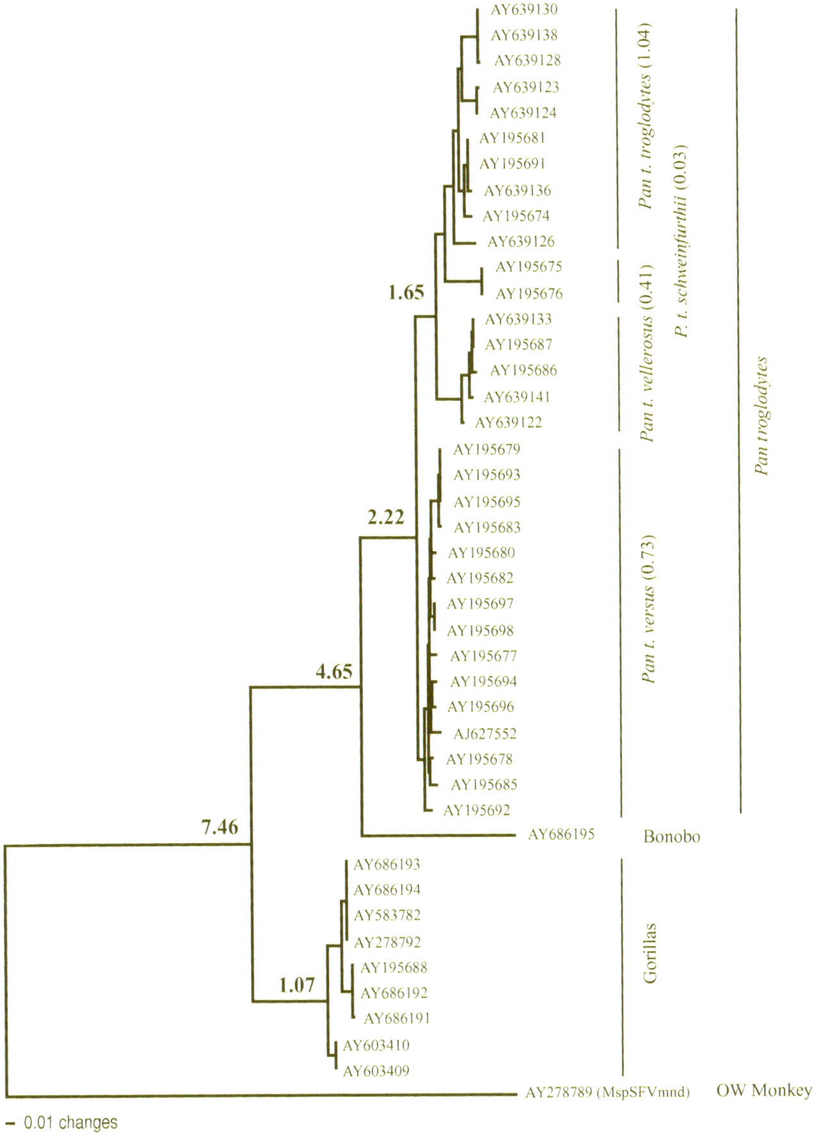

Figure 10.3. Maximum likelihood phylogeny of simian foamy virus based on integrase gene sequences downloaded from GenBank (accession numbers given at each terminus). Taxonomic identifications are given to the right of the tree. Ages of most recent common ancestors are given above the nodes, and ages of taxonomic groups are given in parentheses after taxonomic name (in millions of years).

that it is useful to highlight a few of the studies that are particularly interesting with regard to the inference of host evolutionary history.

Helicobacter pylori

One of the best-studied and most effective pathogens for the study of recent human population dynamics is the bacterium *Helicobacter pylori*, the causative agent of human peptic ulcers. *Helicobacter pylori* is useful for tracking human migrations because it has low pathogenicity, occurs in nearly every human population, does not spread epidemically, and it is genetically very diverse (Wirth *et al.*, 2005). Falush *et al.* (2003) analyzed 370 strains of *H. pylori* in 27 different geographical, ethnic, or linguistic groups using a Bayesian approach. They identified four modern populations (hpAfrica1, hpAfrica2, hpEastAsia, and hpEurope) that could be further subdivided. HpAfrica1 can be further divided into hspWAfrica, which was found in West Africa, the Americas, and South Africa, and hspSAfrica, which was found only in South Africa. HpEastAsia could be further divided into hspAmerind (found in Inuits and Amerindians), hspEAsia, and hspMaori (found in Maoris and Polynesians in New Zealand). HpEurope contained a large assemblage of ethnic groups, including Europeans, Turks, Israelis, Bangladeshis, Ladakhis, Sudanese, Americans, Australians, and South Africans. However, hpEurope could not be divided easily into subpopulations. Overall, the distinction among ethnic and geographic groups shows the level of detail with which *H. pylori* can serve as a marker of host groups. Several interesting findings arose from the Falush *et al.* (2003) study, such as the strong signal of genetic drift in HspMaori compared to hspEastAsia, which indicates a long separation from the *H. pylori* source population. However, HspAmerind showed significantly less drift, indicating that either large numbers of Asians migrated to the Americas, or there were several waves of migration from Asia to the Americas to keep the genetic structure similar (Falush *et al.*, 2003). These findings present testable hypotheses for the examination of human data or more importantly data from additional parasites.

Another recent study of *H. pylori* tested the hypothesis that the bacterium could provide greater resolution of host population genetics than could host data (Wirth *et al.*, 2004). Wirth *et al.* (2004) examined *H. pylori* samples collected from Ladakh, India where two major ethnic communities reside together (Buddhists and Muslims). The two communities have remained distinct for 500–1000 years, yet fast-evolving mitochondrial DNA and microsatellite loci failed to detect genetic differentiation between them (Wirth *et al.*, 2004). DNA sequence data from *H. pylori* populations did detect differences between the two populations as well as differences in their ancestry showing the utility of host-specific parasites and pathogens for inferring host population genetics.

Tapeworms

Three host-specific species of tapeworm infect humans, *Taenia saginata*, *T. asiatica*, and *T. solium* with the obligatory intermediate hosts for *T. saginata* being cattle and *T. asiatica* and *T. solium* being swine. Given that the intermediate hosts are domesticated livestock it was long held that the domestication of cattle and pigs by modern humans permitted these parasites to infect humans through a host switch from an African carnivore (the suspected original definitive host). However, Hoberg *et al.* (2001) used phylogenetic and co-phylogenetic methods to demonstrate that there were two distinct acquisitions of tapeworms in humans (both from carnivore definitive hosts) and that one of these acquisitions predated the origin of modern humans and certainly the domestication of cattle and pigs. Hoberg *et al.* (2001) propose that humans acquired tapeworms by scavenging on bovids long before humans began to domesticate them. Ironically, it seems that domesticated cattle and pigs acquired their tapeworms from humans rather than the other way around.

Human polyomavirus JC virus

The human JC virus (JCV) is part of the polyomavirus group, which is a group of double-stranded DNA viruses that are generally quite host specific. Although JCV is slow evolving among viruses it evolves at a rate fast enough to track human migrations. JC virus occurs in high frequency in human populations, is readily excreted in urine, and is primarily transmitted among cohabitating family members. First infections usually occur in early childhood, and JCV is generally benign to its host unless they become immunocompromised. JC virus is estimated to occur in 70–90% of the human population (Ault & Stoner, 1992), which permits large-scale sampling via simple urine samples (Yogo *et al.*, 2004).

Sugimoto *et al.* (2002) constructed a phylogenetic tree from 65 complete genome sequences of viral DNA using neighbor-joining, maximum parsimony, and maximum likelihood. All three methods revealed three ancestral clusters of viruses subsequently designated A, B, and C (Sugimoto *et al.*, 2002). Using the geographic distributions of the ancestral populations of JCVs, it can be inferred that the population carrying type A gave rise to Europeans, the population carrying type C gave rise to Africans, and the population carrying type B gave rise to several populations of Asians and small populations of Europeans and Africans. Sugimoto *et al.* (2002) suggest that two exoduses occurred from Africa: one directly to Europe (type A) and one to Eurasia (type B). The population carrying type A split into EU-a and EU-b and migrated to present day Europe. It is hypothesized that EU-a and EU-b were at one time settled into separate regions, but eventually enough admixture blurred any distinction. EU-a is also found in northern Japan, Siberia, and Arctic North America. It is possible that Eu-a split into several small populations, some of which migrated east from Europe. The first subtype to split from type B is Af2, which is found in Africa. The second split that occurred within type B is B1-c, a minor European subtype. All subsequent splits produced Asian and Oceanic genotypes (B1-a, B1-b, B1-d, MY, B2, CY, and SC). Type C split into two distinct African subtypes, Af1 and Af3. The three major clades identified by Sugimoto *et al.* (2002) are largely consistent geographically with the three clades seen in human lice (Figure 4 of Reed *et al.*, 2004). Further study could determine whether clade ages are similar between the human parasite (lice) and pathogen (JCV).

Conclusions

Parasites (broadly defined) contain valuable information about their hosts. By examining each parasite in turn, and by examining several at once, we can learn much about the hosts that carry these parasites through time. When studying a permanent parasite one can capture, study, and describe 100% of the niche of that parasite (its host), and the ecological interactions between the host and parasite can be finely studied. When studying an obligate parasite, one knows that the parasite is bound to a particular host in ecological time, and examining that parasite likely will add information to our understanding of host evolutionary history. Those parasites that are both obligate and permanent, especially those that cospeciate and coevolve with their hosts, are likely to contain rich sources of data about host evolution. However, it is interesting to note that several parasites described within this chapter are merely temporary parasites (e.g. pinworms and *Plasmodium*), and yet these parasites still appear to be very useful for inferring host evolutionary history. More studies are needed to fully realize the potential of parasites and pathogens for inferring host evolutionary history. Once a critical mass of studies is available, then perhaps we can adequately triangulate host evolutionary history using a wide diversity of associated parasites, pathogens, mutualists, and commensals.

Acknowledgements

This work was supported by grants to DLR from the University of Florida Research Opportunity SEED Fund and the National Science Foundation (DEB 0445712 and DEB 0555024).

References

Allen, J. M., Reed, D. L., Perotti, M. A. & Braig, H. R. (2007). Evolutionary relationships of *Candidatus* Riesia spp., endosymbiotic *Enterobacteriaceae* living within hematophagous primate lice. *Applied and Environmental Microbiology*, **73**, 1659–1664.

Ault, G. S. & Stoner, G. L. (1992). Two major types of JC virus defined in progressive multifocal leukoencephalopathy brain by early and late coding region DNA-sequences. *Journal of General Virology*, **73**, 2669–2678.

Barker, S. C. (1991). Evolution of host-parasite associations among species of lice and rock-wallabies: coevolution? (J. F. A. Sprent Prize lecture, August 1990). *International Journal for Parasitology*, **21**, 497–501.

Brooks, D. R. (1979). Testing the context and extent of host-parasite coevolution. *Systematic Zoology*, **28**, 299–307.

Brooks, D. R. & Glen, D. R. (1982). Pinworms and primates: a case study in coevolution. *Proceedings of the Helminth Society of Washington*, **49**, 76–85.

Buchner, P. (1965). *Endosymbiosis of Animals with Plant Microorganisms*. New York, NY: Interscience.

Calattini, S., Nerrienet, E., Mauclere, P. *et al.* (2006). Detection and molecular characterization of foamy viruses in Central African chimpanzees of the *Pan troglodytes troglodytes* and *Pan troglodytes vellerosus* subspecies. *Journal of Medical Primatology*, **35**, 59–66.

Cameron, T. W. M. (1929). The species of *Enterobius* Leach, in primates. *Journal of Helminthology*, **7**, 161–182.

Delson, E. (1980). Fossil macaques, phyletic relationships and a scenario of deployment. In *The Macaques, Studies in Ecology, Behavior and Evolution*, ed. D. G. Lindburg. New York, NY: Van Nostrand Reinhold Co., pp. 10–29.

Demanche, C., Berthelemy, M., Petit, T. *et al.* (2001). Phylogeny of *Pneumocystis carinii* from 18 primate species confirms host specificity and suggests coevolution. *Journal of Clinical Microbiology*, **39**, 2126–2133.

di Rienzo, A. & Wilson, A. C. (1991). Branching pattern in the evolutionary tree for human mitochondrial DNA. *Proceedings of the National Academy of Sciences, USA*, **88**, 1597–1601.

Disotell, T. R. (2003). Discovering human history from stomach bacteria. *Genome Biology*, **4**, Issue 5.

Douglas, A. E. (1989). Mycetocyte symbiosis in insects. *Biological Reviews of the Cambridge Philosophical Society*, **64**, 409–34.

Ehrlich, P. R. & Raven, P. H. (1964). Butterflies and plants: a study in coevolution. *Evolution*, **18**, 586–608.

Escalante, A. A., Freeland, D. E., Collins, W. E. & Lal, A. A. (1998). The evolution of primate malaria parasites based on the gene encoding cytochrome *b* from the linear mitochondrial genome. *Proceedings of the National Academy of Sciences, USA*, **95**, 8124–8129.

Escalante, A. A., Cornejo, O. E., Freeland, D. E. *et al.* (2005). A monkey's tale: the origin of *Plasmodium vivax* as a human malaria parasite. *Proceedings of the National Academy of Sciences, USA*, **102**, 1980–1985.

Falush, D., Wirth, T., Linz, B. *et al.* (2003). Traces of human migrations in *Helicobacter pylori* populations. *Science*, **299**, 1582–1585.

Goodman, M., Porter, C. A., Czelusniak, J. *et al.* (1998). Toward a phylogenetic classification of primates based on DNA evidence complemented by fossil evidence. *Molecular Phylogenetics and Evolution*, **8**, 585–598.

Guillot, J., Demache, C., Hugot, J. P. *et al.* (2001). Parallel phylogenies of *Pneumocystis* species and their mammalian hosts. *Journal of Eukaryotic Microbiology*, **48**, 113S–115S.

Gysin, J. (1998). Animal models: primates. In *Malaria: Parasite Biology, Pathogenesis and Protection*, ed. I. W. Sherman. Washington, DC: ASM Press, pp. 419–441.

Hafner, M. S., Sudman, P. D., Villablanca, F. X. *et al.* (1994). Disparate rates of molecular evolution in cospeciating hosts and parasites. *Science*, **265**, 1087–1090.

Harpending, H. C., Sherry, S. T., Rogers, A. R. & Stoneking, M. (1993). The genetic structure of ancient human populations. *Current Anthropology*, **34**, 483–496.

Hasegawa, H. (1999). Phylogeny, host-parasite relationship and zoogeography. *Korean Journal of Parasitology*, **37**, 197–213.

Hayasaka, K., Fugii, K. & Horai, S. (1996). Molecular phylogeny of macaques: implications of nucleotide sequences from an 896-base pair region of mitochondrial DNA. *Molecular Biology and Evolution*, **13**, 1044–1053.

Hoberg, E. P., Alkire, N. L., de Queiroz, A. & Jones, A. (2001). Out of Africa: origins of the *Taenia* tapeworms in humans. *Proceedings of the Royal Society of London B*, **268**, 781–787.

Holmes, E. C. (2004). The phylogeography of human viruses. *Molecular Ecology*, **13**, 745–756.

Hooke, R. (1664). Micrographia: or some physiological description of minute bodies made by magnifying glasses with observations and inquiries thereupon. *Council of the Royal Society of London for Improving Natural Knowledge London*, **244**.

Hugot, J. (1999). Primates and their pinworm parasites: the Cameron hypothesis revisited. *Systematic Biology*, **48**, 523–546.

Hugot, J. P., Demanche, C., Barriel, V., Dei-Cas, E. & Guillot, J. (2003). Phylogenetic systematics and evolution of primate-derived *Pneumocystis* based on mitochondrial or nuclear DNA comparison. *Systematic Biology*, **52**, 735–744.

Iniguez, A. M., Reinhard, K. J., Araujo, A., Ferreira, L. F. & Vicente, A. C. P. (2003). *Enterobius vermicularis*: ancient DNA from North and South American human coprolites. *Memorias do Instituto Oswaldo Cruz*, **98**, 67–69.

Johnson, K. P. & Clayton, D. H. (2003). Coevolutionary history of ecological replicates: comparing phylogenies of wing and body lice to columbiform hosts. In *Tangled Trees: Phylogeny, Cospeciation and Coevolution*, ed. R. Page. Chicago, IL: University of Chicago, pp. 262–286.

Kittler, R., Kayser, M. & Stoneking, M. (2003). Molecular evolution of *Pediculus humanus* and the origin of clothing. *Current Biology*, **13**, 1414–1417.

Kittler, R., Kayser, M. & Stoneking, M. (2004). Molecular evolution of *Pediculus humanus* and the origin of clothing (Vol. 13, p. 1414, 2003). *Current Biology*, **14**, 2309.

Leal, E. S. & Zanotto, P. M. A. (2000). Viral diseases and human evolution. *Memorias do Instituto Oswaldo Cruz*, **95**, 193–200.

Leclerc, M. C., Durand, P., Gauthier, C. *et al.* (2004). Meager genetic variability of the human malaria agent *Plasmodium vivax. Proceedings of the National Academy of Sciences, USA*, **2004**, 14455–14460.

Leo, N. P. & Barker, S. C. (2005). Unravelling the evolution of the head lice and body lice of humans. *Parasitology Research*, **98**, 44–47.

Light, J. E. (2005). Host-parasite cophylogeny and rates of evolution in two rodent-louse assemblages. PhD thesis. Baton Rouge, LA: Louisiana State University, p. 271.

Moran, N. A., von Dohlen, C. D. & Baumann, P. (1995). Faster evolutionary rates in endosymbiotic bacteria than in cospeciating insect hosts. *Journal of Molecular Evolution*, **41**, 727–731.

Mu, J., Joy, D. A., Duan, J. *et al.* (2005). Host switch leads to emergence of *Plasmodium vivax* malaria in humans. *Molecular Biology and Evolution*, **22**, 1686–1693.

Munson, M. A., Baumann, P., Clark, M. A. *et al.* (1991). Evidence for the establishment of aphid-eubacterium endosymbiosis in an ancestor of 4 aphid families. *Journal of Bacteriology*, **173**, 6321–6324.

Page, R. D. M. (1996). Temporal congruence revisited: comparison of mitochondrial DNA sequence divergence in cospeciating pocket gophers and their chewing lice. *Systematic Biology*, **45**, 151–167.

Page, R. D. M. (2003). *Tangled Trees: Phylogenies, Cospeciation, and Coevolution.* Chicago, IL: University of Chicago Press.

Page, R. D., Lee, P. L., Becher, S. A., Griffiths, R. & Clayton, D. H. (1998). A different tempo of mitochondrial DNA evolution in birds and their parasitic lice. *Molecular and Phylogenetic Evolution*, **9**, 276–293.

Page, S. L. & Goodman, M. (2001). Catarrhine phylogeny: noncoding DNA evidence for a diphyletic origin of the mangabeys and for a human-chimpanzee clade. *Molecular Phylogenetics and Evolution*, **18**, 14–25.

Pavesi, A. (2003). African origin of polyomavirus JC and implications for prehistoric human migrations. *Journal of Molecular Evolution*, **56**, 564–572.

Pavesi, A. (2005). Microbes coevolving with human host and ancient human migrations. *Journal of Anthropological Sciences*, **83**, 9–28.

Percy, D. M., Page, R. D. M. & Cronk, Q. C. B. (2004). Plant-insect interactions: double-dating associated insect and plant lineages reveals asynchronous radiations. *Systematic Biology*, **53**, 120–127.

Posada, D. & Crandall, K. A. (1998). MODELTEST: testing the model of DNA substitution. *Bioinformatics*, **14**, 817–818.

Reed, D. L., Smith, V. S., Hammond, S. L., Rogers, A. R. & Clayton, D. H. (2004). Genetic analysis of lice supports direct contact between modern and archaic humans. *Public Library of Science, Biology*, **2**, e304.

Reed, D. L., Light, J. E., Allen, J. M. & Kirchman, J. J. (2007). Pair of lice lost or parasites regained: the evolutionary history of anthropoid primate lice. *BMC Biology*, **5**, Issue 7.

Rogers, A. R. & Harpending, H. (1992). Population growth makes waves in the distribution of pairwise genetic differences. *Molecular Biology and Evolution*, **9**, 552–569.

Sanderson, M. J. (2003). r8s: inferring absolute rates of molecular evolution and divergence times in the absence of a molecular clock. *Bioinformatics*, **19**, 301–302.

Sasaki-Fukatsu, K., Koga, R., Nikoh, N. *et al.* (2006). Symbiotic bacteria associated with stomach discs of human lice. *Applied Environmental Microbiology*, **72**, 7349–7352.

Stauffer, R. L., Walker, A., Ryder, O. A., Lyons-Weiler, M. & Hedges, S. B. (2001). Human and ape molecular clocks and constraints on paleontological hypotheses. *Journal of Heredity*, **92**, 469–474.

Sugimoto, C., Hasegawa, M., Kato, A. *et al.* (2002). Evolution of human polyomavirus JC: implications for the population history of humans. *Journal of Molecular Evolution*, **54**, 285–297.

Switzer, W. M., Salemi, M., Shanmugam, V. *et al.* (2005). Ancience co-speciation of simian foamy viruses and primates. *Nature*, **434**, 376–380.

Swofford, D. L. (2006). *PAUP**: Phylogenetic Analysis using Parsimony* (*and Other Methods). Sunderland, MA: Sinauer.

Thompson, J. D., Gibson, T. J., Plewniak, F., Jeanmougin, F. & Higgins, D. G. (1997). The ClustalX windows interface: flexible strategies for multiple sequence alignment aided by quality analysis tools. *Nucleic Acids Research*, **24**, 4876–4882.

Weckstein, J. D. (2004). Biogeography explains cophylogenetic patterns in toucan chewing lice. *Systematic Biology*, **53**, 154–164.

Wirth, T., Wang, X. Y., Linz, B. *et al.* (2004). Distinguishing human ethnic groups by means of sequences from *Helicobacter pylori*: lessons from Ladakh. *Proceedings of the National Academy of Sciences, USA*, **101**, 4746–4751.

Wirth, T., Meyer, A. & Achtman, M. (2005). Deciphering host migrations and origins by means of their microbes. *Molecular Ecology*, **14**, 3289–3306.

Yoder, A. D., Olson, L. E., Hanley, C. *et al.* (2005). A multidimensional approach for detecting species patterns in Malagasy vertebrates. *Proceedings of the National Academy of Sciences, USA*, **102**, 6587–6594.

Yogo, Y., Sugimoto, C., Zheng, H. Y. *et al.* (2004). JC virus genotyping offers a new paradigm in the study of human populations. *Reviews in Medical Virology*, **14**, 179–191.

11 *Cryptic species and biodiversity of lice from primates*

NATALIE P. LEO

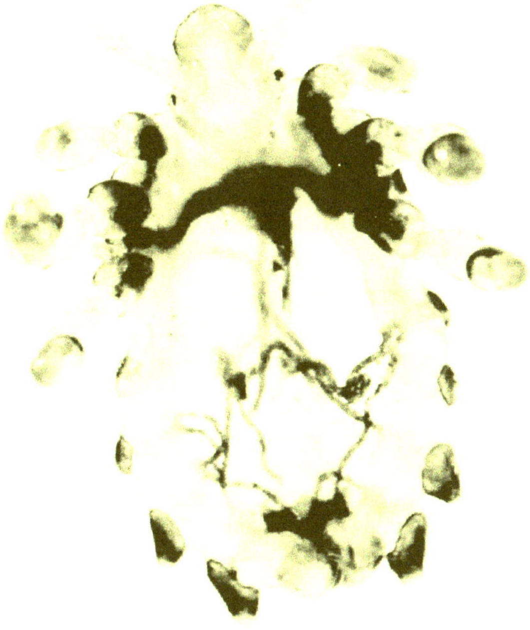

Photograph by Natalie Leo (*Pthirus gorillae*, host: *Gorilla beringei*)

Primate Parasite Ecology. The Dynamics and Study of Host–Parasite Relationships, ed. Michael A. Huffman and Colin A. Chapman. Published by Cambridge University Press.

Introduction

Cryptic species are species that are reproductively isolated, but morphologically indistinguishable. Usually, cryptic species occur under conditions where there is strong selection to keep a particular morphology. Many cryptic species that are known today are arthropods of medical importance, since cryptic species differ in behaviors that can affect their ability to transmit pathogens. For example, different species of anopheline mosquitoes differ in their ability to transmit the protozoan parasites that cause malaria and thus identification of these cryptic species is desirable (Singh *et al.*, 2004). Once cryptic mosquito species are identified, they can be studied for differences in their feeding preferences (human or animal), resting preferences (indoors or outdoors), biting habits (if they are likely to bite multiple people), and insecticide resistance profiles. This information can then be used to identify the likely vectors of malaria in a particular area, and to implement a targeted control program.

Another arthropod of medical importance is the head louse and body louse of humans. Body lice are known to transmit harmful bacterial pathogens, but the role of head lice during outbreaks of disease is less certain (Robinson *et al.*, 2003). The species status of these lice has been an issue of debate for many decades. This is because, among other things, they differ in where they live (head hair and clothes), in their biology (e.g. frequency of feeding, number of eggs laid per day, longevity), and apparently in their ability to transmit pathogens. However, they are so similar in their morphology that a single louse picked at random cannot be identified with confidence as a head or body louse. More recently, researchers turned to molecular methods for an answer. The results from phylogenetic analyses showed that gene sequences from head lice and body lice do not form two distinct lineages, which was interpreted as evidence that head and body lice are the same species (Leo *et al.*, 2002). However, a population genetics approach showed that head lice had not interbred with body lice, even when they infested a person simultaneously (Leo *et al.*, 2005). Since sympatric populations of head lice and body lice do not interbreed, this indicates that they are cryptic species. The phylogenetic results do not contradict this result; if head and body lice diverged so recently that the gene tree does not yet match the species tree (otherwise known as incomplete lineage sorting). This seems likely since the divergence of head and body lice has been estimated to be as recent as 100 000 years ago (Kittler *et al.*, 2004).

The presence of cryptic species of lice on humans raises the question of whether there are cryptic species of lice on other primates. An examination of the literature reveals that there is a case for cryptic species of lice from primates. A taxonomic checklist of the sucking lice and their hosts shows that many of the louse species from primates are found on multiple hosts (Durden & Musser,

1994). This is at odds with the axiom that lice are host-specific. Interestingly, a look at the original species descriptions shows that many of these multi-host species of lice were initially described as different louse species. These were later collapsed into larger species groups on the justification that the original descriptions were of insufficient detail and based on obscure character differences (Ferris, 1935). Thus it is plausible that there might be some, as yet unidentified, cryptic species of primate lice.

A determined effort to check for the presence of cryptic species of lice will not only improve current understanding of primate louse taxonomy, it will provide a solid basis for studies of louse–primate coevolution. While alpha taxonomy cannot tell us about evolutionary relationships (phylogeny), it provides a starting point for investigations into phylogeny. This is because before we can ask how taxa A and B are related, we first need to know that taxa A and B are different. If a species has not been described in the alpha taxonomy, it will not be sampled or included in the phylogeny. Coevolutionary studies compare host–parasite phylogenies, so the absence of a host or parasite species can change the interpretation of coevolutionary events such as cospeciation.

Lice have been used as evolutionary markers for humans (Kittler *et al.*, 2003; Reed *et al.*, 2004) and it seems only a matter of time before attention is turned towards the lice from non-human primates. Lice can reveal events in host history that would otherwise not be recorded in the host DNA. For example, the date for when humans first began to wear clothing was estimated from the divergence date of head and body lice (Kittler *et al.*, 2003), since body lice presumably evolved after the habitat of clothing was invented. Similarly, the date of divergence of two ancient lineages of head lice was used as evidence of direct contact between modern humans and *Homo erectus* (see Reed *et al.*, 2004 and Reed *et al.*, Chapter 10, this volume). It seems likely that the lice from non-human primates could give similar insight into the evolutionary history of their hosts. However, the taxonomy of primate lice is poorly understood (Ferris, 1951; Kim & Emerson, 1968; Durden & Adams, 2005) and is thus not ideal for coevolution studies that require accurate taxonomies and phylogenies of host and parasite. The reasons for the poor state of taxonomy for these lice are varied, but perhaps the greatest obstacle has been a paucity of specimens.

For a robust taxonomy we need lice, and lice from primates are particularly difficult to collect. This is because lice are permanent parasites and thus must be collected directly from the body of the host. Parasitologists do not have this sort of access to primates and must rely on those who work with primates to supply them with specimens. So taxonomy of these lice requires collaboration between parasitologists and those who work on primates: veterinarians, sanctuary staff, and field primatologists. However, this collaboration does not occur often since lice from non-human primates do not usually cause disease in their hosts and are thus not a high priority for most primate researchers or caretakers. I have found that veterinarians are the most likely to reply positively to my requests for lice. Perhaps this is because veterinarians learn about lice as part of their veterinary training, have experience with lice as part of their work, and as a result have a reason to learn more. Among people who have not had the opportunity to learn about lice, a common attitude is that lice should be eradicated as soon as possible. To these people, it may come as a surprise that lice are perceived positively in some human cultures, where infestations are deliberately started and maintained (Trigger, 1981). Since lice are so difficult to collect from primates, a complete picture of their distribution and biodiversity can only be achieved with the active participation of those who work with primates. The most recent taxonomic checklist of the sucking lice shows that fewer than 30 species of lice have been described from primates (Durden & Musser, 1994). There are approximately 350 species of primates (Groves, 2001), yet only 63 of these are known to be principal hosts of lice (Durden & Musser, 1994). While a few primate lice have been described since this checklist (e.g. Mey, 1994), it is fair to say that the current number of louse species known from primates today is an underestimate of their true biodiversity.

Apart from providing a solid taxonomy for coevolution studies, a thorough investigation of the lice from primates is expected to reveal cryptic species and overall patterns of biodiversity. Traditional studies of coevolution compare host and parasite phylogenies to infer coevolutionary events such as cospeciation and host-switching (Banks *et al.*, 2006). More recent studies have compared louse phylogeny to events in the evolutionary history of the host (Kittler *et al.*, 2003; Reed *et al.*, 2004; Leo & Barker, 2005). It is hoped that patterns of biodiversity will provide a third perspective of louse and primate coevolution. Patterns in the biodiversity of lice from primates can be used to generate hypotheses about the evolution of primate behavior and morphology. For example, why is it that humans have relatively high louse biodiversity (three species), while other ape species have only one or no louse species? For permanent parasites such as lice, changes in host morphology and behavior can effect changes in louse evolution. Uncovering the true biodiversity of lice from primates has the potential to help us to better understand how events in the evolution of primates have influenced the evolution of louse host-specificity and biodiversity.

Taxonomy

The beginnings of primate louse taxonomy were far from ideal. Initial descriptions were often based on obscure characters, and species names were taken from trivial details which did not aid identification (Ferris, 1935). A major revision of all sucking lice concluded that many of the lice species described from monkeys were in fact members of the same species, and so these were condensed into fewer louse species (Ferris, 1935). This revision solved many of the problems of the nomenclature for these lice, and was intended to provide a solid basis for future studies. Yet the taxonomy of lice from some primates, such as neotropical cebid monkeys, is still acknowledged to be poorly understood (Ferris, 1951; Kim & Emerson, 1968; Durden & Adams, 2005). The reasons for this apparent lack of progress are addressed here.

The collection of lice from primates is usually opportunistic and thus patchy. Sometimes only some life stages of the lice, such as eggs or nymphs, are found. Ideally, species descriptions are taken from many specimens, so as to account for intraspecific variation. However, this is not always possible for lice. For example, the louse from gorillas (*Pthirus gorillae*) has only been described from nymphs and adult females. To my knowledge, a male *P. gorillae* has never been found. This might indicate that gorilla lice are parthenogenetic, that is, females are able to reproduce without males. However, it is much more likely that the failure to find male lice is due to a sex bias in these lice (Buxton, 1941; Rozsa, 1997; Perotti *et al.*, 2004), exacerbated by the fact that very few lice have ever been collected from gorillas.

Taxonomists can only work with what they have, and in the case of primate lice this is sometimes not very much. In the early days of taxonomy, species descriptions of lice often had little detail and were sometimes based on small differences in morphology from few, even a single louse (Ewing, 1926). Yet the actions of these early taxonomists are understandable given that taxonomy was still a new field, there was a paucity of specimens available, and there was a belief at the time that lice were highly host-specific. Now we know that lice are not always highly host-specific (Barker, 1994). Many of the species characters from initial descriptions of lice were later dismissed as intraspecific variation and the lice collapsed into larger species groups (Ferris, 1935). However, the presence of cryptic species of lice on humans raises the question of whether some of these initial descriptions represent cryptic species. Those small differences that were deemed to be intraspecific variation might instead be differences between cryptic species of lice. While this would be nearly impossible to prove with the small sample sizes of lice that are usually collected from primates, molecular tools allow us to identify cryptic species from just a few lice and from any life stage. Although researchers today have the advantage of molecular tools to help them identify new species, the value of a morphological description for a new species should not be overlooked. The aim should be to include both whenever possible.

Since wild primates are often difficult to access and not easily searched for lice, many louse specimens have come from the skins of primates that died in captivity. However, this in itself presents problems for taxonomy. While lice are generally thought to be host-specific, the truth is that there are many exceptions to this rule (Barker, 1994). Indeed, an experimental study of lice on captive doves indicated that the hosts, not the lice, are largely responsible for louse "host-specificity" (Clayton *et al.*, 2003). This study showed that birds restricted from preening were more susceptible to colonization by an alien species of lice, and that these lice had high rates of survival (Clayton *et al.*, 2003). If lice are not necessarily host-specific, and if host-specificity is largely influenced by the ability of the host to groom effectively, then we should look more critically upon those lice collected from captive primates. It seems plausible that these lice might not occur naturally on these animals and might instead represent instances of host-switching events.

In the first half of the twentieth century, primates were often transported overseas and placed in zoological parks. These conditions were likely to be highly stressful for the animals. New louse species were sometimes described from lice whose primate hosts had died within days or weeks of reaching a zoological park (Ewing, 1926). A primate kept in bad conditions is likely to be less efficient at grooming and thus more susceptible to colonization and establishment by an alien louse species than under natural conditions. Stress, sickness, restriction of movement, as well as close proximity to unfamiliar animals (e.g. humans), might all contribute to the successful establishment of an alien louse species. So there are three possibilities that need to be considered when studying multi-host lice from captive primates: they are multi-host species, they are as yet unidentified cryptic species, or they are alien species that do not normally occur on these hosts (called stragglers in the louse literature).

The fewer lice there are to work with, the more difficult it is to distinguish real differences in morphology from artifacts of the preparation process. Microscopic preparations can cause expansion, collapse, distortion, and parallax error of the exoskeleton of lice, even in areas of the body that are relatively hard (Ferris, 1935). The method used to preserve the animal skin might have distorted the lice. Moreover, comparison of small numbers of lice makes it more difficult to distinguish interspecific variation from intraspecific variation. This has led to incorrect identification of new species in the past (Ferris, 1935). Finally, sometimes only eggs or nymphal stages can be collected. Lice that look quite similar in gross morphology can often be easily distinguished by male genitalia (Mey, 1994), so the absence of male specimens can greatly increase the difficulty of finding characters that can be used to easily identify the species.

Fortunately, the problems associated with unidentified cryptic species and working with a limited number of specimens can be largely overcome with molecular tools. Researchers using molecular methods should be able to identify cryptic species and host-switch events with relative ease, giving particular attention to louse species found from multiple primate hosts and especially those from captive primates. Today, the greatest obstacle towards a robust taxonomy of these lice is a lack of specimens. This is an area in which those working on primates can make a great contribution. While it might be too much to expect non-parasitologists to think lice deserve to be conserved in their own right, it has been pointed out that lice and other parasites might be saved by their ability to serve as indicators of biology and evolution of their hosts (Ashford, 2000). While the taxonomy of lice seems academic, it is a necessary first step for studies of lice–primate coevolution. In addition to providing new insight into events in host evolution, knowledge of which lice are multi-host species and which are alien species for a particular primate could be useful in the care of captive primates. In developed countries, lice are feared and reviled, and measures are taken to exterminate them. This is understandable since lice in humans are known to transmit some very serious diseases, such as epidemic typhus (Gross, 1996; Raoult & Roux, 1999). However, lice from primates are not known to transmit any diseases, and lice in small to moderate numbers do not in themselves cause serious illness. Clinical observations of humans with continued exposure to moderate numbers of head lice showed they had a skin reaction at the point of the bite, followed by itching (Mumcuoglu *et al.*, 1991). Indeed, lice in small numbers might indirectly have a positive effect on the social interaction of primates. In Japanese macaques, grooming has the primary function of hygiene and the secondary function of social bonding (Tanaka, 1995; Zamma, 2002). In some indigenous human cultures, louse infestations are deliberately maintained so that they can be groomed out by family and close friends (Trigger, 1981). Thus, a greater understanding of which lice occur naturally on a given primate species might cause caretakers to think twice about eradicating lice. For caretakers of captive primates, much time and energy could be saved if lice were only treated when alien species of lice were present, or if the lice were present in numbers that pose a threat to the health of the primate.

Pediculus mjobergi – an enigma

The genus *Pediculus* consists of the body louse *P. humanus*, and head louse *P. capitis* from humans, *P. schaeffi* from chimpanzees and bonobos, and *P. mjobergi* from some New World monkeys (*Ateles* spp., *Alouatta* spp., and *Cebus* spp.). Thus *P. mjobergi* does not follow expectations of lice and host cospeciation, where closely related lice are usually found on closely related hosts.

The taxonomic status of *P. mjobergi* is currently unresolved. At one time, four species of lice in the genus *Parapediculus* were recognized from spider monkeys (Ewing, 1926). These were later revised and it was concluded that all but one of these species were actually lice from humans (*P. humanus* and *P. capitis*) that had apparently transferred to monkeys held in captivity (Ferris, 1935). The exception was renamed *P. mjobergi*. This was interesting since *P. mjobergi* is found on monkeys of the genera *Ateles*, *Alouatta*, and *Cebus*, which are not closely related to man. Furthermore, this louse species has only ever been found from captive monkeys, despite efforts to find them on wild monkeys (Ferris, 1935). It was tentatively given species status because it shows very little variation in morphology within the species, despite the fact that its morphology lies within the range of variation known for human lice (Ferris, 1935). It is expected that future studies will clarify the species status of *P. mjobergi* (Ferris, 1935). Certainly, there have been efforts by parasitologists to solve this mystery but these are hampered by the inability to collect specimens, from wild or captive animals. Thus, despite the advantages that molecular methods give modern-day researchers, a lack of specimens means the taxonomic validity of *P. mjobergi* remains unresolved.

Cryptic species or multi-host species of lice?

Many of the lice from monkeys are found on more than one primate species, with 9 of 14 *Pedicinus* spp. listed with two or more principal hosts (Durden & Musser, 1994; Figure 11.1). Explanations for the presence of multi-host parasites include unidentified cryptic parasite species, misclassified (over-split) hosts, recent host-switches, failure of parasites to speciate following host speciation, and incomplete host-switching where lice switch to a novel host but remain a single species on two different host species (Banks & Paterson, 2005). Of these, misclassification of both lice and hosts are of concern for studies of louse–primate coevolution. This is because the accurate inference of coevolutionary events such as cospeciation and host-switching relies on accurate knowledge of the taxonomy and phylogeny of host and parasite. For example, the louse *Pediculus schaeffi* is found on two host species: *Pan troglodytes* (the common chimpanzee) and *Pan paniscus* (the bonobo). A comparison of the primate and louse phylogenies would result in the inference that *P. schaeffi* failed to speciate when the host species did. However, if *P. schaeffi* were found to be comprised of two cryptic species (bonobo lice have not yet been examined molecularly), then the inference would change to that of louse and primate cospeciation.

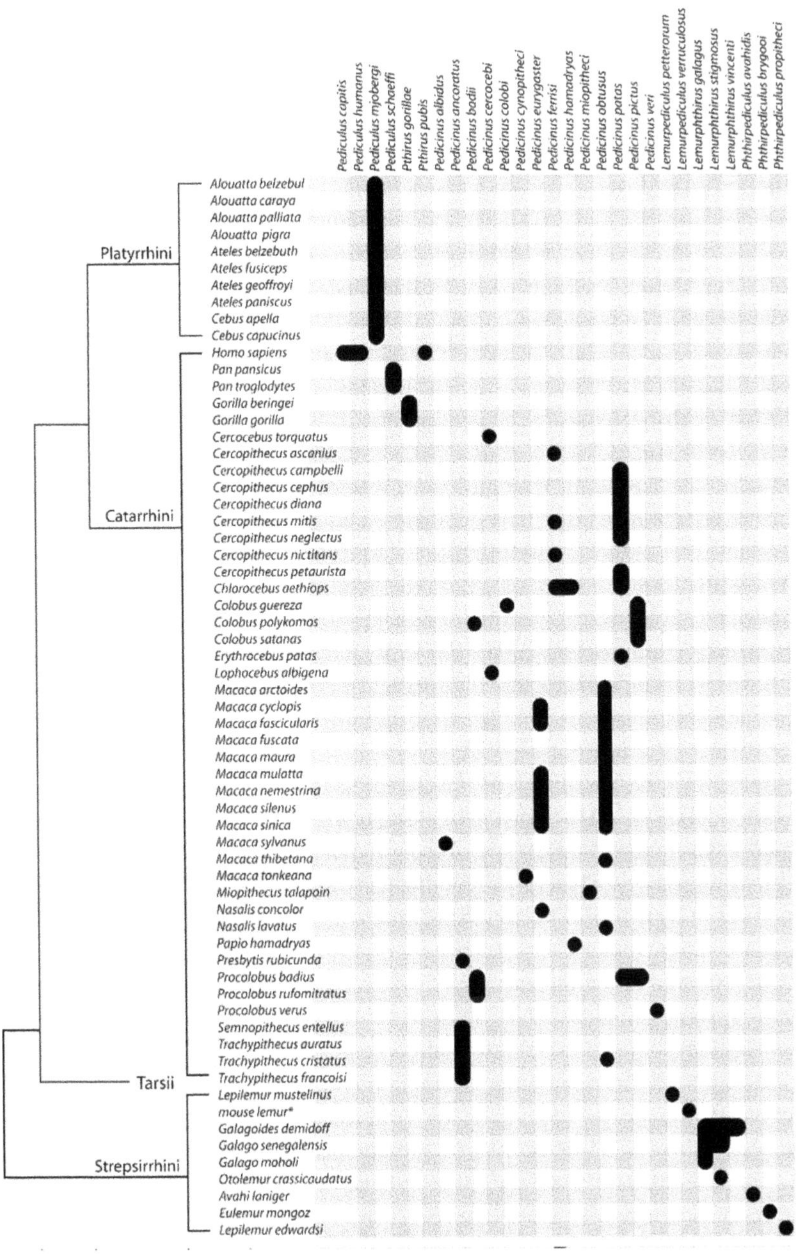

Figure 11.1. The distribution of louse species from primates among their principal hosts (adapted from Durden & Musser, 1994), with the four groups of primates and their relationships to each other indicated (adapted from Purvis, 1995). **Cheirogaleus* sp. or *Microcebus* sp.

There is good reason to suspect that some of the multi-host lice from primates may be cryptic species. On paper, lice from primates make good candidates for cryptic species. Cryptic species are commonly found in lineages that have diversified under strong morphological and physiological constraints (Whittall *et al.*, 2004). These conditions apply to lice. Cryptic species of lice have been reported from doves and from humans (Johnson *et al.*, 2002; Leo *et al.*, 2005). The need for lice to avoid detection and removal by the host likely imposes strong constraints on louse morphology. There is evidence that chewing lice on doves experience constraints on their size, since the body sizes of lice and doves were found to be positively correlated (Clayton *et al.*, 2003). In addition, lice that were experimentally placed on novel hosts of a different size to their usual host were easily removed. This led the authors to suggest that host defense reinforces cospeciation by preventing host-switches between unrelated birds of different size (Clayton *et al.*, 2003). So, there may be selective pressure for lice to maintain an ancestral morphology if they have diverged on two closely related hosts that have maintained similar size and grooming ability. In plants, taxonomic descriptions of cryptic taxa are necessarily based on small differences in morphology, on the assumption that some morphological characters have weaker constraints on them and have become fixed for a species by chance (Whittall *et al.*, 2004). Therefore it might not be coincidence that one of the most reliable characters for distinguishing closely related lice species is the male genitalia. This is a character that presumably would face little or no selective constraint from the host.

As well as the possibility of misclassification in lice from cryptic species, researchers also need to be aware of possible inaccuracies in primate taxonomy. Apart from changes in the name of recorded hosts for lice, the host might have been incorrectly or only loosely identified. For example, the louse *Lemurpediculus verruculosus* was collected from what was described as a "mouse lemur," which could be either *Cheirogaleus* sp. or *Microcebus* sp. (Durden & Musser, 1994). Another potential source of inaccuracy might be a trend towards over-splitting some primate species. Recently, molecular data from *Pan troglodytes* showed a level of genetic differentiation among chimpanzee subspecies that is consistent with that seen among human populations, raising questions about the (validity of) subspecies status for these chimpanzees (Fischer *et al.*, 2006). Similarly, there is no consensus on the number of species and subspecies of gorillas (e.g. Groves, 2001; Vigilant & Bradley, 2004). The policy of conservation of organisms at the species level might also bias the classification of endangered primates towards over-splitting of species. Animals spread across geographical and biological ranges may be easier to conserve if divided into distinct species (Mace, 2004). In determining whether some multi-host species of lice might instead be cryptic species, researchers need to be wary of possible

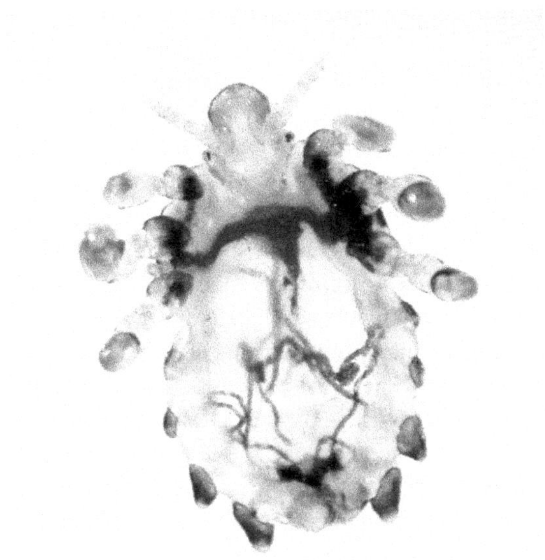

Figure 11.2. *Pthirus gorillae* from a mountain gorilla. Is this a cryptic species?

errors in the recorded hosts for lice, and of possible bias towards over-splitting of primate hosts (Figure 11.2).

Distribution and biodiversity of lice

The number of louse species currently described from primates is likely to be just a fraction of the true biodiversity of these lice. Of approximately 350 species of primates (Groves, 2001), only 63 were known as principal hosts of lice in 1994 (Durden & Musser, 1994). Figure 11.1 shows the known louse species and their principal hosts, adapted from Durden & Musser (1994). It should be noted that the majority of primate species that are absent in Figure 11.1 are absent because of a lack of data and not a known absence of lice. Of the four primate groups, Platyrrhini, Catarrhini, and Strepsirrhini are represented, but Tarsii are not. I could not find any evidence from the literature of lice ever being searched for or found from tarsiers. If tarsiers have lice, it could shed light on some of the questions that surround their phylogenetic position within the order Primates. While one study provides evidence that the Tarsii are a sister group to Platyrrhini and Catarrhini (Purvis, 1995), another study indicates that the robustness of this relationship was dependent on the characters used in the analysis (Poux & Douzery, 2004). Lice from tarsiers might be able to provide an additional source of information to help resolve this uncertainty.

Figure 11.1 also illustrates the patterns of distribution of lice among primate species, and the differences in biodiversity among lice from different groups of primates. Since many primates are not represented, it is only a sketchy picture of the true biodiversity of these lice. Even so, some patterns are apparent. First, lice are quite commonly found on multiple species of primates. Second, primates commonly have multiple species of lice. Third, different species of lice might share some hosts, but not others (e.g. *Pedicinus eurygaster* and *P. obtusus*). The distribution and biodiversity patterns seen in Figure 11.1 raise some questions. Is there something inherent in the behavior or morphology of *Macaca* species that makes them more likely to harbor multiple species of lice than other catarrhines? Why is there only one species of louse found on multiple genera of platyrrhines? For primatologists that are familiar with the biology and evolution of different primate groups, there may be questions about how these patterns in lice might be explained by events in the evolution of primates, or even by behavioral characteristics of certain primate groups. Once a more complete picture of the biodiversity and distribution of these lice is known, it should be easier to form testable hypotheses. The answers are likely to be of interest to both parasitologists and primatologists, since patterns in the biodiversity and distribution of lice will likely be linked to the evolution of new behaviors and morphology in lice and primates (Figure 11.3).

Lice complete their life cycle on their host and their primary mode of transmission is direct contact between hosts. Thus, their habitat is their host and all ecological interactions occur there. Just as habitat disturbance can have major consequences for primate species, similarly changes in host behavior or morphology can have major consequences for lice. For example, when humans began to wear clothing, a new niche on the human body was created. Subsequently a new species of louse *Pediculus humanus* (body lice) evolved from *P. capitis* (head lice; Leo & Barker, 2005). This is interesting because it has been suggested that hair loss was an adaptation to help humans reduce their parasite load, since lice, ticks, and fleas would be easier to detect on naked skin than in hair (Rantala, 1999; Pagel & Bodmer, 2003). Ironically, it seems that loss of hair might have allowed humans to harbor a greater diversity of lice. Humans have three species of lice whereas other ape species have only one or no louse species. Whatever the reason, the retraction of hair to the head and pubic area created two niches for lice, head hair for *P. capitis* and pubic hair for *Pthirus pubis*. Furthermore, the evolution of nakedness later led to the invention of clothing, which created the third niche of clothing for *P. humanus.* So it might be that one change in morphology of humans, hair loss, is responsible for the relatively high biodiversity of lice found on humans today.

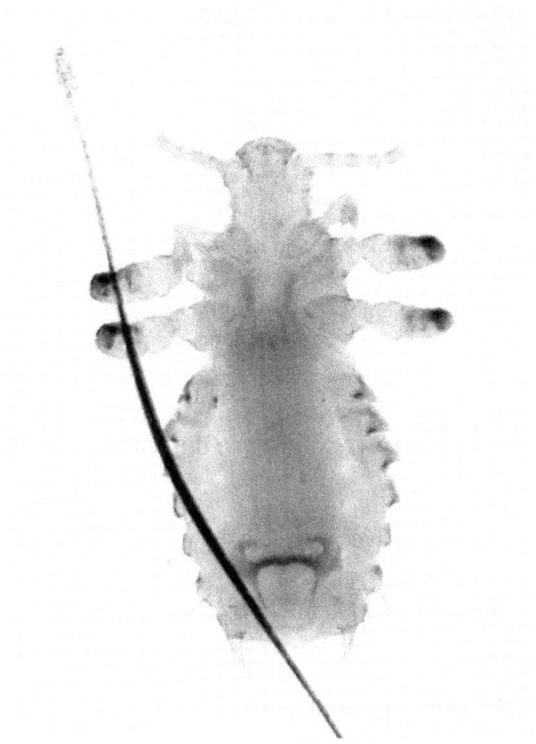

Figure 11.3. A possible new species of *Phthirpediculus* from the sifaka *Propithecus edwardsi*.

In contrast to humans, orangutans and gibbons have no known lice. Since all other apes have lice (humans, chimpanzees, bonobos, gorillas), it seems likely that the lice from gibbons and orangutans have become extinct or "missed the boat." "Missing the boat" is an expression used to describe when a founding host population splits off from the parent population but, by chance, none of the founding host population has lice from the parent population. Another possible explanation is the difference in the behavior of gibbons and orangutans from other apes. Compared with the group living practiced by humans, chimpanzees, and gorillas, the relatively solitary lifestyle of orangutans and small group size of gibbons (Buckley *et al.*, 2006) likely reduces the opportunities for louse transmission among hosts. If louse transmission is too low, the lice die with their host instead of transferring to a younger host and the species will eventually become extinct. This might have occurred with the lice from orangutans and gibbons. However, it is noteworthy that other solitary animals such as pocket gophers show long evolutionary histories of cospeciation with their lice (Nadler &

Hafner, 1993). So perhaps the gibbon and orangutan lost their lice first, which then allowed them to more freely adopt a more solitary lifestyle. Many other primate species groom each other frequently, with the primary function thought to be that of hygiene (Zamma, 2002). Since most primates have at least some areas on their body that they are not able to efficiently groom themselves, there is a benefit to living in a group where other individuals can groom those areas for them. Lorises have been reported to spend about half of their active time on grooming (Nekaris, 2001). An absence of lice, or the ability to effectively self-groom, might remove one of the benefits of group living. Interestingly, the greater flexibility of movement enjoyed by orangutans than by other apes (Isler, 2005), means that there are few areas on their body that they cannot groom themselves. This increase in grooming ability could decrease their dependence on others to help them groom effectively. This hypothesis is very difficult to test, but a comprehensive knowledge of the number and nature of lice on primate species might be used to look for significant correlations between the ability of a host to self-groom, a solitary or group lifestyle, and the number of lice species. This is one example of how observed patterns of lice biodiversity among primates can provide a perspective from which new hypotheses about the evolution of host behavior and morphology can be generated.

The influence of lice on primate behavior

The behavior of hosts has undoubtedly had important consequences for lice, but it is noteworthy that lice have also had an important effect on the behavior of their hosts. Unlike endoparasites, lice are visible to their primate hosts, as evidenced by the deliberate removal of lice and eggs from hair and fur. Where lice bite the host itches, and where lice lay their eggs, nits leave visible evidence that remains months after an active infestation has been eradicated. Not surprisingly this can have an effect on the psyche of the host. Kindergarten children that were asked to draw lice produced drawings that reflected a sense of fear and anxiety (Mumcuoglu, 1991). Frantic parents have applied kerosene or sprayed insecticide directly onto their child's head in a desperate attempt to eradicate lice. Yet in some human cultures, lice do not necessarily have a negative association. In northern Siberia, throwing lice at someone is considered a sign of love (Zinsser, 1935). Some indigenous groups in Australia deliberately maintain lice infestations so that they can be removed by friends and family; it is a relaxing experience that has been likened to a scalp massage (Trigger, 1981).

It should not be assumed that primates consider their lice in a negative light. Lice in small numbers might play a positive role for some primate species. Some primates rely on other individuals in their group to help them remove ectoparasites from areas of the body that they cannot access themselves. In Japanese macaques, it was observed that louse eggs were found mostly on the outer areas of the body, not easily accessible to the infested monkey. The outer areas were also the areas groomed most frequently by other monkeys (Zamma, 2002). If it is beneficial to have others remove our parasites, perhaps primates have evolved to enjoy the feeling of being groomed. Is it a result of chance or positive selection that humans enjoy the feeling of another person stroking our hair? It could be that lice are indirectly responsible for the evolution of behaviors that have a secondary function that continues even in their absence. Why do some women "flick" their hair over their shoulder when flirting with a man? It has been suggested that this might serve the purpose of advertising fitness since this action brings attention to the hair behind the ear, where head lice prefer to lay their eggs (Anita Bailey, personal communication). More obviously, lice have influenced our language through words like "nit-wit" and "lousy." The influence of lice on their primate hosts should not be underestimated.

Conclusion

The taxonomy of lice from primates as it stands today is built from relatively few specimens collected opportunistically. In addition, that many specimens of lice were collected from the bodies of primates that died in captivity raises doubts about whether lice are found on these hosts naturally. Many of the lice from primates are multi-host species, but examination of the evidence shows that some of these lice might instead be cryptic species or a chance transfer to a novel host under captive conditions. While cryptic species are by definition difficult to distinguish with morphological characters, molecular tools make it relatively easy to identify them from small numbers of lice. Molecular tools should also make cases of host-transfers much easier to identify. However, the current museum specimens are not suitable for genetic studies and new specimens, preferably from wild primates, will need to be collected.

The true biodiversity of lice from primates is likely to be much greater than what is currently known, since only about one fifth of known primate species are described as principal hosts of lice. A concerted effort by primatologists and parasitologists to learn about these lice will have many beneficial outcomes. It will provide a robust taxonomy from which accurate phylogenies can be built. From accurate phylogenies, we can more confidently infer coevolutionary events. This has the potential to reveal events that could not be determined from host genetics alone. In the longer term, an accurate picture of the distribution and biodiversity of lice from different primate species will provide a rich source of information from which to draw inspiration for new research.

Here collaboration between parasitologists and primatologists will be the most fruitful. Parasitologists can contribute their expertise on the factors that affect the distribution and biodiversity of lice, and primatologists can contribute their expertise on the evolution of morphology and behavior of the various primate groups. Identifying cryptic species and describing the biodiversity of lice from primates will give us the opportunity to learn how the evolution of primate behavior, morphology and ecology, has influenced the evolution of lice host-specificity and biodiversity.

Acknowledgements

I thank Djaisi Robinson and the anonymous reviewers for their helpful comments on this chapter. The gorilla louse pictured in this chapter was kindly donated by Biological Resource Center of the Mountain Gorilla Veterinary Project, and Patricia C. Wright collected the sifaka louse. This book chapter contains research that was funded by the Japan Society for the Promotion of Science.

References

Ashford, R. W. (2000). Parasites as indicators of human biology and evolution. *Journal of Medical Microbiology*, **49**, 771–772.

Banks, J. C. & Paterson, A. M. (2005). Multi-host parasite species in cophylogenetic studies. *International Journal for Parasitology*, **35**, 741–746.

Banks, J. C., Palma, R. L. & Paterson, A. M. (2006). Cophylogenetic relationships between penguins and their chewing lice. *Journal of Evolutionary Biology*, **19**, 156–166.

Barker, S. C. (1994). Phylogeny and classification, origins, and evolution of host associations of lice. *International Journal for Parasitology*, **24**, 1285–1291.

Buckley, C., Nekaris, K. A. & Husson, S. J. (2006). Survey of *Hylobates agilis albibarbis* in a logged peat-swamp forest: Sabangau catchment, Central Kalimantan. *Primates*, **47**, 327–335.

Buxton, P. A. (1941). Studies on populations of head-lice (*Pediculus humanus capitis*: Anoplura). *Parasitology*, **33**, 224–242.

Clayton, D. H., Bush, S. E., Goates, B. M. & Johnson, K. P. (2003). Host defense reinforces host-parasite cospeciation. *Proceedings of the National Academy of Sciences, USA*, **100**, 15694–15699.

Curnoe, D. & Thorne, A. (2003). Number of ancestral human species: a molecular perspective. *Homo*, **53**, 201–224.

Durden, L. A. & Adams, N. E. (2005). Primary type specimens of sucking lice (Insecta: Phthiraptera: Anoplura) in the U.S. National Museum of Natural History, Smithsonian Institution. *Zootaxa*, **1047**, 21–60.

Durden, L. A. & Musser, G. G. (1994). The sucking lice (Insecta: Anoplura) of the world: a taxonomic checklist with records of mammalian hosts and geographical distributions. *Bulletin of the American Museum of Natural History*, **0**, 1–90.
Ewing, H. E. (1926). A revision of the American lice of the genus *Pediculus*, together with a consideration of the significance of their geographical and host distribution. *Proceedings of the United States National Museum*, **68**, 1–30.
Ferris, G. F. (1935). *Contributions Toward a Monograph of the Sucking Lice*. London: Humphrey Milford Oxford University Press.
Ferris, G. F. (1951). *The Sucking Lice*. San Francisco, CA: The Pacific Coast Entomological Society.
Fischer, A., Pollack, J., Thalmann, O., Nickel, B. & Pääbo, S. (2006). Demographic history and genetic differentiation in apes. *Current Biology*, **16**, 1133–1138.
Gross, L. (1996). How Charles Nicolle of the Pasteur Institute discovered that epidemic typhus is transmitted by lice: reminiscences from my years at the Pasteur Institute in Paris. *Proceedings of the National Academy of Sciences, USA*, **93**, 10539–10540.
Groves, C. (2001). *Primate Taxonomy*. London: Smithsonian Institution Press.
Isler, K. (2005). 3D-kinematics of vertical climbing in hominoids. *American Journal of Physical Anthropology*, **126**, 66–81.
Johnson, K. P., Williams, B. L., Drown, D. M., Adams, R. J. & Clayton, D. H. (2002). The population genetics of host specificity: genetic differentiation in dove lice (Insecta: Phthiraptera). *Molecular Ecology*, **11**, 25–38.
Kim, K. C. & Emerson, K. C. (1968). Descriptions of two species of Pediculidae (Anoplura) from great apes (Primates Pongidae). *Journal of Parasitology*, **54**, 690–695.
Kittler, R., Kayser, M. & Stoneking, M. (2003). Molecular evolution of *Pediculus humanus* and the origin of clothing. *Current Biology*, **13**, 1414–1417.
Kittler, R., Kayser, M. & Stoneking, M. (2004). Erratum. Molecular evolution of *Pediculus humanus* and the origin of clothing. *Current Biology*, **14**, 2309.
Leo, N. P. & Barker, S. C. (2005). Unravelling the evolution of the head lice and body lice of humans. *Parasitology Research*, **98**, 44–47.
Leo, N. P., Campbell, N. J. H., Yang, X., Mumcuoglu, K. & Barker, S. C. (2002). Evidence from mitochondrial DNA that head lice and body lice of humans (Phthiraptera: Pediculidae) are conspecific. *Journal of Medical Entomology*, **39**, 662–666.
Leo, N. P., Hughes, J. M., Yang, X. *et al.* (2005). The head and body lice of humans are genetically distinct (Insecta: Phthiraptera, Pediculidae): evidence from double infestations. *Heredity*, **95**, 34–40.
Mace, G. (2004). The role of taxonomy in species conservation. *Philosophical Transactions of the Royal Society of London B: Biological Sciences*, **359**, 711–719.
Mey, E. (1994). *Pedicinus*-Formen (Insecta, Phthiraptera, Anoplura) seltener Schlankaffen (Mammalia, Primates, Colobinae) aus Vietnam. *Rudolstädter Naturhistorische Schriften*, **6**, 83–92.
Mumcuoglu, K. Y. (1991). Head lice in drawings of kindergarten children. *Israel Journal of Psychiatry and Related Sciences*, **28**, 25–32.

Mumcuoglu, K. Y., Klaus, S., Kafka, D., Teiler, M. & Miller, J. (1991). Clinical observations related to head lice infestation. *Journal of the American Academy of Dermatology*, **25**, 248–251.

Nadler, S. A. & Hafner, M. S. (1993). Systematic relationships among pocket gopher chewing lice (Phthiraptera: Trichodectidae) inferred from electrophoretic data. *International Journal for Parasitology*, **23**, 191–201.

Nekaris, K. A. (2001). Activity budget and positional behavior of the Mysore slender loris (*Loris tardigradus lydekkerianus*): implications for slow climbing locomotion. *Folia Primatologica (Basel)*, **72**, 228–241.

Pagel, M. & Bodmer, W. (2003). A naked ape would have fewer parasites. *Proceedings of the Royal Society of London. Series B*, **270**, S117–S119.

Perotti, M. A., Catala, S. S., Ormeno Adel, V. *et al.* (2004). The sex ratio distortion in the human head louse is conserved over time. *BioMed Central Genetics*, **5**, 10.

Poux, C. & Douzery, E. J. (2004). Primate phylogeny, evolutionary rate variations, and divergence times: a contribution from the nuclear gene IRBP. *American Journal of Physical Anthropology*, **124**, 1–16.

Purvis, A. (1995). A composite estimate of primate phylogeny. *Philosophical Transactions of the Royal Society of London B: Biological Sciences*, **348**, 405–421.

Rantala, M. J. (1999). Human nakedness: adaptation against ectoparasites? *International Journal for Parasitology*, **29**, 1987–1989.

Raoult, D. & Roux, V. (1999). The body louse as a vector of reemerging human diseases. *Clinical Infectious Diseases*, **29**, 888–911.

Reed, D., Smith, V. S., Hammond, S. L., Rogers, A. R. & Clayton, D. H. (2004). Genetic analysis of lice supports direct contact between modern and archaic humans. *PLoS Biology*, **2**, e340.

Robinson, D., Leo, N. P., Prociv, P. & Barker, S. C. (2003). Potential role of head lice, *Pediculus humanus capitis*, as vectors of *Rickettsia prowazekii. Parasitology Research*, **90**, 209–211.

Rozsa, L. (1997). Adaptive sex-ratio manipulation in *Pediculus humanus capitis*: possible interpretations of Buxton's data. *Journal of Parasitology*, **83**, 543–544.

Singh, O. P., Chandra, D., Nanda, N. *et al.* (2004). Differentiation of members of the *Anopheles fluviatilis* species complex by an allele-specific polymerase chain reaction based on 28S ribosomal DNA sequences. *American Journal of Tropical Medicine and Hygiene*, **70**, 27–32.

Tanaka, I. (1995). Matrilineal distribution of louse egg-handling techniques during grooming in free-ranging Japanese macaques. *American Journal of Physical Anthropology*, **98**, 197–201.

Trigger, D. S. (1981). Blackfellows, whitefellows, and head lice. *Australian Institute of Aboriginal Studies Newsletter*, **15**, 63–72.

Vigilant, L. & Bradley, B. J. (2004). Genetic variation in gorillas. *American Journal of Primatology*, **64**, 161–172.

Whittall, J. B., Hellquist, C. B., Schneider, E. L. & Hodges, S. A. (2004). Cryptic species in an endangered pondweed community (*Potamogeton*, Potamogetonaceae) revealed by AFLP markers. *American Journal of Botany*, **91**, 2022–2029.

Wildman, D. E., Uddin, M., Liu, G., Grossman, L. I. & Goodman, M. (2003). Implications of natural selection in shaping 99.4% nonsynonymous DNA identity between humans and chimpanzees: enlarging genus *Homo*. *Proceedings of the National Academy of Science, USA*, **100**, 7181–7188.
Zamma, K. (2002). Grooming site preferences determined by lice infection among Japanese macaques in Arashiyama. *Primates*, **43**, 41–49.
Zinsser, H. (1935). *Rats, Lice and History*, 1st edn. London: George Routledge & Sons, Ltd.

12 *Prevalence of* Clostridium perfringens *in intestinal microflora of non-human primates*

SHIHO FUJITA, ASAMI OGASAWARA, AND
TAKASHI KAGEYAMA

Photograph by Michael A. Huffman (*Pan troglodytes schweinfurthii*)

Primate Parasite Ecology. The Dynamics and Study of Host–Parasite Relationships, ed. Michael A. Huffman and Colin A. Chapman. Published by Cambridge University Press.

Introduction

Relationship between normal intestinal microflora and the host animal

The mammalian intestinal microflora consists of over 400–500 bacteria species (Moore & Holdeman, 1974), and plays an important role in maintaining health and preventing disease. Intestinal bacteria aid in the digestion of food, metabolize drugs and foreign compounds, and produce vitamins (Mitsuoka, 1982; Willett, 1992). In addition, the normal microbial flora inhibits the establishment of extraneous pathogens by stimulating the host's immune system, which helps to prevent infectious disease (Mitsuoka, 1982; Ryan, 2004).

In a healthy host, a balance in the proportions of different bacterial species present in the intestine is maintained, and the composition of microflora is stable over relatively long periods (Drasar *et al.*, 1976; Granato, 2003). However, the bacterial proportions of the microflora depend on many factors including animal species, age, sex, diet, exogenous microorganisms, as well as emotional stress (Mitsuoka, 1982). The composition of the microflora reflects lifestyle, physiology, and/or health status of a host. Thus, monitoring the intestinal microflora not only provides information about host–microorganism interactions, but also about the health status of the host.

Clostridium perfringens

Clostridium perfringens is a normal member of the intestinal microflora in domestic animals and humans, and is a ubiquitous bacterium in the natural environment (Saito, 1990; Allen *et al.*, 2003). However, it has been reported that when the intestinal microflora is disturbed, populations of pathogenic bacteria, including *C. perfringens*, increase in proportion and cause autogenous infections (Mitsuoka, 1982). *Clostridium perfringens* is classified into several types each of which produce different types of toxins (McDonel, 1980; Smedley *et al.*, 2004) and cause a range of occasionally fatal diseases in a diversity of animals, including humans (Niilo, 1980; Hatheway, 1990; Borriello, 1995; Songer, 1996). Since hyper-proliferation of this bacterium in the intestinal tract can be a disease-causing agent even in a healthy individual, there are good reasons to check the prevalence of this bacterium as part of a strategy for assessing health status.

The polymerase chain reaction (PCR) method has been widely used for the rapid detection of bacteria. In this chapter, we describe the prevalence of *C. perfringens* in fecal specimens of chimpanzees (*Pan trogrodytes verus, P. t. schweinfurthii*) and Japanese macaques (*Macaca fuscata fuscata*) using the nested PCR assay, which is composed of a two-step PCR. Furthermore, we assess the difference in the intestinal microflora among populations living under different conditions.

Methods and materials

Subject animals

Chimpanzees

We collected fecal specimens from captive and wild chimpanzees. Captive chimpanzees were reared in a social group (three adult males, eight adult females, and three infants under a year of age) at the Primate Research Institute (PRI) of Kyoto University, Japan (35°22′N, 136°56′E). Details of the housing facility were given by Matsuzawa (2003). The chimpanzees were fed commercial monkey-chow (Oriental Yeast Co. Ltd., Tokyo, Japan), various fruits, and vegetables three times a day. They had access to water *ad libitum.*

We also collected fecal specimens from two wild populations at Mahale and at Bossou. The Mahale Mountains National Park is located on the eastern shores of Lake Tanganyika, Tanzania (6°11′S, 29°74′E). Vegetation in the home range of chimpanzees is composed of semideciduous forests and woodland (Uehara & Ihobe, 1998). The history of the study group and the flora of this research site were given by Nishida (1990). In Mahale, where eastern chimpanzees (*P. t. schweinfurthii*) inhabit, the subject group consisted of 50–60 members during the study period. We collected feces from Mahale chimpanzees in two wet seasons (wet season I: February through March 2001, and II: January through February 2005), and a dry season (September through October 2001).

Bossou is located on the southeastern edge of the Republic of Guinea (7°65′N, 8°50′W). Vegetation in the home range of chimpanzees here consists of deciduous forests surrounded by savanna areas and cultivated fields. The history of the study group and the flora of this research site were given by Sugiyama (1981). In Bossou, where western chimpanzees (*P. t. verus*) inhabit, the subject group consisted of 10–20 members during the study period. We collected feces from the Bossou chimpanzees in a dry season (March 2002) and a wet season (July through August 2004). The main food item of the wild chimpanzees at both sites was fruits, consumed for about 60% of total feeding time. Other food consumed included leaves, pith, flowers, seeds, insects, mammals, and other minor food items (Fujita, 2003).

Japanese macaques

The captive monkeys were reared in three social groups (Takahama, and Arashiyama-1 and 2) in corral enclosures (about 500 m^2) at the PRI. All monkeys were fed commercial monkey-chow every day (Oriental Yeast Co.

Ltd., Tokyo, Japan) supplemented with sweet potatoes three times a week. They had access to water *ad libitum*.

We also collected fecal specimens from wild monkeys at Kinkazan and at Ichikawa. Kinkazan Island is located on Ishinomaki, Miyagi, Japan (38°16′N, 141°35′E). Vegetation at this site is composed of deciduous broad-leaved forests and coniferous forests, and a general description of this island was given by Tsuji & Takatsuki (2004). In Kinkazan, the size of the subject troop varied from 30–40 individuals during the study. The monkeys there eat fruits (including nuts and seeds), leaves, flowers, bark, herbaceous plants, fungi, animal matter (including insects, spiders, limpets, and frogs), and other minor items (Tsuji *et al.*, 2006). The other study site, Ichikawa, is located in Shinshiro, Aichi, Japan (34°53′N, 137°29′E). Vegetation at this site consists of deciduous broad-leaved forests and coniferous forests, and detailed information about the monkeys there is not available.

Fecal collection and sample preparation

From the Mahale chimpanzees, we collected 16 samples from 16 individuals during the dry season, 15 samples from 15 individuals during wet season I, and 50 samples from 40 individuals (1–3 sample(s)/individual) during wet season II. At Bossou, we collected 23 samples from 15 individuals (1–3 sample(s)/individual) during the dry season and 30 samples from 13 individuals (1–4 sample(s)/individual) during the wet season. From the Japanese monkeys, we collected fecal samples from 36, 38, 28, 38, and 28 individuals (1 sample/individual) in Takahama, Arashiyama-1 and -2, Ichikawa, and Kinkazan, respectively.

For the captive chimpanzees and monkeys, we collected fecal samples on the floor of individual cages (for the macaques), indoor experimental booths and sleeping rooms (for the chimpanzees). A small amount of feces (*c.* 1.0 g) was suspended in the same volume of glycerol (50%) and kept at –30 °C until use.

For the wild chimpanzees and monkeys, we scooped feces into an appropriately labeled plastic bag immediately after the chimpanzees defecated, and kept them at ambient temperature until the end of the day's observation. Upon returning to the field station at the end of the day, we weighted 1.0 g of fecal specimens and suspended them in the cooked meat medium (Becton Dickinson Co. Ltd., Tokyo, Japan). All subject animals in captivity and in the wild appeared healthy and did not have diarrhea.

A small aliquot of the fecal specimen (frozen feces for the captive chimpanzees and fresh ones in medium for the wild chimpanzees) was suspended in 1.2 ml of GAM broth (Nissui-Pharmaceutical Co. Ltd., Tokyo, Japan)

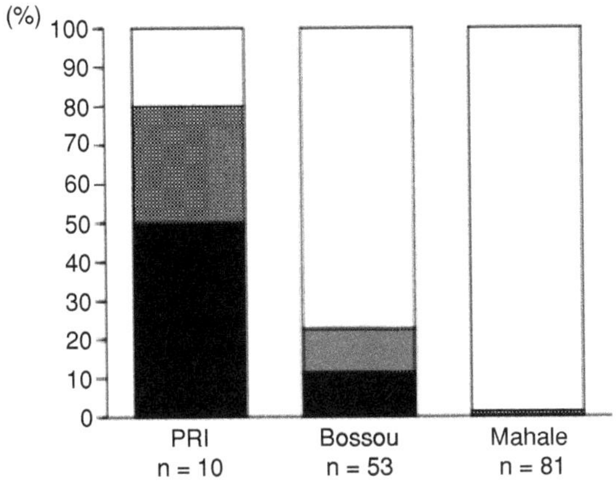

Figure 12.1. Detection rates of *Clostridium perfringens* in feces of captive (PRI) and wild (Mahale and Bossou) chimpanzees (redrawn from Fujita & Kageyama, 2007). The black and shaded area of each bar shows percentage of animals in which *C. perfringens* was detected by the first PCR and by the nested PCR, respectively. Although more than 10^4 colony-forming units (CFU) of bacterial DNA were necessary for product detection with the first PCR, DNA from 10 CFUs were sufficient with the nested PCR (Fujita & Kageyama, 2007, Fig. 1), showing the sensitivity to have increased 10^2 to 10^3-fold. The open area shows percentage of animals in which the bacterium was not detected.

followed by anaerobic incubation at 37 °C for 24 h. Bacterial DNA was isolated according to Kageyama *et al.* (2002).

Nested PCR for the detection of C. perfringens

Two pairs of primers for α-toxin (*plc*) gene of *C. perfringens* were used for the detection by nested PCR. Five toxinotypes (type A, B, C, D, and E) of *C. perfringens* have been recognized. The type A strain is most common in the intestinal tract of animals and in the natural environment, and produces α-toxin in large amounts (Hatheway, 1990). The sequences of primers and the expected sizes of PCR products are given in a previous report (Fujita & Kageyama, 2007). Procedures of nested PCR and electrophoresis were performed as described by Kageyama *et al.* (2002).

Results and discussion

Chimpanzees

The results of bacterial DNA detection from the captive and wild chimpanzees are given in Figure 12.1. Animals in which the bacterium is detected by the first

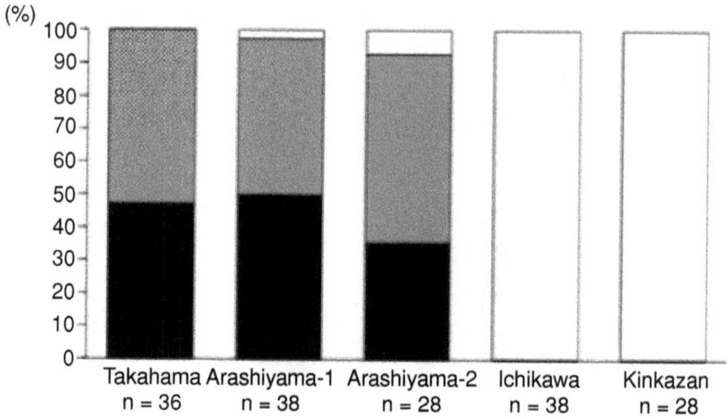

Figure 12.2. Detection rates of *C. perfringens* in feces of captive (Takahama, and Arashiyama-1 and 2) and wild (Ichikawa and Kinkazan) Japanese monkeys. See Figure 12.1 for the explanation of each bar.

and the following nested PCR are carriers of a larger amount of the bacterium than those in which bacteria are detected only by the nested PCR. For the captive chimpanzees, 5 of 10 samples were positive in the first PCR and the following nested PCR assay, whereas 3 of 10 samples were so only in the nested PCR assay, indicating that 80% (8/10) of samples were positive in total. In contrast, the detection rate in wild chimpanzees was low. At Bossou, *C. perfringens* was detected in 11.3% (6/53) of fecal samples by the first PCR and in 11.3% (6/53) only by the nested PCR. At Mahale, the bacterium was detected in only one of 81 fecal samples (1.3%) by the nested PCR, and the bacterium was not detected at all by the first PCR.

Japanese macaques

The results of bacterial DNA detection from the captive and wild Japanese macaques are given in Figure 12.2. For the captive groups, *C. perfringens* was detected in 35–50% (44.3% on average) of fecal samples by the first PCR and the following nested PCR, and in 47–57% (52.4% on average) only by the nested PCR. This result indicates that almost all of the captive monkeys were carriers of the bacterium. In contrast, no bacterium was detected from the feces of the wild monkeys at either Ichikawa or Kinkazan.

Contributing factors to the differences in detection rates of C. perfringens

Detection rates of *C. perfringens* were higher in the feces of captive animals than in those of wild ones. The supply of artificial diets to animals in captivity might have been a causative factor in the increased prevalence of intestinal *C. perfringens*. That is, the low fiber content of their diets might have predisposed them to the proliferation of the bacterium in the intestine. This speculation is supported by the results of previous studies, where the numbers of several bacterial species, including *Clostridium* spp., have been shown to be low in bark-eating monkeys (Benno *et al.*, 1987). In human populations, when fed larger amounts of dietary fiber, the incidence of fecal *C. perfringens* has been reported to be low (Finegold *et al.*, 1977; Benno *et al.*, 1986, 1989a, b). Benno *et al.* (1989b) suggested that high intake of dietary fiber prevented intestinal putrefaction, where saprogens, including *C. perfringens*, proportionally decreased in the microflora. On the other hand, a high calorie diet is known to increase the number and incidence of *C. perfringens* in the intestinal microflora. In rhesus monkeys that were fed with high fat and high calorie diets, a large population of *C. perfringens* was observed in the intestinal microflora, coinciding with a rise in blood cholesterol levels (Graber & Boltjes, 1968). Since the captive chimpanzees and monkeys at PRI are fed with monkey-chow, which contains high levels of energy and low levels of dietary fiber, *C. perfringens* might have been able to easily colonize and increase in their intestinal tracts.

The diet composition of wild populations differed markedly between habitats and between seasons. In the present study, detection rates were higher in the chimpanzees at Bossou than in those at Mahale. Bossou chimpanzees usually feed on several cultivated plant species in arable fields and this high calorie food might explain the increase in the proportion of *C. perfringens* in the Bossou group. A previous study reported that the reproductive potential of Bossou females is higher than that of Mahale females, suggesting that the nutritional status of chimpanzees is better at Bossou than at Mahale (Fujita, 2003). In addition, at Bossou, the detected rates of *C. perfringens* were higher in the wet season than in the dry season. In the wet season 16.7% (5/30) of fecal samples scored positive by the first PCR and the following nested PCR and 13.3% (4/30) did only by the nested PCR, whereas the values were 4.3% (1/23) and 8.7% (2/23), respectively, in the dry season. Seasonal differences in the detection rates at Bossou were not significant within the sample base ($DF = 2$, $\chi^2 = 2.5$, $P = 0.29$). However, within the individual base, 20% of the subjects (3/15) carried *C. perfringens* in the dry season, whereas 46% of the subjects (6/13) did so in the wet season. It has been reported that the availability of fruit (a main food item for chimpanzees) is lower in the wet season than in the dry season (Takasaki *et al.*, 1990), and that the body weight of chimpanzees decreases over the wet season (Uehara & Nishida, 1998). Lower nutritional values of the ingested food during the wet season might weaken the immune response of the hosts (Drasar, 1974), resulting in an increase of the *C. perfringens* population in the microflora.

Not only diet but also environmental factors may affect intestinal microflora. The bacterium is known to be stable for a long time in a natural environment because of sporulation (Allen *et al.*, 2003), and it has been reported that this bacterium was isolated from surfaces in a hospital facility (Borriello *et al.*, 1985). Therefore, various environmental factors such as soil, water, and surface of the facility could be natural reservoirs of this bacterium. In the present study, the facility for keeping animals in captivity can be a source of infection because human and domestic animals have *C. perfringens* in their intestine as a component of the normal microflora. The captive animals, therefore, have more chance of being infected with this bacterium in the artificial environment. In addition, the captive animals might have more contact with feces from other group members than the wild ones, leading to rapid dissemination and maintenance of *C. perfringens* in the guts of these populations. A similar explanation might apply in the case of the wild chimpanzees. At Mahale, local people, other than rangers, tour guides, researchers, and field assistants, are restricted from the National Park, resulting in less frequent contact between humans and chimpanzees. On the other hand, at Bossou, the chimpanzees' habitat is close to villages of local people, and the chimpanzees pass a road and forage on agricultural crops regularly. In addition, the local people defecate in the undergrowth around the village and in the farmland. These indirect contacts between humans and chimpanzees are anticipated to increase the possibility of infection with *C. perfringens* in chimpanzees (see Gasser *et al.*, Chapter 3, this volume for a discussion of transmission of pathogens between primates and humans).

Mental stress is another major contributing factor that causes disorders of the intestinal microflora (Holdeman *et al.*, 1976). It was reported that the isolation rate of *C. perfringens* was higher from samples in corralled animals than in free-ranging animals (dog: Balish *et al.*, 1977; reindeer: Aschfalk *et al.*, 2003). These results indicate that captive conditions easily cause tension among animals because of space restrictions, leading to a disturbance of their intestinal microflora. Since chimpanzees and macaques are highly social animals, psychological stress may increase the proportion of this bacterium in their microflora.

In conclusion, the prevalence of *C. perfringens* in the feces of chimpanzees and Japanese monkeys differed greatly between animals in captivity and those in the wild. In the wild chimpanzees, moreover, the prevalence of *C. perfringens* differed between populations and between seasons. Diet and local environment were probably major contributing factors to these differences. For further understanding of ecological background of microbial proportions in the intestine, information on the intestinal microflora in various species and in various populations is necessary. In the present study, although all the samples were derived from clinically healthy chimpanzees, the prevalence of *C. perfringens* could reflect a disorder of the intestinal microflora. The simple and non-invasive technique used in this study might be useful in the assessment of health risks in both captive and wild primates.

Acknowledgements

We sincerely thank the Tanzania Commission for Science and Technology, the Tanzania Wildlife Research Institute, and the Tanzania National Parks, for permission to work in the Mahale Mountains National Park; the Direction Nationale de la Recherche Scientifique et Technique, the Republic of Guinea, and the Institut Recherche Environnementale de Bossou for permission to work in Bossou.

References

Allen, S. D., Emery, C. L. & Lyerly, D. M. (2003). *Clostridium*. In *Manual of Clinical Microbiology*, ed. P. R. Murray, E. J. Baron, J. H. Jorgensen, M. A. Pfaller & R. H. Yolken, 8th edn. Washington, DC: ASM Press, pp. 835–356.

Aschfalk, A., Kemper, N. & Höller, C. (2003). Bacteria of pathogenic importance in faeces from cadavers of free-ranging or corralled semi-domesticated reindeer in northern Norway. *Veterinary Research Communications*, **27**, 93–100.

Balish, E., Cleven, D., Brown, J. & Yale, C. E. (1977). Nose, throat, and fecal flora of beagle dogs housed in "locked" or "open" environments. *Applied and Environmental Microbiology*, **34**, 207–221.

Benno, Y., Suzuki, K., Narasiwa, K. *et al.* (1986). Comparison of the fecal microflora in rural Japanese and urban Canadians. *Microbiology and Immunology*, **30**, 521–532.

Benno, Y., Itoh, K., Miyao, Y. & Mitsuoka, T. (1987). Comparison of fecal microflora between wild Japanese monkeys in a snowy area and laboratory-reared Japanese monkeys. *Japan Journal of Veterinary Science*, **49**, 1059–1064.

Benno, Y., Endo, K., Miyoshi, H. *et al.* (1989a). Effect of rice fiber on human fecal microflora. *Microbiology and Immunology*, **33**, 435–440.

Benno, Y., Endo, K., Mizutani, T. *et al.* (1989b). Comparison of fecal microflora of elderly persons in rural and urban areas of Japan. *Applied and Environmental Microbiology*, **55**, 1100–1105.

Borriello, S. P. (1995). Clostridial disease of the gut. *Clinical Infectious Disease*, **20** (Suppl. 2), S242–S250.

Borriello, S. P., Barclay, F. E., Welch, A. R. *et al.* (1985). Epidemiology of diarrhoea caused by enterotoxigenic *Clostridium perfringens*. *Journal of Medical Microbiology*, **20**, 363–372.

Drasar, B. S. (1974). Some factors associated with geographical variations in the intestinal microflora. In *The Normal Microbial Flora of Man*, ed. F. A. Skinner. London: Academic Press, pp. 187–196.

Drasar, B. S., Jenkins, D. J. A. & Cummings, J. H. (1976). The influence of a diet rich in wheat fibre on the human faecal flora. *Journal of Medical Microbiology*, **9**, 423–431.

Finegold, S. M., Sutter, V. L., Sugihara, P. T. *et al.* (1977). Fecal microbial flora in Seventh Day Adventist populations and control subjects. *American Journal of Clinical Nutrition*, **30**, 1781–1792.

Fujita, S. (2003). Reproductive biology in wild female primates: variability in hormonal profiles, behavior and reproductive parameters. PhD thesis, Kyoto University.

Fujita, S. & Kageyama, T. (2007). Polymerase chain reaction detection of *Clostridium perfringens* in feces from captive and wild chimpanzees, *Pan troglodytes. Journal of Medical Primatology*, **36**, 25–32.

Graber, C. D. & Boltjes, B. H. (1968). Blood cholesterol levels, autochthonous intestinal flora and *Clostridium perfringens* recovery in rhesus monkeys fed a high sucrose diet. *Journal of Choronic Disease*, **21**, 255–264.

Granato, P. A. (2003). Pathogenic and indigenous microorganisms of humans. In *Manual of Clinical Microbiology*, 8th edn, ed. P. R. Murray, E. J. Baron, J. H. Jorgensen, M. A. Pfaller & R. H. Yolken, Washington, DC: ASM Press, pp. 44–54.

Hatheway, C. L. (1990). Toxigenic *Clostridia. Clinical Microbiology Reviews*, **3**, 66–98.

Holdeman, L. V., Good, I. J. & Moore, W. E. C. (1976). Human fecal flora: variation in bacterial composition within individuals and a possible effect of emotional stress. *Applied and Environmental Microbiology*, **31**, 359–375.

Kageyama, T., Ogasawara, A., Fukuhara, R. *et al.* (2002). *Yersinia pseudotuberculosis* infection in breeding monkeys: detection and analysis of strain diversity by PCR. *Journal of Medical Primatology*, **31**, 129–135.

Matsuzawa, T. (2003). The Ai project: historical and ecological contexts. *Animal Cognition*, **6**, 199–211.

McDonel, J. L. (1980). *Clostridium perfringens* toxins (type A, B, C, D, E). *Pharmacology and Therapeutics*, **10**, 617–655.

Mitsuoka, T. (1982). Recent trends in research on intestinal flora. *Bifidobacteria and Microflora*, **1**, 3–24.

Moore, W. E. C. & Holdeman, L. V. (1974). Human fecal flora: the normal flora of 20 Japanese-Hawaians. *Applied Microbiology*, **27**, 961–979.

Niilo, L. (1980). *Clostridium perfringens* in animal diseases: a review of current knowledge. *Canadian Veterinary Journal*, **21**, 141–148.

Nishida, T. (1990). A quarter century of research in the Mahale Mountains: an overview. In *The Chimpanzees of the Mahale Mountains: Sexual and Life History Strategies*, ed. T. Nishida. Tokyo: University of Tokyo Press, pp. 3–35.

Ryan, K. J. (2004). Normal microbial flora. In *Sherris Medical Microbiology: An Introduction to Infectious Diseases*, ed. K. J. Ryan & C. G. Ray, 4th edn. New York, NY: The MacGraw-Hill Companies, pp. 141–148.

Saito, M. (1990). Production of enterotoxin by *Clostridium perfringens* derived from human, animals, foods, and the natural environment in Japan. *Journal of Food Protection*, **53**, 115–118.

Schlapp, T., Blaha, I., Bauerfeind, R. *et al.* (1995). Synthesis and evaluation of a non-radioactive gene probe for the detection of *C. perfringens* alpha toxin. *Molecular and Cellular Probes*, **9**, 101–109.

Smedley III, J. G., Fisher, D. J., Sayeed, S., Chakrabarti, G. & McClane, B. A. (2004). The enteric toxins of *Clostridium perfringens. Reviews of Physiology, Biochemistry and Pharmacology*, **152**, 183–204.

Songer, J. G. (1996). Clostridial enteric diseases of domestic animals. *Clinical Microbiology Reviews*, **9**, 216–234.

Sugiyama, Y. (1981). Observations on the population dynamics and behavior of wild chimpanzees at Bossou, Guinea, in 1979–1980. *Primates*, **22**, 435–444.

Takasaki, H., Nishida, T., Uehara, S. *et al.* (1990). Summary of meteorological data at Mahale research camps, 1973–1988. In *The Chimpanzees of the Mahale Mountains: Sexual and Life History Strategies*, ed. T. Nishida. Tokyo: University of Tokyo Press, pp. 291–300.

Tsuji, Y. & Takatsuki, S. (2004). Food habits and home range use of Japanese macaques on an island inhabited by deer. *Ecological Research*, **19**, 381–388.

Tsuji, Y., Fujita, S., Sugiura, H., Saito, C. & Takatsuki, S. (2006). Long-term variation in fruiting and the food habits of wild Japanese macaques on Kinkazan Island, Northern Japan. *American Journal of Primatology*, **68**, 1068–1080.

Uehara, S. & Ihobe, H. (1998). Distribution and abundance of diurnal mammals, especially monkeys, at Kasoje, Mahale Mountains, Tanzania. *Anthropological Science*, **106**, 349–369.

Uehara, S. & Nishida, T. (1987). Body weights of wild chimpanzees (*Pan troglodytes schweinfurthii*) of the Mahale Mountains National Park, Tanzania. *American Journal of Physical Anthropology*, **72**, 315–321.

Willett, H. P. (1992). Normal flora and opportunistic infections. In *Zinsser Microbiology*, ed. W. K. Joklik, H. P. Willett, D. B. Amos & C. M. Wilfert, 20th edn. Norwolk: Appleton & Lange, pp. 393–400.

13 *Intestinal bacteria of chimpanzees in the wild and in captivity: an application of molecular ecological methodologies*

KAZUNARI USHIDA

Photograph by Kazunari Ushida (*Pan troglodytes verus*)

Primate Parasite Ecology. The Dynamics and Study of Host–Parasite Relationships, ed. Michael A. Huffman and Colin A. Chapman. Published by Cambridge University Press.

Importance of intestinal bacteria

The importance of intestinal bacteria for humans has been stressed with respect to health promotion and nutrient supply to the host (Gibson & Macfarlane, 1995; Hooper, 2004; Kelly *et al.*, 2005). In the human colon, bacterial microflora consists of several hundred species of which some are present at very high density (over 10^{11}/g). Therefore, intestinal bacteria represent over 90% of cells associated with the human body. Interestingly, the predominant intestinal bacteria are the same for human subjects whose ethnicity and foods differ (Finegold *et al.*, 1983). Moreover, it is noteworthy that the intestinal ecosystem of the adult human is very stable and there is little temporal variation within the same subject. Indeed, dietary change has little influence on the matured intestinal ecosystem (Mitsuoka & Ohno, 1977; Tannock, 1983). It is therefore suggested that there may be a human-associated, or coevolved, intestinal bacterial ecosystem (Hooper, 2004).

Colonization of intestinal commensal bacteria has great impact on the development and maintenance of the mucosal immune system that is a primary defense against pathogenic penetration. A great deal of information has become available about the so-called protective intestinal flora over the last 30 years and this has led us to understand one of the major beneficial functions of the intestinal commensal bacteria (Hentges *et al.*, 1985; Hooper, 2004; Kelly *et al.*, 2005). In the case of humans, there have been reports that suggest the importance of bifidobacteria, a kind of lactic acid bacteria, as an indicator of gut disorders (Mitsuoka & Kaneuchi, 1977; Mitsuoka, 1990; Leahy *et al.*, 2005). In this context, the theory of Metchinikoff (1907) has been revived, and as a result there are many lactic acid bacteria now commercialized as probiotics, or beneficial bacteria, that improve the "intestinal balance" (Fuller, 1989). Probiotics were thought to be effective in preventing diarrhea in young chimpanzees in the Tacugama chimpanzee sanctuary in Sierra Leone (Garriga and Kabasawa, personal communication).

From a nutritional point of view, the importance of intestinal bacteria is clear for all herbivores. All great apes, including humans, have well-developed intestines that meet their dietary needs (Stevens & Hume, 1995). Chimpanzees, gorillas, and orangutans consume plant material as a principal food and its fibrous components require bacterial fermentation to supply energy to the host. The retrieved energy is in the form of short chain fatty acids such as acetate, propionate, and butyrate, which typically contribute a significant part of the energy requirements of these animals. An unbalanced intestinal microflora may lead to non-pathogenic diarrhea that is caused by abnormal intestinal fermentation and abnormal accumulation of lactate and succinate (Tsukahara & Ushida, 2001; Tsukahara & Ushida, 2002; Hashizume *et al.*, 2003). Therefore it is important to understand the role of intestinal bacteria in promoting great ape health. Such information is particularly important for maintaining apes in captivity. With this in mind, we compare the intestinal microflora of wild chimpanzees at Bossou, Guinea and Mahale, Tanzania with those in the captive Primate Research Institute group in Inuyama, Japan.

A culture-independent approach to the bacteriology of wild animals

Little information is available about the intestinal microbiota of the wild apes, although much is known about that of humans, livestock, and experimental rodents. Our knowledge of the intestinal microbiota of wild apes is constrained by methodological limitations. Traditional bacteriological approaches are based on culturing bacteria, which requires fresh samples and considerable experimental equipment for anaerobic culturing. This is obviously difficult to undertake under field conditions. The recent development of so-called "molecular microbial ecology" now makes it possible to assess the intestinal microbiota of wild apes when fresh fecal samples are available. This new methodology is culture-independent and determines the bacterial genomic DNA of the sample. Bacterial DNA in the feces is easily preserved by placing the sample in ethanol or acetone (Fukatsu, 1999). This makes it possible to safely preserve and transport the samples to the laboratory, although ethanol-fixed samples are no longer suitable for culturing. We routinely use ethanol, because ethanol is available in most small cities in Africa.

This new molecular approach mostly amplifies bacterial 16S rRNA genes by polymerase chain reaction (PCR). Bacterial genomic DNA is extracted from the fixed fecal samples to serve as DNA templates for PCR. There are several primer sets available to amplify bacterial 16S rRNA in general, which are called universal primers for the bacterial 16S rRNA gene. Since each bacterial species has its own particular sequences in the 16S rRNA gene, the sequence data of this gene is suitable for classification and phylogenetic analyses.

To assess the structure of a particular microbial community, the 16S rRNA gene amplified with universal primers can be analyzed in a sequence-specific manner, namely, amplified rDNA restriction analysis (ARDRA) (Inoue & Ushida, 2003b), denaturing gradient gel electrophoresis (DGGE), or temperature gradient gel electrophoresis (TGGE) (Müyzer *et al.*, 1993). The analytical ranges of these molecular techniques are, however, believed to be limited to 10, or a few more, of the most prevalent species (over 10^5/g) (Itoh, 2005).

A particular bacterial family, genera, or species can be detected directly by specific PCR. The recent development of quantitative PCR using a light cycler system has made it possible to quantify particular bacteria in a community (Matsuki *et al.*, 2004). Quantification of particular bacteria can also be assessed by fluorescent in situ hybridization (FISH) technology. In this method, a probe specific to the particular sequence of genomic DNA of the target bacteria is required. Under the microscope, it is possible to distinguish target bacteria from others by fluorescence. In this methodology, paraformaldehyde-fixed bacteria are much better than those fixed with ethanol, because the bacterial cells shrink in ethanol.

Importance of sampling protocol

Feces of the subjects must be collected just after defecation for the study of intestinal bacteria. Rapid collection is important to avoid a shift of bacterial composition in feces due to the penetration of air and to avoid eventual contamination by environmental bacteria. This precaution is easily achieved in the laboratory. However, in the field animals must be continuously followed to collect samples immediately after defecation. In our studies at Bossou and Mahale, chimpanzees were recognized individually and feces were collected just after defecation (Ushida *et al.*, 2006; Fujita & Kageyama, 2007; Uenishi *et al.*, 2007). The correct identification of individuals is essential to avoid redundant sampling and to measure individual variation of intestinal microflora. However, this precaution is not always possible depending on field conditions (Frey *et al.*, 2006; S. Fujita, personal communication).

Application of ARDRA and TGGE analyses of Bossou chimpanzees

According to the ARDRA of Bossou individuals, 596 operational taxonomic units (OTUs) were detected (Uenishi *et al.*, 2007). Among them, 119 OTUs were detected in at least two individuals. We conducted sequence analysis on the 119 OTUs which are considered as predominant bacterial groups in Bossou chimpanzees. Among them, 58 OTUs harbored a partial sequence of 16S rDNA of Firmicutes, such as *Clostridium*, *Eubacteria*, *Lactobacillus*, *Ruminococcus*, and *Bacillus*. The next most abundant group harbored a partial sequence of 16S rDNA of Bacteroidetes (48 OTUs), such as *Prevotella* and *Bacteroides*. Partial sequences of 16S rDNA of Proteobacteria (8 OTUs), such as *Succinimonas*, *Succinivibrio*, and *Acinetobacter*, and of *Actinobacteria* (4 OTUs), such as *Bifidobacterium* and *Slackia*, were detected. One OTU harbored a partial sequence of 16S rDNA of Spirochetes, genus *Treponema*. Each OTU of *E. coli* clones,

Table 13.1. *Similarity of bacterial composition between chimpanzees in Bossou and Human – summary of ARDRA results*

	% of phylogenetic lineage			
Bacterial phylum	Chimpanzee	Human[1]	Gorilla[2]	Pig[3]
Actinobacteria	1.7	0.2	5.3	1.1
Bacteriodetes	32.7	47.7	1.1	11.2
Firmicutes	60.3	50.8	71	81.3
Flexistipes				0.3
Fusobacteria		0.08		
Lentisphaerae			3.2	
Planctomycetes			1.1	0.3
Proteobacteria	4.8	0.6		5.3
Spirochetes	0.6		1.1	0.5
Verrucomicrobia		0.6	17.2	
Unclassified		0.02		

[1]Eckburg *et al.* (2005).
[2,3]See Frey *et al.* (2006).

of which the cloning plasmids harbor the same 16S rDNA. Assuming that the number of *E. coli* clones belonging to a particular OTU represents the relative contribution of this OTU to the whole gut ecosystem, the values for representative bacterial phylum found in chimpanzees, humans, gorillas, and pigs are presented in Table 13.1. The values are calculated from 594 clones of 119 OTUs. The results show the predominant presence of Firmicutes and Bacteroidetes, phylums commonly classified as plant polymer and sugar-fermenting bacteria. The most prevalent OTUs were *Eubacterium* sp.-like bacterium, *Ruminococcus obeum*-like bacterium, *Anaeroplasma varium*-like bacterium, *Ruminococcus* sp.-like bacterium, and *Prevotella* sp.-like bacterium (Uenishi *et al.*, 2007). Among them, one OTU close to *Eubacterium* sp. and the other OTU close to *R. obeum* were detected from all individuals. This is also confirmed by TGGE analysis (Figure 13.1) on fecal bacteria of Bossou chimpanzees. Sequence analyses on the bands in TGGE commonly shared by chimpanzees in Bossou suggest that they belong to the genera *Clostridium*, *Lactobacillus*, and *Bifidobacterium* (Uenishi *et al.*, 2007).

The predominant presence of the plant polymer and sugar-fermenting bacteria in Bossou chimpanzees is expected, because these bacteria are usually detected from a wide range of herbivorous and omnivorous animals, including humans (Holdeman *et al.*, 1977; Salyers *et al.*, 1977). The detection of the *C. coccoides*-group (*Clostridium* rRNA cluster XIVa), such as *Eubacterium* sp., and *Ruminococci*, such as *R. obeum*, in human feces has also been reported

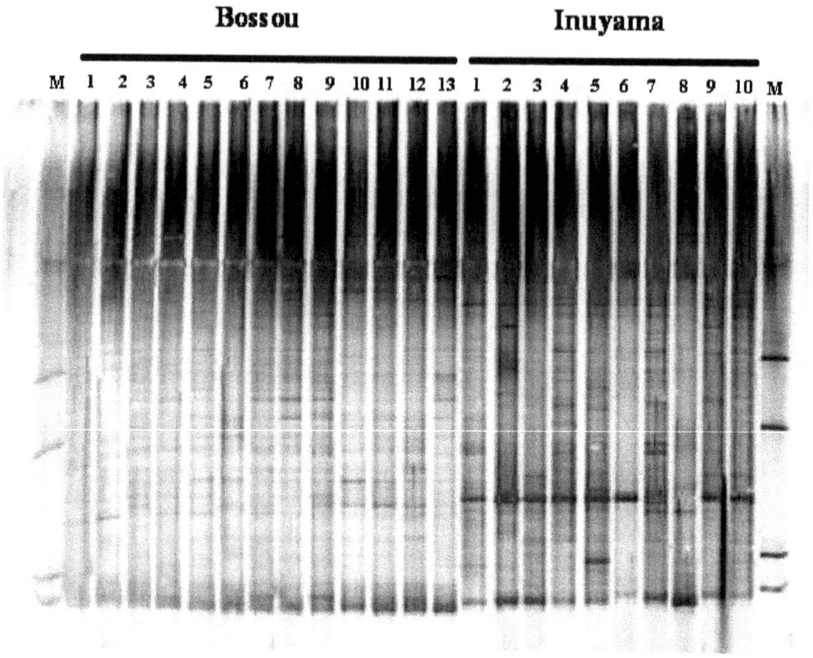

Figure 13.1. Gel image of temperature gradient gel electrophoresis on partial 16S rDNA of intestinal bacteria of chimpanzees. M, marker. From left to right, Jire, Foaf, Pama, Peley, Fotayu, Fokaiye, Yolo, Tua, Yo, Fanle, Fana, Jeje, and Velu for Bossou individuals. For Inuyama individuals, Ayumu, Cleo, Puchi, Popo, Reo, Pan, Pal, Pico, Ai, and Chloe.

(Suau *et al.*, 1999; Akkermans *et al.*, 2000; Zoetendal *et al.*, 2002). Three sequences with the highest homology to *R. obeum* and *Eub. hallii* and *Faecalibacterium prausnitzii* (*C. leptum*-group Clostridium rRNA cluster IV) were present in all human subjects, and their universal role in the human gastrointestinal tract was suggested (Akkermans *et al.*, 2000). A comparison with human data (Eckburg *et al.*, 2005) suggested that chimpanzees in Bossou harbor a type of intestinal microflora similar to that of humans at the phylum level (Table 13.1). Since the analytical methods used were not the same for all of these studies cited in Table 13.1, comparison should be made cautiously.

Differences and similarities in composition of intestinal microflora between wild and captive chimpanzees

The TGGE band profiles from Bossou chimpanzees and those in captivity at Inuyama differed (Figure 13.2), although there were several commonly shared

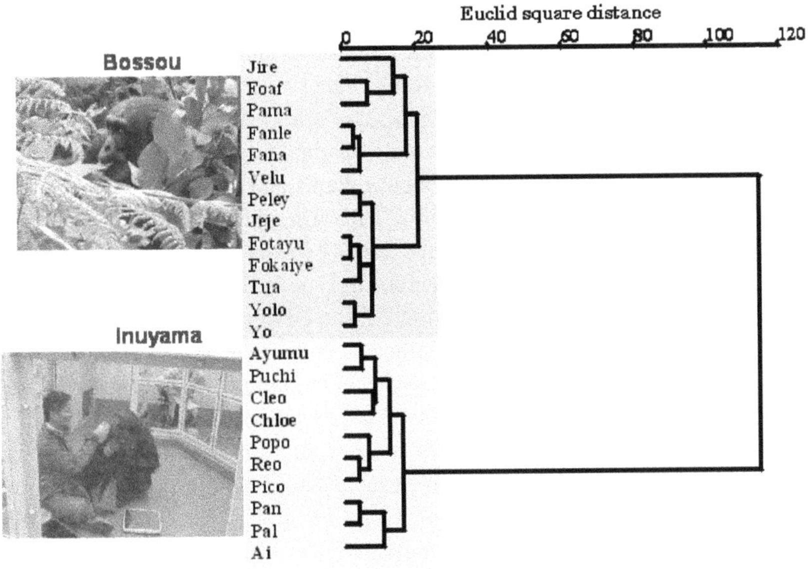

Figure 13.2. Clustering analysis on the community structure of intestinal microflora of chimpanzees (Uenishi *et al.*, 2007).

bands. It is noteworthy that the compositions of intestinal microflora of Inuyama chimpanzees were relatively similar, as evidenced by the small Euclid square distance (<20). Unexpectedly, the same is true for Bossou chimpanzees. In the case of humans, variation of the intestinal microflora is large (Mitsuoka & Ohno, 1977). In our experiment involving eight students from the laboratory, the Euclid square distance for eight individuals was 40 to 60 units (Tokunaga and Ushida, unpublished data), much larger than that observed in chimpanzees presented here. It is easy to understand individual variation in human intestinal microflora, because variation is known to be produced by a number of environmental factors (Conway, 1995). Therefore, it was unexpected to find such high similarity in the intestinal microflora of free-ranging Bossou chimpanzees.

Why do wild Bossou chimpanzees have relatively similar intestinal microflora? Development of intestinal microflora is affected by such factors as the intensity of transmission of maternal bacteria and dietary conditions. In Figure 13.2, individual chimpanzees harboring similar bacterial composition are placed side by side. For example, in the case of Bossou chimpanzees, each of five combinations, Pama and Foaf, Jeje and Peley, Yo and Yolo, Fotayu and Fokaiye, and Fana and Fanle are placed in this manner. In the case of Inuyama chimpanzees, there were four combinations, Ayumu and Petit, Leo and Pico, Chloe and Cleo, and Pan and Pal. Among these combinations, Fana and Fanle, Fotayu and Fokaiye, Yo and Yolo, Cloe and Cleo, and Pan and Pal, were mother–offspring pairs. The late weaning of chimpanzees (4–5 years of age; Clarke, 1977) may contribute to the intensive transmission of maternal intestinal bacteria in chimpanzees. In the case of rodents and pigs, feces is one of the most possible sources of maternal intestinal bacteria, because pups and piglets consume a considerable amount of dam's and sow's feces (Sansom & Gleed, 1981; Inoue & Ushida, 2003b). Therefore, the similarity in intestinal microflora between a dam and her pups can be explained by intensive transmission of maternal intestinal bacteria (Inoue & Ushida, 2003b), probably due to coprophagy. Coprophagy is not so often observed in wild chimpanzees, however, and the direct consumption by infants of their mother's feces has not been observed (S. Fujita and R. Garriga, personal communication). The offspring may have an increased chance of mother fecal contamination if they are touched by the mother's hand after she performs coprophagy. However, if this form of contact contamination occurs, than we should also expect fecal contamination between non-kin and horizontal transfer within a group.

Dietary conditions can affect the construction of intestinal microflora. Since in Inuyama, chimpanzees were basically offered the same meal, this may explain the high similarity in their intestinal microflora. The same may be true for Bossou chimpanzees at least to some extent, because the size of the group is small and they live in a relatively limited area (Matsuzawa *et al.*, 2006), where they consume the same types of natural foods (Yamakoshi, 1998; Takemoto, 2004).

Detection of particular bifidobacteria

Temperature gradient gel electrophoresis analyses indicate that *Bifidobacterium* spp. were a common intestinal bacteria in chimpanzees. We focused here on bifidobacteria, because these bacteria have been recognized as human-associated lactic acid bacteria and are believed to contribute to human health (Mitsuoka & Kaneuchi, 1977; Mitsuoka, 1990; Leahy *et al.*, 2005). *Bifidobacterium adolescentis*, *B. angulatum*, *B. bifidum*, *B. breve*, the *B. catenulatum* group (*B. catenulatum* and *B. pseudocatenulatum*), and the *B. longum* group (*B. longum* and *B. infantis*) are common in the human intestine (Matsuki *et al.*, 1999). When we checked the presence of bifidobacteria in chimpanzees using bifidobacteria-specific PCR (Matsuki *et al.*, 1998, 2004), we detected bifidobacteria close to *B. angulatum* (12 OTUs over 14) from all chimpanzees in Bossou (Figure 13.3). Bifidobacteria close to *B. pseudolongum* (6 OTUs over 9) and *B. dentium* (3 OTUs over 9) were detected from five young chimpanzees in Inuyama. These bifidobacteria are common in humans as well. Interestingly,

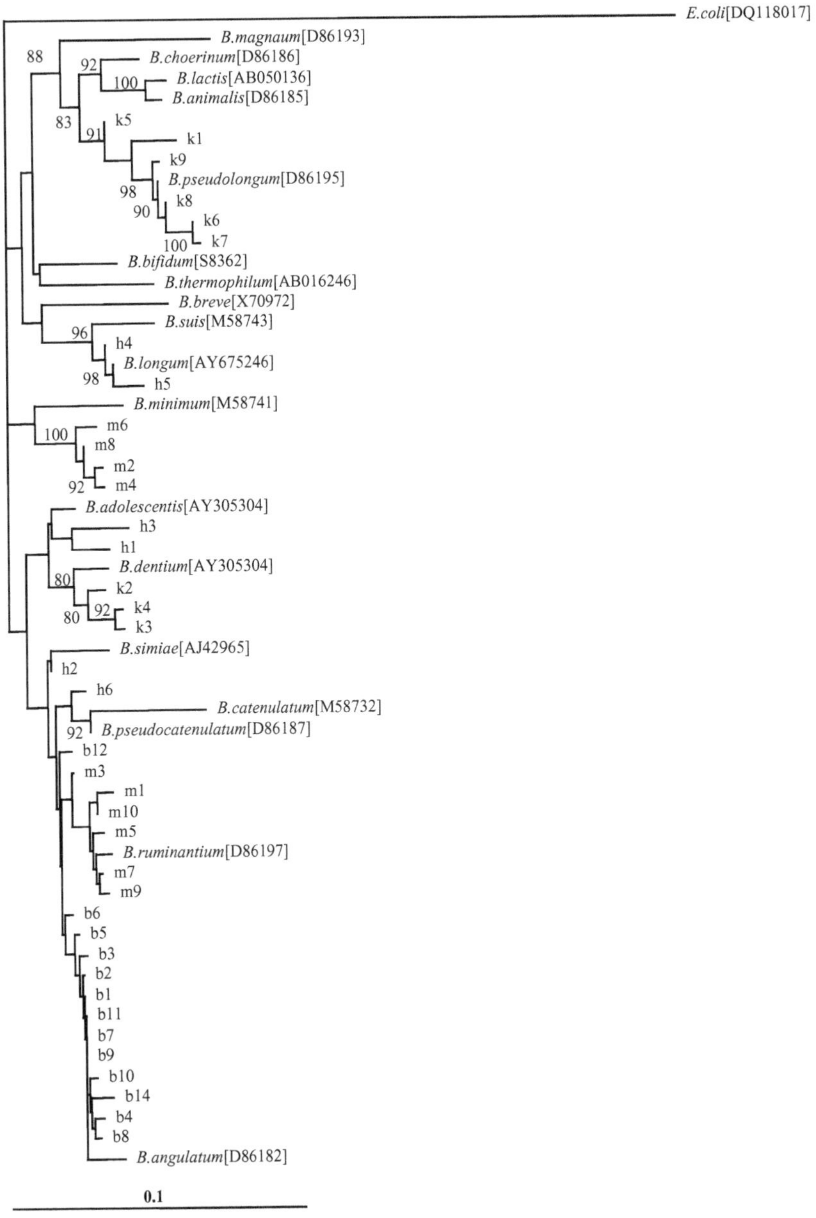

Figure 13.3. Phylogenetic tree of bifidobacterial SSU rDNA sequences detected from chimpanzees. The tree was constructed by the neighbor-joining method on the partial sequence of 16S rDNA amplified with bifidobacteria-specific primers (Matsuki *et al.*, 2004). k, samples from individuals of Kyoto University Primate Research Institute at Inuyama; m, samples from Mahale individuals; b, samples from Bossou individuals; h, retrieved from human feces (Tokunaga and Ushida, unpublished data). The figures on the node indicate the bootstrap value for 1000 calculations. Values under 80% were omitted.

B. dentium was detected only from one young chimpanzee that had been raised by artificial feeding formula instead of chimpanzee breast milk.

On the other hand, we have detected bifidobacteria close to *B. ruminantium* (6 OTUs over 10) and *B. minimum* (4 OTUs over 10) from 10 chimpanzees in Mahale (Figure 13.3). The former was isolated from bovine rumen (Biavati & Mattarelli, 1991), and the latter was from non-human sources (Simpson *et al.*, 2004). The detection of human-type bifidobacteria from Bossou and Inuyama chimpanzees and non-human-type bifidobacteria from chimpanzees in Mahale is interesting because it may reflect the extent of human activity at Bossou and Inuyama on the chimpanzees' intestinal microflora.

The study of human intestinal bacteria has a long history, dating to the discovery of bifidobacteria (Tissier, 1900). Recently, the beneficial roles of intestinal commensal bacteria have been recognized and given its potential role in health maintenance a comprehensive study on the intestinal bacteria of great apes, particularly in the wild, is needed. The study of intestinal bacteria in great apes has only recently begun and the present chapter briefly reports findings from the first step of our project.

Acknowledgements

A part of this study was supported by the HOPE project from the Japan Society for Promoting Science. The author thanks Dr. T. Matsuzawa, Primate Research Institute, Kyoto University (KUPRI) for the opportunity to work on chimpanzees in Bossou and KUPRI. Sincere gratitude is expressed to Dr. S. Fujita (Department of Veterinary Medicine, Yamaguchi University). Thanks are also due to Dr. A. Kato, Dr. N. Maeda, Dr. K. Kumazaki, Dr. A. Kaneko, Dr. T. Kageyama, and Mr. G. Ohashi in KUPRI for their help with sample collection. The microbiological works were indebted to Dr. Y. Ohashi, Mr. G. Uenishi, and Miss M. Hiraguchi of the author's laboratory. Information about the weaning behavior of chimpanzees was supplied by Dr. T. Matsusaka (School of Human Cultures, The University of Shiga Prefecture). Important information was supplied from Dr. R. M. Garriga and Ms. A. Kabasawa at Tacugama Chimpanzee Sanctuary in Sierra Leone. The author is indebted to Messrs. P. Goumy, P. Cherif, and J. Doré for their assistance with the collection of chimpanzee feces. We thank L'Institut de Recherche Environnementale de Bossou (IREB) and La Direction Nationale de la Recherche Scientifique et Technique (DNRST) of the government of the Republic of Guinea, the Tanzania Commission for Science and Technology, the Tanzania Wildlife Research Institute, the Tanzania National Parks, the Mahale Wildlife Research Centre, and the Mahale Mountains National Park for their support.

References

Akkermans, A. D. L., Zoetendal, E. G., Favier, C. F. *et al.* (2000). Temperature and denaturing gradient gel electrophoresis analysis of 16S rRNA from human fecal samples. *Bioscience Microflora*, **19**, 93–98.

Biavati, B. & Mattarelli, P. (1991). *Bifidobacterium ruminantium* sp. nov. and *Bifidobacterium merycicum* sp. nov. from the rumens of cattle. *International Journal of Systematic Bacteriology*, **41**, 163–168.

Clarke, C. B. (1977). A preliminary report on weaning among chimpanzees of the Gombe National Park, Tanzania. In *Primate Bio-social Development*, ed. S. Chevallier-Skolnikof & F. E. Poirier. New York, NY: Garland, pp. 235–260.

Conway P. L. (1995). Microbial ecology of the human large intestine. In *Human Colonic Bacteria: Role in Nutrition, Physiology, and Pathology*, ed. G. R. Gordon & G. T. Macfarlane. London: CRC Press, pp. 1–24.

Eckburg, P. B., Bik, E. M., Bernstein, C. N. *et al.* (2005). Diversity of the human intestinal microbial flora. *Science*, **308**, 1635–1638.

Finegold, S. M., Sutter, V. L. & Mathiesen, G. E. (1983). Normal indigenous flora. In *Human Intestinal Microflora in Health and Disease*, ed. D. J. Hentges. New York, NY: Academic Press, pp. 3–31.

Frey, J. C., Rothman, J. M., Pell, A. N. *et al.* (2006). Fecal bacterial diversity in a wild gorilla. *Applied and Environmental Microbiology*, **72**, 3788–3792.

Fujita, S. & Kageyama, T. (2007). Polymerase chain reaction detection of *Clostridium perfringens* in feces from captive and wild chimpanzees, *Pan troglodytes. Journal of Medical Primatology*, **36**, 25–32.

Fukatsu, T. (1999). Acetone preservation: a practical technique for molecular analysis. *Molecular Ecology*, **8**, 1935–1945.

Füller, R. (1989). Probiotics in man and animals. *Journal of Applied Bacteriology*, **66**, 365–378.

Gibson, G. R. & Macfarlane, G. T. (1995). *Human Colonic Bacteria: Role in Nutrition, Physiology, and Pathology*. London: CRC Press.

Godon, J. J., Zumstein, E., Dabert, P., Habouzit, F. & Moletta, R. (1997). Molecular microbial diversity of an anaerobic digestor as determined by small-subunit rDNA sequence analysis. *Applied and Environmental Microbiology*, **63**, 2802–2813.

Hashizume, K., Tsukahara, T., Yamada, K., Koyama, H. & Ushida, K. (2003). *Megasphaera elsdenii* JCM1772 T normalizes hyperlactate production in the large intestine of fructooligosaccharide-fed rats by stimulating butyrate production. *Journal of Nutrition*, **133**, 3187–3190.

Hentges, D. J., Stein, A. J., Casey, S. W. & Que, J. U. (1985). Protective role of intestinal flora against infection with *Pseudomonas aeruginosa* in mice: influence of antibiotics on colonization resistance. *Infection and Immunity*, **47**, 118–122.

Holdeman, L. V., Cato, E. P. & Moore, W. E. C. (1977). *Anaerobe Laboratory Manual*, 4th edn. Blacksburg, VA: Southern Printing.

Hooper, L. V. (2004). Bacterial contributions to mammalian gut development. *Trends in Microbiology*, **12**, 129–134.

Inoue, R. & Ushida, K. (2003a). Development of the intestinal microflora in rats and its possible interaction with the evolution of the luminal IgA in the intestine. *FEMS Microbiology Ecology*, **45**, 147–153.

Inoue, R. & Ushida, K. (2003b). Vertical and horizontal transmission of intestinal commensal bacteria in the rat model. *FEMS Microbiology Ecology*, **46**, 213–219.

Itoh, K. (2005). *Probiotics and Biogenics*, 1st edn. (In Japanese) Tokyo: NTS.

Kelly, D., Conway, S. & Aminov, R. (2005). Commensal gut bacteria: mechanisms of immune modulation. *Trends in Microbiology*, **26**, 326–333.

Leahy, S. C., Higgins, D. G., Fitzgerald, G. F. & van Sinderen, D. (2005). Getting better with bifidobacteria. *Journal of Applied Microbiology*, **98**, 1303–1315.

Matsuki, T., Watanabe, K., Tanaka, R. & Oyaizu, H. (1998). Rapid identification of human intestinal bifidobacteria by 16S rRNA-targeted species- and group-specific primers. *FEMS Microbiology Letters*, **167**, 113–121.

Matsuki, T., Watanabe, K., Tanaka, R., Fukuda, M. & Oyaizu, H. (1999). Distribution of bifidobacterial species in human intestinal microflora examined with 16S rRNA-gene-targeted species-specific primers. *Applied and Environmental Microbiology*, **65**, 4506–4512.

Matsuki, T., Watanabe, K., Fujimoto, J. *et al.* (2004). Quantitative PCR with 16S rRNA-gene-targeted species-specific primers for analysis of human intestinal bifidobacteria. *Applied and Environmental Microbiology*, **70**, 167–173.

Matsuzawa, T., Tomonaga, M. & Tanaka, M. (2006). *Cognitive Development in Chimpanzees*. Tokyo: Springer-Verlag.

Metchnikoff, E. (1907). *The Prolongation of Life*, 1st edn. New York, NY: GP Putman's Sons.

Mitsuoka, T. (1990). Bifidobacteria and their role in human health. *Journal of Industrial Microbiology*, **6**, 263–268.

Mitsuoka, T. & Kaneuchi, C. (1977). Ecology of the bifidobacteria. *American Journal of Clinical Nutrition*, **30**, 1799–1810.

Mitsuoka, T. & Ohno, K. (1977). Die Eaekalflora bei Menschen. V. Mitteiling: Die Schwankungen in der Zusammensetsung der Faekalflora gesunder Erwachsener. *Zentralblatt für Bakteriologie, Parasitenkunde, Infektionskrankheiten und Hygiene. Erste Abteilung Originale. Reihe A: Medizinische Mikrobiologie und Parasitologie*, **238**, 228–236.

Müyzer, G., Waal, E. C. & Uitterlinden, A. G. (1993). Profiling of complex microbial populations by denaturing gradient gel electrophoresis analysis of polymerase chain reaction-amplified genes coding for 16S rRNA. *Applied and Environmental Microbiology*, **59**, 695–700.

Salyers, A. A., West, S. E., Vercellotti, J. R. & Wilkins, T. D. (1977). Fermentation of mucins and plant polysaccharides by anaerobic bacteria from the human colon. *Applied and Environmental Microbiology*, **34**, 529–533.

Sansom, B. F. & Gleed, P. T. (1981). The ingestion of sow's faeces by suckling piglets. *British Journal of Nutrition*, **46**, 451–456.

Simpson, P. J., Ross, R. P., Fitzgerald, G. F. & Stanton, C. (2004). *Bifidobacterium psychraerophilum* sp. nov. and *Aeriscardovia aeriphila* gen. nov., sp. nov., isolated from a porcine caecum. *International Journal of Systematic and Evolutionary Microbiology*, **54**, 401–406.

Stevens, C. E. & Hume, I. D. (1995). *Comparative Physiology of the Vertebrate Digestive System*, 2nd edn. Cammbridge, UK: Cambridge University Press.

Suau, A., Bonnet, R., Sutren, M. *et al.* (1999). Direct analysis of genes encoding 16S rRNA from complex communities reveals many novel moleclar species within human gut. *Applied and Environmental Microbiology*, **65**, 4799–4807.

Takemoto, H. (2004). Seasonal change in terrestriality of chimpanzees in relation to microclimate in the tropical forest. *American Journal of Physical Anthropology*, **124**, 81–92.

Tannock, G. W. (1983). Effect of dietary and environmental stress on the gastrointestinal tract. In *Human Intestinal Microflora in Health and Disease*, ed. D. J. Hentges. New York, NY: Academic Press, pp. 517–539.

Tissier, H. (1900). *Recherche sur la flore intestinale des nourissons (Etat normal et pathologique)*. Thèse, Université de Paris.

Tsukahara, T. & Ushida, K. (2001). Organic acid profiles in feces of pigs either with pathogenic or non-pathogenic diarrhea. *Journal of Veterinary Medical Science*, **63**, 1351–1354.

Tsukahara, T. & Ushida, K. (2002). Succinate accumulation in pig large intestine during antibiotic-associated diarrhea and the constitution of succinate-producing flora. *Journal of General and Applied Microbiology*, **48**, 143–154.

Uenishi, G., Fujita, S., Ohashi, G. *et al.* (2007). Molecular analyses of the intestinal microbiota of chimpanzees in the wild and in captivity. *American Journal of Primatology*, **69**, 367–376.

Ushida, K., Fujita, S. & Ohashi, G. (2006). Nutritional significance of the selective ingestion of *Albizia zygia* gum exudate by wild chimpanzees in Bossou, Guinea. *American Journal of Primatology*, **68**, 143–151.

Yamakoshi, G. (1998). Dietary responses to fruit scarcity of wild chimpanzees at Bossou: possible implications for ecological importance of tool use. *American Journal of Physical Anthropology*, **106**, 283–295.

Zoetendal, E. G., Ben-Amor, K., Harmsen, H. J. *et al.* (2002). Quantification of uncultured *Ruminococcus obeum*-like bacteria in human fecal samples by fluorescent *in situ* hybridization and flow cytometry using 16S rRNA-targeted probes. *Applied and Environmental Microbiology*, **68**, 4225–4232.

14 *Gastrointestinal parasites of bonobos in the Lomako Forest, Democratic Republic of Congo*

JOZEF DUPAIN, CARLOS NELL, KLÁRA JUDITA PETRŽELKOVÁ, PAOLA GARCIA, DAVID MODRÝ, AND FRANCISCO PONCE GORDO

Photograph by Craig Sholley (*Pan paniscus*)

Primate Parasite Ecology. The Dynamics and Study of Host–Parasite Relationships, ed. Michael A. Huffman and Colin A. Chapman. Published by Cambridge University Press.

Introduction

Habitat destruction, human population growth, and increased encroachment in formerly inaccessible forest blocks intensifies contact between humans and our closest relatives, the great apes. This increases the risk of disease transmission that might have devastating effects on the health status and survival of the species. The recent event of emerging viral diseases such as Ebola has proven the potential danger of still little-known threats (Bermejo *et al.*, 2006). Parasites influence the health status of populations and are candidates for inter-species transmission. The parasites of the common chimpanzee (*Pan troglodytes*) have been extensively studied in captivity and also in the wild (File *et al.*, 1976; McGrew *et al.*, 1989; Kawabata & Nishida, 1991; Landsoud-Soukate *et al.*, 1995; Huffman *et al.*, 1997; Ashford *et al.*, 2000; Murray *et al.*, 2000; Lilly *et al.*, 2002). A checklist of parasites and commensals reported for the chimpanzee (*Pan*) was compiled by Myers & Kuntz (1972). However, there is a lack of information about intestinal parasites of bonobo (*Pan paniscus*) even in captive conditions (Stam, 1960; Vuylsteke, 1964). Only one study dealt with bonobo intestinal parasites in their natural habitat (Hasegawa *et al.*, 1983). They examined 390 feces collected under nests of bonobos from four groups living in Wamba region, Democratic Republic of Congo during October to December 1981.

Our aim is to contribute to the knowledge about parasites of bonobos in the wild and present some suggestions for future research.

Materials and methods

This study was conducted in the Lomako Forest, Equator Province, Democratic Republic of Congo (Dupain *et al.*, 2000) (Figure 14.1) from October to November 1998, which is a heavy rain season (monthly rainfall is > 200 mm). Fresh feces were collected in the morning (at dawn or soon after it) under the night nests of parties of the semi-habituated community of bonobos studied at the research site Iyema-Lomako (N 00°55′33″E 21°05′21″). Bonobos of one community are typically organized in parties of temporary and changing composition. This study focuses on one community. The habituation of this community started in 1995 and continued on and off for approximately 15 months. The first author could identify individually almost 10 members of this community of potentially 30–50 individuals. The members of the bonobo community accepted the researchers' presence rather well which allowed good observation in the morning when bonobos typically wake up and defecate/urinate from their night nests. As habituation was not complete, in most cases, once the bonobos

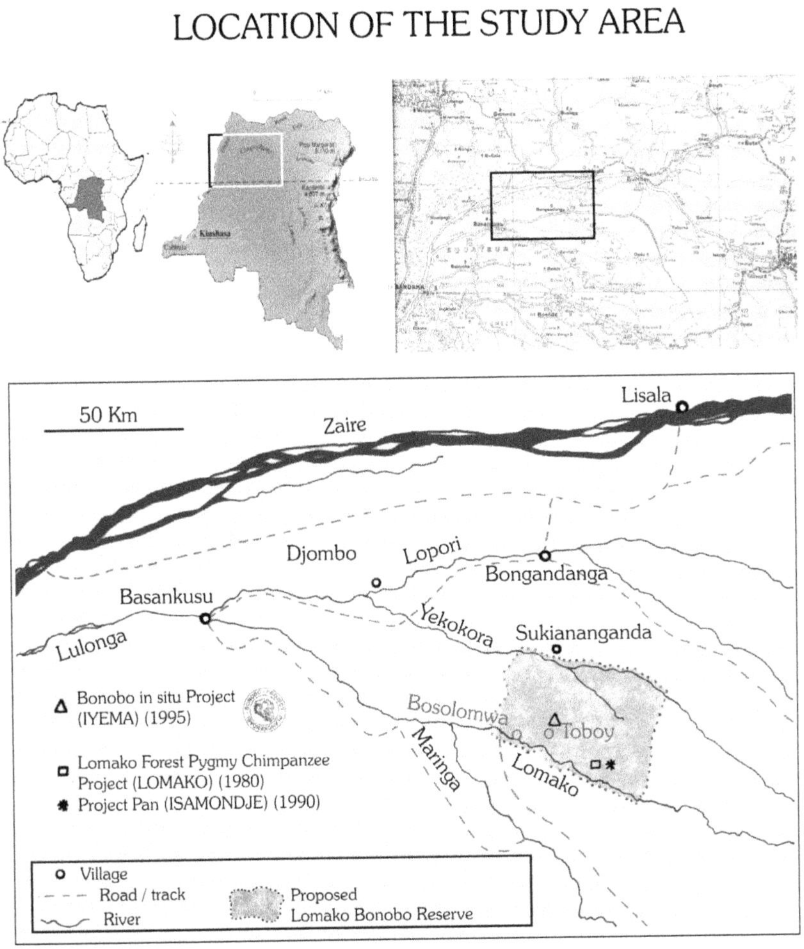

Figure 14.1. Location of the study area.

started traveling, they could not be followed. Only a few samples could be attributed to specific individuals, but care was given on any day not to sample the same individual twice. Each sample was kept in a hermetic 50 ml flask, adequately labeled, and transported to the campsite in less than 4 hours after collection. Upon arrival at the camp, approximately 10 ml of each faecal sample was immediately fixed with 10% formalin and subsequently transferred to the Departamento de Parasitología, Facultad de Farmacia, Universidad Complutense in Spain, where the parasitological analyses were carried within 3 months of the collection. About half of each sample was kept in the original flask for possible further analyses/repetitions. The other half was diluted 1:1

with formalin and filtered through standard stainless sieves (about 100 μm porosity). The material retained in the sieves were resuspended in 20 ml phosphate buffered saline (PBS) and observed under stereomicroscope at 16–64× magnification. The filtered material was concentrated using the simplified Ritchie's method (Ritchie, 1948), as modified by Young *et al.* (1979), and the sediment was observed using temporary wet mounts stained with Lugol's iodine under light/phase contrast microscopy at 100–400× magnification.

Results

Eighty-seven fecal samples of bonobos were examined. No samples were diarrheic, with traces of blood or with mucus. No macroscopic parasite structures were found. Five species of intestinal parasites were identified in the bonobo samples, including one entodiniomorph ciliate (*Trogodytella* sp.) and five nematodes (*Trichuris* sp., hookworms, *Strongyloides fuelleborni*, *Oesophagostomum* sp., and *Ascaris* sp.); besides, four different types of unidentified eggs were found, three of them were trematodes and one was a nematode (Figures 14.2–14.10).

Exclusion of the ciliate *Troglodytella* which is a possible symbiont of African apes (Reichenow, 1920; Collet *et al.*, 1984) leaves *Oesophagostomum* sp. as the most prevalent real parasite. Morphometric data, basic descriptions of helminth eggs, and prevalences are given in Table 14.1. Most of the animals were infected with 1–3 parasite species (1 species – 10.3%, 2 – 25.3%, 3 – 27.6%, 3 – 24.1%, 4 – 10.3%, 5 – 2.3%). If we exclude the suggested symbiont ciliate, most animals were infected by two or less parasite species (1 species – 27.6%, 2 – 29.9%, 3 – 11.5%, 4 – 2.3%).

Discussion

The bonobo is the closest relative of the common chimpanzee, *P. troglodytes*. However, in contrast to the latter species, the information about the biology (including parasites) of bonobos is limited. The only study focused on intestinal parasites of wild bonobos is that by Hasegawa *et al.* (1983; Table 14.2). The spectrum of parasites found during our study is similar to that reported in *P. troglodytes* (*Troglodytella, Trichuris, Oesophagostomum*, ancylostomatid nematodes, *Strongyloides,* and *Ascaris*).

Troglodytella spp. and other entodiniomorph ciliates occur in African apes in high prevalences and a symbiotic role of these ciliates has been suggested (Collet *et al.*, 1984; Stahl, 1984). Prevalence of *Troglodytella* sp. recorded

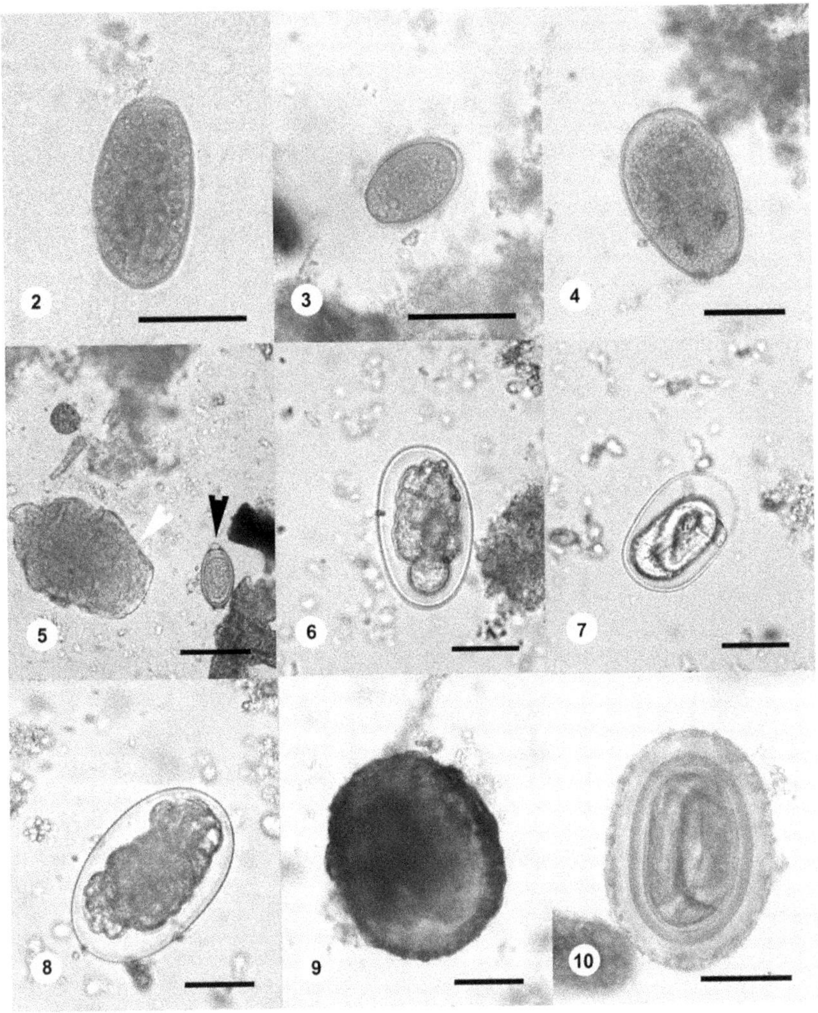

Figures 14.2–14.10. Parasites recovered from the feces of bonobos (*Pan paniscus*). 2. Trematode 1. 3. Trematode 2. 4. Trematode 3. 5. *Troglodytella* trophozoite (white arrowhead) and *Trichuris* egg (black arrowhead). 6. Hookworm egg. 7. Embryonated egg of *Strongyloides fuelleborni*. 8. *Oesophagostomum* egg. 9. *Ascaris* sp. 10. Nematode 1 egg, note thick egg wall and coiled larva inside. Scale bars: 2, 3, and 5 = 50 μm; all other = 25 μm.

in this study (75%) was lower than in Wamba (99%; Hasegawa *et al.*, 1983). Available data suggest the species of the genus *Troglodytella* to be host specific. *Troglodytella abrassarti* is reported from chimpanzees, while *T. gabonensis* and *T. gorillae* occur in gorillas (Reichenow, 1920; Swezey, 1935; Imai *et al.*, 1991).

Table 14.1. *Prevalence and characteristics of the bonobo parasites found. Mean and range sizes were obtained from n = 10 measurements except when otherwise stated*

Parasite	Prevalence % (n = 87)	Characteristics	Compatible species described in chimpanzees
Troglodytella	74.7	Trophozoites 144.6 (100–190) × 84.5 (70–100) μm, oval in shape, with a truncated anterior end; oral aperture apical; with skeletal plates	*Troglodytella abrassarti* Brumpt & Joyeux, 1912
Trematode 1	18.4	Eggs 77.8 (60–90) × 46.8 (35–58) μm, operculate, containing a sparse, yellowish morule	
Trematode 2	3.4	Eggs 44.9 (32–55) × 27.7 (20–35) μm, operculate	*Dicrocoelium dendriticum* (Rudolphi, 1818)
Trematode 3	1.1	Eggs 48.8 (47–51) × 22.0 (19–24) μm (n = 4), operculate	*Concinnum brumpti* (Railliet, Henry & Joyeux, 1912)
Trichuris	2.3	Eggs 55.2 (52–60) × 25.5 (24–27) μm (n = 6), yellowish, ovoid with acute extremes with a salient plug in each. Egg shell smooth and thin	*Trichuris trichiura* (Linnaeus, 1771)
Hookworms	16.1	Eggs 65.1 (63–67) × 40.4 (39–42) μm, ellipsoidal, with rounded extremes and a thin shell, containing a sparse morule with less than 16 cells	*Necator americanus* (Stiles, 1902) *N. congolensis* (Geodelst, 1916) *Trichostrongylus colubriformis* (Giles, 1892)
Strongyloides	35.6	Eggs 52.2 (45–60) × 37.0 (28–45) μm, ellipsoidal, with rounded extremes and a thin shell, containing a dense morule or a larvae	*Strongyloides fuelleborni* (von Linstow, 1905)
Oesophagostomum	50.6	Eggs 71.1 (67–75) × 42.3 (39–47) μm, ellipsoidal, with rounded extremes and a thin shell, containing a dense morule	*Oesophagostomum bifurcum* (Creplin, 1849) *O. stephanostomum* (Stossich, 1904)
Ascaris	3.4	Eggs 67.6 (55–80) × 54.0 (40–75) μm (n = 5), brown, subspherical, with a thick shell covered by a globular albuminoid layer, containing a developing embryo	*Ascaris lumbricoides* (Linnaeus, 1758)
Nematode 1	3.4	Eggs 63.0 (59–68) × 48.7 (48–49) μm (n = 3), subspherical to ovoid, with a thick shell covered by an albuminoid-like layer with surface pitted, containing larvae	*Physaloptera* and other spirurids

Table 14.2. *A comparison of intestinal parasites of wild bonobos from Wamba; Hasegawa* et al. *(1983) and Lomako (this study)*

	Prevalence (%)	
	Wamba	Lomako
Protozoa		
Troglodytella sp.	99.0	74.7
Trematoda		
Dicrocoeliidae gen. sp (Trematode 2, 3)	45.1	3.4 + 1.1
Trematode 1		18.4
Nematoda		
Capillaria sp.	21.0	
Trichuris sp.	3.3	2.3
Strongyloides sp.	52.9	35.6
Oesophagostomum sp.	17.9	50.6
Strongylida fam. gen. sp. (hookworms)	21.0	16.1
Oxyuridae gen. sp.	6.2	
Ascaris sp.		3.4
Nematode 1		3.4

Based on this tentative assumption, *Troglodytella* in bonobos might represent a separate species.

The genus *Trichuris* possesses rather complex taxonomy, with more than 60 species being described from various hosts (Anderson, 2000). The eggs identified as *Trichuris* sp. in this study resembled those of the human whipworm, *T. trichiura*, although the dimensions of the eggs (52–60 × 24–27 μm) we recorded are greater than those from *T. trichiura* usually reported from humans. However, the variability of reported dimensions of eggs of *T. trichiura* from humans is remarkable: i.e. 50–56 × 21–25 (Kotlán, 1960), 49–65 × 20–29 (Burrows, 1965), 47–56 × 22–25 (Jíra, 1998). Generally, the reports and taxonomy of *Trichuris* from apes are controversial in several aspects. *Trichuris trichiura* is commonly recorded from primates, including chimpanzees, and cross-transmissions are considered to be possible (Ruch, 1959). Cross-infections with *T. trichiura* between humans and habituated Virunga mountain gorillas were suggested by Mudakikwa *et al.* (1998) and Sleeman *et al.* (2000), but without direct proof. Ooi *et al.* (1993) identified both studied whipworms from man and primates (macaque, baboon) as *T. trichiura* despite the slight variation observed in the length of the spicule. On the other hand, the eggs of *Trichuris* found in Bwindi gorillas were morphologically dissimilar to *T. trichiura* (Nkurunungi, 1999) and certain morphological differences between *Trichuris* from primates and *T. trichiura* from man were recorded by

Healy & Myers (1973). Also Hasegawa *et al.* (1983) reported the eggs of *Trichuris* in Wamba bonobos to be slightly larger (54–57 × 23–26 μm) than those of *T. trichiura*, which is congruent with our results. Evidently, only methods of molecular taxonomy can elucidate the diversity and phylogeny of *Trichuris* in primates and to solve the question of possible conspecificity of *Trichuris* in man and wild apes. Both Lomako and Wamba bonobos had low prevalence of *Trichuris* infection (2.3%, 3.3%), which is similar to most other populations of common chimpanzee (Landsoud-Soukate *et al.*, 1995; Ashford *et al.*, 2000; Murray *et al.*, 2000).

Rhabditid nematodes of the genus *Strongyloides* are common and prevalent in wild common chimpanzees (Gombe: Murray *et al.*, 2000; Mahale: Huffman *et al.*, 1997; Kibale: Muehlenbein, 2005). The eggs or larvae of *Strongyloides* do not have morphological features that allow the identification of the species. However, the presence of eggs, larvae, or both in feces is a specific character: *Strongyloides stercoralis* is the only *Strongyloides* whose eggs hatch in the intestine and then only larvae (but not eggs) are released in feces, while *S. fuelleborni* releases only eggs in fresh feces, and larvae can be observed only if enough time (6–10 hours) has passed since the deposition of the feces to its fixation/analysis (Cordi & Otto, 1934; Speare, 1989). In the present study, only the eggs were found in all positive samples, allowing their identification as *S. fuelleborni*. Similarly, Hasegawa *et al.* (1983) reported *Strongyloides* in 53% of positive samples of Wamba bonobos and referred to it as *S. fuelleborni*.

Oesophagostomum is the most prevalent parasite in the vast majority of studied chimpanzee populations (Gombe: Murray *et al.*, 2000, Mahale: Huffman *et al.*, 1997; Kibale: Krief *et al.*, 2005; Muehlenbein, 2005) and it was the most prevalent parasite also in our study. Vuylsteke (1964) recorded *O. stephanostomum* in the cecum of bonobo. The size of *Oesophagostomum* sp. eggs found in Wamba bonobos was identical to those of *O. stephanostomum* (60–80 × 40–45 μm) (Hasegawa *et al.*, 1983). However, it is impossible to distinguish between *Oesophagostomum* species on the basis of egg size (Healy & Myers, 1973). A recent study focused on self-medicative abilities of bonobos at Lomako revealed that *Oesophagostomum* infections increased after the onset of the rainy season (Dupain *et al.*, 2002), and the same was observed in Mahale common chimpanzees by Huffman *et al.* (1997). At both sites, a presence of leaf-swallowing behavior, a special type of self-medicative behavior (Huffman & Caton, 2001) was stimulated by increased *Oesophagostomum* infections.

Similarly to *Oesophagostomum,* also the identification of *Ancylostoma* and *Necator* (and the other Ancylostomatidae) cannot be done on the basis of their egg morphology, and coprocultures are needed to obtain the L3 larvae which allow the identification to the genus level. The term "hookworm" is usually used to refer to these parasites and we follow this convention throughout this study. Several species of *Necator* and *Ancylostoma* have been recorded from common chimpanzees (Myers & Kuntz, 1972), and *N. americanus* was reported from bonobo by Stam (1960). Prevalence of hookworms in Wamba bonobos (21%, Hasegawa *et al.*, 1983) is slightly higher than Lomako's (16%). Some authors present hookworms (*Ancylostoma* and *Necator*) and nodular worms (*Oesophagostomum*) together (e.g. Kibale: Ashford *et al.*, 2000; Mondika: Lilly *et al.*, 2002). In general, parasites from the order Strongylida seem to be common for chimpanzees with the exception of the Mt. Assirik population (McGrew *et al.*, 1989).

The characteristics of the *Ascaris* eggs are compatible with those of *A. lumbricoides.* This species was found by Landsoud-Soukate *et al.* (1995) in few chimpanzees in Lope (Gabon). There are no previous records of *Ascaris* species from wild bonobo, but Stam (1960) described a case from a captive bonobo. *Ascaris* sp. was recorded in 14% of chimpanzees, 22% of gorillas and also in local people and researchers in Mondika, Central African Republic (Lilly *et al.*, 2002) and in both sympatric gorillas and chimpanzees in Dja Faunal reserve surroundings in Cameroon (Petrželková *et al.*, 2003). In Gombe chimpanzees, Murray *et al.* (2000) found a high prevalence of nematode eggs, which closely resembled ascarid eggs. Nejsum *et al.* (2006) described the occurrence of *A. suum* in captive chimpanzees. However, deeper studies are needed to clarify the status of *Ascaris* occurring in the wild and evaluate the possibility of cross-infections.

The overall morphology of the nematode sp. 1 eggs strongly suggests its classification within the spirurids. Generally, spirurid nematodes of genera *Physaloptera*, *Abbreviata*, *Gongylonema*, and *Streptopharagus* are common parasites of African primates (Myers & Kuntz, 1965; Hahn *et al.*, 2003; Gillespie *et al.*, 2004) and the egg-based determination should always be considered to be a tentative one. So far, findings of spirurid eggs in chimpanzees were reported as *Abbreviata* (= *Physaloptera*) *caucasica* (File *et al.*, 1976; Muehlenbein, 2005), *Physaloptera* sp. (Murray *et al.*, 2000), and *Gongylonema*-like (Landsoud-Soukate *et al.*, 1995). Moreover, *Streptopharagus* sp. and *Abbreviata* (= *Physaloptera*) *caucasica* were reported from chimpanzees based on the identification of adults (Healy & Myers, 1973; File *et al.*, 1976). Very recently, *Protospirura muricola* was also detected in chimpanzees based on the presence of eggs and adults in feces (Petrželková *et al.*, 2006). So far, spirurid nematodes are not reported from bonobos. The life cycle of most of the spirurid nematodes requires the ingestion of insect intermediate host (Anderson, 2000) and its presence in Lomako bonobos is then somehow congruent with dicrocoelids discussed below.

Trematode sp. 1 eggs do not correspond to any trematode found in chimpanzees. The trematode sp. 2 and sp. 3 eggs resemble the morphology of those of *Dicrocoelium dendriticum* and *Concinnum brumpti* respectively, both known from common chimpanzee (Myers & Kuntz, 1972). However, the egg morphology is closely similar among the dicrocoelid species and their size ranges usually overlap (Yamaguti, 1971). Thus, the trematode identification made in this work on the basis of the egg morphology should be considered as tentative. At Wamba, 45% of the individuals were infected by unidentified dicrocoelids. In comparison, at Lomako, 23% of individuals were infected. *Dicrocoelium dendriticum* itself was reported only in few populations of common chimpanzee and in low prevalences: 1–2% (Huffman *et al.*, 1997), 3% (Landsoud-Soukate *et al.*, 1995). Recent data about *C. brumpti* are missing. It is surprising that both studied bonobo populations were infected by trematodes in remarkable prevalences. Generally, dicrocoelid trematodes utilize the insects as a second intermediate host (Yamaguti, 1971), but the exact transmission route to the great apes remains unknown. Although bonobos probably do not consume invertebrates as regularly as chimpanzees do (Bermejo *et al.*, 1994), they still eat larvae, termites, and ants and can gain the infection through this way. The insectivory has been observed also in Lomako bonobos (Dupain, personal observation). Additionally, bonobos might be infected by smaller insects swollen unintentionally with food, which is the common transmission trait of *Dicrocoelium* spp. in ruminants.

Bonobos host a rich community of parasites, but not all of them could be fully identified to species (or even genus) by coprological examination, as used in our study. In cases of unusual findings (mainly in those of low prevalences), one should always consider the possibility of pseudoparasitic origin of observed eggs. Bonobos were reported to consume various vertebrates (reptiles, shrews, squirrels, infant duikers, bats) (de Waal & Lanting, 1997) and in such cases the eggs of the helminth parasites infecting the digested prey could pass unaltered through the digestive tract of the bonobos and then could be detected in their feces. Moreover, the intentional or unintentional coprophagy can lead to the accidental diagnosis of the parasite eggs from the ingested feces. The examination of serial samples collected from habituated bonobos and chimpanzees during several days can greatly help to distinguish between pseudoparasites and real parasites.

The data about the host specificity of parasites of African primates are missing or at least very limited. The host specificity of *Trichuris* is discussed above. In Lomako, the studied bonobo groups share their habitat not only with humans, but also with *Cercocebus aterrimus*, *Colobus angolensis*, *Allenopithecus nigroviridis*, and three species of *Cercopithecus* (*C. ascanius*, *C. wolfi*, and *C. neglectus*) (Dupain, personal observation). To understand the epidemiological and ecological background of the parasitic infections found, further research should be focused also to the parasitofauna of these primate species.

Most of the animals were infected with one to three parasite species and we found no negative samples. Recently, Muehlenbein (2005), working with chimpanzees in Kibale, clearly demonstrated the increase of the mean parasite species richness when more samples were collected from each animal and stressed a need of serial sampling on multiple non-consecutive days for accurate parasitological diagnoses. Logistically, the opportunity of serial sampling in non-habituated populations of apes is limited, but it is probable that further intense research will increase the spectrum of bonobo parasites. We hope that our study should encourage such studies. The recent finding of *Protospirura* in chimpanzees shows how limited is the knowledge about parasites of our closest relatives (Petrželková *et al.*, 2006). The risk of disease transmission from humans to apes is considered to be a serious threat (Butynski, 2001). However, in many cases, our data about transmission of parasites from humans to the apes (and vice versa) are purely speculative, lacking any direct (e.g. molecular) evidence for such a transmission (see also Gasser *et al.*, Chapter 3, this volume). Evidently, joint efforts of field ecologists, conservation biologists, parasitologists, and molecular biologists, and the wider implementation of methods of molecular taxonomy, is necessary to solve questions about the diversity and host specificity of bonobo parasites.

Acknowledgements

We are indebted to F. Moravec and V. Baruš for the taxonomic opinion about the spirurids. The final stage of the study was supported, in part, by the grant from the Grant Agency of the Czech Republic No. 524–06-0264 to K. J. P. and D. M. We thank the Ministère de l'Environnement, Protection de la Nature et Tourisme and the Ministère de l'Enseignement Supérieur, Recherche Scientifique et Technologies, Kinshasa, République Démocratique du Congo, for providing authorizations and mission orders. All fieldwork was made possible through the financial support of the KBC. Logistical support was provided by the Belgian Embassy and Cooperation in Kinshasa (Dem. Rep. Congo), Philip Heuts, Eric De Bock, CDI-Bwamanda, Claudine Minesi (ABC), Paul DePetter and J. Cl. Hoolans (Nocafex), Father Paul (Procure Saint-Anne, Kinshasa), and the missionaries of Mill Hill (Basankusu). The research would not have been possible without the financial and other help of the Royal Zoological Society of Antwerp. We thank the Flemish Government for structural support to the Centre for Research and Conservation of the RZSA. We thank Lourdes Trujillo for all logistical help.

References

Anderson, R. C. (2000). *Nematode Parasites of Vertebrates. Their Development and Transmission*. Wallingford, UK: CABI Publishing.

Ashford, R. W., Reid, G. D. F. & Wrangham R. W. (2000). Intestinal parasites of the chimpanzees *Pan troglodytes* in Kibale Forest, Uganda. *Annals of Tropical Medicine and Parasitology*, **94**, 173–179.

Bermejo, M., Illera, G. & Pi, J. S. (1994). Animals and mushrooms consumed by bonobos (*Pan paniscus*) – new records from Lilungu (Ikela), Zaire. *International Journal of Primatology*, **15**, 879–898.

Bermejo, M., Rodriguez-Teijeiro, J. D., Ilera, G. *et al.* (2006). Ebola outbreak killed 5000 gorillas. *Science*, **314**, 1564.

Burrows, R. B. (1965). *Microscopic Diagnosis of the Parasites of Man*. New Haven, NJ and London: Yale University Press.

Butynski, T. M. (2001). Africa's Great Apes. In *Great Apes and Humans: The Ethics of Coexistence*, ed. B. B. Beck, T. S. Stoinski, H. Hutchin *et al.* Washington, DC: Smithsonian Institution Press, pp. 133–149.

Collet, J. Y., Bourreau, E., Cooper, R. W., Tutin, C. E. G. & Fernandez, M. (1984). Experimental demonstration of cellulose digestion by *Troglodytella gorillae*, an intestinal ciliate of lowland gorillas (*Gorilla gorilla gorilla*). *International Journal of Primatology*, **5**, 328–328.

Cordi, J. M. & Otto, G. F. (1934). The effect of various temperatures on the eggs and larvae of *Strongyloides*. *American Journal of Hygiene*, **19**, 103–114.

Dupain, J., Van Krunkelsven, E., Van Elsacker, L. & Verheyen, R. F. (2000). Current status of the bonobo (*Pan paniscus*) in the proposed Lomako Reserve (Democratic Republic of Congo). *Biological Conservation*, **94**, 265–272.

Dupain, J., Van Elsacker, L., Nell, C. *et al.* (2002). New evidence for leaf swallowing and *Oesophagostomum* infection in bonobos (*Pan paniscus*). *International Journal of Primatology*, **23**, 1053–1062.

File, S. K., McGrew, W. C. & Tutin, C. E. G. (1976). The intestinal parasites of a community of feral chimpanzees, *Pan troglodytes schweinfurthii. Journal of Parasitology*, **62**, 259–261.

Gillespie, T. R., Greiner, E. C. & Chapman, C. A. (2004). Gastrointestinal parasites of the guenons of western Uganda. *Journal of Parasitology*, **90**, 1356–1360.

Hahn, N. E., Proulx, D., Muruthi, P. M., Alberts, S. & Altmann, J. (2003). Gastrointestinal parasites in free-ranging Kenyan baboons (*Papio cynocephalus* and *P. anubis*). *International Journal of Primatology*, **24**, 271–279.

Hasegawa, H., Kano, T. & Mulavwa, M. (1983). A parasitological survey on the feces of pygmy chimpanzees, *Pan paniscus*, at Wamba, Zaire. *Primates*, **24**, 419–423.

Healy, G. R. & Myers, B. J. (1973). Intestinal helminths. *Chimpanzee*, **6**, 265–296.

Huffman, M. A. & Caton, J. M. (2001). Self-induced increase of gut motility and the control of parasitic infections in wild chimpanzees. *International Journal of Primatology*, **22**, 329–346.

Huffman, M. A., Gotoh, S., Turner, L. A., Hamai, M. & Yoshida, K. (1997). Seasonal trends in intestinal nematode infection and medicinal plant use among chimpanzees in the Mahale, Tanzania. *Primates*, **38**, 111–125.

Imai, S, Ikeda, S. I., Collet, J. Y. & Bonhomme, A. (1991). Entodiniomorphid ciliates from the wild lowland gorilla with the description of a new genus and 3 new species. *European Journal of Protistology*, **26**, 270–278.
Jíra, J. (1998). *Lékařská helmintologie*. Praha: Galén.
Kawabata, M. & Nishida, T. (1991). A preliminary note on the intestinal parasites of wild chimpanzees in the Mahale Mountains, Tanzania. *Primates*, **32**, 275–278.
Kotlán, A. (1960). *Helminthologie*. Budapest: Akadémia Kiadó.
Krief, S., Huffman, M. A., Sevenet, T. *et al.* (2005). Noninvasive monitoring of the health of *Pan troglodytes schweinfurthii* in the Kibale National Park, Uganda. *International Journal of Primatology*, **26**, 467–490.
Landsoud-Soukate, J., Tutin, C. E. G. & Fernandez, M. (1995). Intestinal parasites of sympatric gorillas and chimpanzees in the Lopé Reserve, Gabon. *Annals of Tropical Medicine and Parasitology*, **89**, 73–79.
Lilly, A. A., Mehlman, P. T. & Doran, D. (2002). Intestinal parasites in gorillas, chimpanzees, and humans at Mondika research site, Dzanga-Ndoki National Park, Central African Republic. *International Journal of Primatology*, **23**, 555–573.
McGrew, W. C., Tutin, C. E. G., Collins, D. A. & File, S. K. (1989). Intestinal parasites of sympatric *Pan troglodytes* and *Papio* spp., at two sites – Gombe (Tanzania) and Mt. Assirik (Senegal). *American Journal of Primatology*, **17**, 147–155.
Mudakikwa, A. B., Sleeman, J. M., Foster, J. W., Madder, L. L. & Patton, S. (1998). An indicator of human impact: gastrointestinal parasites of mountain gorillas (*Gorilla gorilla beringei*) from the Virunga Volcanoes Region, Central Africa. In *Proceedings of the American Association of Zoo Veterinarians/American Association of Wild Veterinarians Joint Conference*, ed. C. K. Baer. Philadelphia, PA: American Association of Zoo Veterinarians, pp. 436–437.
Muehlenbein, M. P. (2005). Parasitological analyses of the male chimpanzees (*Pan troglodytes schweinfurthii*) at Ngogo, Kibale National Park, Uganda. *American Journal of Primatology*, **65**, 167–179.
Murray, S., Stem, C., Boudreau, B. & Goodall, J. (2000). Intestinal parasites of baboons (*Papio cynocephalus anubis*) and chimpanzees (*Pan troglodytes*) in Gombe National Park. *Journal of Zoo and Wildlife Medicine*, **31**, 176–178.
Myers, B. J. & Kuntz, R. E. (1965). A checklist of parasites reported for the baboon. *Primates*, **6**, 137–194.
Myers, B. J. & Kuntz, R. E. (1972). A checklist of parasites and commensals reported for the chimpanzee (*Pan*). *Primates*, **13**, 433–471.
Nejsum, P., Grondahl, C. & Murrell, K. D. (2006). Molecular evidence for the infection of zoo chimpanzees by pig *Ascaris*. *Veterinary Parasitology*, **139**, 203–210.
Nkurunungi, J. B. (1999). A survey of the gastro-intestinal helminths of the wild mountain gorilla (*Gorilla gorilla beringei Matschie*) and man in Bwindi Impenetrable National Park – Southwestern Uganda. In *Proceedings of the Ecological Monitoring Programme Workshop*: *Research as an Important Tool of Ecological Monitoring in Bwindi Impenetrable National Park, Uganda*. Bwindi Impenetrable National Park, pp. 9–11.
Ooi, H. K., Tenora, F., Itoh, K. & Kamiya, M. (1993). Comparative study of *Trichuris trichiura* from nonhuman primates and from man, and their difference with *T. suis*. *Journal of Veterinary Medical Science*, **55**, 363–366.

Petrželková, K. J., Dupain, J., Van Dyck, S. *et al.* (2003). Possible medicinal plants in the diet of the western lowland gorilla (*Gorilla gorilla gorilla*) and the central chimpanzee (*Pan troglodytes troglodytes*) in the Dja Faunal Reserve surroundings in Cameroon – a pilot study. *Proceeding of Zoological Days. February*, 13–14, 2003. Brno, Czech Republic.

Petrželková, K. J., Hasegawa, H., Moscovice, L. R. *et al.* (2006). Parasitic nematodes in the chimpanzee population on Rubondo Island, Tanzania. *International Journal of Primatology*, **27**, 767–777.

Reichenow, E. (1920). Den Wiederkäuern-Infusorien verwandte Formen aus Gorilla und Schimpanse. *Archive für Protistenkunde*, **41**, 1–33.

Ritchie, L. S. (1948). An ether sedimentation technique for routine stool examinations. *Bulletin of the U.S. Army Medical Department*, **8**, 326.

Ruch, T. C. (1959). *Diseases of Laboratory Primates*. Philadelphia, PA: W. B. Saunders Co.

Sleeman, J. M., Meader, L. L., Mudakikwa, A. B., Foster, J. W. & Patton, S. (2000). Gastrointestinal parasites of mountain gorillas (*Gorilla gorilla beringei*) in the Parc National des Volcans, Rwanda. *Journal of Zoo and Wildlife Medicine*, **31**, 322–328.

Speare, R. (1989). Identification of species of *Strongyloides*. In *Strongyloidiasis: A Major Roundworm Infection of Man*, ed. D. I. Grove. London: Taylor and Francis, pp. 11–83.

Stahl, A. B. (1984). Hominid dietary selection before fire. *Current Anthropology*, **25**, 151–168.

Stam, A. B. (1960). Un cas mortel d'ascaridio (*Ascaris lumbricoides*) chez le chimpanzee nain *(Pan paniscus Schweir). Annales de Parasitologie*, **35**, 675.

Swezey, W. W. (1935). Cultivation of *Troglodytella abrassarti*, a parasitic ciliate of the chimpanzee. *Journal of Parasitology*, **21**, 10–17.

Vuylsteke, C. (1964). Mission de zoologie medicale au Maniema (Congo, Leopoldville). 3. Vermes. Nematoda. *Annals Koninklijk Museum voor Midden-Africa, Tervuren, series B, Zoologische Wetenschappen*, **132**, 41–66.

de Waal, F. & Lanting, F. (1997). *Bonobo: The Forgotten Ape*. Berkeley, CA: University of California Press.

Yamaguti, S. (1971). *Synopsis of Digenetic Trematodes of Vertebrates*, Vol. I. Tokyo: Keigaku Publishing Co.

Young, K. H., Bullock, S. L., Melvin, D. M. & Spruill, C. L. (1979). Ethyl acetate as a substitute for the ethyl ether in the formalin-ether sedimentation technique. *Journal of Clinical Microbiology*, **10**, 852–853.

15 *Habitat disturbance and seasonal fluctuations of lemur parasites in the rain forest of Ranomafana National Park, Madagascar*

PATRICIA C. WRIGHT, SUMMER J. ARRIGO-NELSON, KRISTINA L. HOGG, BRIAN BANNON, TONI LYN MORELLI, JEFFREY WYATT, A. L. HARIVELO, AND FELIX RATELOLAHY

Photograph by Summer J. Arrigo-Nelson (*Propithecus edwardsi*)

Primate Parasite Ecology. The Dynamics and Study of Host–Parasite Relationships, ed. Michael A. Huffman and Colin A. Chapman. Published by Cambridge University Press.

Introduction

As early as 1986, understanding the effects of parasites and disease on animal populations in disturbed and disappearing habitats was targeted as an important research goal of conservation biology (Dobson & May, 1986). With some notable exceptions, parasites and their hosts are considered to be in an ecological or evolutionary balance such that parasites are able to survive and reproduce effectively without killing the host, thus maintaining their ability to continue reproducing in the host in the future (Dobson & May, 1986). However, human habitat disturbance appears to be disrupting this delicate balance. Today, as the devastation of natural habitats forces primate populations to become more concentrated in protected areas and nature reserves, conservationists are becoming more aware of the impact of disease and parasites on wild animal populations (Stuart *et al.*, 1990, 1993; Myers & Rothman, 1995; Stuart & Strier, 1995; Stoner, 1996; Lilly *et al.*, 2003; Gillespie *et al.*, 2004, 2005; Chapman *et al.*, 2005; Nunn & Altizer, 2006; Gillespie & Chapman, 2006).

Madagascar is an island renowned for its endemism, effected by over 100 million years of isolation (Krause *et al.*, 1997). At the same time, Madagascar is also well known for its growing habitat loss, having lost over 85% of its forests to human impact (Green & Sussman, 1990). This forest destruction continues today, especially in the form of slash-and-burn agricultural practices, and is a major reason for the endangerment of many of Madagascar's species – especially its lemurs (Mittermeier *et al.*, 2006). Given their evolutionary uniqueness and the growing threat of primate extinction on the island, Madagascar has been the focus of many behavioral studies of lemurs throughout the last two decades (Wright, 1999; Gould & Sauther, 2007a). While recent studies on the impact of predation and feeding competition have increased our understanding of lemur evolution (Arrigo-Nelson, 2006; Gould & Sauther, 2007b; Karpanty & Wright, 2007), few studies have addressed the effects of parasites on lemur populations, a subject which will increase in importance as habitats continue to be disturbed by human intervention (Holmes, 1996).

Although initial laboratory and zoological studies of lemur parasites date back to the early 1900s (see Irwin, in press), there are few studies of these parasites from lemurs living in naturalistic conditions. Some of the earliest of these surveys of wild lemur populations have reported infection by ectoparasites and intestinal worms, but focused on incidences of the blood parasites that cause trypanosomiasis, malaria, and filariasis (Landau *et al.*, 1989; Rabetafika *et al.*, 1989; Garell & Meyers, 1995; Junge & Garell, 1995; Rabetafika, 1995). In the past five years, however, biomedical health assessments of wild lemur populations have begun to include parasite identification (Dutton *et al.*, 2003; Junge & Louis, 2002, 2005; Muehlenbein, *et al.*, 2003; O'Connor, 2003; Sauther *et al.*,

2006), and the endoparasite diversity of nine species of lemurs in Ranomafana National Park has recently been described by Hogg *et al.* (in press).

The symbiosis of lemur/parasite interactions is unique among primates for three principal reasons. First, geophagy (the intentional consumption of soil) is common among this group, with the Milne-Edwards' sifaka (*Propithecus edwardsi*) consuming mouthfuls of soil an average of two times per week (Arrigo-Nelson *et al.*, in press). This behavior is important to understanding both endo- and ectoparasite infections, as it allows for the transmission of parasites from the ground to the sifaka without the need for intermediate hosts. Pinworms and some whipworms can be ingested by eating the eggs distributed in soil, and exposure to mites or ticks may also be increased by contact with the ground. Second, unlike the higher primates, lemurs do not typically use their hands or lips to groom themselves or their group mates. Instead, they rely almost exclusively on their specially evolved tooth combs to part and clean their fur, creating a direct pathway from the fur to the digestive tract. Additionally, the morphology of the tooth comb is such that parasite eggs can be scraped from the fur and swallowed (Ankel-Simons, 2006).

In this chapter, we take an ecological approach to examining parasite infections among lemurs by focusing on the Milne-Edwards' sifaka, a species whose behavior, ecology, and demography have been intensively studied for over 21 years at Ranomafana National Park (Wright, 1995; Pochron *et al.*, 2004; Arrigo-Nelson, 2006). Beyond parasite species identification, we have documented the species richness and infection intensity of both endo- and ectoparasites in individually identifiable sifakas, over a 5-year period. Here, we will investigate how habitat disturbance and seasonality shape parasite infection and the natural host/parasite equilibrium of wild populations. While acknowledging the complexities of the ecology, we tested the following predictions:

(1) Habitat disturbance will decrease the diversity of parasites that are dependent on intermediate hosts (i.e. some nematodes utilize beetles, or orthopterans as intermediate hosts) living in tropical forests while the relative prevalence of parasites may increase or remain the same (Anderson & May, 1982).

(2) Sifakas living in disturbed habitats will have higher parasite infection intensities, given that disturbance can increase food stress, ranging patterns, population densities (Stoner, 1996), and, possibly, immunosuppression (Dobson, 1988).

(3) Parasite species richness and infection intensity will be higher within all populations, during the humid, wet season, when conditions are most favorable for the parasites (Stuart & Strier, 1995; Huffman *et al.*, 1997; Altizer *et al.*, 2006).

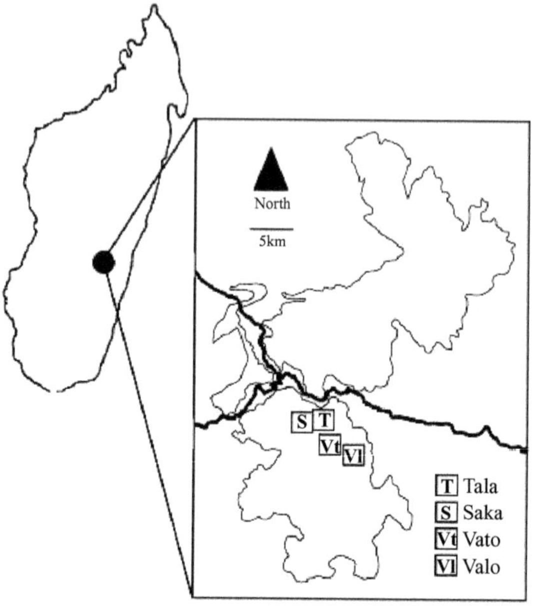

Figure 15.1. Schematic map of study sites within Ranomafana National Park.

Methods

Study site

The research was conducted within Ranomafana National Park (RNP) in south-eastern Madagascar, located at 21°E 16′S latitude and 47°E 20′E longitude (Figure 15.1). The park was created in 1991 and encompasses 43 500 ha of montane rain forest (Wright, 1997); it ranges in altitude from 600–1500 m and rainfall averages 3000 mm per year. This is a seasonal environment, with a warm/wet season lasting from December through March (average temperature: 26 °C, average rainfall: 357 mm/month) and a cool/drier season from June to September (average temperature: 15 °C, average rainfall: 168 mm/month) (RNP weather station records).

The history of human disturbance within the boundaries of RNP has been determined through both quantitative botanical surveys of each research site (Lowry *et al.*, 1997; Balko & Underwood, 2005; Arrigo-Nelson, 2006) and through interviews with elderly village residents living in RNP's peripheral zone and the local officials of Madagascar's Department of Water and Forests (Wright, unpublished data). The Talatakely (Tala) and Sakaroa (Saka) sites were classified as "disturbed," having areas that were clear-cut before 1947

(Tala: ~10% of site area; Saka: ~30%, as well as primary forest which was heavily timbered from 1986–1990 (~35% of site area). The Valohoaka (Valo) and Vatoharanana (Vato) sites were classified as "undisturbed," as they contain primary forest with minimum evidence of human disturbance. Note that although ~1% of the Vatoharanana (Vato) site was also selectively logged from 1986–1990, logging intensity was much less severe than at the sites classified as disturbed (due primarily to the site's distance from the road and its steep gradient). Botanical analyses of the site clearly support its inclusion in the "undisturbed" habitat category (Balko & Underwood, 2005; Tecot, 2008).

There are at least 12 sympatric species of primates (seven diurnal and five nocturnal) in the Ranomafana National Park, of which *Propithecus edwardsi* (the Milne-Edwards' sifaka) is the largest (5.8 kg; Glander *et al.*, 1992). The sifaka social system is variable, with group size ranging from three to nine individuals (Pochron & Wright, 2003), and adult females are consistently dominant to adult males (Pochron *et al.*, 2003). *Propithecus edwardsi* have a long lifespan (~30 years), with 29% of the females over 18 years of age (King *et al.*, 2005; Wright *et al.*, 2008).

Sampling protocol

Endoparasites

For the endoparasite assessment, fecal samples were collected during the cool/drier season from selected marked lemurs at the Tala (n = 10 individuals) and Vato (n = 8) research sites, from June 6 to July 19, 2001 (Hogg, 2002). All samples were collected opportunistically, during the early morning hours and placed in vials filled with 10% neutral-buffered formalin and were marked with the identity of the lemur who produced the sample, the date, and the precise time when the sample was produced. Upon returning to the USA, each fecal sample was mixed with zinc sulfate solution, centrifuged, and examined under a light microscope (Rabetafika, 1995) by KLH. Following standard protocol, four slides were observed from each fecal sample and all parasite eggs or larva were counted and recorded using a McMaster grid.

Ectoparasites

Following an established capture protocol (Glander *et al.*, 1992; Wright, 1995), study animals were captured and searched for ectoparasites in the warm/wet (2002, 2003, 2005, 2006) and cool/drier (2003, 2004, 2005, 2007) seasons at the selectively logged site, Tala (n = 73 individuals). Additional individuals were captured from 2002–2007 at the more undisturbed site of Valo (n = 24) and the disturbed site of Saka (n = 18). While anesthetized, each sifaka was examined for ectoparasites following a standardized procedure. Moving from head to tail, each animal's fur was parted and both the fur and skin were inspected. A human lice comb and small forceps were used to pick up and remove lice, ticks, fleas, leeches, flies, and their eggs, with special attention paid to the eye, ear, underarm, and genital regions. Data were recorded by a second individual, who listed the number of ticks, mites, lice, fleas, flies, or leeches and area of the body where the ectoparasites were found. When necessary, the combed materials were also examined under a light microscope by PCW, to aid in identification.

Data analysis

The endo-and ectoparasite data sets were assessed on two levels. First, overall parasite richness was considered (see Greiner and McIntosh, Chapter 1, this volume for comparison). As species-level parasite identifications were beyond the scope of this project, "classes" were created (usually at the genus level for individually identifiable parasite morphotypes) which were then used as the unit of analysis. For both endo- and ectoparasites, richness was measured on a scale from 0–5, where a score of 0 would indicate that no parasites were present, a 1 would indicate that one class of parasite was detected, and so on. Second, parasite infection intensity was examined to score each individual's overall health; low intensity could have a minimal effect, but high intensity might reduce vigor and consequently fitness. Intensity scores for ectoparasites were calculated based on number of individuals of each parasite "class" found on each sifaka sampled during a given capture event. For both of these measures, endo- and ectoparasite infections were examined and comparisons were made between sifakas living in disturbed and undisturbed forest habitats. Additionally, inter-seasonal comparisons were also carried out for the ectoparasite data set. Due to unequal sample sizes and non-normal distribution of means, non-parametric analyses were employed (Sokal & Rohlf, 1995). First, to examine differences in endoparasite parasite "class" richness and infection intensity, Mann–Whitney U tests were performed to determine the effect of site (disturbed vs. undisturbed). Second, to examine differences in ectoparasite species richness and infection intensity, Kruskal–Wallis tests compared the Tala (disturbed), Saka (disturbed), and Valo (undisturbed) study sites. Finally, Mann–Whitney U tests were used to compare parasite infection intensity and parasite "class" richness between Tala and Valo during the hot/wet and cool/drier seasons.

Table 15.1a. *Endoparasite richness. The proportion of individuals at each site infected with 0–5 of the endoparasite classes discussed in this study. Endoparasite classes: tapeworms (*Anoplocephala *and* Monezia*), strongyles, pinworms, and other roundworms*

		Endoparasite richness score					
Site	n	0	1	2	3	4	5
Tala	10	0	30	30	20	20	0
Vato	8	25	62.5	12.5	0	0	0

Results

Behavioral observations (PCW and SAN) and veterinary assessments (JW) show no indication that sifakas are distressed by parasitic infections, with diarrhea, gastrointestinal upset, anemia, and weight changes (all characteristic symptoms of acute endoparasite infection) observed only rarely. However, it should be noted that long-term study has revealed that sifakas living in undisturbed habitats tend to weigh more than those living in disturbed areas (Arrigo-Nelson, 2006). Infection by lice and mites, which were found on the majority of sifakas examined, did not appear to cause any obvious health problems, even on the individuals with large numbers. The biting midge flies swarmed around the heads of most individuals, with males having the largest swarms. Behavioral strategies to combat the flies included head shaking and batting at them occasionally.

Parasite prevalence and richness

Endoparasites

A total of six classes of endoparasites were identified in the feces of the animals sampled: nematodes (*Lemuricola* sp., plus those nematodes which could not be assigned to a lower taxonomic level), tapeworms (*Monezia* sp. and *Anoplocephala* sp.), *Strongylus* sp., and pinworms (*Physocephalus* sp.). Although no adult intestinal parasites were observed, parasite eggs were detected in 17 of the 19 individuals sampled. Of the ten individuals sampled within the disturbed forest site (Tala), parasite richness ranged from 1–4 taxa per individual (mean = 2.30, SD = 1.160), while of the eight individuals sampled in the pristine forest site (Vato), six were found to be infected with endoparasites (Table 15.1a).

Table 15.1b. *Ectoparasite, biting insect, and leech richness. The proportion of individuals at each site infested with 0–5 of the ectoparasite classes discussed in this study. Ectoparasite classes: flies, lice, ticks, other mites, and leeches*

		Ectoparasite richness score					
Site	n	0	1	2	3	4	5
Tala	73	9.6	32.9	23.3	19.2	12.3	2.7
Saka	18	0	22.2	22.2	33.3	22.2	0
Valo	24	4.2	20.8	45.8	20.8	4.2	4.2

Within this population, parasite richness ranged from 0–2 (mean = 0.88, SD = 0.641), as no *Stronglyus* sp., *Anoplocephala* tapeworm, *Lemuricola* sp., nor *Physocephalus* infections were recorded for this population. Sifakas in the disturbed forest site had a higher parasite richness than those in the undisturbed forest site (Mann–Whitney U test, Z = −2.623, P = 0.009).

Ectoparasites, biting insects, and leeches

Five main taxa of ectoparasites were detected on the sifakas sampled: flies (Hippoboscidae and Ceratopogonidae), lice (*Phthirpediculus* sp.), mites (*Makialges* sp. or *Gaudalges* sp. (Bochkov & O'Conner, 2006), plus another species of Laelipid mite, leeches, and ticks (*Haemaphysalis* sp.?), along with three organisms which we were unable to identify. Two genera of blood-sucking flies were identified: the larger, dark, louse flies (Hippoboscidae) with reduced wings and smaller black biting midges which had functional wings and flew from the animal. Leeches, *Malagobdella fallax* and possibly *Malagobdella niarchosorum* (Borda, 2006) were occasionally found on the ears, eyes, or the underside of the sifaka. Tiny mites and ticks were sometimes attached around the rim of the eye skin. No fleas were seen on the sifakas. One adult mosquito (Culicidae) was identified.

Ectoparasite richness ranged from 0–5 (out of a possible 6) in the Tala (disturbed forest) and Valo (undisturbed forest) sifaka populations, and from 1–4 in the Saka (disturbed forest) population (Table 15.1b). Eight individuals possessed no ectoparasites (one from Valo and seven from Tala). Average parasite type richness between the disturbed forest sites was 1.85 at Tala (n = 73; SD = 1.24) and 2.47 at Saka (n = 18; SD = 1.12). Valo, the undisturbed forest site, had an average parasite richness of 2.13 (n = 24; SD = 1.08) (Table 15.3). There is no significant difference in parasite richness between these sites (Kruskal–Wallis, H = 3.451, P = 0.178).

Table 15.2a. *Endoparasite infection intensity. The average number (mean and SD) of parasites identified within a sifaka fecal sample at each site, for each endoparasite class discussed. Mann–Whitney U tests revealed that only* Monezia *tapeworm infection intensity differed significantly between sites*

	Site			
Endoparasite class	Tala	Vato	Z	*P*
Monezia tapeworm	5.90 ± 10.18	0.25 ± 0.70	−1.975	0.048*
Anoplocephala tapeworm	0.10 ± 0.31	0	−0.894	0.371
Strongylus sp.	0.60 ± 1.07	0	−1.641	0.101
Pinworms	0.30 ± 0.67	0	−1.302	0.193
Other roundworms	21.60 ± 23.33	9.13 ± 7.22	−1.619	0.106

*Statistical significance at $P < 0.05$.

Infection intensity

Endoparasites

Across individuals, nematodes (unidentified roundworms) were the most abundant parasite found in samples from individuals at both sites, followed by the tapeworm *Monezia* (Table 15.2a). At Tala, 100% of the individuals sampled were infected with unidentified roundworm eggs and 60% were infected with *Monezia* tapeworm. Infection rates were 50% of the animals sampled at this site, nematode (unidentified roundworm) eggs were found to be "frequent" (having more than 20 eggs on the four slides taken from the standardized 0.60 ml sample) and "common" (more than 20 eggs on one slide taken from the standard sample) in 10% of the animals sampled. Individuals at the Tala site were also found to be infected with strongyles, *Anoplocephala* tapeworms, and pinworms in much lower abundances (never more than three per individual). At Vato, 75% of animals were found to be infected with unidentified roundworms and 12.5% infected with *Monezia* tapeworms. No infections of strongyles, *Anoplocephala* tapeworms, or pinworms were identified in this population. Comparisons of the infection intensity of each parasite class revealed significant between-site differences only for *Monezia* tapeworm infection (Mann–Whitney U test, $Z = -1.975, P = 0.048$).

Ectoparasites, biting insects, and leeches

Ectoparasite infection intensity ranged from 0 to over 175 individuals, depending on the parasite taxon sampled and sifaka individual infected (Table 15.2b). Lice and mites were consistently most abundant among sifakas at all three sites, although not all individuals had high abundance scores for both of these taxa. Fly and tick abundance was uniformly lower than that for lice and mites, but

Table 15.2b. *Ectoparasite, biting insect, and leech infestation intensity. The average number (mean and SD) of parasites detected on a captured sifaka at each site, for each ectoparasite class discussed. Kruskal–Wallis tests revealed significant between-site differences in the intensity of louse, mite, and leech infestation*

	Site				
Ectoparasite class	Tala	Saka	Valo	H	*P*
Flies	2.48 ± 5.31	2.33 ± 3.36	1.25 ± 1.57	0.359	0.836
Lice	19.10 ± 34.78	52.83 ± 43.88	20.38 ± 38.46	20.019	< 0.001**
Ticks	1.45 ± 2.47	1.11 ± 2.03	2.96 ± 5.91	0.616	0.735
Other mites	3.73 ± 8.63	43.44 ± 60.21	6.13 ± 24.34	12.138	0.002**
Leeches	0.38 ± 0.97	0	0.04 ± 0.21	6.855	0.032*

*Statistical significance at $P < 0.05$; **Statistical significance at $P < 0.001$.

were also present at all three sampled sites. Only leech abundance was unevenly distributed among sites, with no leeches found at Saka and only one found at Valo. When each ectoparasite class was examined individually, the infection intensities of lice, mites, and leeches were found to differ significantly between sites (Kruskal–Wallis tests, lice: H = 20.019. $P < 0.001$; mites: H = 12.138, $P = 0.002$; leeches: H = 6.855, $P = 0.032$, flies: H = 6.855, $P = 0.836$; ticks: H = 0.616, $P = 0.735$). However, these differences do not appear to follow an undisturbed/disturbed forest dichotomy as infection intensities of lice and mites were more similar between Valo and Tala than either was to Saka, and leech infection intensity was more similar between Valo and Saka than either was to Tala.

Seasonal variation in ectoparasite, biting insects, and leech infection

To determine whether climate plays a role in ectoparasite infection intensity and/or parasite richness, data from the warm/wet and cool/drier seasons at both the Tala (disturbed forest) and Valo (undisturbed forest) sites were compared. Statistical comparisons at the Saka site were not possible, as only two individuals were sampled during the cool/drier season. Ectoparasite richness was found to differ significantly at Tala but not at Valo (Mann–Whitney U tests, Tala: $Z = -2.318$, $P = 0.020$; Valo: $Z = -1.533$, $P = 0.125$), with richness scores higher in the warm/wet season than in the cool/drier season (Figure 15.2).

When ectoparasite infection intensity was compared between seasons, for each parasite class, lice infection intensity was found to vary significantly

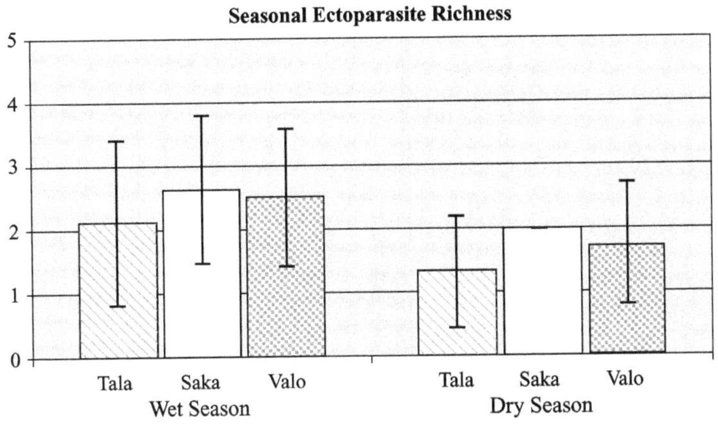

Figure 15.2. Seasonal ectoparasite richness. Average (mean and SD) of ectoparasite richness scores in both the warm/wet and cool/drier seasons. Tala: $n = 50$ (warm/wet) and 23 (cool/drier); Saka: $n = 16$ (warm/wet) and 2 (cool/drier); Valo: $n = 12$ (warm/wet) and 12 (cool/drier). Significant interseasonal differences were found in richness at the Tala site. Due to the small sample size during the dry season, Saka was excluded from the statistical analyses.

at both sites (Mann–Whitney U tests, Tala: $Z = -2.505$, $P = 0.012$; Valo: $Z = -2.261$, $P = 0.024$), with more lice being counted during captures in the warm/wet season (Table 15.3). Additionally, tick infections in the undisturbed forest (Valo) ($Z = -3.257$, $P = 0.001$) and fly infections in the disturbed forest (Tala) ($Z = -2.949$, $P = 0.003$) also were found to be significantly more intense during the warm/wet season. No significant differences were found in the seasonal intensity of other mites (Tala: $Z = -0.320$, $P = 0.749$; Valo: $Z = -0.499$, $P = 0.618$) or leeches (Tala: $Z = -1.470$, $P = 0.141$; Valo: $Z = -1.000$, $P = 0.317$) infections at either site, tick infections at Tala ($Z = -0.673$, $P = 0.501$), and fly infections at Valo ($Z = -0.245$, $P = 0.806$).

Discussion

Long-term study of the Milne-Edwards' sifaka (*Propithecus edwardsi*), has documented high infant and juvenile mortality rates (Pochron *et al.*, 2004), with most deaths attributable to either predator attack or infanticide (Wright, 1995, 1998; Karpanty & Wright, 2007). To date, mortality from disease and loss of condition due to parasite infection has not been observed as a cause of death. However, given the role that parasitism has been found to play in other primate species (e.g. Milton, 1996), we examined the richness and intensity of parasite

Table 15.3. *Seasonal ectoparasite, biting insect, and leech infestation intensities. Average (mean and SD) infestation intensity scores in both the warm/wet and cool/drier seasons for both the Tala [n = 50 (warm/wet) and 23 (cool/drier)] and Valo [n = 12 (warm/wet) and 12 (cool/drier)] study sites. Significant inter-seasonal differences were found in lice infestation intensity at both sites*

Ectoparasite class	Talatakely (Tala)				Valohoaka (Valo)			
	Warm/wet	Cool/Drier	Z	P	Warm/wet	Cool/Drier	Z	P
Flies	3.30 ± 6.16	0.70 ± 1.74	−2.949	0.003**	1.25 ± 1.76	1.25 ± 1.42	−0.245	0.806
Lice	25.94 ± 39.58	4.22 ± 11.50	−2.505	0.012*	39.25 ± 48.09	1.50 ± 1.88	−2.261	0.024*
Ticks	1.26 ± 2.22	1.87 ± 2.94	−0.673	0.501	5.75 ± 7.46	0.17 ± 0.58	−3.257	0.001**
Other mites	4.34 ± 9.85	2.39 ± 5.02	−0.32	0.749	1.25 ± 2.34	11.00 ± 34.36	−0.499	0.618
Leeches	0.44 ± 0.93	0.26 ± 1.05	−1.47	0.141	0.08 ± 0.29	0	−1.000	0.317

*Statistical significance at $P < 0.05$; **Statistical significance at $P < 0.001$.

infections within these rain-forest lemurs to gain insights into the possible effects of parasites on condition, survival, and reproductive success. While earlier studies of lemur parasitism have been limited to descriptions of parasite species detected during one sampling period (Garell & Meyers, 1995), an increase in field work and new methods for capturing lemurs in the wild have made it possible to study ectoparasites *in situ* over a longer time period (Takahata *et al.*, 1998; Junge & Louis, 2002, 2005; Muehlenbein *et al.*, 2003; Sauther *et al.*, 2006). While preliminary, these data offer us the first systematic investigation of host/parasite dynamics within the sifakas of Ranomafana National Park (RNP) and will serve as a starting point for further research in this area.

Specifically, this study was designed to elucidate the relationship between habitat disturbance, seasonality, and parasitism, by testing three hypotheses. Contrary to hypothesis 1, we found that parasite richness did not decrease, due to a reduction in vector availability, with increasing habitat disturbance. Instead, endoparasite richness was greater within the disturbed forest site. Ectoparasite diversity remained constant between the undisturbed forest and the lesser of the two disturbed sites (Tala) and then increased significantly in the most disturbed site (Saka). In part, these findings may suggest that habitat disturbance within the Tala region of the RNP has not been great enough in magnitude to have altered parasite ecology at the site. Or perhaps it reflects the all-vegetarian diet of sifakas, since Orthopterans or beetles, which may be intermediate hosts for many endoparasites, have not been seen to be ingested by *Propithecus edwardsi*.

Similarly, hypothesis 2, which predicted that sifakas living in disturbed habitats would have higher parasite infection intensities, was only weakly supported by the data. Of the five classes of endoparasites examined here, only the intensity of *Monezia* tapeworm infections were found to differ significantly between sifakas living in the disturbed and undisturbed habitats. Additionally, as with parasite diversity, analyses revealed that ectoparasite infection intensities remained consistently similar between the undisturbed forest and the lesser of the disturbed forest sites, and were found to be significantly greater in the most disturbed site for louse and mite infections. Given the marked habitat disturbance (both intensive timber exploitation and large-scale clear cutting) that took place in Saka, and the resulting ecological stress to the animals (*sensu* Stoner, 1996), it is not surprising that the sifakas living there were found to have high parasite infection intensities. However, why the fly and tick infections were not found to follow the same pattern as the lice and mites is unknown at present. Furthermore, that ectoparasite infection intensities at Tala and Valo were so similar is quite surprising, given the fact that long-term study of these sifaka populations has been able to document significant differences in forest structure, food resource availability and abundance, and sifaka diet and ranging patterns between the two sites (Balko & Underwood, 2005; Arrigo-Nelson, 2006); conditions which have, in other studies, been linked to increased energetic costs and the lowering of the host's immune system to parasite infection (Santa Cruz, 2000). As mentioned above, it is possible that the type and degree of disturbance experienced at Tala was simply not sufficient to disrupt the natural host/parasite balance at the site to such a degree that it was still detectable 20 years following the cessation of all disturbance activities, while the extent of the initial disturbance at Saka was great enough to create a much longer (if not permanent) imbalance in host/parasite dynamics. We look forward to the continued monitoring of these sites, to better understand the relationship between disturbance level and parasite infection and to determine how this relationship changes over time.

Finally, data on ectoparasite richness and infection intensity appear to show seasonal fluctuations in support of hypothesis 3. While all three sites showed a trend for greater species richness scores during the warm/wet season, seasonal differences in ectoparasite richness at Tala were found to differ significantly. Furthermore, fly (Tala), louse (Tala and Valo), and tick (Valo) infection intensities were all found to be greater during the warm/wet season (although it should also be noted that other mites and leeches did not show this same pattern). While this hypothesis could not be tested for endoparasite infections at this point, due to a lack of comparative data, we look forward to doing so in the near future. Given the marked climatic seasonality seen at RNP, it is not surprising to see that parasite ecology shows similar fluctuations here, to those seen at other sites (Stuart & Strier, 1995; Huffman *et al.*, 1997; Huffman *et al.*, Chapter 16, this volume). Many parasite species with direct life cycles show strong seasonality of re-infection, making it important to sample in both wet and dry seasons to appreciate parasite ecology and its effect on the health of hosts (see Huffman *et al.*, Chapter 16, this volume, but see Vitazkova & Wade, 2006, for contrasting results). It is therefore suggested that researchers consider this fact, when establishing sampling protocols.

Although our primary goal was to examine the impacts of habitat disturbance and seasonality on parasite infection, there is an additional conservation implication from this research. Recently, several mainland African studies have suspected links between human presence in an area and the magnitude of parasite infections in the non-human primates living there (Lilly *et al.*, 2003; Gillespie *et al.*, 2005; Weyher *et al.*, 2006). As within RNP, one of the study sites (Tala) experiences heavy tourism from July–November (~30 000 tourists/year). Because the other three sites are visited by few or no tourists, we were interested to see whether or not we would detect any difference within the sifaka population that may be attributable to tourist presence (Wallis & Lee, 1999). We find that, in terms of endoparasite infections, the data presented here may be cause for concern, as sifakas living in the disturbed forest were found to suffer from infections of strongyles, *Anoplocephala* tapeworms, and pinworms, which at least at the generic level are known to infect either humans or their commensals (Kightlinger *et al.*, 1995; Lehtonen *et al.*, 2001; Laakkonen, 2003) and which were never found in fecal samples from the undisturbed forest population. However, because here we have not identified the endoparasites definitively to species, and it cannot be inferred if these endoparasites can be spread from one primate species to another, this evidence is only suggestive (see Gasser *et al.*, Chapter 3, this volume, for a discussion of how difficult it is to evaluate whether or not transmission has occurred). Our results may be an additional factor encouraging modern human sanitation facilities within the areas of the park frequented by humans (see also Vitazkova, Chapter 18, this volume). In terms of ectoparasite richness and infection intensity, sifakas at the site with many tourists showed no increase over the undisturbed forest baseline and significantly lower infections than the disturbed but non-touristed (Saka) site. Although further examination is needed, this finding is likely attributable to some combined effect of ectoparasite ecology (including but not limited to host specificity) and the mechanisms of ectoparasite transmission (Anderson & May, 1978).

As habitats become increasingly disturbed, and fragmented, parasites may have an increasing effect on the fitness of primate populations (Anderson & May, 1982; Roberts *et al.*, 2002; Altizer *et al.*, 2003). Baseline data on patterns of parasitic infections in wild primate populations may provide an index of population health and information to assess and manage disease risks (Stuart & Strier, 1995; Gillespie, 2006). Our results offer the first opportunity to examine host–parasite relationships in a well-studied population of rain-forest lemurs. This study suggests that the richness and intensity of endoparasites may increase with forest disturbance, although the pattern was less clear with relation to the effects of habitat disturbance on ectoparasite infection. However, whether increased parasite richness and/or intensity affects either longevity or reproductive success remains to be seen, and until we have these data, it is premature to advise conservation managers on actions to help populations of this endangered lemur species.

Acknowledgements

We received authorization to do this research in Madagascar from the Ministry of Environment, Water and Forests, and the CAFF/CORE committee. This research would not have been possible without the logistical assistance of MICET, especially Benjamin Andriamihaja, ANGAP, and the Central ValBio and its Director Anna Feistner. Special appreciation goes to the *Propithecus* Project team, including the late Georges Rakotonirina, Raymond Ratsimbazafy, Remy Rakotovao, Georges Rene Randrianirina, Laurent Randrianasolo, and Albert Telo, and Ed Louis' lemur capture team led by Richard Randriapoina and funded by the Henry Doorly Zoo. We gratefully acknowledge the assistance of Michael Stuart who inspired and trained KLH to identify parasites. Special thanks to Jennifer Legon and Zeph Pendleton, and Rita Riewerts for their careful diligence in ectoparasite photos and counts, to Natalie Leo for identification of the lice, and to Pascal Rabeson and Stan Vaughn for identification of the flies. We are grateful to Mark Siddall and Liz Borda for leech identification. We thank Jukka Jernvall for his advice and wisdom on earlier drafts of this manuscript. Randy Junge is thanked for his expert comments. We are grateful to the editors Colin Chapman and Mike Huffman for inviting us to contribute to this volume. We also thank our research sponsors: Wenner-Gren Foundation for Anthropological Research, Earthwatch Institute, Margot Marsh Biodiversity Fund, Leakey Foundation, Norman and Lucile Packard Foundation, Saint Louis Zoo (FRC committee), Stony Brook University, National Science Foundation, USA, the National Institutes of Health, USA, and the Seneca Park Zoo Docent organization.

References

Altizer, S., Nunn, S. L, Trall, P. H. *et al.* (2003). Social organization and parasite risk in mammals: integrating the theory and empirical studies. *Annual Review of Ecology, Evolution and Systematics*, **34**, 517–547.

Altizer, S., Dobson, A., Hosseini, P., Pascual, M. & Rohani, P. (2006). Seasonality and the dynamics of infectious diseases. *Ecology Letters*, **9**, 467–484.

Anderson, R. M. & May, R. M. (1978). Regulation and stability of host-parasite population interactions. I. Regulatory processes. *Journal of Animal Ecology*, **47**, 219–247.

Anderson, R. M. & May, R. M. (1982). Coevolution of hosts and parasites. *Parasitology*, **85**, 411–426.

Ankel-Simons, F. (2006). *An Introduction to Primate Anatomy*, 3rd edn. San Diego, CA: Elsevier.

Arrigo-Nelson, S. J. (2006). The effects of habitat disturbance on ecology and behavior of *Propithecus edwardsi*, the Milne-Edwards' sifaka in Ranomafana National Park, Madagascar. Unpublished PhD thesis, Stony Brook University, New York.

Arrigo-Nelson, S. J., Baden, A. L. & Wright, P. C. (in press) Testing hypotheses for geophagy in a Malagasy rainforest lemur. *Biotropica.*

Balko, E. A. & Underwood, H. B. (2005). Effects of forest structure and composition on food availability for *Varecia variegata* at Ranomafana National Park, Madagascar. *American Journal of Primatology*, **66**, 45–70.

Bochkov, A. V. & O'Conner, B. M. (2006). Revision of the genus *Gaudalges* (Acari Psoroptidae). Parasites of Malagasy lemurs. *Acarina*, **14**, 3–20.

Borda, E. (2006). A revision of the Malagabdellinae (Arhynchobdellida, Domanibdellidae), with a description of a new species, *Malagabdella niarchosorum*, from Ranomafana National Park, Madagascar. *American Museum Novitates*, **3531**, 1–13.

Chapman, C. A., Gillespie T. R. & Speirs, M. L. (2005). Parasite prevalence and richness in sympatric Colobines: effects of host density. *American Journal of Primatology*, **67**, 259–266.

Dobson, A. P. (1988). The population biology of parasite-induced changes in host behavior. *Quarterly Review of Biology*, **63**, 139–165.

Dobson, A. P. & May, R. M. (1986). Disease and conservation. In *Conservation Biology: The Science of Scarcity and Diversity*, ed. M. Soule. Sunderland, MA: Sinauer Associates, Inc., pp. 345–365.

Dutton, C. J., Junge, R. E. & Louise, E. E. Jr. (2003). Biomedical evaluation of free-ranging ring-tailed lemurs (*Lemur catta*) in Tsimannapetsosoa Strict Nature Reserve, Madagascar. *Journal of Wildlife Medicine*, **34**, 16–24.

Garell, D. M. & Meyers, D. M. (1995). Hematology and serum chemistry values of free-ranging golden crowned sifaka (*Propithecus tattersalli*). *Journal of Zoo and Wildlife Medicine*, **26**, 382–386.

Gillespie, T. R. (2006). Noninvasive assessment of gastrointestinal parasite infections in free-ranging primates. *International Journal of Primatology*, **27**, 1129–1143.

Gillespie, T. R. & Chapman, C. A. (2006). Predictions of parasite infection dynamics in primate metapopulations based on attributes of forest fragmentation. *Conservation Biology*, **20**, 441–448.

Gillespie, T. R., Greiner, E. C. & Chapman, C. A. (2004). Gastrointestinal parasites of the guenons of western Uganda. *Journal of Parasitology*, **90**, 1356–1360.

Gillespie, T. R., Chapman, C. A. & Greiner, E. C. (2005). Effects of logging on gastrointestinal parasite infections and infection risk in African primates. *Journal of Applied Ecology*, **42**, 699–707.

Glander, K., Wright, P. C., Daniels, P. S. & Merenlender, A. (1992). Morphometrics and testicle size of rainforest lemur species from southeastern Madagascar. *Journal of Human Evolution*, **22**, 1–17.

Gould, L. & Sauther, M. L. (2007a). Lemuriformes. In *Primates in Perspective*, ed. C. J. Campbell, A. Fuentes, K. C. MacKinnon, M. Panger & S. K. Bearder. New York, NY: Oxford University Press, pp. 46–72.

Gould, L. & Sauther, M. L. (2007b). Anti-predator strategies in a diurnal prosimian, the ring-railed lemur (*Lemur catta*), at the Beza Mahafaly Special Reserve, Madagascar. In *Primate Anti-Predator Strategies*, ed. S. L. Gursky & K. A. I. Nekaris. New York, NY: Springer, pp. 275–288.

Green, G. M. & Sussman, R. W. (1990). Deforestation history of the eastern rain forests of Madagascar from satellite images. *Science*, **248**, 212–215.

Hogg, K. L. (2002). Effect of habitat disturbance on parasite load and diversity in two species of lemur (*Eulemur rubriventer* and *Propithecus diadema edwardsi*) at Ranomafana National Park, Madagascar. Unpublished M.A. thesis, Stony Brook University, New York.

Hogg, K. L., Wade, S. & Wright, P. C. (in press). Parasites of nine lemur species at Ranomafana National Park, Madagascar. *Journal of Zoo and Wildlife Medicine*.

Holmes, J. C. (1996). Parasites as threats to biodiversity in shrinking ecosystems. *Biodiversity and Conservation*, **5**, 975–983.

Huffman, M. A., Gotoh, S., Turner, L. A., Hamai, M. & Yoshida, K. (1997). Seasonal trends in intestinal nematode infection and medicinal plant use among chimpanzees in the Mahale Mountains, Tanzania. *International Journal of Primatology*, **38**, 111–125.

Irwin, M. T. (in press). A review of the parasites of the lemurs of Madagascar in companion and exotic animal parasitology, ed. D. D. Bowman. International Veterinary Information Service.

Junge, R. E. & Garell, D. (1995). Veterinary evaluation of ruffed lemurs (*Varecia variegata*) in Madagascar. *Primate Conservation*, **16**, 44–46.

Junge, R. E. & Louis, E. E. (2002). Medical evaluation of free-ranging primates in Betampona Reserve, Madagascar. *Lemur News*, **7**, 23–25.

Junge, R. E. & Louis, E. E. (2005). Biomedical evaluation of two sympatric lemur species (*Propithecus verreauxi deckeni* and *Eulemur fulvus rufus*) in Tsiombokibo classified forest, Madagascar. *Journal of Zoo and Wildlife Medicine*, **36**, 581–589.

Karpanty, S. M. & Wright, P. C. (2007). Predation on lemurs in the rainforest of Madagascar by multiple predator species: observations and experiments. In *Primate Anti-Predator Strategies*, ed. S. L. Gursky & K. A. I. Nekaris. New York, NY: Springer, pp. 77–99.

Kightlinger, L. K., Seed, J. R. & Kightlinger, M. B. (1995). The epidemiology of *Ascaris lumbricoides, Trichuris trichiura* and hookworms in children in the Ranomafana rainforest, Madagascar. *Journal of Parasitology*, **81**, 159–169.

King, S. J., Arrigo-Nelson, S. J., Pochron, S. T. *et al.* (2005). Dental senescence in a long-lived primate links infant survival to rainfall. *Proceedings of the National Academy of Sciences*, USA, **102**, 16579–16583.

Krause, D. W., Harman, J. H. & Wells, N. A. (1997). Late Cretaceaous vertebrates from Madagascar: implications for biotic change in deep time. In *Natural and Human Induced Change in Madagascar*, ed. S. M. Goodman & B. D. Patterson. Washington, DC: Smithsonian Institution Press, pp. 3–43.

Laakkonen, J. (2003). Trypomastigotes and potential flea vectors of the endemic rodents and the introduced *Rattus rattus* in the rainforests of Madagascar. *Biodiversity and Conservation*, **12**, 1775–1783.

Landau, I., Lepers, J. P., Rabetafika, L., Baccam, D., Peters, W. & Coulanges, P. (1989). Plasmodies des lemuriens malagaches. *Annales Parasitologie Humaine et Comparée*, **64**, 171–184.

Lehtonen, J. T., Mustonen, O., Ramiarinjanahary, H., Niemelä, J., & Rita, H. (2001). Habitat use by endemic and introduced rodents along gradient of forest disturbance in Madagascar. *Biodiversity and Conservation*, **10**, 1185–1202.

Lilly, A. A., Mehlman, P. M. & Doran, D. (2003). Intestinal parasites in gorillas, chimpanzees and humans at Mondika Research Site, Dzanga-Ndoza Park, Central African Republic. *International Journal of Primatology*, **23**, 555–573.

Lowry, P. P. II, Schatz, G. E. & Phillipson, P. B. (1997). The classification of natural and anthropogenic vegetation in Madagascar. In *Natural and Human Induced*

Change in Madagascar, ed. S. M. Goodman & B. D. Patterson. Washington, DC: Smithsonian Institution Press, pp. 381–405.

Milton, K. (1996). Effects of bot fly (*Alouattamyia baeri*) parasitism in a free ranging howler monkey (*Alouatta palliata*) population in Panama. *Journal of Zoological Society, London*, **239**, 39–63.

Mittermeier, R. A., Konstant, W. R., Hawkins, F. *et al.* (2006). *The Lemurs of Madagascar*. Washington, DC: Conservation International.

Muehlenbein, M. P., Schwartz, M. & Richard, A. (2003). Parasitologic analyses of the sifaka (*Propithecus verreauxi verreauxi*) at Beza Mahafaly, Madagascar. *Journal of Zoo and Wildlife Medicine*, **34**, 274–277.

Myers, J. H. & Rothman, L. E. (1995). Virulence and transmission of infectious diseases in humans and insects: evolutionary and demographic patterns. *Trends in Ecology and Evolution*, **10**, 194–198.

Nunn, C. & Altizer, S. (2006). *Infectious Diseases in Primates: Behaviour, Ecology and Evolution.* Oxford: Oxford University Press.

O'Connor, B. M. (2003). Acariformes: parasitic and commensal mites of vertebrates. In *The Natural History of Madagascar*, ed. S. M. Goodman & J. P. Benstead. Chicago, IL: University of Chicago Press, pp. 593–602.

Pochron, S. T. & Wright, P. C. (2003). Variability in adult group compositions of a prosimian primate. *Behavioral Ecology and Sociobiology*, **54**, 285–293.

Pochron, S. T., Fitzgerald, J., Gilbert, C. C. *et al.* (2003). Patterns of female dominance in *Propithecus diadema edwardsi* of Ranomafana National Park, Madagascar. *American Journal of Primatology*, **61**, 173–185.

Pochron, S. T., Tucker, W. T. & Wright, P. C. (2004). Demography, life history and social structure in *Propithecus diadema edwardsi* from 1986–2000 in Ranomafana National Park, Madagascar. *American Journal of Physical Anthropology*, **125**, 61–72.

Rabetafika, L. (1995). *Nouveaux Sporozoaires de la Faune Malgache*. Unpublished PhD thesis, University of Antananarivo, Madagascar.

Rabetafika, L. Landau, I., Baccam, D, Lepers, J.-P., Coulanges, P. & Rakotofiringa, S. (1989). Sporogonie et schizogonie pre-erythrocytaire de plasmodies de lemuriens malgaches. *Annales de Parasitologie Humaine et Comparée*, **64**, 243–250.

Roberts, M. G., Dobson, A. P., Arneberg, P., de Leo, G. A. & Krecek, R. C. (2002). Parasite community ecology and biodiversity. In *The Ecology of Wildlife Diseases*, ed. P. J. Hudson, A. Rizzoli, B. T. Grenfell, H. Heesterbeek & A. P. Dobson. Oxford, UK: Oxford University Press, pp. 63–82.

Santa Cruz, A. C. M. (2000). Habitat fragmentation and parasitism in howler monkeys (*Alouatta caraya*). *Neotropical Primates*, **8**, 146–148.

Sauther, M. L., Fish, K. D., Cuozzo, F. P. *et al.* (2006). Patterns of health, disease and behavior among wild ring-tailed lemurs, *Lemur catta*: effects of habitat and sex. In *Ringtailed Lemur Biology: Lemur catta in Madagascar*, ed. A. Jolly, R. W. Sussman, N. Koyama & H. Rasamimanana. New York, NY: Springer, pp. 313–331.

Sokal, R. R. & Rohlf, F. J. (1995). *Biometry*, 3rd edn. New York, NY: W. H. Freeman and Company.

Stoner, K. E. (1996). Prevalence and intensity of intestinal parasites in mantled howling monkeys (*Alouatta palliata*) in Northeastern Costa Rica: implication for conservation biology. *Conservation Biology*, **10**, 539–546.

Stuart, M. D. & Strier, K. B. (1995). Primates and parasites: a case for a multidisciplinary approach. *International Journal of Primatology*, **16**, 577–593.

Stuart, M. D., Greenspan, L. L., Glander, K. E. & Clarke, M. R. (1990). A coprological survey of parasites of wild mantled howling monkeys, *Alouatta palliata palliata. Journal of Wildlife Diseases*, **26**, 547–549.

Stuart, M. D., Strier, K. B. & Pierberg, S. M. (1993). A coprological survey of parasites of wild muriquis, *Brachyteles arachnoides*, and brown howling monkeys, *Alouatta fusca. Proceedings of the Helminthological Society of Washington*, **60**, 111–115.

Takahata, Y., Kawamoto, Y., Hirai, K. *et al.* (1998). Ticks found among the wild ring-tailed lemurs at the Berenty Reserve, Madagascar. *African Study Monographs*, **19,** 217–222.

Tecot, S. (2008). Reproduction and ecology of *Eulemur rubriventer* in Ranomafana National Park, Madagascar. University of Texas, Austin.

Vitazkova, S. K. & Wade, S. E. (2006). Parasites of free-ranging black howler monkeys (*Alouatta pigra*) from Belize and Mexico. *American Journal of Primatology*, **68**, 1089–1097.

Wallis, J. and Lee, D. (1999). Primate conservation: the prevention of disease transmission. *International Journal of Primatology*, **20**, 803–826.

Weyher, A. H., Ross, C. & Semple, S. (2006). Gastrointestinal parasites in crop raiding and wild foraging *Papio anubis* in Nigeria. *International Journal of Primatology*, **27**, 1519–1534.

Wright, P. C. (1995). Demography and life history of free-ranging *Propithecus diadema edwardsi* in Ranomafana National Park, Madagascar. *International Journal of Primatology*, **16**, 835–854.

Wright, P. C. (1997). The future of biodiversity in Madagascar: a view from Ranomafana National Park. In *Natural and Human Induced Change in Madagascar*, ed. S. M. Goodman & B. D. Patterson. Washington, DC: Smithsonian Institution Press, pp. 381–405.

Wright, P. C. (1998). Impact of predation risk on the behavior of *Propithecus diadema edwardsi* in the rain forest of Madagascar. *Behaviour*, **135**, 483–512.

Wright, P. C. (1999). Lemur traits and Madagascar ecology: coping with an island environment. *Yearbook of Physical Anthropology*, **42**, 31–72.

Wright, P. C., King, S. J., Baden, A. L. & Jernvall, J. (2008). Aging in wild female lemurs: sustained fertility with increased infant mortality. In *Aging in Primates*, ed. S. W. Margulis & S. Atsalis. Basel, Switzerland: Karger Press.

16 *Chimpanzee–parasite ecology at Budongo Forest (Uganda) and the Mahale Mountains (Tanzania): influence of climatic differences on self-medicative behavior*

MICHAEL A. HUFFMAN, PAULA PEBSWORTH, CHRIS BAKUNEETA, SHUNJI GOTOH, AND MASSIMO BARDI

Photograph by Michael A. Huffman (*Pan troglodytes schweinfurthii*). Leaf swallowing behavior

Primate Parasite Ecology. The Dynamics and Study of Host–Parasite Relationships, ed. Michael A. Huffman and Colin A. Chapman. Published by Cambridge University Press.

Introduction

Long-term monitoring of host populations from habitats of differing climatic regimes over multiple seasons is important to better understand the dynamics of host–parasite relationships and their potential effects on host health and behavior. Varying environmental factors, such as temperature and rainfall, can affect the development and maturation of parasites within a given habitat, and can in turn influence temporal patterns of their transmission and their incidence of infection within their definitive hosts (Taylor & Michel, 1953; Fabiyi *et al.*, 1988; Anderson, 1992; Huffman *et al.*, 1997; Marquardt *et al.*, 2000). For example, to persist in host populations within a given habitat, parasite life cycles generally demand that progeny spend some time exposed to external environmental factors if the process of moving between definitive host individuals, i.e. parasite transmission, is direct. Alternatively, if a parasite's method of transmission is indirect because it includes stages within intermediate hosts, the parasites are also indirectly exposed to external environmental factors affecting these hosts. Thus, environmental variation between habitats can potentially result in differences in the health and behavior of host species infected with parasite species that have direct and/or indirect life cycles, and long-term monitoring is expected to enhance our ability to understand and predict differences across regions.

Self-medication by primates to prevent or control parasite infection in the wild is a host behavior that can differ across habitats (Huffman, 1997, 2001). Leaf swallowing, a self-medicative behavior that utilizes the folding and swallowing of rough, bristly leaves without chewing them, is known to occur among all the African great ape species at various sites across their distribution (Huffman, 1997, 2001). Leaf swallowing acts to stimulate rapid gut motility, which assists in flushing out helminth parasites during peak periods of infection (Huffman *et al.*, 1996, 1997). Based on the ratio of leaves swallowed to the number of adult nematodes of *Oesophagostomum stephanostomum* (nodular worms) expelled per dung (2:1) at Mahale National Park (western Tanzania), it was estimated that repeated treatment could have a significant impact on a chimpanzee's worm burden (Huffman & Caton, 2001). Reduction in worm numbers has been observed after repeated bouts of leaf-swallowing and within 24 hours after bitter pith chewing of *Vernonia amygdalina*, another form of self-medication studied in detail at Mahale (Huffman *et al.*, 1993, 1996). At Kibale National Park (southwestern Uganda) a significant relationship between leaf swallowing and the presence of *Bertiella studeri* tapeworm proglottids (reproductive segment) in the dung has been reported (Wrangham, 1995). *Oesophagostomum stephanostomum* and *Bertiella studeri* (tapeworm) are the only two parasite species associated with leaf-swallowing behavior in chimpanzees to date. These studies suggest that there is an adaptive value for hosts in self-medicative behavior, and given the wide range of African great ape habitats in which leaf swallowing occurs, it is important to understand the factors responsible for regional differences in parasite species associated with these host behaviors.

Environmental factors such as rainfall can influence infection dynamics in primates known to have self-medicative behaviors. For example, for *Oesophagostomum* at Mahale, the peak period of infection is the rainy season months of November through May, during which leaf swallowing most frequently occurs (Kawabata & Nishida, 1991; Huffman *et al.*, 1997). Recent evidence from three additional chimpanzee study sites, Iyema-Lamako in central Zaire, Gashaka in eastern Nigeria, and Fongoli in southeastern Senegal, have shown similar trends in increased frequency of leaf swallowing during the rainy season and concurrent increased prevalence of *Oesophagostomum* infections (Dupain *et al.*, 2002; Pruetz & Johnson-Fulton, 2003; Fowler *et al.*, 2007). At Gombe, a direct relationship between parasite infection and leaf swallowing has not been demonstrated, but peaks in the frequency of leaf swallowing have been observed during the rainy season of some years, but not others (Wrangham & Goodall, 1989). Nonetheless, these data generally support the idea that climate can be a trigger for the completion of *Oesophagostomum*'s life history (re-infection) and for the incidence of host's self-medicative behaviors. Less is known about the importance of environmental factors in the transmission biology of *B. studeri*. At Kibale, nothing has been published which suggests that a relationship between rainfall and proglottid expulsion or between rainfall and leaf swallowing exists (Wrangham, 1995). Indeed, until now nothing has been reported about seasonal effects on infection by *Bertiella* in any host species (e.g. Brack, 1987). At Budongo, a forest reserve in northwestern Uganda, observations of leaf swallowing were first made by Chris Bakuneeta and colleagues while conducting research on the feeding ecology of the Sonso and Pabidi chimpanzee communities. The presence of *Bertiella* infection has also been observed in the Sonso chimpanzees in unpublished independent surveys by G. Kalema-Zikusoka (Masters thesis, North Carolina State University, USA) and G. Renehan (Masters thesis, Tufts University, USA). However, nothing is known about the relationship between leaf-swallowing behavior and *Oesophagostomum* infections at any of these Ugandan study sites.

The above information suggests that chimpanzees in two forests in Tanzania and Uganda may be using the same behaviors (i.e. leaf swallowing) in response to two different parasite species with very distinct life cycles (*Oesophagostomum* – direct life cycle, *Bertiella* – indirect life cycle). Clear differences in seasonal patterns of rainfall and the life cycles of these two parasites present in chimpanzees at sites in both regions provide an excellent opportunity to investigate the role of climate on parasites and its effect on the health and behavior of the host.

The aim of this study was to assess the effect of seasonal and longer-term climatic variation on the nature of parasite infection patterns and their impact on host health and anti-parasitic behaviors among chimpanzee populations living under different climatic conditions. An inter-site comparison was conducted on two populations of chimpanzees, the Sonso group living in the Budongo forest (northwestern Uganda) and the M group of the Mahale Mountains (western Tanzania). This is an important step towards understanding the ecological dynamics of host–parasite interactions and their effect on self-medicative behavior of the host. Results are discussed with reference to climatic influences on infection parameters and its impact on leaf-swallowing behavior to gain insights into regional level effects and their potential use in predicting the occurrence of this and other self-medicative behaviors across equatorial Africa.

Methods and materials

Study sites

The Sonso study site (1°44′N, 3°33′E) is located within the Budongo Forest Reserve in western Uganda and was established in 1990 by Vernon Reynolds and the Budongo Forest Project staff (Reynolds, 1992). The Budongo Forest is approximately 428 km^2 and consists of moist semi-deciduous tropical forest ranging in elevation from 914 to 1094 m asl (Eggeling, 1947; Synnott, 1985; Plumptre, 1996).

The Mahale M-group study site (6°07′S, 29°44′E) is located within the Mahale Mountains National Park, western Tanzania on Lake Tanganyika and was established in 1965 by Junichiro Itani (Nishida, 1990). The national park was established in 1985, and is approximately 1613 km^2. It ranges between an elevation of 773 m at the lakeshore to the highest peak 2515 m asl, and is a mixture of tropical semi-evergreen forest and Miombo woodland (Nishida, 1990). Data from Mahale presented here for comparison with Budongo, were collected between 1989 and 1994 (Huffman & Seifu, 1989; Huffman *et al.*, 1993, 1996, 1997).

Observation periods and goals

The Budongo study was divided into three periods. Period I (October 1995–January 1997) was initiated by MAH with the help of the Budongo Forest Project on-site co-directors. Baseline data were collected to assess the seasonality of parasite infection and presence of leaf-swallowing behavior. Field support-staff were trained to record the presence or absence of whole leaves and/or adult worms (whole or fragments) in dung encountered while searching for or following individuals or groups of chimpanzees.

During period II (February 1998–December 1998) behavioral, health, and parasitological parameters were measured and correlated between February and October by PP using focal-animal and *ad libitum* behavioral observations. When possible, chimpanzees were followed from dawn to their night nests ($n = 341.5$ hour, 14 individuals, 106 focal sessions). *In situ* macroscopic monitoring and microscopic analysis of dung encountered during animal follows, and sample collection from known individuals, was continued ($n = 295$).

In period III (January 1999–May 1999) monthly monitoring of a subset of five individual chimpanzees previously observed by PP was initiated ($n = 116$). This protocol was designed to obtain longitudinal data on the dynamics of parasite infection in individuals, similar to studies at Mahale by Huffman *et al.* (1997). Routine surveillance for the presence of visible parasites and whole leaves in dung encountered while following chimpanzees continued.

MAH and colleagues collected the Mahale data analyzed in this chapter during three study periods between 1989 and 1994. Period I was between August 1989 and March 1990, period II was between August 1990 and February 1992, and period III was between October 1993 and March 1994 (Huffman *et al.*, 1993, 1996, 1997).

Climatological analysis

Rainfall data was collected at Sonso between January 1995 and December 1999 by the Budongo Forest Project staff and was used here for analysis on the effects of rainfall on parasite prevalence and chimpanzee leaf-swallowing behavior. Researchers and staff of the Mahale Mountains Chimpanzee Research Project collected rainfall data at Kansyana camp between January 1989 and December 1994. Months were classified as "wet" if cumulative rainfall exceeded 100 mm, "transient" if it was between 99 and 51 mm, and "dry" if it was 50 mm or less (Newton-Fisher, 1999).

Parasitological analysis

Difficulties in inter-population comparisons of parasite infection due to differences in sampling techniques can occur (Ashford *et al.*, 2000). We minimize this problem by using the same techniques in comparisons of the Mahale and Budongo samples. Dung samples collected in the field were taken back to camp and measured out into three 1 g replicates. One replicate was preserved with 5% neutral formalin and sent to the Primate Research Institute, Kyoto University for analysis by S. Gotoh. The remaining two sets were kept as backup samples.

Parasite species prevalence was quantified using the McMaster flotation and formalin-ether concentration techniques (Huffman *et al.*, 1993, 1997).

Fecal samples used for the analysis of parasite prevalence (% of positive samples, individuals) were collected during two different periods at both sites and each period was characterized by variable monthly rainfall distributions. Rainfall for each collection periods is expressed as a ratio of wet:transient and dry months according to the definitions above (e.g. 6:5 = an 11-month period including 6 wet months and 5 transient to dry months). This comparison is useful as an indicator of the relative effects of climatic variability on parasite prevalence.

Results

Inter-site variation in seasonal patterns of rainfall

Yearly fluctuation in cumulative rainfall was apparent for both Budongo and Mahale (Figure 16.1A, B). However, the average annual rainfall varied little between these two sites (Budongo: 1761 mm, SD 280, n = 5 years; Mahale: 1534 mm, SD 237, n = 3 years). Nonetheless, a significant difference was found in the proportion of dry months between the two sites. In 39% of the 62 months monitored at Mahale between 1989 and 1994, less than 50 mm of rainfall was recorded, while only 18% of the 61 months monitored at Budongo between 1995 and 1999 had less than 50 mm of rainfall (Figure 16.1A, B, $\chi^2_{(1)} = 6.5, P = 0.01$).

There is a striking difference in the seasonality of rainfall between the two sites (Figure 16.1). At Mahale, dry periods with 0–50 mm of rainfall typically range from 4 to 5 months, and occur between June and November (Takasaki *et al.*, 1990) when the ground becomes hard and dry. In contrast, at Budongo and other study sites in western Uganda such as Kibale, rainfall is more widely distributed throughout the year, and there are typically few if any prolonged periods of dryness (Struhsaker, 1997).

Inter-site comparison of parasite prevalence and infection intensity

Prevalence

Twenty-three parasite taxa are reported to infect *Pan* species across their natural distribution (Table 16.1). Of the taxa listed in either one or both of the Mahale and Budongo groups (n = 15), chimpanzees share 60% in common. The three most widely represented nematode species were *O. stephanostomum*,

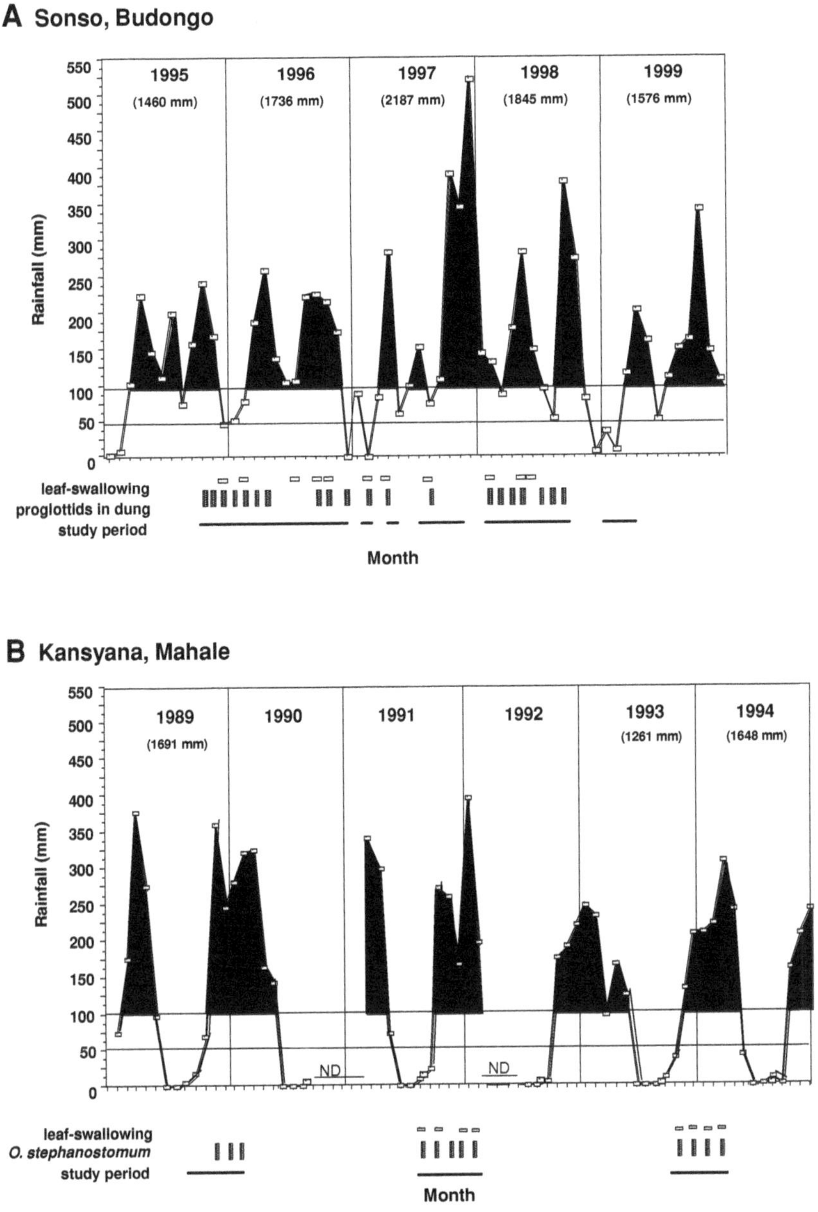

Figure 16.1. Monthly distribution of rainfall and the occurrence of leaf swallowing and the shedding of *Bertiella* tapeworm proglottids among chimpanzees at Sonso, Budongo and Kansyana, Mahale for a 5-year period.

Table 16.1. *Parasites identified in free-ranging chimpanzee and bonobo populations*

	Study site	Chimpanzees	Bonobos
Nematoda			
Trichuris trichiura	**M, Bd**	X	
Trichuris sp.	G, Kb, Wb	X	X
Strongyloides fulleborni	**M, Bd**	X	
Strongyloides sp.	G, Kb	X	X
Oesophagostomum stephanostomum	**M, Bd,** Iy	X	X
Oesophagostomum sp.	G, Kb, Wb	X	X
Ternidens deminutus	Bd	X	
Necator sp.	G, **M**	X	
Ancylostoma sp.	Iy		X
Ascaris sp.	G, Iy	X	X
Probstmayria gombensis	G	X	
Probstmayria sp.	Kb	X	
Physaloptera sp.	G	X	
Prosthenorchis sp.	**M**	X	
Oxyuridae gen. sp.	Wb		X
Cestoda			
Bertiella studeri	**M**, **Bd**, G, Kb	X	
Trematoda			
Dicrocelium lanceatum	M	X	
Dicrocelium sp.	Iy		X
Protozoa			
Troglodytella sp.	**M**, **Bd**, G, Kb, Wb, Iy	X	X
Giardia lamblia	**M, Bd**, Kb	X	
Entamoeba coli	**M, Bd**, G, Kb	X	
E. chattoni	Kb	X	
E. hartmanni	Kb	X	
Endolimax nana	**M, Bd**	X	
Iodamoeba buetschlii	**M, Bd**	X	
Iodamoeba sp.	Kb	X	
Balantidium coli	G	X	
Blastocystis hominis	**Bd**	X	
Chilomastix mesnili	**Bd**	X	

Site: M, Mahale (Tanzania); G, Gombe (Tanzania); Bd, Budongo (Uganda); Kb, Kibale (Uganda); Wb, Wamba (Democratic Republic of Congo); Iy, Iyema-Lomako (DRC). References quoted for Budongo (Barrows, 1996; this study); Gombe (File *et al.*, 1976; McGrew *et al.*, 1989); Iyema-Lomako (Dupain *et al.*, 2002); Mahale (Huffman *et al.*, 1997; Kawabata & Nishida, 1991); Kibale (Ashford *et al.*, 2000); and Wamba (Hasegawa *et al.*, 1983). Not all studies made identification to species level.

S. fulleborni, and *T. trichiura*. At Budongo, Mahale, Gombe, and Kibale, the cestode (tapeworm) *B. studeri* is also present. The parasites *O. stephanostomum* and *B. studeri* are most important to our present analysis on leaf-swallowing behavior at Mahale and Budongo.

These four most common parasite species were analyzed for variation in prevalence (% chimpanzees infected in each period) between and within the Budongo and Mahale sites (Table 16.2). There was a statistically significant difference in prevalence of *O. stephanostomum* over the two periods at Budongo (Fisher's exact test, $P = 0.003$). Prevalence was higher in the 1996 period (79%) than the 1998–1999 period (40%). The collection period of 1996 consisted of all wet months, while that of 1998–1999 were half wet and half transient and dry months. At Mahale, no significant difference in *O. stephanostomum* prevalence between the two collection periods was noted (42%, 59%). The ratio of wet to transient and dry months were nearly equal (5: 3; 89–90; 6: 291–292). When the two collection periods for each site were combined, *O. stephanostomum* prevalence at Budongo and Mahale were nearly the same (50%, 58%), with no statistical difference in prevalence detected during the periods sampled (Table 16.2). The ratio of wet to transient and dry months when combined were nearly equal at both sites.

There is a marked difference in the prevalence of *B. studeri* between Budongo and Mahale, but statistical tests could not be performed because of the extremely low prevalence at Mahale. *Bertiella studeri* proglottids were found in only 0.5% (2/403 samples) of the dung visually inspected at Mahale between 1987 and 1996 (this occurred outside of current study and not included in Table 16.2). Similarly, Kawabata & Nishida (1991) report a low prevalence (0.7%) for the 7 months sampled between 1975 and 1976 in both the K and M groups.

Inter-site differences in the prevalence of *S. fulleborni* and *T. trichiura* were statistically significant (Fisher's exact test, $P = 0.0001$ for *S. fulleborni*; 0.002 for *T. trichiura* in Table 16.2). The prevalence of *S. fulleborni* was higher at Budongo than Mahale, while the reverse was true for *T. trichiura*. For these two species, prevalence did not change within the same site over different sampling periods, suggesting that within-site variation in rainfall patterns had little influence on prevalence, but that inter-site differences in the distribution of rainfall did. Neither species has been implicated in leaf swallowing or other forms of self-medication at the two sites.

Intensity

The intensity (eggs per gram feces, EPG) of *O. stephanostomum* infections in a sub-population of individuals was tracked over time at both Mahale and Budongo (Figure 16.2). Distinct inter-site differences were noted. At Budongo, the intensity of infection is generally low (0–100), with individual peaks during

Table 16.2. *Prevalence of infection with intestinal parasites common to the Mahale M Group (Tanzania) and Budongo Sonso group (Uganda) of chimpanzees (*Pan troglodytes schweinfurthi*)*

	Mahale		Budong		Mahale	Budong
Year of study	1989–1990	1991–1992	1996	1998–1999	Periods combined for each group	
Months studied	Aug.–Feb.	Aug.–Feb.	March–August	June–April		
Ratio of wet: transient and dry months	5:3	6:2	6:0	6:5	11:5	12:5
Methods of parasitological analysis	FSE, McM	FSE, McM, H-M	FSE, McM, H-M	FSE, McM, H-M		
No. samples (individuals) analyzed	$n = 161$ ($n = 48$)	$n = 156$ ($n = 49$)	$n = 68$ (24)	$n = 150$ ($n = 33$)	($n = 97$)	($n = 57$)
Reference	Huffman *et al.*, 1997	Huffman *et al.*, 1997	Barrows, 1996	This study	This study	This study
% Samples (individuals) found positive for:						
Oesophagostomum stephanostomum	18(42)	37(59) NS	82(79)	23(40) *	(50)	(58) NS
Strongyloides fulleborni	14(35)	13(31) NS	29(67)	62(81) NS	(33)	(75) ***
Trichuris trichiura	7(16)	24(27) NS	1(4)	0.6(3) NS	(22)	(4) *
Bertiella studeri	0(0)	0(0) nt	38(62)	9(24) **	(0)	(43) nt

Methods of analysis used in the studies were formalin ether sedimentation (FSE), McMasters quantitative flotation (McM), and Harada-Mori coproculture (H-M) for *O. stephanostomum* larvae I.D. Levels of statistical significance using Fisher's exact test, two tailed: * $0.002 \geq P \geq 0.003$; ** $P = 0.006$; *** $P = 0.0001$, NS = not significant, nt: not testable.

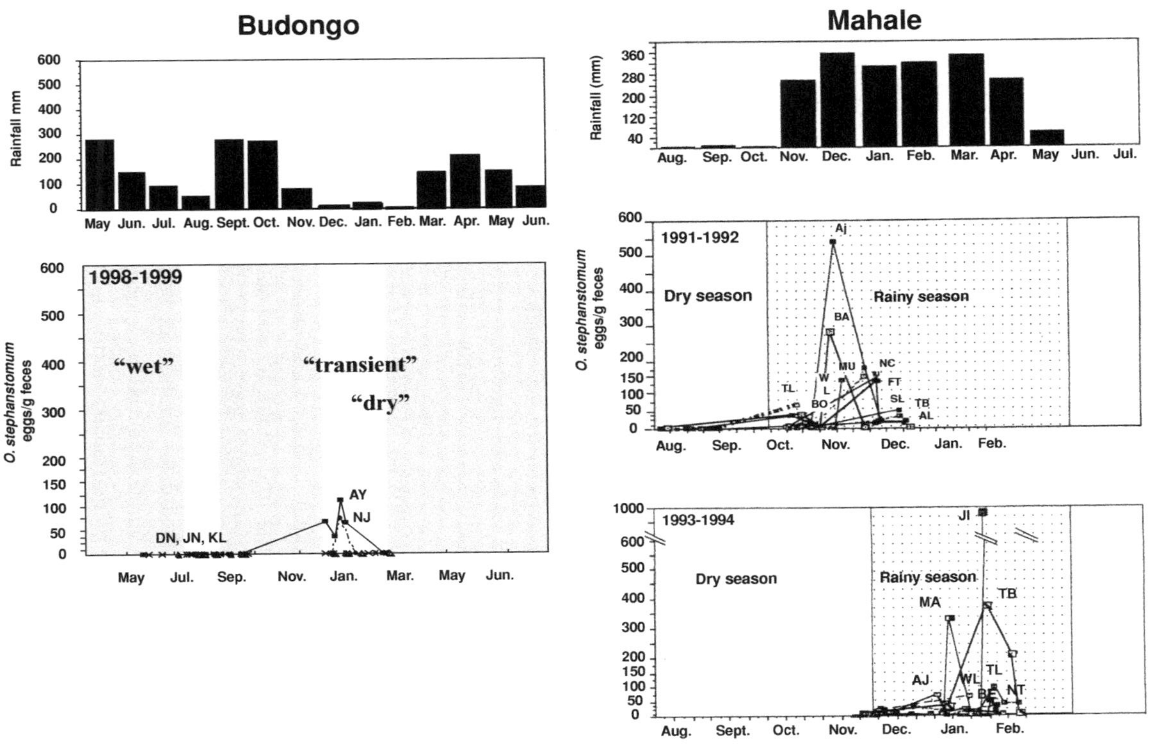

Figure 16.2. Comparison of monthly rainfall patterns and the intensity of *Oesophogostomum* infection in individual chimpanzees at Budongo and Mahale.

some, but not all dry and transient months of rainfall. This is in sharp contrast to the pattern previously noted for Mahale (Huffman *et al.*, 1997), in which sharp individual increases in intensity (50–550) occur 1–2 months after the onset of the rainy season, otherwise remaining low during the dry season (Figure 16.2).

Relationship between leaf swallowing and *Bertiella* proglottid expulsion

At Budongo, *B. studeri* proglottids were found in 8.6% (n = 535) of all dung collected and visually inspected during the study. The distribution of proglottid shedding was spread out into four major periods over the three years (Figure 16.1a). In 55% of the 36 months that dung samples were collected, proglottid shedding was observed. No significant correlation was found between monthly rainfall and the number of dung samples found to contain proglottids (Spearman rank correlation, $r_s = -0.0404$, $n = 36$, $P = 0.8152$). However, proglottids were shed more frequently than expected in transient and dry months ($\chi^2_{(1)} = 3.8$, $n = 36$, $P = 0.05$). Though not statistically significant, proglottids were found in fewer months of the wettest year, 1997, than in either 1996 or 1998 (Figure 16.1a). There was a trend for leaf swallowing of *Aneilema aquinoctiale* (Commelinaceae) to coincide significantly more frequently in months of proglottid shedding than expected (Figure 16.2a, $\chi^2_{(1)} = 4.4$, $n = 36$, $P = 0.03$).

In sharp contrast, the occurrence of *B. studeri* from microscopic investigation or from the presence of proglottids in the dung of Mahale chimpanzees is extremely rare. Proglottids and eggs from this species have only been found in 0.7% of 1769 dung samples collected and inspected macro- and or microscopically during different periods in the dry and rainy seasons of 1976–1977, 1983–1984, 1989–1990, 1991–1992, 1993–1994, 1995, 1996 (Uehara, 1987, personal communication; Kawabata & Nishida, 1991; Huffman *et al.*, 1997; M. A. Huffman, unpublished data). Both cases occurred outside of the current study period. In one sample collected in October 1995, a 2-cm portion of proglottid, an adult *O. stephanostomum* worm, and leaves of *A. aquinoctiale* were all found together in the same dung of an M group chimpanzee.

Discussion

The effect of rainfall on parasite infection patterns

The mode of transmission for *O. stephanostomum* and *B. studeri* are different, but both must undergo part of their life cycle outside of the host, making them equally vulnerable to climatic fluctuations. However, regional variation in climatic patterns affects prevalence and intensity of infection differently, and according to each species typical biological and ecological traits.

Bertiella studeri

Adult *B. studeri* range in size from 10 to 30 cm long by 1 cm wide. The head (scolex) of this species has no true hooks but instead attaches to the intestinal wall of the host with its four prominent shaped suckers. The adult causes no apparent lesions from attachment and little overt symptoms of ill health from infection (Brack, 1987). A known obligate intermediate host is the oribatid mite which picks up the small (38–45 micrometer) eggs transmitted from gravid proglottids periodically shed in the host's feces. Eggs can also be shed directly into the environment without proglottids. The eggs take less than 2 months to develop into cysticercoids that are then capable of infecting their definitive host (non-human primates and occasionally humans) upon incidental ingestion of the mite (Brack, 1987).

At Budongo, proglottids were being shed in more than half of the months surveyed, with a tendency for this to occur in drier months. In contrast, to the south in Kibale, Wrangham (1995) reported just one isolated and prolonged period of proglottid shedding that lasted 7 consecutive months during a 6.5-year period of monitoring the Kanyawara chimpanzee group. This period of proglottid shedding was spread out across wet, transient, and dry months. Proglottids were found in 5.2% (n = 271) of the dung collected and visually inspected (Wrangham, 1995). The difference in rate of occurrence between these two Ugandan sites was not significant, but shedding of proglottids appears to occur under more limited conditions and less periodically in Kibale than at Budongo. The possible reasons for this remain unknown, as the ecology of *Bertiella* remains largely unstudied. Further inter-site investigation into factors such as temperature, humidity, and intermediate host activity are needed.

Oesophagostomum stephanostomum

Oesophagostomum is a common genus found in a wide range of primate species (Brack, 1987). Three species have been reported in chimpanzees; *O. stephanostomum, O. bifurcum,* and *O. polydentatum* (Yamashita, 1963; Brack, 1987; Polderman & Blotkamp, 1995). Adult worms of *O. stephanostomum* are 2–3 cm long and readily visible when present. *Oesophagostomum* spp. eggs are shed in the feces and must undergo development outside the host before re-infection can take place via the ingestion of minute infective stage L3 larvae. Larval stages developing in the outside environment during the rainy season, when conditions are moist and humid, have a higher probability of surviving and infecting hosts than those shed in the dry season. This is supported by two years of data at Mahale showing high peaks in infection during the rainy season

(Figure 16.2). After successful infection, the larvae form cysts in the intestinal mucosa for a period when they mature into juvenile and adult forms prior to emergence back into the lumen. It has been suggested that the encysted L4 stage may be prolonged in highly seasonal environments to maximize reproductive potential by avoiding the shedding of eggs that would otherwise perish if expelled into the hot dry soil (Anderson, 1992; Polderman & Blotkamp, 1995). The low levels of infection detected in the dry season at Mahale suggest that a prolonged L4 stage period occurs here during this time (Huffman *et al.*, 1997).

Significant pathology can occur at the encysted L4 stage and *O. stephanostomum* is said to be the most hazardous nodular worm species in the great apes (Rousselot & Pelissier, 1952). Symptoms of moderate to heavy infection range from weight loss, enteritis, diarrhea, anemia, and lethargy to intense pain. Pathology includes hemorrhagic cysts containing larvae, bacterial invasion of lesions, blockage of the colon, and epigastric or periumbilical masses easily detected by physical examination. Gross submucosal or subserosal nodules (5–20 mm diameter) are associated with the larvae. These nodules contain caseous, necrotic centers in a fibrous capsule. Histological findings typically include intense inflammatory cell reaction accompanied by high levels of neutrophils and eosinophils associated with the nodules (McClure & Guilloud, 1971; Brack, 1987; Polderman & Blotkamp, 1995).

At Mahale the presence of adult worms of *O. stephanostomum* in the dung of chimpanzees is rare. Worms were found in 3.7% (n = 245) of the dung inspected over 9 months, and significantly so together with whole leaves (Huffman *et al.*, 1997). Worms were never observed in the dung of Budongo chimpanzees, despite the fact that this species is the most prevalent nematode infection at Budongo (Table 16.2).

Oesophagostomum stephanostomum larvae in the highly seasonal regions of Togo and Ghana, and Trychostrongyle nematodes in general, have the robust ability to undergo larval arrestment (anahydrobiosis) in the outside environment, surviving for long periods of drought until the next rains (Polderman & Blotkamp, 1995; Lettini & Sukhdeo, 2006). At Budongo, rainfall is distributed rather evenly throughout the majority of the year (Figure 16.1). This lack of a distinct and prolonged dry season may exclude the necessity of prolonged larval arrestment in the intestinal mucosa of hosts, and of anahydrobiosis outside of the host at Budongo. If true, a relatively high turnover of encysted larvae into the lumen of chimpanzees in Budongo may reduce the characteristic pathology caused by such cysts, and preclude the use of leaf swallowing to control infection by decreasing the number of adults in the lumen. Further research into the possible relationship between climatic factors and region-specific variation in larval arrestment and anahydrobiosis capabilities may shed light into the selective pressure of climate on *Oesophagostumum* biology and host pathology.

Seasonal infection patterns and the occurrence of leaf-swallowing behavior

At Budongo no significant association between the occurrence of leaf swallowing and monthly rainfall was found. This is in accordance with the absence of a significant correlation between proglottids shedding and rainfall. At Kibale too, no seasonal relationship between leaf swallowing and *Bertiella* proglottids expulsion was reported (Wrangham, 1995).

When adult *O. stephanostomum* are found in the dung at Mahale, their presence is significantly associated with leaf-swallowing behavior (Huffman *et al.*, 1996). The apparent absence of *O. stephanostomum* larval arrestment in Budongo, possibly due to the relatively equitable distribution of rain throughout the year and no obvious pathology in the host, might help explain why, unlike at Mahale, leaf swallowing is not associated with *O. stephanostomum* infections at either Budongo or Kibale.

At Mahale, the presence of adult *O. stephanostomum* worms in the dung is highly seasonal, occurring most noticeably during the rainy season. Here the onset of leaf swallowing is significantly correlated with marked seasonal increases in infection levels of *O. stephanostomum* (Huffman *et al.*, 1997).

Site-specific impact of parasite infection on self-medicative behavior

Chimpanzees of Budongo and Mahale were found to share many common parasite species. *Oesophagostomum stephanostomum* was the most prevalent parasite occurring at both sites. *Bertiella studeri* was also recognized to occur at both sites, but prevalence was markedly higher at Budongo than at Mahale. Leaf swallowing was associated with seasonal peaks in *O. stephanostomum* infection (increased prevalence, elevated intensity) in Mahale chimpanzees, but not at Budongo (Table 16.1; Figure 16.2). Seasonal rainfall patterns, larval arrestment, adult reproductive dormancy during the dry period, and an apparent surge of infection at the beginning of the rainy season in Mahale may explain why chimpanzees there purge adult worms from the lumen most frequently during the rainy season. In contrast, this was not the case at Budongo where infection appears to be constant and intensity comparatively low throughout the year. The pattern noted for Budongo may reflect a homeostasis in the host–parasite relationship brought about by favorable conditions for parasite life-cycle events (transmission, reproduction) and low pathological impact (low intensity infections) on the host. At Mahale, the seasonally induced surge in infection by the parasite and the resultant increased intensity of infection and associated increase in related symptoms may be responsible for the combined therapeutic practice of leaf swallowing and bitter pith chewing, another self-medicative behavior characteristic of this site (Huffman & Seifu, 1989; Huffman *et al.*, 1993; Huffman, 1997, 2007; Huffman & Caton, 2001).

Regional climatic differences, particularly in rainfall patterns between Tanzania and Uganda seem to be associated with regional differences in host–parasite associations and related self-medicative practices. A relationship between leaf swallowing and the shedding of *B. studeri* proglottids was recognized at Budongo, resembling the trend described for Kibale chimpanzees, a site exhibiting similar ecological and climatic characteristics. In spite of the similarly high levels of prevalence of *Oesophagostomum* at Budongo and Mahale, no evidence currently exists for a similar relationship between leaf swallowing and *Oesophagostomum* worm expulsion at Budongo.

It is without doubt that climate is a driving force behind the transmission biology of many gastrointestinal parasites. Seasonality can both directly and indirectly affect host–parasite interactions (Altizer *et al.*, 2006; Lass & Ebert, 2006; Stone *et al.*, 2007). While this study focuses on the effect of rainfall in host–parasite relations, we do not exclude the potential role of other factors and alternative or complementary explanations for the patterns we see at these two sites. For example, host population density, social organization, relative abundance of intermediate hosts in the habitat, differences in host food resource quality and abundance could possibly inhibit or enhance transmission (Anderson, 1992; Altizer *et al.*, 2006; Nunn & Altizer, 2006; Chapman *et al.*, Chapter 21, this volume). In the future such alternative hypotheses need to be investigated at these and other study sites across the geographic distribution of African great apes.

Predicting host–parasite relationships and the occurrence of leaf swallowing in African great apes

The patterns discussed above suggest that chimpanzees in different regions respond with similar anti-parasitic strategies to infections with different gastrointestinal parasite species, in this case a nematode and a cestode. Elucidating factors responsible for inter-regional differences affecting parasite transmission and pathology should further our understanding of the emergence of host self-medicative behavior. A pattern emerges in the relationship between the monthly distribution of rainfall, *Bertiella* and *Oesophagostomum* infection ecology and leaf-swallowing behavior that allows us to form predictions about what the state of host–parasite relationships may be at other sites across the distribution of the African great apes. One prediction is that, given the presence of both parasite species in the ape host, those hosts living in highly seasonal environments will be more likely to face increased levels of pathology from *Oesophagostomum* infections and respond with leaf swallowing, than hosts living in less strongly seasonal, moist environments. Conversely, ape hosts living in less seasonal, moist environments will be more prone to *Bertiella* infections than those hosts living in strongly seasonal environments. This prediction is partially supported by recent observations of leaf swallowing and its association with parasite infection at other sites. A correlation between the presence of whole leaves in the dung and oesophagostomiasis has been noted in one bonobo population and two chimpanzee populations in central and western Africa (Dupain *et al.*, 2002; Pruetz & Johnson-Fulton, 2003; Fowler *et al.*, 2007). These sites exhibit distinct dry and rainy seasons and leaf swallowing was most prevalent during the rainy season months.

There are a number of other chimpanzee and gorilla study sites across Africa where leaf swallowing has been noted to occur (Huffman, 1997, 2001), but from which detailed research into parasite ecology and ape self-medication has yet to be initiated. In particular, information is still scarce about the eastern and western lowland gorillas and chimpanzees of western and central Africa. Further research is expected to fill in gaps in our understanding of host–parasite ecology and to support or further expand current views of anti-parasite strategies of self-medication.

Recent advances in the use of such tools as GIS, remote sensing, and epidemiology have made it possible to predict the influence of climate, temperature, vegetation, water bodies, human settlement, and other factors on the severity of vector-borne diseases of consequence to humans and livestock (Arambulo & Astudillo, 1991; Hay *et al.*, 1998; Bavia *et al.*, 2001; Anyamba *et al.*, 2002; Beck *et al.*, 2002; Nunn & Altizer, 2006). In the same way, research in these areas can be expected to provide useful insight into regional variability of parasite infection patterns, the impacts on primate health, and their implications for the evolution of self-medicative behavior in primates.

Acknowledgements

MAH and PP give their sincere thanks to the J. William Fulbright Foreign Scholarship Board, the United States Information Agency for their support of PP and to the Leakey Foundation for their support of MAH during fieldwork on this project in Uganda. MAH's fieldwork in Tanzania was funded by grants under the Monbusho International Scientific Research Program to T. Nishida (No. 62041021, 63043017, 0304146) and from the Plant Sciences Research Foundation of the Faculty of Agriculture, Kyoto University. Sincere thanks go to Prof. Vernon Reynolds (founder of the Budongo Forest Project), and previous on-site directors of Budongo, Andy Plumptre, Jeramie Lindsell, Lucy Beresford-Stooke, Mark Atwater, and Fred Babweteera. Without their support this work would never have continued as long as it did. Our heartfelt gratitude goes to the people of Sonso and Mahale whose dedication and expert knowledge made it all work in the field, and to their families who made life there immensely enjoyable.

References

Altizer, S., Dobson, A., Hosseini, P. *et al.* (2006). Seasonality and the dynamics of infectious diseases. *Ecology Letters*, **9**, 467–484.

Anderson, R. C. (1992). *Nematode Parasites of Vertebrates. Their Development and Transmission*. Wallingford, UK: CABI Publishing.

Anyamba, A., Lithicum, K. J., Mahoney, R., Tucker, C. J. & Kelley, P. W. (2002). Mapping potential risk of Rift Valley fever outbreaks in African savannas using vegetation index time series data. *Photogrammetric Engineering and Remote Sensing*, **68**, 137–145.

Arambulo, P. V. & Astudillo, V. (1991). Perspectives on the application of remote sensing and geographic information system to disease control and health management. *Preventative Veterinary Medicine*, **11**, 345–352.

Ashford, R. W., Reid, G. D. F. & Wrangham, R. W. (2000). Intestinal parasites of the chimpanzee *Pan troglodytes* in Kibale Forest, Uganda. *Annals for Tropical Medicine and Parasitology*, **94**, 173–179.

Barrows, M. (1996). A survey of the intestinal parasites of the Primates in Budongo forest, Uganda. Masters thesis, Department of Medicine, University of Glasgow, Scotland.

Bavia, M. E., Malone, J. B., Hale, L. *et al.* (2001). Use of thermal and vegetation index data from earth observing satellites to evaluate the risk of schistosomiasis in Bahia, Brazil. *Acta Tropica*, **79**, 79–85.

Beck, L. R., Lobitz, B. M. & Wood, B. L. (2002). Remote sensing and human health: new sensors and new opportunities. *Emerging Infectious Diseases*, **6**, 217–226.

Brack, M. (1987). *Agents Transmissible from Simians to Man*. Berlin: Springer-Verlag.

Dupain, J., van Elsaker, L., Nell, C. *et al.* (2002). New evidence for leaf swallowing and *Oesophagostomum* infections in bonobos (*Pan paniscus). International Journal of Primatology*, **23**, 1053–1062.

Eggeling, W. J. (1947). Observations on the ecology of the Budongo Rain Forest, Uganda. *Journal of Ecology*, **34**, 20–87.

Fabiyi, J. B., Copeman, D. B. & Hutchinson, G. W. (1988). Abundance and survival of infective larvae of the cattle nematodes *Cooperia punctata, Haemonchus placei* and *Oesophagostomum radiatum* from fecal pats in a wet tropical climate. *Australian Veterinary Journal*, **65**, 229–231.

File, S. K., McGrew, W. C. & Tutin, E. C. (1976). The intestinal parasites of a community of feral chimpanzees, *Pan troglodytes schweinfurthii. Journal of Parasitology*, **62**, 259–261.

Fowler, A., Koutsioni, Y. & Sommer, V. (2007). Leaf-swallowing behavior in Nigerian chimpanzees: evidence for assumed self-medication. *Primates*, **48**, 73–76.

Hasegawa, H., Kano, T. & Mulavwa, M. (1983). A parasitological survey on the feces of pygmy chimpanzees, *Pan paniscus,* at Wamba, Zaire. *Primates*, **24**, 419–423.

Hay, S. I., Snow, R. W. & Rogers, D. J. (1998). From predicting mosquito habitat to malaria seasons using remotely sensed data: practice, problems and perspectives. *Parasitology Today*, **14**, 306–313.

Huffman, M. A. (1997). Current evidence for self-medication in primates: a multidisciplinary perspective. *Yearbook of Physical Anthropology*, **40,** 171–200.

Huffman, M. A. (2001). Self-medicative behavior in the African Great Apes: an evolutionary perspective into the origins of human traditional medicine. *BioScience*, **51**, 651–661.

Huffman, M. A. (2007). Primate self-medication. In *Primates in Perspective*, ed. C. J. Campbell, A. Fuentes, K. C. Mackinnon, M. Panger & S. K. Bearder. New York, NY: Oxford University Press, pp. 677–690.

Huffman, M. A. & Caton, J. M. (2001). Self-induced increase of gut motility and the control of parasitic infections in wild chimpanzees. *International Journal of Primatology*, **22**, 329–346.

Huffman, M. A. & Seifu, M. (1989). Observations on the illness and consumption of a possibly medicinal plant *Vernonia amygdalina* (DEL.), by a wild chimpanzee in the Mahale Mountains National Park, Tanzania. *Primates*, **30**, 51–63.

Huffman, M. A., Gotoh, S., Izutsu, D., Koshimizu, K. & Kalunde, M. S. (1993). Further observations on the use of *Vernonia amygdalina* by a wild chimpanzee, its possible effect of parasite load, and its phytochemistry. *African Study Monographs*, **14**, 227–240.

Huffman, M. A., Page, J. E., Sukhdeo, M. V. K. *et al.* (1996). Leaf-swallowing by chimpanzees, a behavioral adaptation for the control of strongyle nematode infections. *International Journal of Primatology*, **72**, 475–503.

Huffman, M. A., Gotoh, S., Turner, L. A., Hamai, M. & Yoshida, K. (1997). Seasonal trends in intestinal nematode infection and medicinal plant use among chimpanzees in the Mahale Mountains, Tanzania. *Primates*, **38**, 111–125.

Kawabata, M. & Nishida, T. (1991). A preliminary note on the intestinal parasites of wild chimpanzees in the Mahale Mountains, Tanzania. *Primates*, **32**, 275–278.

Lass, S. & Ebert, D. (2006). Apparent seasonality of parasite dynamics: analysis of cyclic prevalence patterns. *Proceedings of the Royal Society of London B*, **273**, 199–206.

Lettini, S. E. & Suhkdeo, M. V. K. (2006). Anhydrobiosis increases survival of trichostrongyle nematodes. *Journal of Parasitology*, **92**, 1002–1009.

Marquardt, W. C., Demaree, R. S. & Grieve, R. B. (2000). *Parasitology and Vector Biology*, 2nd edn. San Diego, CA: Harcourt Academic Press.

McClure, M. & Guilloud, N. B. (1971). Comparative pathology of the chimpanzee. In *The Chimpanzee*, Vol. 4, ed. G. H. Bourne. Basel: Karger. pp. 103–272.

McGrew, W. C., Tutin, C. E. G., Collins, D. A. & File, S. K. (1989). Intestinal parasites of sympatric *Pan troglodytes* and *Papio* spp. at two sites: Gombe (Tanzania) and Mt. Assirik (Senegal). *American Journal of Primatology*, **17**, 147–155.

Newton-Fisher, N. E. (1999). The diet of chimpanzees in the Budongo Forest Reserve, Uganda. *African Journal of Ecology*, **37**, 344–354.

Nishida, T. (1990). A quarter century of research in the Mahale Mountains. In *The Chimpanzees of the Mahale Mountains: Sexual and Life History Strategies*, ed. T. Nishida. Tokyo: University of Tokyo Press, pp. 237–255.

Nunn, C. L. & Altizer, S. (2006). *Infectious Diseases in Primates – Behavior, Ecology and Evolution*. Oxford: Oxford University Press.

Plumptre, A. J. (1996). Changes following 60 years of selective timber harvesting in the Budongo Forest Reserve, Uganda. *Forest Ecology and Management*, **89**, 101–113.

Polderman, A. M. & Blotkamp, J. (1995). *Oesophagostomum* infections in humans. *Parasitology Today*, **11**, 451–456.

Pruetz, J. D. & Johnson-Fulton, S. (2003). Evidence for leaf swallowing behavior by savanna chimpanzees in Senegal – a new site record. *Pan Africa News*, **10**, 14–16.

Reynolds, V. (1992). Chimpanzees in the Budongo Forest, 1962–92. *Journal of Zoology*, **228**, 695–699.

Rousselot, R. & Pellissier, A. (1952). Pathologie du gorille. III. Oesophagostomose nodulaire a *Oesophagostomum stephanostomum* du gorille et du chimpanze. *Bulletin de la Société de Pathologie Exotique et de ses Filiales*, **45**, 568–574.

Stone, L., Olinky, R. & Huppert, A. (2007). Seasonal dynamics of recurrent epidemics. *Nature*, **446**, 533–536.

Struhsaker, T. (1997). *Ecology of an African Rain Forest: Logging in Kibale and the Conflict between Conservation and Exploitation*. Gainesville, FL: University Press of Florida.

Synnott, T. J. (1985). *A Checklist of the Flora of the Budongo Forest Reserve, Uganda with Notes on Ecology and Phenology*. CFI Occasional Paper. Oxford: Forestry Institute, Oxford Press.

Takasaki, H., Nishida, T., Uehara, S. *et al.* (1990). Appendix. Summary of meteorological data at Mahale Research Camps, 1973–1988. In *The Chimpanzees of the Mahale Mountains: Sexual and Life History Strategies,* ed. T. Nishida. Tokyo: University of Tokyo Press, pp. 291–300.

Taylor, E. L. & Michel, J. F. (1953). The parasitological and pathological significance of arrested development in nematodes. *Journal of Helminthology*, **27**(3/4), 199–205.

Uehara, S. (1987). Sex and group differences in feeding on animals by wild chimpanzees in the Mahale Mountains National Park, Tanzania. *Primates*, **27**, 1–13.

Wrangham, R. W. (1995). Relationship of chimpanzee leaf swallowing to a tapeworm infection. *American Journal of Primatology*, **37**, 297–303.

Wrangham, R. W. & Goodall, J. (1989). Chimpanzee use of medicinal leaves. In *Understanding Chimpanzees*, ed. P. G. Heltne & L. A. Marquardt. Cambridge, MA: Harvard University Press, pp. 22–37.

Yamashita, J. (1963). Ecological relationships between parasites and primates. *Primates*, **4**, 1–96.

Part III

The ecology of primate–parasite interactions

17 *Primate exposure and the emergence of novel retroviruses*

NATHAN D. WOLFE AND WILLIAM M. SWITZER

Photograph by Nathan D. Wolfe (*Cercopithecus cephus*)

Primate Parasite Ecology. The Dynamics and Study of Host–Parasite Relationships, ed. Michael A. Huffman and Colin A. Chapman. Published by Cambridge University Press.

Introduction

Closely related species generally share susceptibility to the same groups of microorganisms (Wolfe *et al.*, 2000). Primates, including humans, share broadly similar physiologic and genetic characteristics and thus are susceptible to many of the same viruses, bacteria, fungi, protozoa, helminths, and ectoparasites (Ruch, 1959; Brack, 1987, but see Gasser *et al.*, Chapter 3, this volume). Behaviors that involve direct contact between different species of primates can allow for the transmission of microorganisms across species boundaries. A variety of behaviors involving direct contact can facilitate the transmission of microorganisms from non-human primates (NHPs) to humans (Wolfe *et al.*, 1998), with consequences for human health; from humans to NHPs, with consequences for wildlife conservation (Wallis & Lee, 1999); and between distinct species of NHPs.

Among humans, a range of ancient and contemporary behaviors facilitate such cross-species transmission. Among contemporary behaviors, the care of captive NHPs has led to the transmission of a range of infections, including simian foamy virus (SFV) (Heneine *et al.*, 1998), herpesvirus B (HBV) (Huff & Barry, 2003), primate malaria (Coatney *et al.*, 1997), and tuberculosis (Kalter *et al.*, 1978) to humans. Non-human primates ecotourism (e.g. gorilla watching) has been associated with the possible transmission from humans to NHPs of diseases that include scabies (*Sarcoptes scabiei*) (Graczyk *et al.*, 2001), intestinal parasites (Sleeman *et al.*, 2000), and measles (Butynski & Kalina, 1998). Laboratory handling of tissues or fluids of NHPs has led to the transmission of a range of infections to humans, including simian immunodeficiency virus (SIV) (Khabbaz *et al.*, 1994) and SV40, which moved from macaque cell cultures to vaccine lots and was subsequently distributed through oral polio vaccine to millions of people (Shah, 1989). Additionally, keeping NHP pets has been linked to transmission of a variety of microorganisms (Renquist & Whitney, 1987).

Nevertheless, while contemporary activities have changed the ways in which humans are exposed to NHPs, they are by no means unique in permitting exposure to sources of infection from NHPs. Because of the broad range of body fluid and tissue types to which humans may become exposed during hunting and butchering, and because such behaviors are very widespread, they appear to be particularly important for cross-species transmission (Wolfe *et al.*, 2000) (although other less common activities, such as wildlife necropsy, may have similar per incident risks; Le Guenno *et al.*, 1995). Hunting of primates is an ancient practice that we share with our closest relative, the chimpanzee (*Pan troglodytes*). While little is known about hunting by bonobos (*Pan paniscus*), it appears likely that hunting by humans dates to the time of the common ancestor with chimpanzees and bonobos (5–7 MYA) (Stanford, 1996, 1999). Hunting of NHPs, like exposure to the body fluids of NHPs in laboratory and zoo settings, provides an important portal for the entry of novel agents into human populations.

In this chapter we will review aspects of the biology of the primate retroviruses and discuss in detail how interspecies primate interaction, namely human contact with NHPs, leads to the ongoing cross-species transmission of retroviruses.

The emergence of pandemic human retroviruses

Retroviruses are a large and diverse group of enveloped RNA viruses in the family Retroviridae that replicate in a unique way, using a viral reverse transcriptase (RT) enzyme to transcribe the RNA genome into linear double-stranded DNA. Retroviruses can be either exogenous in nature, replicating independent of the host genome, or they can exist endogenously as proviral DNA integrated into the germ line of the host. As such, retroviruses can be transmitted as either infectious virions (exogenously) or as endogenous proviral DNA passed from parent to offspring.

Taxonomically, retroviruses are divided into two sub-families: the Orthoretrovirinae, composed of six genera (Alpha-, Beta-, Gamma-, Delta-, and Epsilonretroviruses and the Lentiviruses) and the Spumaretrovirinae, composed of only the Spumavirus (foamy virus) genus (Linial *et al.*, 2004). Exogenous retroviruses of simian origin and potential public health significance are found in both retrovirus sub-families, including the type D simian retrovirus (SRV, Betaretrovirus), simian and human T-lymphotropic viruses (STLV and HTLV, respectively; Deltaretroviruses); simian and human immunodeficiency viruses (SIV and HIV, respectively; Lentiviruses); and simian foamy virus (SFV, Spumavirus) (Linial *et al.*, 2004). Retroviruses typically cause lifelong, persistent infections with long periods of clinical latency prior to disease development.

The emergence of new viruses into humans can be conceptualized as a series of steps leading from exposure to global spread (Figure 17.1). Exposure to the viruses of other animals is a frequent event, with the majority of such exposure events not leading to successful infection. For viruses capable of causing zoonotic infections (primary infections resulting from direct animal exposure), only some will be capable of secondary spread. Even viruses capable of causing both zoonotic infections as well as some secondary cases will often be of limited public health significance, due to the fact that they are poorly adapted to the new host (Antia *et al.*, 2003). Emergence can only occur following successful

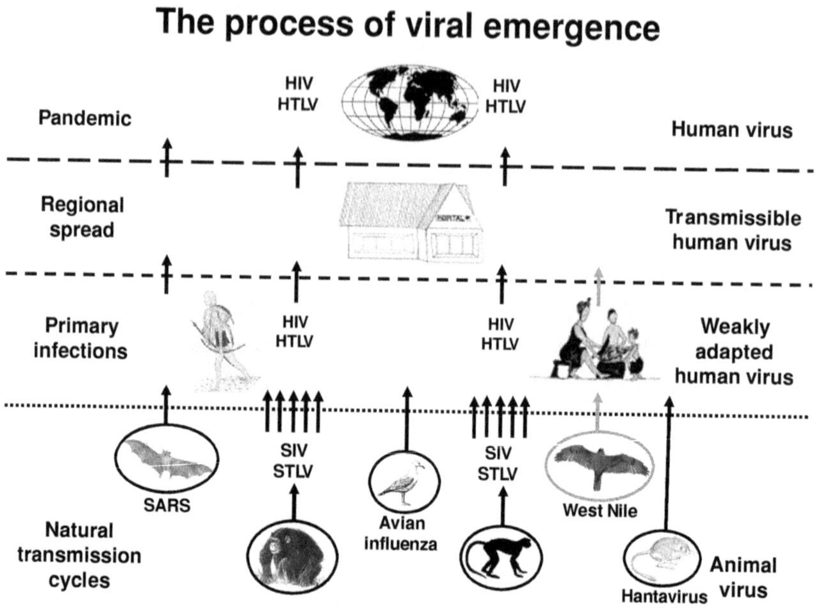

Figure 17.1. A diagrammatic representation of the process of viral emergence. Step 1, primary infection or successful cross-species transmission from the wild animal reservoir to humans; 2, local secondary transmission between humans to regional spread; step 3, pandemic spread from a regional epidemic into the global population. Examples of various zoonotic infections, including simian retroviruses, are shown with arrows indicating the level of transmission at each step (represented by horizontal dashed lines).

cross-species transmission of a virus, defined here as sustained transmission in the new host. Identifying the frequency of virus transmission at each of these steps, as well as the factors which permit movement from each step is the ultimate goal of the study of disease emergence, and holds the potential to improve our ability to predict disease emergence.

It has been clear for some time that the emergence of the pandemic human retroviruses, HIV and HTLV, resulted following the introduction of multiple independent viruses from NHPs to humans (Figure 17.1). Phylogenetic studies of HIV-1 and HIV-2, alongside their counterpart SIVs, have shown that HIV is the result of as many as eight independent introductions from African monkeys and apes although only two of these zoonotic events resulted in pandemics (Hahn *et al.*, 2000; Apetrei *et al.*, 2004). Similarly, HTLV-1 and HTLV-2, the two known variants of pandemic human HTLV, originated independently and are related to distinct lineages of STLV-1 and STLV-2, respectively, with HTLV-1 having resulted from multiple independent introductions of STLV-1 into human populations (Vandamme *et al.*, 1988; Mahieux *et al.*, 1998; Salemi *et al.*, 1999, 2000; Slattery *et al.*, 1999; Gessain & Mahieux, 2000; Salemi *et al.*, 2000). Clearly then, NHP retroviruses have the potential for successful cross-species transmission. Nevertheless, the factors that allow retroviruses present in NHP reservoirs to successfully cross species boundaries and emerge in human populations remain largely unknown and unstudied. The approach advocated in this chapter focuses on the study of persons who are highly exposed to the blood and body fluids of primates, either through contact in laboratories and primate housing facilities or through the hunting and butchering of wild NHP game, with detailed follow-up of individuals with zoonotic infections and their contacts for evidence of secondary transmission (Sotir *et al.*, 1997; Wolfe *et al.*, 2004a, b). Such studies have the potential to elucidate factors that allow for retrovirus zoonosis, secondary spread, and successful cross-species transmission. This, in turn, has the potential to clarify not only how the existing pandemic retroviruses emerged, but also, and perhaps more importantly, how to prevent the emergence of future retroviruses.

Human infections with simian foamy viruses

Spumaviruses are also known as foamy viruses because they display an unusual cell culture cytopathology of increased intracellular vacuolization giving the infected cells a "foamy appearance" by electron microscopic examination. Foamy viruses have been isolated from many different species of mammals (Meiering & Linial, 2001). Foamy viruses from NHPs are referred to as simian foamy viruses (SFVs) (Meiering & Linial, 2001) and tend to be widespread across species and infect many Old and New World monkeys, apes, and prosimians at high prevalence (Meiering & Linial, 2001; Hussain *et al.*, 2003). In captivity, more than 70% of adult NHPs are infected with SFV (Meiering & Linial, 2001; Hussain *et al.*, 2003). Less is known about the prevalence of SFV in wild-living primates, but rates as high as 62% have been observed in some species (Hussain *et al.*, 2003; Calattini *et al.*, 2004b; Liu *et al.*, 2005). The wide distribution of SFV among a variety of NHPs has been shown recently to be the result of cospeciation of SFV with the primate host, suggesting an ancient infection (Switzer *et al.*, 2005).

Simian foamy virus has a broad host range and can infect many types of cells from a variety of animal species in vitro, including humans, resulting in cytopathology and cell death. Persistent infection of cell lines with SFV has also been reported (Meiering & Linial, 2001). Although SFV infection was reported in one orangutan with encephalopathy there have been no clinical diseases associated with foamy virus infection in other species of NHPs to date (McClure *et al.*, 1994). The pathogenicity of SFV in many species appears unclear and no direct association between infection and disease has been proven (Meiering & Linial, 2001). The persistent and non-pathogenic nature of SFV infection may be related to the ancient cospeciation of this virus with NHPs (Switzer *et al.*, 2005). Although cross-species transfer of SFV has been reported between NHP species, it is unclear if these infections will lead to disease in the new host, as occurs with SIV and STLV (Gessain & Mahieux, 2000; Hahn *et al.*, 2000; Araujo & Hall, 2004; Switzer *et al.*, 2005).

Latent SFV provirus DNA has been found in most cells and tissues of persistently infected animals with infectious isolates obtained mainly from the oral mucosa and blood (Falcone *et al.*, 1999; Meiering & Linial, 2001; Hussain *et al.*, 2003; Murray *et al.*, 2006). Contact with these two body fluids has been implicated in horizontal transmission of SFV, such as occurs with biting and licking, although sexual transmission is also suspected to occur (Broussard *et al.*, 1997; Blewett *et al.*, 2000; Meiering & Linial, 2001). More recently, viral RNA was found in the feces of 75% of wild-living chimpanzees suggesting that contact with feces, especially mucocutaneously, may also increase the risk of SFV infection (Liu *et al.*, 2008). Evidence of vertical transmission has been reported in a mother and offspring chimpanzee pair, although additional data are needed to confirm this route of transmission (Switzer *et al.*, 2004). Newborn and infant primates test negative upon losing passively acquired maternal antibodies, but can get infected as juveniles, presumably by contact with infected adults (Broussard *et al.*, 1997; Blewett *et al.*, 2000; Meiering & Linial, 2001).

Early studies described a relatively high rate of seroreactivity to foamy viruses among human populations, but these studies lacked definitive evidence of human infection and were not subsequently confirmed (Weiss, 1988; Schweizer *et al.*, 1995; Ali *et al.*, 1996; Meiering & Linial, 2001; Heneine *et al.*, 2003). Improved diagnostic assays have not documented evidence of foamy virus infection in large numbers of persons in the general population (Schweizer *et al.*, 1995; Ali *et al.*, 1996). In contrast, screening of primate handlers and researchers exposed to NHPs and NHP origin retroviruses using validated molecular and serologic tests revealed a substantial prevalence of SFV in this population and thus pointed out the high risk of cross-species transmission of SFV to humans exposed to primates (Heneine *et al.*, 2003; Switzer *et al.*, 2004). Studies at the Centers for Disease Control and Prevention (CDC) which provides voluntary testing for simian retroviruses for persons working at zoos and primate centers has identified to date 14 of 418 (3.35%) workers tested to be infected with SFV (Switzer *et al.*, 2004). The infected persons were both men and women working at zoos and research institutions with different occupations including veterinarians, animal handlers, and scientists, and were infected with a range of divergent SFVs (Switzer *et al.*, 2004).

The SFV-infected workers generally report working with the primate species as the source of their SFV and in many cases recall receiving injuries from these primate species such as bites and scratches (Heneine *et al.*, 2003; Switzer *et al.*, 2004). Some workers, however, did not report any specific injuries, and therefore it is unclear if transmission of SFV to humans from exposure to NHP bodily fluids may occur via other routes (Switzer *et al.*, 2004).

The high seroprevalence of SFV infection documented in workers exposed occupationally to NHPs in zoos and primate centers has raised questions about whether SFV infects human populations in natural settings. Recently, a study of 1099 bushmeat hunters in Cameroon who reported a history of hunting, butchering, and/or keeping primates as pets documented a 1% SFV seroprevalence. Simian foamy virus infection in these studies was determined by phylogenetic analysis to have originated from mandrills (*Mandrillus sphinx*), De Brazzas's monkeys (*Cercopithecus neglectus*), and gorillas (*Gorilla gorilla*), which are all NHP species found in Cameroon and commonly hunted in this region (Wolfe *et al.*, 2004a, b). Additional recent evidence of SFV infection in primate hunters documented gorilla- and chimpanzee-type SFV in two Bantous and one Baka pygmy from southern Cameroon (Calattini *et al.*, 2004a). The SFV gorilla-infected persons in this study reported having received significant bite wounds from gorillas (Calattini *et al.*, 2004a). Another study of 82 workers exposed to Asian macaques around religious temples in Indonesia reported an SFV infection of a cynomolgus macaque origin consistent with the prevalence of this NHP species in the area (Jones-Engel *et al.*, 2005). Overall these results show that SFV is actively crossing into humans, and they document human susceptibility to infection with at least seven different SFV variants. They also demonstrate the wide geographic distribution of cross-species infection among exposed humans in North America, Europe, Central Africa, and Asia. Although NHPs are also found in South America and SFV infection of these New World primates has been demonstrated (Meiering & Linial, 2001), there is currently no information available to determine if zoonotic SFV infection also occurs in this region.

While several studies have now documented primary SFV infection in persons exposed to NHPs (Brooks *et al.*, 2002; Heneine *et al.*, 2003; Calattini *et al.*, 2004a; Switzer *et al.*, 2004; Wolfe *et al.*, 2004b), less is known of the ability of this virus to transmit secondarily among humans and cause disease. Data available from the CDC study of primate workers showed that spouses of six SFV-infected men remained uninfected after 9–19 years suggesting that this virus may not be easily transmitted sexually from males to females (Switzer *et al.*, 2004). However, more data are needed to fully assess transmission of SFV. Specimens were not available from spouses and close contacts of SFV-infected women identified in the CDC study to determine if transmission could occur from female to male or from mother to child (Switzer *et al.*, 2004).

The consistent finding of SFV in the peripheral blood mononuclear cells (PBMCs) of persistently infected persons raises questions about the possibility of spread of these viruses following exposure via blood donations from infected persons. It is noteworthy that 11 cases in the CDC study reported being blood donors and six of these persons were confirmed to be SFV-infected at the time of their donation by testing of archived sera (Switzer *et al.*, 2004). A retrospective study of recipients from a blood donor infected with chimpanzee-like SFV failed to identify evidence of SFV infection in two recipients of red cells, one recipient of filtered red cells, and one recipient of platelets (Boneva *et al.*, 2002). However, these blood products were all leuokocyte-reduced which may help explain the absence of transmission seen in this study since leukocytes are reservoirs for SFV in the blood. Thus, more data are needed to better define the risks for SFV transmission through donated blood.

Although SFV is non-pathogenic in naturally infected NHPs (Meiering & Linial, 2001), the significance of SFV infection in humans is poorly known. The introduction of SFV infections into humans is of concern because changes in the pathogenicity of simian retroviruses following cross-species infection are well documented, as evidenced by the fact that both HIV-1 and HIV-2 evolved from SIVs that benignly infect their natural primate hosts (Hahn *et al.*, 2000; Apetrei *et al.*, 2004). Published findings from different studies of SFV-infected humans suggest that these are asymptomatic infections which are consistent with natural SFV infection of NHP (Heneine *et al.*, 2003; Switzer *et al.*, 2004). However, the limited number of cases, short duration of follow-up, and, more importantly, selection biases in the enrollment of healthy workers to identify cases all limit the ability to identify potential disease associations (Heneine *et al.*, 2003; Switzer *et al.*, 2004). Incidence of disease in SFV-infected persons may be low, may follow long latency periods, or may be associated with specific SFV clades that have not yet been identified. Additional studies such as long-term follow-up of SFV-infected humans are needed to better assess clinical outcomes of SFV infection and to define the public health implications of these infections.

Identification of HTLV-3 and HTLV-4

The transmission of SFV to humans exposed to NHPs raises questions about whether other simian retroviruses are also being transmitted to persons with primate exposures. We have recently examined the diversity of HTLV among primate bushmeat hunters in Cameroon who had documented SFV infections and who thus may be at risk for infection with additional simian retroviruses. We then looked for evidence of HTLVs of possible simian origin in these primate hunters. HTLV-1 and HTLV-2 endemically infect humans (Yamashita *et al.*, 1996; Gessain & Mahieux, 2000; Araujo & Hall, 2004). HTLVs have spread globally to at least 22 million persons sexually, from mother to child, and by exposure to contaminated blood through transfusions and injection drug use (Yamashita *et al.*, 1996; Gessain & Mahieux, 2000; Araujo & Hall, 2004). HTLV-1 causes adult T-cell leukemia and HTLV-1–associated myelopathy/tropical spastic paraperesis (HAM/TSP) and other inflammatory diseases in about 2–5% of those infected (Yamashita *et al.*, 1996; Gessain & Mahieux, 2000). HTLV-2 is less pathogenic than HTLV-1 and has been associated with a neurologic disease similar to HAM/TSP (Araujo & Hall, 2004). HTLVs are antigenically and genetically closely related to STLV, which are composed of three major groups, STLV-1, -2, and -3 (Fultz, 1994; Goubau *et al.*, 1994; Digilio *et al.*, 1997; Gessain & Mahieux, 2000). STLV has been found in more than 33 species of African and Asian Old World primates, both in captivity and the wild (Fultz, 1994; Digilio *et al.*, 1997; Mahieux *et al.*, 1998; Vandamme *et al.*, 1998; Meertens *et al.*, 2002, 2003; Takemura *et al.*, 2002; Meertens & Gessain, 2003; Courgnaud *et al.*, 2004; Leendertz *et al.*, 2004; Van Dooren *et al.*, 2004). STLV-1 has been implicated in the development of T-cell lymphomas and leukemia in NHPs (Hubbard *et al.*, 1993; Voevodin *et al.*, 1996). Previous studies showing close phylogenetic relationships between HTLV-1 and STLV-1 in the same geographic region have suggested that HTLV-1 originated from multiple cross-species infections of STLV-1 (Mahieux *et al.*, 1998; Salemi *et al.*, 1998, 1999, 2000; Vandamme *et al.*, 1998; Gessain & Mahieux, 2000).

Our recently reported study of HTLV diversity among primate hunters from Cameroon led to the identification of two new HTLVs and thus demonstrated that HTLV diversity is far greater than previously thought (Wolfe *et al.*, 2005). In this study, 930 samples were screened for HTLV by serologic testing and PCR amplification. Two new viruses that are distinct from HTLV-1 and HTLV-2 were discovered (Wolfe *et al.*, 2005). The first HTLV clustered phylogenetically within the diversity of STLV-3 and represents the first human member in this group. This virus was named HTLV-3 because of its genetic similarity to STLV-3 (Wolfe *et al.*, 2005). The second unique HTLV discovered in this population was distinct from HTLV-1, -2, and -3 and all STLVs and appears to have evolved independently over a long period of time (Wolfe *et al.*, 2005). This virus was designated HTLV-4. While viruses similar to HTLV-4 are likely to be present in NHPs, more work is needed with improved diagnostic assays to identify the simian counterpart of this new virus. Recently, a second HTLV-3 infection was identified in a Bakola pygmy from southern Cameroon (Calattini *et al.*,

2005). This person also reported frequent exposure to NHPs and is infected with an HTLV-3 most similar to an STLV-3 found in a red-capped mangabey from Cameroon (Meertens *et al.*, 2002; Calattini *et al.*, 2005). Combined with a history of STLVs crossing from primates to humans and a wide geographic distribution of STLV-3 across Africa (Goubau *et al.*, 1994; Salemi *et al.*, 1999; Takemura *et al.*, 2002; Courgnaud *et al.*, 2004; Van Dooren *et al.*, 2004), these results suggest that HTLV-3 infection may be more widespread than suspected.

Consistent with previous results, many STLV-1-like infections were also identified in our study confirming the active and frequent transmission and spread of STLV-1 viruses among humans (Wolfe *et al.*, 2005). The viruses found in these individuals included strains genetically similar to STLV-1 from mandrills (*Mandrillus sphinx*), gorillas (*Gorilla gorilla gorilla*), chimpanzee (*Pan troglodytes*), colobus (*Piliocolobus badius*), and crested mona monkeys (*Cercopithecus pogonias*) (Wolfe *et al.*, 2005).

Health exams of participants and collection of information regarding person to person contact were not obtained in either study that identified the new HTLV-3 and HTLV-4 viruses, thus an assessment of either disease associations or secondary transmission with these novel HTLV infections was not possible (Calattini *et al.*, 2005; Wolfe *et al.*, 2005). Therefore, clinical evaluations and longitudinal epidemiologic studies are needed for persons infected with HTLV-3, HTLV-4, and STLV-1-like viruses to determine if these viruses cause disease and/or are transmissible among humans, as occurs with the other HTLVs (Yamashita *et al.*, 1996; Gessain & Mahieux, 2000; Araujo & Hall, 2004). Although STLV-3 has been reported to transform or immortalize human CD4+ lymphocytes in vitro (Goubau *et al.*, 1994), further studies are needed to determine both the cellular tropism of HTLV-3 and HTLV-4 and the ability of these viruses to transform lymphocytes.

HTLV is screened for in blood banks in the USA and Europe but is not typically screened for in Africa, thus further spread of these viruses among West and Central Africans may be facilitated by blood donations from infected persons. The finding that HTLV-4 and HTLV-3 are serologically indistinguishable from HTLV-1 and HTLV-2 by current assays may explain why these viruses were not previously identified, highlighting the need for improved diagnostic assays to detect these viruses reliably and accurately (Wolfe *et al.*, 2005). Although plasma samples from the HTLV-3- and HTLV-4-infected persons were detected by an ELISA containing purified HTLV-1 and HTLV-2 viral lysates (Wolfe *et al.*, 2005) or by an HTLV-1 immunofluorescent assay (Calattini *et al.*, 2005), the overall sensitivity of these screening methods for detecting these new viruses is not known. Similarly, the sensitivity of detecting HTLV-3 and HTLV-4 by using screening assays employing recombinant proteins or synthetic peptides is also not known.

Other factors influencing retrovirus emergence

A number of factors have the potential to influence retrovirus zoonosis (i.e. the entry of retroviruses into persons exposed to blood and body fluids of NHPs), secondary spread, and subsequent population level emergence. Exposure to the blood and body fluids of NHPs is required for zoonosis. Such exposure can be readily studied both in captive and wild animal settings (Sotir *et al.*, 1997; Wolfe *et al.*, 2004). Exposure to the blood and body fluids of NHPs also represents a possible point of intervention in attempting to limit retrovirus zoonosis. Nevertheless, exposure alone is insufficient for retrovirus emergence. Of the persons who are exposed to the blood and body fluids of NHPs, only a small percentage are infected with zoonotic retroviruses (Brooks *et al.*, 2002, Heneine *et al.*, 2003; Switzer *et al.*, 2004; Wolfe *et al.* 2004a, 2005; Jones-Engel *et al.*, 2005). In addition, some individuals who show clear evidence of exposure – for example through a positive Western blot result – have negative viral PCR results (Calattini *et al.*, 2004a; Switzer *et al.*, 2004; Wolfe *et al.*, 2004a, b). Such discordant results may be due to many factors, including PCR diagnostics which are either too specific or not sufficiently sensitive.

Host genetic factors may also play a role in determining whether exposed individuals go on to have active zoonotic infections. During the past few years a number of cellular viral defense mechanisms have been documented, some of which appear to inhibit viral zoonosis. One gene, the APOBEC host gene, causes lentivirus hypermutation and can thereby limit viral infection (Harris & Liddament, 2004). The lentivirus Vif protein deactivates APOBEC, allowing infection, but only in a species-specific manner. For example, HIV Vif deactivates human APOBEC (i.e. thereby allowing infection in humans) but HIV Vif does not deactivate the APOBEC of other primate species, effectively limiting the potential for successful cross-species infections in novel hosts (Harris & Liddament, 2004; Song *et al.*, 2005). One possible explanation for the occurrence of low levels of zoonosis in particular retrovirus taxa could be the presence of polymorphisms in APOBEC and other host restriction genes such as Trim5alpha which may make particular individuals in a novel host species more susceptible to cross-species infection (Harris & Liddament, 2004; Song *et al.*, 2005). Such a pattern might be expected to contribute to sporadic low prevalence zoonotic infections that may not be likely to cause secondary spread. Recently, SFVs have also been shown to be inhibited by APOBEC cytidine deaminases which may help explain both the low level of viral replication in persistently infected NHPs and humans and the apparent low frequency of secondary spread from zoonotically infected humans (Russell *et al.*, 2005; Delebecque *et al.*, 2006).

Virus evolution and adaptation are also likely to play important roles in determining whether primary zoonotic infections have the potential to cause secondary infections. Mutation rates and propensity for recombination, for example, are mechanisms which may allow viruses to generate genetic diversity that provides the fuel for adaptation to novel host species (Moya *et al.*, 2004) and that may increase the probability that a zoonotic infection has the potential for secondary spread. In addition to playing a direct role in retrovirus emergence, studying patterns of host–virus evolution may also provide clues about the potential that viruses have for successful cross-species transmission (Moya *et al.*, 2004). Successful cross-species transmission will also depend on other factors such as viral load and route of exposure, which are known to be important for transmission of other retroviruses like HIV and HTLV.

Conclusion

In this chapter we have discussed the implications of exposure to NHP blood and body fluids on the entry and establishment of novel retroviruses into humans. The work demonstrates that a much broader range of retroviruses than previously thought is capable of causing primary zoonotic infections among humans, and that such zoonotic transmission may be occurring on a regular basis. Three new retroviruses previously undocumented in humans including SFVs, HTLV-3, and HTLV-4 have all been identified in persons exposed to the blood and body fluids of NHPs. But the results also suggest that there are many natural hurdles that prevent primary human infections from spreading and becoming permanently established in a new species. The regular occurrence of primary primate retrovirus infections suggests that zoonosis, per se, may not be the rate limiting step in pandemic retrovirus emergence, and that other factors such as viral adaptation and evasion of cellular level host viral defense probably play an important role in successful cross-species transmission and pandemic human retrovirus emergence.

Host behavior can play an important role in facilitating cross-species transmission of microorganisms. For example, the majority of primates are not carnivorous, but hominoids evolved the ability to hunt (with humans, chimpanzees, and bonobos likely sharing a common ancestor who hunted). The evolution of hunting among hominoids is likely to have played an important role in the emergence of retroviruses. For example, chimpanzee SIV (SIVcpz) resulted from the recombination of SIV from two monkey species (red-capped mangabeys and greater spot nosed guenons) (Bailes *et al.*, 2003). It seems probable that chimpanzee propensity for hunting monkeys contributed to simultaneous multiple infections, which provided the opportunity for the recombination of SIV required for successful cross-species transmission. And, while we have focused on retroviruses in this chapter, it is important to note that hunting by chimpanzees has implications for the cross-species transmission of other agents. A study of Ebola hemorrhagic fever among chimpanzees in the Taï forests showed that the primary risk factor for contracting Ebola among wild chimpanzees was hunting behavior, which showed a stronger association with infection than other acknowledged risk factors, such as "touching dead bodies" (Formenty *et al.*, 1999).

Of course, the evolution of hunting 5–7 MYA or more among hominoids was not the end of the story. Ongoing changes in behavioral repertoire can have complicated consequences for disease emergence. The discovery of cooking some 1.9 mya (Wrangham *et al.*, 1999) probably served to decrease the risks associated with human ancestors' consumption of NHPs, since cooking inactivates a number of infectious agents. And a range of more recent behaviors, such as blood transfusions and intravenous drug use, provide novel opportunities for secondary transmission, again increasing the opportunities for establishment of novel zoonotic agents. Such "artificial" chains of secondary transmission have the potential of prolonging the period in which a virus can adapt to novel hosts or recombine with previously existing viruses, potentially increasing the frequency at which zoonotic infections can emerge. Furthermore, increases in human population density, improved hunting techniques (e.g. the invention and proliferation of guns), globalized trade, and road building all increase contact between humans and NHPs, and are likely increasing the frequency of cross-species transmission. Ongoing studies at the ecological interface between humans and NHPs (e.g. among humans hunting NHPs), as well as studies at the interface between NHPs (e.g. among chimpanzees hunting monkeys) have the potential to help define how novel agents successfully establish themselves in a new host species, while simultaneously providing applied human health and conservation benefits by characterizing the diversity of potentially pathogenic agents that are making these transitions.

Note

Use of trade names is for identification only and does not imply endorsement by the U.S. Department of Health and Human Services, the Public Health Service, or the Centers for Disease Control and Prevention. The findings and conclusions in this report are those of the authors and do not necessarily represent the views of the Centers for Disease Control and Prevention.

References

Ali, M., Taylor, G. P., Pitman, R. J. *et al.* (1996). No evidence of antibody to human foamy virus in widespread human populations. *AIDS Research in Human Retroviruses*, **12**, 1473–1483.

Antia, R., Regoes, R. R., Koella, J. C. & Bergstrom, C. T. (2003). The role of evolution in the emergence of infectious diseases. *Nature*, **426**, 658–661.

Apetrei, C., Robertson, D. L. & Marx, P. (2004). The history of SIVs and AIDS: epidemiology, phylogeny, and biology of isolates from naturally SIV infected non-human primates (NHP) in Africa. *Frontiers Bioscience*, **9**, 225–254.

Araujo, A. & Hall, W. W. (2004). Human T-lymphotropic virus type II and neurological disease. *Annals of Neurology*, **56**, 10–19.

Bailes, E., Gao, F., Bibollet-Ruche, F., Courgnaud, V. *et al.* (2003). Hybrid origin of SIV in chimpanzees. *Science*, **300**, 1713.

Blewett, E. L., Black, D. H., Lerche, N. W., White, G. & Eberle, R. (2000). Simian foamy virus infections in a baboon breeding colony. *Virology*, **278**, 183–193.

Boneva, R. S., Grindon, A., Horton, S. *et al.* (2002). Simian foamy virus infection in a blood donor. *Transfusion*, **42**, 886–891.

Brack, M. (1987). *Agents Transmissible from Simians To Man*. Berlin: Springer-Verlag.

Brooks, J. I., Rudd, E. W., Pilon, R. G. *et al.* (2002). Cross-species retroviral transmission from macaques to human beings. *Lancet*, **360**, 387–388.

Broussard, S. R., Comuzzie, A. G., Leighton, K. L. *et al.* (1997). Characterization of new simian foamy viruses from African nonhuman primates. *Virology*, **237**, 349–359.

Butynski, T. M. & Kalina, J. (1998). Gorilla tourism: a critical look. In *Conservation of Biological Resources*, ed. E. J. Milner-Gulland & R. Mace. Oxford: Blackwell Science, pp. 294–366.

Calattini, S., Mauclere, P., Tortevoye, P. *et al.* (2004a). Interspecies transmission of simian foamy viruses from chimpanzees and gorillas to Bantous and Pygmy hunters in southern Cameroon. *Abstracts of the 5th International Foamy Virus Conference*, pp. 7–8.

Calattini, S., Nerrienet, E., Mauclere, P. *et al.* (2004b). Natural simian foamy virus infection in wild-caught gorillas, mandrills and drills from Cameroon and Gabon. *Journal of General Virology*, **85**, 3313–3317.

Calattini, S., Chevalier, S. A., Duprez, R. *et al.* (2005). Discovery of a new human T-cell lymphotropic virus (HTLV-3) in Central Africa. *Retrovirology*, **2**, 30.

Courgnaud, V., Van Dooren, S., Liegeois, F. *et al.* (2004). Simian T-cell leukemia virus (STLV) infection in wild primate populations in Cameroon: evidence for dual STLV type 1 and type 3 infection in agile monkeys. *Journal of Virology*, **78**, 4700–4709.

Coatney, G. R., Collins, W. E., Warren, M. & Contacos, P. G. (1997). *The Primate Malarias*. Washington, DC: US Government Printing Office.

Delebecque, F., Suspene, R., Calattini, S. *et al.* (2006). Restriction of foamy viruses by APOBEC cytidine deaminases. *Journal of Virology*, **80**, 605–614.

Digilio, L., Giri, A., Cho, N. *et al.* (1997). The simian T-lymphotropic/leukemia virus from *Pan paniscus* belongs to the type 2 family and infects Asian macaques. *Journal of Virology*, **71**, 3684–3692.

Falcone, V., Leupold, J., Clotten, J. *et al.* (1999). Sites of simian foamy virus persistence in naturally infected African green monkeys: latent provirus is ubiquitous, whereas viral replication is restricted to the oral mucosa. *Virology*, **257**, 7–14.

Formenty, P., Boesch, C., Wyers, M. *et al.* (1999). Ebola virus outbreak among wild chimpanzees living in a rain forest of Côte d'Ivoire. *Journal of Infectious Diseases*, **179** (Suppl. 1), S120–S126.

Fultz, P. N. (1994). Simian T-lymphotropic virus type 1. In *The Retroviridae, Vol. 3*, ed. J. A. Levy. New York, NY: Plenum Press, pp. 111–131.

Gessain, A. & Mahieux, R. (2000). Epidemiology, origin and genetic diversity of HTLV-1 retrovirus and STLV-1 simian affiliated retrovirus. *Bulletin de la Societe de Pathologie Exotique*, **93**, 163–171.

Goubau, P., Van Brussel, M., Vandamme, A. M., Liu, H. F. & Desmyter, J. (1994). A primate T-lymphotropic virus, PTLV-L, different from human T-lymphotropic viruses types I and II, in a wild-caught baboon (*Papio hamadryas*). *Proceedings of the National Academy of Sciences, USA*, **91**, 2848–2852.

Graczyk, T. K., Mudakikwa, A. B., Cranfield, M. R. & Eilenberger, U. (2001). Hyperkeratotic mange caused by *Sarcoptes scabiei* (Acariformes: *Sarcoptidae*) in juvenile human-habituated mountain gorillas (*Gorilla gorilla beringei*). *Parasitology Research*, **87**, 1024–1028.

Hahn, B. H., Shaw, G. M., De Cock, K. M. & Sharp, P. M. (2000). AIDS as a zoonoses: scientific and public health implications. *Science*, **287**, 607–614.

Harris, R. S. & Liddament, M. T. (2004). Retroviral restriction by APOBEC proteins. *Nature Review, Immunology*, **4**, 868–877.

Heneine, W., Switzer, W. M., Sandstrom, P. *et al.* (1998). Identification of a human population infected with simian foamy viruses. *Nature Medicine*, **4**, 403–407.

Heneine, W., Schweizer, M., Sandstrom, P. & Folks, T. (2003). Human infection with foamy viruses. *Current Topics in Microbiology and Immunology*, **277**, 181–196.

Hubbard, G. B., Mone, J. P., Allan, J. S. *et al.* (1993). Spontaneously generated non-Hodgkin's lymphoma in twenty-seven simian T-cell leukemia virus type 1 antibody-positive baboons. *Laboratory Animal Science*, **43**, 301–309.

Huff, J. L. & Barry, P. A. (2003). B-virus (Cercopithecine herpesvirus 1) infection in humans and macaques: potential for zoonotic disease. *Emerging Infectious Diseases* (serial on the internet). Feb. 2003 (cited Jan. 13, 2003). Available from http://www. cdc.gov/ncidod/EID/vol9no2/02- 0272.htm

Hussain, A. I., Shanmugam, V., Bhullar, V. B. *et al.* (2003). Screening for simian foamy virus infection by using a combined antigen Western blot assay: evidence for a wide distribution among Old World primates and identification of four new divergent viruses. *Virology*, **309**, 248–257.

Jones-Engel, L., Engel, G. A., Schillaci, M. A. *et al.* (2005). Primate to human retroviral transmission in Asia. *Emerging Infectious Diseases*, **11**, 1028–1035.

Kalter, S. S., Millstein, C. H., Boncyk, L. H. & Cummins, L. B. (1978). Tuberculosis in nonhuman primates: a threat to humans. *Developments in Biological Standardization*, **41**, 85–91.

Khabbaz, R. F., Heneine, W., George, J. R. *et al.* (1994). Brief report: infection of a laboratory worker with simian immunodeficiency virus. *New England Journal of Medicine*, **330**, 172–177.

Le Guenno, B., Formentry, P., Wyers, M. *et al.* (1995). Isolation and partial characterization of a new strain of Ebola virus. *Lancet*, **345**, 1271–1444.

Leendertz, F. H., Junglen, S., Boesch, C. *et al.* (2004). High variety of different simian T-cell leukemia virus type 1 strains in chimpanzees (*Pan troglodytes verus*) of the Taï National Park, Cote d'Ivoire. *Journal of Virology*, **78**, 4352–4356.

Linial, M. L., Fan, H., Hahn, B. *et al.* (2004). In *Virus Taxonomy, 7th Rep. International Committee on the Taxonomy of Viruses*, ed. C. M. Fauquet, M. A. Mayo, J. Maniloff, U. Desselberger & L. A. Ball. London: Elsevier/Academic Press, pp. 421–440.

Liu, W., Worobey, M. *et al.* (2008). Molecular ecology and natural history of simian foamy virus infection in wild-living chimpanzees. *PLoS Pathogens*, **4(7)**, 1–22.

Mahieux, R., Chappey, C., Georges-Courbot, M. C. *et al.* (1998). Simian T-cell lymphotropic virus type 1 from *Mandrillus sphinx* as a simian counterpart of human T-cell lymphotropic virus type 1 subtype D. *Journal of Virology*, **72**, 10316–10322.

McClure, M. O., Bieniaz, P. D., Schulz, T. F. *et al.* (1994). Isolation of a new foamy retrovirus from orangutans. *Journal of Virology*, **68**, 7124–7130.

Meertens, L. & Gessain, A. (2003). Divergent simian T-cell lymphotropic virus type 3 (STLV-3) in wild-caught *Papio hamadryas papio* from Senegal: widespread distribution of STLV-3 in Africa. *Journal of Virology*, **77**, 782–789.

Meertens, L., Mahieux, R., Mauclere, P., Lewis, J. & Gessain, A. (2002). Complete sequence of a novel highly divergent simian T-cell lymphotropic virus from wild caught red-capped mangabeys (*Cercocebus torquatus*) from Cameroon: a new primate T-lymphotropic virus type 3 subtype. *Journal of Virology*, **76**, 259–268.

Meertens, L., Shanmugam, V., Gessain, A. *et al.* (2003). A novel, divergent simian T-cell lymphotropic virus type 3 in a wild-caught red-capped mangabey (*Cercocebus torquatus torquatus*) from Nigeria. *Journal of General Virology*, **84**, 2723–2727.

Meiering, C.D. & Linial, M. L. (2001). Historical perspective of foamy virus epidemiology and infection. *Clinical Microbiology Review*, **14**, 165–176.

Moya, A., Holmes, E. C. & Gonzalez-Candelas, F. (2004). The population genetics and evolutionary epidemiology of RNA viruses. *Nature Review, Microbiology*, **2**, 279–288.

Murray, S. M., Picker, L. J., Axthelm, M. K. & Linial, M. L. (2006). Expanded tissue targets for foamy virus replication with simian immunodeficiency virus-induced immunosuppression. *Journal of Virology*, **80**, 663–670.

Renquist, D. M. & Whitney, R. A. (1987). Zoonoses acquired from pet primates. *Veterinary Clinics of North America Small Animal Practice*, **17**, 219–240.

Ruch, T. C. (1959). *Diseases of Laboratory Primates*. Philadelphia, PA: W. B. Saunders Company.

Russell, R. A., Wiegand, H. L., Moore, M. D. *et al.* (2005). Foamy virus Bet proteins function as novel inhibitors of the APOBEC3 family of innate antiretroviral defense factors. *Journal of Virology*, **79**, 8724–8731.

Salemi, M., Van Dooren, S., Audenaert, E. *et al.* (1998). Two new human T-lymphotropic virus type I phylogenetic subtypes in seroindeterminates, a Mbuti pygmy and a Gabonese, have closest relatives among African STLV-I strains. *Virology*, **246**, 277–287.

Salemi, M., Van Dooren, S. & Vandamme, A. M. (1999). Origin and evolution of human and simian T-cell lymphotropic viruses. *AIDS Review*, **1**, 131–139.

Salemi, M., Desmyter, J. & Vandamme, A. M. (2000). Tempo and mode of human and simian T-lymphotropic virus (HTLV/STLV) evolution revealed by analyses of full-genome sequences. *Molecular Biology and Evolution*, **17**, 374–386.

Schweizer, M., Turek, R., Hahn, H. *et al.* (1995). Markers of foamy virus (FV) infections in monkeys, apes, and accidentally infected humans: appropriate testing fails to confirm suspected FV prevalence in man. *AIDS Research and Human Retroviruses*, **11**, 161–170.

Shah, K. V. (1989). A review of the circumstances and consequences of simian virus SV40 contamination of human vaccines. *Symposium on Continuous Cell Lines as Substrates for Biologicals. Developments in Biological Standardization, Vol. 70.* Arlington, VA: International Association of Biological Standardization.

Slattery, J. P., Franchini, G. & Gessain, A. (1999). Genomic evolution, patterns of global dissemination, and interspecies transmission of human and simian T-cell leukemia/lymphotropic viruses. *Genome Research*, **9**, 525–540.

Sleeman, J. M., Meader, L. L., Mudakikwa, A. B., Foster, J. W. & Patton, S. (2000). Gastrointestinal parasites of mountain gorillas (*Gorilla gorilla beringei*) in the Parc National des Volcans, Rwanda. *Journal of Zoo Wildlife Medicine*, **31**, 322–328.

Song, B., Javanbakht, H., Perron, M. *et al.* (2005). Retrovirus restriction by TRIM5 alpha variants from Old World and New World primates. *Journal of Virology*, **79**, 3930–3937.

Sotir, M., Switzer, W., Schable, C. *et al.* (1997). Risk of occupational exposure to potentially infectious NHP materials and to simian immunodeficiency virus. *Journal of Medical Primatology*, **26**, 233–240.

Stanford, C. B. (1996). The hunting ecology of wild chimpanzees: implications for the behavioral ecology of Pliocene hominids. *American Anthropologist*, **98**, 96–113.

Stanford, C. B. (1999). *The Hunting Apes: Meat-eating and the Origins of Human Behavior*. Princeton, NJ: Princeton University Press.

Switzer, W. M., Bhullar, V., Shanmugam, V. *et al.* (2004). Frequent simian foamy virus infection in persons occupationally exposed to NHPs. *Journal of Virology*, **78**, 2780–2789.

Switzer, W. M., Salemi, M., Shanmugam, V. *et al.* (2005). Ancient co-speciation of simian foamy virus and primates. *Nature*, **434**, 376–380.

Takemura, T., Yamashita, M., Shimada M. K. *et al.* (2002). High prevalence of simian T-lymphotropic virus type L in wild Ethiopian baboons. *Journal of Virology*, **76**, 1642–1648.

Vandamme, A. M., Salemi, M. & Desmyter, J. (1998). The simian origins of the pathogenic human T-cell lymphotropic virus type 1. *Trends in Microbiology*, **6**, 477–483.

Van Dooren, S., Shanmugam, V., Bhullar, V. *et al.* (2004). Identification in gelada baboons (*Theropithecus gelada*) of a distinct simian T-cell lymphotropic virus 3 with a broad range of Western blot reactivity. *Journal of General Virology*, **85**, 507–551.

Voevodin, A., Samilchuk, E., Schatzl, H., Boeri, E. & Frachini, G. (1996). Interspecies transmission of macaque simian T-cell leukemia/lymphoma virus type 1 in baboons results in an outbreak of malignant lymphoma. *Journal of Virology*, **70**, 1633–1639.

Wallis, J. & Lee, D. R. (1999). Primate conservation: the prevention of disease transmission. *International Journal of Primatology*, **20**, 803–826.

Weiss, R. A. (1988). Foamy retroviruses: a virus in search of a disease. *Nature*, **333**, 497–498.

Wolfe, N. D., Escalate, A. A., Karesh, W. B. *et al.* (1998). Wild primate populations in emerging infectious disease: the missing link. *Emerging Infectious Diseases*. Available from http://www. cdc.gov/ncidod/eid/vol4no2/wolfe.htm.

Wolfe, N. W., Mpoudi-Ngole, E., Gockowski, J. *et al.* (2000). Deforestation, hunting and the ecology of microbial emergence. *Global Change and Human Health*, **1**, 10–25.

Wolfe, N. D., Prosser, T. A., Carr, J. K. *et al.* (2004a). Exposure to NHPs in rural Cameroon. *Emerging Infectious Diseases*, **10**, 2094–2099.

Wolfe, N. D., Switzer, W. M., Carr, J. K. *et al.* (2004b). Naturally acquired simian retrovirus infections in central African hunters. *Lancet*, **363**, 932–937.

Wolfe, N. D., Heneine, W., Carr, J. K. *et al.* (2005). Emergence of unique primate T-lymphotropic viruses among central African bushmeat hunters. *Proceedings of the National Academy of Sciences, USA*, **102**, 7994–7999.

Wrangham, R. W., Jones, J. H., Laden, G. *et al.* (1999). The raw and the stolen. *Current Anthropology*, **40**, 567–594

Yamashita, M., Ido, E., Miura, T. & Hayami, M. (1996). Molecular epidemiology of HTLV-I in the world. *Journal of Acquired Immune Deficiency Syndrome and Human Retrovirology*, **13** (Suppl. 1), S124–S131.

18 *Overview of parasites infecting howler monkeys,* Alouatta *sp., and potential consequences of human–howler interactions*

SYLVIA K. VITAZKOVA

Photograph by Sylvia K. Vitazkova (*Alouatta pigra*)

Primate Parasite Ecology. The Dynamics and Study of Host–Parasite Relationships, ed. Michael A. Huffman and Colin A. Chapman. Published by Cambridge University Press.

Introduction

Humans and wildlife have interacted for hundreds of thousands of years, but the level at which these two groups currently interact is unprecedented due to such factors as human population growth, changes in agricultural practices, and extraction of natural resources (Daszak *et al.*, 2000; Slingenbergh *et al.*, 2004). Anthropogenic environmental changes have been linked to changes in host–parasite interactions, such as the recent emergence of deadly infectious diseases and the corresponding threat to the conservation of wildlife (Schrag & Wiener, 1995; Wolfe *et al.*, 1998; Daszak, 2000; Patz *et al.*, 2000; Thompson, 2000). The increasing popularity of wildlife tourism and the presence of researchers may contribute to human–wildlife disease exchange. However, it is the encroachment of permanent settlements into areas previously uninhabited by people (or inhabited only by small, nomadic tribes) that poses the major threat to the health of both wildlife and humans (Adams *et al.*, 2001; Graczyk *et al.*, 2002; Nizeyi *et al.*, 2002b; Woodford *et al.*, 2002). The threat of cross-species infection is especially relevant between human and non-human primates because of their similar physiology (Wolfe *et al.*, 1998; Woodford *et al.*, 2002). For example, in Uganda, free-ranging, human-habituated gorillas, and humans and cattle that inhabit adjacent areas, have all been found to be infected with the same assemblage of *Giardia duodenalis* (syn. *G. lamblia, G. intestinalis*) (Graczyk *et al.*, 2002). The authors of the study speculate that humans were probably the source of the gorillas' parasitic infection, as "a large percentage of the local community does not follow park regulations regarding the disposal of their fecal waste, as self-reported in a questionnaire" (Graczyk *et al.*, 2002). A survey of tourists visiting Kibale National Park in Uganda, which is home to chimpanzees and other wildlife, indicated that the visitors had a high prevalence of infectious symptoms, particularly diarrhea (Adams *et al.*, 2001). It is suggested that the parasites causing these infections could be transmitted to the chimpanzee population, particularly since adequate latrines were not provided for the visitors and defecations (which were presumably not buried) sometimes occurred in the bushes where chimpanzees congregated. However, care needs to be taken when suggesting inter-species parasite transmission, as certain generalist parasites that seem to be the same may, in fact, be two or more host species-specific subtypes. For example, *Oesophagostomum bifurcum* found in humans and Mona monkeys was thought to be the same subtype (de Gruijter *et al.*, 2004); however, subsequent high-resolution DNA fingerprinting showed that *O. bifurcum* samples from humans and monkeys were genetically distinct (de Gruijter *et al.*, 2005; Gasser *et al.*, Chapter 3, this volume).

Research that investigates links between human activities and parasitic diseases in non-human primates is just beginning and has focused mostly on African non-human primates (Sleeman *et al.*, 2000; Adams *et al.*, 2001; Graczyk *et al.*, 2001; Lilly *et al.*, 2002; Nizeyi *et al.*, 2002a; Gillespie & Chapman, 2006). Old World apes are our closest relatives and are susceptible to many of the diseases that affect humans (Orihel, 1970; Hoffert, 1998; Muriuki *et al.*, 1998; Adams *et al.*, 2001; Leclerc *et al.*, 2004; Legesse & Erko, 2004). However, neotropical non-human primates are also susceptible to many human diseases, such as Chagas disease (Ziccardi & Lourenço-de-Oliveira, 1997), malaria (Volney *et al.*, 2002; Di Fiore *et al.*, Chapter 7, this volume), and ascariasis (Michaud *et al.*, 2003). For example, several different species of neotropical monkeys have been found to be infected with simian malaria, *Plasmodium brasilianum* (Thoisy *et al.*, 2001). *Plasmodium brasilianum* and *P. malariae*, which infect African primates, are genetically indistinguishable from each other (Escalante *et al.*, 1995). One hypothesis is that the *P. brasilianum* is actually *P. malariae*, which may have been brought to the Americas by humans who had previously been infected with the parasite in Africa, probably by being bitten by mosquitoes that had fed on infected chimpanzees (Coatney, 1971).

To begin examining possible links in human and non-human primate parasitic diseases in the Neotropics, this chapter examines some possible links between the parasites of howler monkeys, genus *Alouatta* (Cebidae), from Central and South America, human diseases, and human activities, including a case study of *A. pigra* and *Giardia* sp. infection.

Alouatta sp. and gastrointestinal parasites

Members of the neotropical genus *Alouatta* are large-bodied and relatively slow-moving non-human primates. Groups defend territories from other groups of *Alouatta*, but may share habitat with other primate species. Howler monkeys range from southern Mexico to Argentina (Cortez-Ortiz *et al.*, 2003) and are energy-minimizing folivores, consuming large quantities of leaves, fruits, and other vegetation and resting up to 80% of the day (Milton, 1980). The primates' common name is reflective of the loud calls made primarily by males, sometimes by females, which are used in demarcating and defending group territories (Horwich & Gebhard, 1983). Black howler monkeys, *Alouatta pigra*, are found in the smallest geographical area of all *Alouatta*, ranging only in parts of Belize, southern Mexico, and northern Guatemala (Horwich & Jones, 1998). These primates prefer riverine habitats in low-lying tropical rain forests under 300 m above sea level (Horwich & Johnson, 1986).

Stuart *et al.* (1998) published a comprehensive review of *Alouatta* parasites in 1998, noting that no parasites had been reported for *A. pigra* at that time. Since then, two studies found that *Controrchis biliophilus*, *Trypanoxyuris minutus*, *Giardia duodenalis* (syn. *G. lambdlia, G. intestinalis*), *Entamoeba coli*, and a strongyle-type nematode infect *A. pigra* (Eckert *et al.*, 2006; Vitazkova & Wade, 2006). Of these parasites, *Giardia duodenalis* and *Entamoeba coli* are known to infect humans (Thompson *et al.*, 2000; Aimpun & Hshieh, 2004). Several other parasites, of which a number have been detected in humans, have been documented in other species of *Alouatta* (Table 18.1).

Human–howler monkey interactions

Alouatta sp. is more tolerant of habitat alteration, such as decreased home ranges due to habitat destruction, than most other neotropical primate species (Bicca-Marques, 2003). *Alouatta's* ability to be largely folivorous and its energy-minimizing dietary strategy (Milton, 1980) are probably the main reasons for its adaptability to habitat alteration (Silver & Marsh, 2003). This adaptability benefits the primates in that they are able to survive in degraded habitats, but it also potentially exposes them to other dangers, including human and domestic animal parasites. Additionally, increased primate density, which often results from habitat fragmentation (Ostro *et al.*, 2001), may lead to increased contamination of the environment with parasite eggs, larvae, or cysts, which could increase the primates' parasite prevalence and intensity of infection, and negatively affect their health. Inbreeding depression, which may occur when primate groups are isolated from each other in forest fragments (James *et al.*, 1997), may further make primate populations vulnerable to parasites due to decreased immune competence.

A 2004 coprological survey of Mayan inhabitants from remote villages in southern Belize found that 66% of those tested were infected with at least one parasite, including the following species: hookworms (55%), *Ascaris lumbricoides* (30%), *Entamoeba coli* (21%), *Trichuris trichiura* (19%), *Giardia lamblia* (12%), *Iodamoeba beutschlii* (9%), *Entamoeba histolytica* (syn. *E. dispar*) (6%), *Entamoeba hartmani* (3%), *Strongyloides stercoralis* (1%), *Endolimax nana* (<1%), *Isospora belli* (<1%), and *Chilomastix mesnili* (<1%) (Aimpun & Hshieh, 2004). Although the southernmost range of *A. pigra* in Belize has not been determined, and, thus, it is unknown whether these Mayan populations come into contact with the monkeys, the results of this study indicate that people could potentially be a source of several parasitic infections for *A. pigra*, particularly parasites with a direct fecal–oral life cycle. A preliminary

Table 18.1. *Endoparasites detected in feces reported from* Alouatta *sp. Species or genera highlighted in* **bold** *have been reported from both* Alouatta *and humans (Taylor* et al.*, 2001; Pillai & Kain, 2003). Parasites indicated with * were detected in free-ranging* Alouatta*; parasites indicated with ** were detected in captive animals; whether the origin of parasites without any asterisk is from a wild or captive howler monkey is unknown*

Parasite type	Species	Reference
Protozoa	Symptoms caused in humans. Protozoa are often asymptomatic, although in some individuals they may cause non-specific symptoms, such as diarrhea, blood in feces, fever, nausea, vomiting, and abdominal cramps (CDC, 2004; Schuster & Visvesvara, 2004; Solaymani-Mohammadi & Petri, 2006). Some parasites, such as *Balantidium coli*, may cause more serious colonic lesions, which allow for secondary bacterial infection (Schuster & Visvesvara, 2004).	
	Balantidium coli**	(Bonilla-Moheno, 2002)
	Blastocystis sp.*	(Phillips *et al.*, 2004)
	Chilomastix sp.**	(Kuntz & Myers, 1972; Stuart *et al.*, 1998; Phillips *et al.*, 2004)
	Cryptosporidium sp.*	(Bonilla-Moheno, 2002)
	Endolimax nana**	(Bonilla-Moheno, 2002)
	*Entamoeba coli**	(Thoisy *et al.*, 2001; Bonilla-Moheno, 2002; Eckert, 2002)
	Entamoeba hartmanni**	(Bonilla-Moheno, 2002)
	Entamoeba histolytica**	(Kuntz & Myers, 1972; Stuart *et al.*, 1998; Thoisy *et al.*, 2001)
	Entamoeba polecki**	(Thoisy *et al.*, 2001, Bonilla-Moheno, 2002)
	Enteromonas sp.*	(Thoisy *et al.*, 2001)
	Giardia intestinalis (= *G. lamblia, G. duodenalis)**	(Gual *et al.*, 1990; Stoner, 1996; Stuart *et al.*, 1998; Bonilla-Moheno, 2002)
	Iodamoeba butschlii**	(Kuntz & Myers, 1972; Eckert, 2002; Phillips *et al.*, 2004)
	Retortamonas (= *Embadomonas*) *intestinalis**	(Kuntz & Myers, 1972; Stuart *et al.*, 1998)
	Trichomonas sp.**	(Stuart *et al.*, 1998)
	Trichomonas hominis***	(Kuntz & Myers, 1972)
Coccidia	Symptoms caused in humans: Coccidia may cause protracted diarrhea and malabsorption, with heavy infections causing inflammation and possible injury to the intestinal lining (Goodgame, 1996; Looney, 1998).	
	Cyclospora sp.*	(Bonilla-Moheno, 2002)
	Isospora arctopitheci**	(Duszynski *et al.*, 1998; Stuart *et al.*, 1998)
	Isospora scorzi (*experimental*)**	(Duszynski *et al.*, 1998)
	*Eimeria**	(Serrano, 1998; Bonilla-Moheno, 2002)

(*cont.*)

Table 18.1 (*cont.*)

Parasite type	Species	Reference
Trematodes	Symptoms caused in humans: The trematodes detected to date in *Alouatta* sp. have not been found to infect humans. Therefore, no symptoms are on record.	
	Controrchis biliophilus (= *Controrchis caballeroi*)*	(Stuart *et al.*, 1998; Jimenez-Quioros & Brenes, 1957; Gomes & Pinto, 1978; Gonzales *et al.*, 1983; Pastor-Nieto, 1991, 1993; Castillejos, 1993; Aceves, 1995; Garcia, 1995; Hermida-Lagunes *et al.*, 1996; Bonilla-Moheno, 2002)
	*Controrchis sp.***	(Kuntz & Myers, 1972)
	*Athesmia foxi**	(Aguirre & Guerrero, 2001)
Cestodes	Symptoms caused in humans: Cestodes may cause intermittent diarrhea, gastric pain, anorexia, and weight loss (Bhagwant, 2004).	
	Bertiella mucronata*	(Stuart *et al.*, 1998; Santa Cruz, 2000)
	Bertiella sp.**	(Kuntz & Myers, 1972)
	Mathevotaenia (= *Oochoristic*) *megastoma* (= *Taenia metastoma, Atriotaenia megastoma, Bertiella fallax*)*	(Stuart *et al.*, 1998)
	Mathevotaenia sp. (= *Atriotaenia* sp.)**	(Kuntz & Myers, 1972)
	*Moniezia rugosa**	(Stuart *et al.*, 1998)
	Moniezia sp.**	(Kuntz & Myers, 1972)
	Raillietina demerariensis*	(Stuart *et al.*, 1998; Aguirre & Guerrero, 2001)
	*Raillietina alouattae**	(Stuart *et al.*, 1998; Aguirre & Guerrero, 2001)
	Raillietina sp.**	(Kuntz & Myers, 1972)
Nematodes	Symptoms caused in humans: Nematodes may cause some of the most severe clinical effects of all gastrointestinal parasites. Different species may cause severe itching and skin eruptions (if larvae migrate under the skin) (Schad, 1994), retarded growth and intellectual development in children and iron deficiency anemia (Georgiev, 2000), blockage of the gastrointestinal tract, invasion and obstruction of the biliary or pancreatic ducts (de la Fuente-Lira *et al.*, 2006), and watery diarrhea, abdominal pain, cramping, and nausea (Ralph *et al.*, 2006).	
	Ancylostoma quadridenata*	(Stuart *et al.*, 1998)
	Ancylostoma sp.**	(Kuntz & Myers, 1972; Stuart *et al.*, 1998)
	Ascaris elongata (?)	(Stuart *et al.*, 1998)
	Ascaris lumbriocoides	(Stuart *et al.*, 1998)
	Ascaris sp.*	(Thoisy *et al.*, 2001; Phillips *et al.*, 2004)

Table 18.1 (*cont.*)

Parasite type	Species	Reference
	Enterobius sp.**	(Kuntz & Myers, 1972; Garcia, 1995; Hermida-Lagunes *et al.*, 1996)
	Filariopsis (= *Filaroides*) *asper*	(Stuart *et al.*, 1998)
	Filariopsis sp. (= *Filaroides* sp.)**	(Kuntz & Myers, 1972)
	Necator americanus*	(Thoisy *et al.*, 2001)
	Longistriata sp.**	(Kuntz & Myers, 1972)
	*Oxyuridae**	(Santa Cruz, 2000)
	Parabronema (*Squamanema*) *bonnie**	(Pastor-Nieto, 1991, 1993; Hermida-Lagunes *et al.*, 1996; Stoner, 1996; Stuart *et al.*, 1998)
	Physaloptera dilatata*	(Stuart *et al.*, 1998)
	Strongyloides sp.*	(Garcia, 1995; Santa Cruz, 2000; Thoisy *et al.*, 2001; Phillips *et al.*, 2004)
	*Strongyloides cebus**	(Stuart *et al.*, 1998)
	Trichostrongylus sp.*	(Thoisy *et al.*, 2001)
	Trichuris dispar	(Stuart *et al.*, 1998)
	Trichuris sp.	(Kuntz & Myers, 1972; Stuart *et al.*, 1998; Phillips *et al.*, 2004)
	*Trypanoxyuris minutus**	(Pastor-Nieto, 1991, 1993; Castillejos, 1993; Hermida-Lagunes *et al.*, 1996; Stuart *et al.*, 1998)
	Trypanoxyuris sp.*	(Kuntz & Myers, 1972; Thoisy *et al.*, 2001; Eckert, 2002)
	Vianella (= *Longistriata*) *dubia*	(Stuart *et al.*, 1998)
Acanthocephalans	Symptoms caused in humans: The acanthocephalans detected to date in *Alouatta* sp. have not been found to infect humans. Therefore, no symptoms are on record.	
	*Prosthenorchis elegans***	(Stuart *et al.*, 1998)

examination of this type of infection potential is presented in the following case study of *A. pigra* and *Giardia* sp. infection.

Alouatta pigra *and* Giardia *sp.*

A study of demographic and ecological factors that affect the prevalence of parasites in *A. pigra* was conducted between 2002 and 2004 (Vitazkova & Wade, 2006, 2007). Four parasites were detected: (1) *Controrchis biliophilus* (Dicrocoeliidae), prevalence 80% (wet season) and 81% (dry season); (2) *Trypanoxyuris minutus* (Oxyuridae), prevalence 8% (wet season) and 27% (dry season); (3) *Giardia* sp. (Hexamitidae), prevalence 40% (wet season) and 27% (dry season); and (4) *Entamoeba* sp. (Endamoebidae) in a single specimen. The most important factor in predicting whether an individual *A. pigra* would be infected with a parasite was the individual's membership in a particular social troop. The prevalence of each parasite showed a different pattern, with *Controrchis biliophilus*, a fluke, tending to be more prevalent in *A. pigra* that inhabited disturbed habitats, and *Trypanoxyuris minutus*, a pinworm, tending to be more prevalent in *A. pigra* from undisturbed habitats. *Giardia* sp. tended to be more prevalent in primates living at high densities. Molecular techniques allowed for more precise identification of *Giardia* sp. in *A. pigra* feces from one of the study sites than would morphological examination alone. DNA extracted from *Giardia duodenalis*-positive samples from *A. pigra* inhabiting the Community Baboon Sanctuary in Belize was found to be from *Giardia duodenalis* (syn. *G. lamblia, G. intestinalis*) Assemblage A and Assemblage B (Vitazkova & Wade, 2006). There are many genetic differences within the *G. duodenalis* species complex (Thompson, 2002). *Giardia duodenalis* Assemblage A and Assemblage B are globally the most widespread of the *G. duodenalis* subtypes (Thompson *et al.*, 2000; Thompson, 2002), and are known to infect other primates, including humans and mountain gorillas, *Gorilla gorilla berengei* (Thompson, 2000; Graczyk *et al.*, 2002). *Giardia duodenalis* Assemblage A subtype usually infects a variety of species, including humans, livestock, cats, dogs, beavers, guinea pigs, and slow lorises (Thompson *et al.*, 2000). *Giardia duodenalis* Assemblage B subtype usually infects similar hosts as Assemblage A, but Assemblage B has not been isolated from livestock (Thompson, 2000).

The original source(s) of *G. duodenalis* infection in *A. pigra* is as yet unknown: did the monkeys become infected through exposure to humans and their domestic animals, to other wildlife, or do *A. pigra* naturally harbor *G. duodenalis* infections? Circumstantial evidence suggests that humans and/or their domestic animals may be the source of the infection in *A. pigra* inhabiting the Community Baboon Sanctuary, a mixture of secondary growth forest fragments, cattle pastures, plantations, and seven villages covering 47 km^2 along the Belize River in northern Belize. Seven out of 14 (50%) *A. pigra* inhabiting the Baboon Sanctuary in 2003 were infected with *Giardia* sp., whereas no infection was detected in *A. pigra* inhabiting the Cockscomb Basin Wildlife Reserve, a 400 km^2 protected area with no permanent human settlements along the eastern slope of the Maya Mountains in southern Belize, during 2003 ($n = 10$) (Vitazkova, 2005). The Cockscomb population is descended from 63 howler monkeys that were translocated from the Baboon Sanctuary in 1992 and 1993 (Koontz *et al.*, 1994). Twenty 1 g fecal samples that had been collected from the captured monkeys during translocation (preserved in 10% formalin and stored in Para-Pak vials) were examined by the author and her colleagues in July 2003. One of the samples was positive for *Giardia* sp. Although the sample size was small and the samples may have become degraded while in storage for 10 years, this indicated that at least one of the founding members of the Cockscomb population had been infected prior to translocation, whereas no infection was detected with repeated sampling at this site 10 years later. These findings suggest that *Giardia* sp. infection in *A. pigra* may be perpetuated at the Baboon Sanctuary by some factor(s) that are not present at Cockscomb, such as the highly altered habitat and presence of humans and domestic animals at the former location.

Alouatta pigra may become infected by walking on ground contaminated with *G. duodenalis* cysts, then transferring the cysts from their hands onto food and ingesting them. Although *A. pigra* rarely come to the ground in their natural state, in fragmented forests, such as the Baboon Sanctuary, the primates have been observed coming to the ground to travel from one forest patch to another (pers. obs.). Possible sources of *G. duodenalis* cysts at the Baboon Sanctuary include exposed animal or human excrement, used toilet paper discarded in the "bush," and, possibly, contaminated water (Woodford *et al.*, 2002). Cattle, horses, and dogs all graze, roam, and defecate freely in the fields, under bush, and paths under the canopy inhabited by *A. pigra* at this site. These domestic animals are rarely treated for parasitic infections. Children, especially prone to *G. duodenalis* infection due to poor hygiene (Thompson, 2000), may sometimes defecate in the forest. Hunters, researchers, and their assistants often spend an entire day following animals, and any of these people may be infected with *G. duodenalis,* and may also defecate in the bush without burying his/her excrement. Latrines used by infected people may contribute to the spread of *G. duodenalis* cysts, particularly during the wet season, when they can overflow due to heavy rains. Water may also become contaminated with *G. duodenalis* in any of the ways described above, although it is unlikely that *A. pigra* become infected with *Giardia* sp. through water, as the primates usually obtain water from foodstuffs or tree holes (Horwich & Lyon, 1990). Other possible routes of transmission are hand feeding of *A. pigra* by local guides and tourists, which was seen to occur frequently, and transportation of *G. duodenalis* from feces to primate food, lips, or hands by house flies, *Musca domestica* (Doiz *et al.*, 2000).

The hypotheses that are presented above are based on circumstantial evidence. More detailed molecular analyses, such as high resolution DNA fingerprinting using random amplified polymorphic DNA analyses (de Gruijter *et al.*, 2004), are needed to definitively determine whether humans and/or their domestic animals are the source of *Giardia* sp. infection in *A. pigra* at the Community Baboon Sanctuary in Belize.

Next steps

Wildlife and humans will increasingly interact as people continue to colonize habitats previously uninhabited by humans. Since it is unlikely that this kind of interaction will be stopped, it is important that steps be taken to record the impact of human activities on wildlife, and to minimize this impact. It is important to conduct research to establish baseline data on parasites of wild animals and to conduct periodic coprological surveys of wildlife near human settlements to monitor for unusual outbreaks of parasitic infections. Periodic coprological surveys of the human populations living near wildlife are also important, as is ensuring that any parasitic diseases are treated in a timely manner. Educating people, especially children, about proper hygiene and the risk of disease transmission from wildlife could reduce the risk of zoonoses passing between humans and non-human animals. Finally, conducting detailed genetic analyses of human and non-human primate parasites to determine whether interspecific infection is taking place will greatly enhance our understanding of human–wildlife disease ecology. The use of molecular techniques, already important tools in conservation, holds the promise of opening an exciting new chapter in the study of wildlife and human disease transmission.

Acknowledgements

None of this work would have been possible without the support of Dr. Susan E. Wade and Dr. Michael A. Huffman, who have for years generously shared their knowledge, resources, and enthusiasm with me. Mr. Michael Wade and Mr. Emiliano Pop were invaluable field assistants. Ms. Jessie Young and Mr. Oswald McFadzean at the Community Baboon Sanctuary provided logistical support and permission to conduct research on their land, while the Belize Audubon Society and the Government of Belize approved research permits. Thanks are also due to Drs. Fred Koontz and Wendy Westrom for donation of fecal samples from *A. pigra* translocated in 1992–1993, and to Dr. Rebecca Traub (molecular analyses) and Ms. Stephanie Schaaf (ELISAs) for assistance with sample processing.

References

Aceves, M. (1995). Identificacion de nematodos en monos aulladores (*Alouatta palliata*) en la isla de Agaltepec. Facultad de Medicina Veterinaria y Zootecnia, Universidad Nacional Autonoma de Mexico. Thesis.

Adams, H. A., Sleeman, J. M., Rwego, I. & New, J. C. (2001). Self-reported medical history survey of humans as a measure of health risk to the chimpanzees (*Pan troglodytes schweinfurthii*) of Kibale National Park, Uganda. *Oryx*, **35**, 308–312.

Aguirre, A. A. & Guerrero, R. (eds.) (2001). *Mexico, Central and South America. Helminths of Wildlife.* Enfield, UK: Science Publishers, Inc.

Aimpun, P. & Hshieh, P. (2004). Survey for intestinal parasites in Belize, Central America. *Southeast Asian Journal of Tropical Medicine & Public Health*, **35**, 506–511.

Bhagwant, S. (2004). Human *Bertiella studeri* (family Anoplocephalidae) infection of probable Southeast Asian origin in Mauritian children and an adult. *American Journal of Tropical Medicine and Hygiene*, **70**, 225–228.

Bicca-Marques, J. C. (2003). How do howler monkeys cope with habitat fragmentation? In *Primates in Fragments: Ecology and Conservation.* ed. L. K. Marsh. New York, NY: Kluwer Academic and Plenum, pp. 283–298.

Bonilla-Moheno, M. (2002). Prevalencia de parásitos gastroentéricos en primates (*Alouatta pigra* y *Ateles geoffroyi yucatanensis*) localizados en hábitat conservado y fragmentado de Quintana Roo, México. Facultad de Ciencias, Cuidad de Mexico, Universidad Nacional Autonoma de Mexico.

Castillejos, M. (1993). Identificacion de parasitos gastrointestinales en monos aulladores (*Alouatta palliata*) en al reserva "El Zapotal" Chiapas, Mexico. Facultad de Medicina Veterinaria y Zootecnia, Universidad Nacional Autonoma de Mexico. Thesis.

Centers for Disease Control and Prevention (2004). *Non-Pathogenic Intestinal Amoebas Fact Sheet.* Atlanta, GA: CDC.

Coatney, G. R. (1971) The simian malarias: zoonoses, anthroponoses, or both? *American Journal of Tropical Medicine and Hygiene*, **20**, 795–803.

Cortez-Ortiz, L., Bermingham, W., Rico, C. *et al.* (2003). Molecular systematics and biogeography of the Neotropical monkey genus, *Alouatta. Molecular Phylogenetics and Evolution*, **26**, 64–81.

Daszak, P. (2000). Emerging infectious diseases: bridging the gap between humans and wildlife. *The Scientist*, **14**, 14–17.

Daszak, P., Cunningham, A. A. & Hyatt, A. D. (2000). Emerging infectious diseases of wildlife – threats to biodiversity and human health. *Science*, **287**, 443–449.

De Gruijter, J. M., Ziem, J., Verweij, J. J., Polderman, A. M. & Gasser, R. B. (2004). Genetic substructuring within *Oesophagostomum bifurcum* (Nematoda) from human and non-human primates from Ghana based on random amplified polymorphic DNA analysis. *American Journal of Tropical Medicine and Hygiene*, **71**, 227–233.

De Gruijter, J. M., Gasser, R. B., Polderman, A. M., Asigri, V. & Dijkshoorn, L. (2005). High resolution DNA fingerprinting by AFLP to study the genetic variation among *Oesophagostomum bifurcum* (Nematoda) from human and non-human primates from Ghana. *Parasitology*, **130**, 229–237.

De La Fuente-Lira, M., Molotla-Xolalpa, C. & Rocha-Guevara, E. R. (2006). Biliary ascariasis. Case report and review of the literature. *Cirugia y cirujanos*, **74**, 195–198.

Doiz, O., Clavel, A., Morales, S. *et al.* (2000). House fly (*Musca domestica*) as a transport vector of *Giardia lamblia*. *Folia Parasitologica*, **47**, 330–331.
Duszynski, D. W., Upton, S. J. & Couch, L. (1998). Coccidia (Eimeriidae) of Primates and Scadentia (monkeys, apes humans, tree shrews). htttp://biology.unm.edu/biology/coccidia/primates.html.
Eckert, K. A. (2002). Endoparasites and defecation patterns of black howler monkeys (*Alouatta pigra*) from Belize. Masters thesis. San Francisco, CA: San Francisco State University.
Eckert, K. A., Hahn, N. E., Genz, A. *et al.* (2006). Coprological surveys of *Alouatta pigra* at two sites in Belize. *International Journal of Primatology*, **27**, 227–238.
Escalante, A. A., Barrio, E. & Ayala, F. J. (1995). Evolutionary origin of human and primate malarias: evidence from the circumsporozoite protein gene. *Molecular Biology and Evolution*, **12**, 616–626.
Garcia, O. (1995). Identificacion de nematodos en monos aulladores (*Alouatta palliata*) cautivos en el estado de Veracruz. Facultad de Medicina Veterinaria y Zootecnia, Universidad Nacional Autonoma de Mexico. Thesis.
Georgiev, V. S. (2000). Necatoriasis: treatment and developmental therapeutics. *Expert Opinion on Investigational Drugs*, **9**, 1065–1078.
Gillespie, T. R. & Chapman, C. A. (2006). Forest fragment attributes predict parasite infection dynamics in primate metapopulations. *Conservation Biology*, **20**, 441–448.
Gomes, D. C. & Pinto, R. M. (1978). Contribuicao ao conhecimento de fauna helmintologica da Ragiao Amazonica – Trematodeos. *Atlas Soc. Biol.*, **19**, 43–46.
Gonzales, B., Paasch, L. H. & Paasch, P. A. (1983). Identificacion de parasitos metazoarios en cortes histologicos. *Veterinaria Mexico*, **14**, 159–174.
Goodgame, R. W. (1996). Understanding intestinal spore-forming protozoa: Cryptosporidia, Microsporidia, Isospora, and Cyclospora. *Annals of Internal Medicine*, **124**, 429–441.
Graczyk, T. K., Dasilva, A. J., Cranfield, M. R. *et al.* (2001). *Cryptosporidium parvum* Genotype 2 infections in free-ranging mountain gorillas (*Gorilla gorilla beringei*) of the Bwindi Impenetrable National Park, Uganda. *Parasitology Research*, **87**, 368–370.
Graczyk, T. K., Bosco-Nizeyi, J., Ssebide, B. *et al.* (2002). Anthropozoonotic *Giardia duodenalis* genotype (Assemblage A) infections in habitats of free-ranging human-habituated gorillas, Uganda. *Journal of Parasitology*, **88**, 905–909.
Gual, F., Guerrero, C. & Quiroz, H. (1990) Determinacion de parasitos gastroentericos en primates del Zoologico de Chapultepec. Facultad de Medicina Veterinaria y Zootecnia, Universidad Nacional Autonoma de Mexico. Thesis.
Hermida-Lagunes, J., Caneles-Espinosa, D., Osorio, D. & Garcia-Serrano, O. (1996). Relationships between parasitism, hematological values and body weight in adult females of *Alouatta palliata mexicana.* IPS/ASP Congress Abstracts, 698.
Hoffert, S. P. (1998). Researchers call for collaboration on wild primates, human diseases. *The Scientist*, **12**, 1.
Horwich, R. & Gebhard, K. (1983). Roaring rhythms in black howler monkeys (*Alouatta pigra*) of Belize. *Primates*, **24**, 290–296.

Horwich, R. & Jones, M. (1998). Wildlife conservation crosses international borders – profile of a community conservation organization. *Journal of Wildlife Rehabilitation*, **21**, 29–45.

Horwich, R. H. & Johnson, E. D. (1986). Geographical distribution of the black howler (*Alouatta pigra*) in Central America. *Primates*, **27**, 53–62.

Horwich, R. H. & Lyon, J. (1990). *A Belizean Rainforest – The Community Baboon Sanctuary*. Gays Mills, WI: Community Conservation, Inc.

James, R. A., Leberg, P. L., Quattro, J. M. & Vrijenhoek, R. C. (1997). Genetic diversity in black howler monkeys (*Alouatta pigra*) from Belize. *American Journal of Physical Anthropology*, **102**, 329–336.

Jimenez-Quioros, O. & Brenes, R. R. (1957). Helmintos de la Republica de Costa Rica. V. Sobre la validez del genero *Controrchis* Price 1928 (Trematoda, Dicrocoeliidae) *y descripcion de* Controrchis caballeroi n. sp. *Revista de Biologia Tropical*, **5**, 103–121.

Koontz, F. W., Horwich, R., Saqui, E. *et al.* (1994). Reintroduction of black howler monkeys (*Alouatta pigra*) into the Cockscomb Basin Wildlife Sanctuary, Belize. American Zoo and Aquarium Annual Conference. Bethesda, MD.

Kuntz, R. E. & Myers, B. J. (1972). *Parasites of South American Primates: South American Primates in Captivity*. Regent's Park, London: The Zoological Society of London.

Leclerc, M. C., Hugot, J. P., Durand, P. & Renaud, F. (2004). Evolutionary relationships between 15 *Plasmodium* species from New and Old World primates (including humans): an 18S rDNA cladistic analysis. *Parasitology*, **129**, 677–684.

Legesse, M. & Erko, B. (2004). Zoonotic intestinal parasites in *Papio anubis* (baboon) and *Cercopithecus aethiops* (vervet) from four localities in Ethiopia. *Acta Tropica*, **90**, 231–236.

Lilly, A. A., Mehlman, P. T. & Doran, D. (2002). Intestinal parasites in gorillas, chimpanzees, and humans at Mondika Research Site, Dzanga-Ndoki National Park, Central African Republic. *International Journal of Primatology*, **23**, 555–573.

Looney, W. J. (1998) *Cyclospora* species as a cause of diarrhea in humans. *British Journal of Biomedical Science*, **55**, 157–161.

Michaud, C., Tantalean, M., Ique, C., Montoya, E. & Gozalo, A. (2003). A survey for helminth parasites in feral New World non-human primate populations and its comparison with parasitological data from man in the region. *Journal of Medical Primatology*, **32**, 341–345.

Milton, K. (1980). *The Foraging Strategy of Howler Monkeys: A Study in Primate Economics*, New York, NY: Columbia University Press.

Muriuki, S. M. K., Murugu, R. K., Munene, E., Karere, G. M. & Chai, D. C. (1998). Some gastro-intestinal parasites of zoonotic (public health) importance commonly observed in old world non-human primates in Kenya. *Acta Tropica*, **71**, 73–82.

Nizeyi, J. B., Cranfield, M. R. & Graczyk, T. K. (2002a). Cattle near the Bwindi Impenetrable National Park, Uganda, as a reservoir of *Cryptosporidium parvum* and *Giardia duodenalis* for local community and free-ranging gorillas. *Parasitology Research*, **88**, 380–385.

Nizeyi, J. B., Sebunya, D., Dasilva, A. J. *et al.* (2002b). Cryptosporidiosis in people sharing habitats with free-ranging mountain gorillas (*Gorilla gorilla beringei*), Uganda. *American Journal of Tropical Medicine and Hygene*, **66**, 442–444.

Orihel, T. C. (1970). The helminth parasites of nonhuman primates and man. *Laboratory Animal Care*, **20**, 395–400.

Ostro, L. E. T., Silver, S. C., Koontz, F. W., Howich, R. H. & Brockett, R. (2001). Shifts in social structure of black howler (*Alouatta pigra*) groups associated with natural and experimental variation in population density. *International Journal of Primatology*, **22**, 733–748.

Pastor-Nieto, R. (1991). Identificacion de helmintos del mono aullador (*Alouatta palliata*). Facultad de Medicina Veterinaria y Zootecnia, Universidad Nacional Autonoma de Mexico. Thesis.

Pastor-Nieto, R. (1993). Preliminary note on the identification of gastrointestinal helminth parasites of a wild troop of howler monkeys (*Alouatta palliata)* in Southern Mexico. *Diplomado en medicina y manejo de fauna silvestre. Modulo IV: Medicina y manejo de primates.* Facultad de Medicina Veterinaria y Zootecnia, Universidad Nacional Autonoma de Mexico.

Patz, J. A., Graczyk, T. K., Geller, N. & Vittor, A. Y. (2000). Effects of environmental change on emerging parasitic diseases. *International Journal of Parasitology*, **30**, 1395–1405.

Phillips, K. A., Haas, M. E., Grafton, B. W. & Yrivarren, M. (2004). Survey of the gastrointestinal parasites of the primate community at Tambopata National Reserve, Peru. *Journal of Zoology*, **264**, 149–151.

Pillai, D. R. & Kain, K. C. (2003). Common intestinal parasites. *Current Treatment Options in Infectious Diseases*, **5**, 207–217.

Ralph, A., O'Sullivan, M. V. N., Sangster, N. C. & Walker, J. C. (2006). Abdominal pain and eosinophilia in suburban goat keepers – trichostrongylosis. *Medical Journal of Australia*, **184**, 467–469.

Santa Cruz, A. C. M. (2000). Habitat fragmentation and parasitism in howler monkeys (*Alouatta caraya*). *Neotropical Primates*, **8**, 146–148.

Schad, G. A. (1994). Hookworms: pets to humans. *Annals of Internal Medicine*, **120**, 434–435.

Schrag, S. J. & Wiener, P. (1995). Emerging infectious diseases: what are the roles of ecology and evolution? *Trends in Ecology and Evolution*, **10**, 319–324.

Schuster, F. L. & Visvesvara, G. S. (2004). Amebae and ciliated protozoa as causal agents of waterborne zoonotic disease. *Veterinary Parasitology*, **126**, 91–120.

Serrano, M. A. (1998). Incidencia de Protozoarios gastrointestinales en en primates del Zoologico de Zacango de Calimaya Estado de Mexico. Facultad de Medicina Veterinaria y Zootecnia, Universidad Autonoma del Estado de Mexico. Thesis.

Silver, S. C. & Marsh, L. K. (2003). Dietary flexibility, behavioral plasticity, and survival in fragments: lessons from translocated howlers. In *Primates in Fragments: Ecology and Conservation*, ed. L. K. Marsh. New York, NY: Kluwer Academic and Plenum, pp. 251–266.

Sleeman, J. M., Meader, L. L., Mudakikwa, A. B., Foster, J. W. & Patton, S. (2000). Gastrointestinal parasites of mountain gorillas (*Gorilla gorilla beringei*) in the

Parc National des Volcans, Rwanda. *Journal of Zoological and Wildlife Medicine*, **31**, 322–328.
Slingenbergh, J., Gilbert, M., De Balogh, K. & Wint, W. (2004). Ecological sources of zoonotic diseases. *Revue Scientifique et Technique de L'Office International Des Epizooties*, **23**, 467–484.
Solaymani-Mohammadi, S. & Petri, W. A. (2006). Zoonotic implications of the swine-transmitted protozoal infections. *Veterinary Parasitology*, **140**, 189–203.
Stoner, K. E. (1996). Prevalence and intensity of intestinal parasites in mantled howling monkeys (*Alouatta palliata*) in northeastern Costa Rica: implications for conservation biology. *Conservation Biology*, **10**, 539–546.
Stuart, M., Pendergast, V., Rumfelt, S. *et al.* (1998). Parasites of wild howlers (*Alouatta* spp.). *International Journal of Primatology*, **19**, 493–512.
Taylor, L. H., Latham, S. M. & Woolhouse, M. E. J. (2001). Risk factors for human disease emergence. *Philosophical Transactions of the Royal Society of London B*, **356**, 983–989.
Thoisy, B. D., Vogel, I., Reynes, J. M. *et al.* (2001). Health evaluation of translocated free-ranging primates in French Guiana. *American Journal of Primatology*, **54**, 1–16.
Thompson, R. C. A. (2000). Giardiasis as a re-emerging infectious disease and its zoonotic potential. *International Journal of Parasitology*, **30**, 1259–1267.
Thompson, R. C. A. (2002). Towards a better understanding of host specificity and the transmission of *Giardia*: the impact of molecular techniques. In *Giardia: The Cosmopolitan Parasite*, ed. B. E. Olson, M. E., Olson & P. M. Wallis. Wallingford, UK: CABI Publishing.
Thompson, R. C. A., Hopkins, R. M. & Homan, W. L. (2000). Nomenclature and genetic groupings of *Giardia* infecting mammals. *Parasitology Today*, **16**, 210–213.
Vitazkova, S. K. (2005). Demographic and ecological factors affecting the parasites of free-ranging black howler monkeys, *Alouatta pigra*, from Belize and Mexico. PhD thesis. New York, Columbia University.
Vitazkova, S. K. & Wade, S. E. (2006). Parasites of free ranging black howler monkeys, *Alouatta pigra*, from Belize and Mexico. *American Journal of Primatology*, **68**, 1089–1097.
Vitazkova, S. K. & Wade, S. E. (2007). The effects of ecology on the endo-parasites of *Alouatta pigra*. *International Journal of Primatology*, **28**, 1327–1343.
Volney, B., Pouliquen, J. F., De Thoisy, B. & Fandeur, T. (2002). A sero-epidemiological study of malaria in human and monkey populations in French Guiana. *Acta Tropica*, **82**, 11–23.
Wolfe, N. D., Escalante, A. A., Karesh, W. B. *et al.* (1998). Wild primate populations in emerging infectious disease research: the missing link? *Emerging Infectious Diseases*, **4**, 149–158.
Woodford, M. H., Butynski, T. M. & Karesh, W. B. (2002). Habituating the great apes: the disease risks. *Oryx*, **36**, 152–160.
Ziccardi, M. & Lourenço-de-Oliveira, R. (1997). The infection rates of trypanosomes in squirrel monkeys at two sites in the Brazilian Amazon. *Memorias do Instituto Oswaldo Cruz*, **92**, 465–470.

19 *Primate parasite ecology: patterns and predictions from an ongoing study of Japanese macaques*

ALEXANDER D. HERNANDEZ, ANDREW J. MACINTOSH, AND MICHAEL A. HUFFMAN

Photograph by Alexander D. Hernandez (*Macaca fuscata yakui*)

Introduction

The behavior, ecology, and evolution of parasites or infectious diseases in primates are exciting emerging fields of study. For over 50 years primatology has clarified our understanding of primate behavior, ecology, and evolution (Smuts

Primate Parasite Ecology. The Dynamics and Study of Host–Parasite Relationships, ed. Michael A. Huffman and Colin A. Chapman. Published by Cambridge University Press.

et al., 1987; Sussman, 2006), and this occurred at the same time that parasitology began to incorporate ecological and evolutionary paradigms after centuries of an exacting biomedical rubric that directed the way parasites were treated in scholarly circles (Sukhdeo & Hernandez, 2005). Thus, it is not surprising that, as academic fields of study, parasite and primate ecology only recently began to explore their commonalities, especially because of the close relationship between non-human primates and humans, and the increasing threat of emerging infectious diseases speculated to have resulted from continuing anthropogenic abuse of natural ecosystems sustaining non-human primates (Wolfe *et al.*, 1998; Wallis & Lee, 1999; Chapman *et al.*, 2005; Gillespie *et al.*, 2005). However, early reviews highlight the lack of empirical data because very few studies have been designed to test predictions on how the ecology of parasites affects the ecology of primates, or how primate ecology affects parasite ecology (Nunn & Altizer, 2006; Chapman *et al.*, Chapter 21, this volume; Huffman *et al.*, Chapter 16, this volume). The time seems ripe then, for studies that explicitly test the relation between parasites and primates in the context of field patterns and theoretical models that have evolved in both fields.

For decades parasite ecology has been concerned with the rules of assembly in parasite populations and communities within their hosts, the roles of parasites in host population regulation, and the evolutionary and ecological implications of parasite mediation in trophic interactions (Anderson & May, 1979; Freeland, 1983; Minchella & Scott, 1991; Hudson *et al.*, 1998; Combes, 2001; Lafferty & Kuris, 2002; Poulin, 2007a). Few rules have actually emerged from these tireless efforts. However, one clear pattern that appears to be nearly universal is the aggregated distribution of a parasite's population within its host's population (Poulin, 2007b).

The pattern of aggregated parasite distributions generally shows that a relatively small number of host individuals in a population harbors a large number of individuals in a parasite's population (Crofton, 1971; Pennycuick, 1971; Shaw & Dobson, 1996). This is such a widespread pattern that before new host populations are sampled for the first time, it is relatively safe to predict that the pattern will emerge, and even predict the level of aggregation to expect with some level of confidence (Poulin, 2007b). Some of the factors thought to determine this aggregated distribution are categorized as being independent of the number of parasites infecting a host, i.e. they are density independent, and can include variables such as host gender and behavior (Pennycuick, 1971; Shaw & Dobson, 1996). Primates in general are highly social animals and it has been suggested that studying their behavior can help us better understand the importance of parasite density independent factors, such as variation in host social dominance, to the dynamics of parasitic diseases including their aggregated distribution within host populations (Nunn & Altizer, 2006).

Relatively few studies have examined the relation between host social dominance and parasitism in wild non-human primates. Data support the prediction that the higher the social-rank status of individuals, the less abundant the parasites; however, support for the exact opposite prediction also exists. For example, high-ranking yellow baboons, *Papio cynocephalus*, shed more nematode eggs, and this pattern was more prominent in males than females (Hausfater & Watson, 1976; Meade, 1984). In contrast, death due to infection with unknown parasites was more likely in low-ranking vervet monkeys, *Cercopithecus aethiops* (Cheney *et al.*, 1988), while female olive baboons, *Papio cynocephalus anubis*, from lower-ranking families were also more frequently infected with lice (Eley *et al.*, 1989). In another study, social rank did not appear to influence parasitism in troops of olive baboons (Muller-Graf *et al.*, 1996). Clearly, social dominance may influence dynamics of parasite populations in wild non-human primates, but further study of questions relevant to this topic is undoubtedly needed.

A vast majority of our early understanding about primate social dominance and behavior has come from studies on cercopithecines, especially Japanese macaques, *Macaca fuscata,* and this has been extensively reviewed in recent times (Gouzoules & Gouzoules, 1987; Melnick & Pearl, 1987; Sussman, 2006; Thierry, 2006). Longitudinal studies on provisioned and non-provisioned populations of this species, and relatively easy identification of individuals in groups, have helped elucidate how behavioral interactions and social relationships can be influenced by kinship, especially among females (Yamada, 1963; Koyama, 1970; Yamagiwa & Hill, 1998). For example, Japanese macaque females within a troop are matrilineal kin that generally do not disperse; they form kin-groups or female-bonded groups where closely related females associate with each other more than with more distant kin (Wrangham, 1980; Yamagiwa & Hill, 1998). Dominance rank among females is also generally linear and stable, and this stability may be re-enforced by the permanent association of females with kin (Yamagiwa & Hill, 1998). However, the relation between dominance rank and parasite infections in these monkeys has never been tested, especially because relatively few studies exist on the interaction between parasites and hosts in wild Japanese macaque populations (Nunn & Altizer, 2005).

In this chapter we use one month of preliminary field data to outline how the study of a subspecies of Japanese macaque, *Macaca fuscata yakui*, on the island of Yakushima can help test some early predictions in the fledgling field of primate parasite ecology. We focus on host social dominance and the distribution of parasite populations within hosts. However, our main objective is to illustrate the merger of classical ideas in parasite and primate ecology.

Materials and methods

Study site description

Research on the ecology of parasites and Japanese macaques in a lowland forest (between 0 and 399 m asl) with mixed subtropical and warm-temperate broad-leaved evergreen trees on the northwest coast of the island of Yakushima was started in December 2006. The island, located in the southwest of Japan (30°N, 131°E), is the southern-most location in the geographic distribution of Japanese macaques, and has a total area of a little over 500 km^2. Yakushima is one of only two unique study sites where wild macaques have been habituated without provisioning, and studied continuously since 1973 (Yamagiwa & Hill, 1998). This provides an opportunity to learn about macaque behaviors that could be important to their interaction with parasites in a relatively undisturbed natural ecosystem, and existing long-term studies provide excellent background data on their ecology.

Study group description

Approximately 10 troops have been identified and their individual members counted in the northwest area of Yakushima since 1974, and for our study we have been focusing on one relatively new study troop in the yearly census, Umi troop. "Umi" is the Japanese word for "sea," and this name may have been chosen because the troop ranges mostly near, and often visits, the rocky coastline of its territory. Presently, Umi troop is composed of 13 adult females, seven adult males, 19 one- to four-year-old juveniles, and two infants. The size of this troop is higher than the average of 32 individuals/troop reported for Japanese macaques on Yakushima, but it is within the range of variation that has been reported over the years, i.e. 25–57 (Kawamura & Itani, 1952; Iwano, 1983).

Behavioral observations and fecal collections

Behavioral observations and fecal collections were started in December 2006 on all 13 adult females and three adult males (ranks 1, 2, 3) in the Umi group and will continue until August 2008. Observations are made during a 4-week block that is followed by a 4-week break, and by the end of the study in 2008 there will be a total of 10 observation blocks. Four 30-minute focal-follows are conducted for each individual in a 4-day period within a week, and their activity is recorded every 2 minutes. When a focal-session ends, switching occurs to a different individual in the vicinity, and who is on a list of focal-animals in need of observation for that week. If no other individuals are spotted nearby within a 5-minute period, a new focal-session is started for the individual that was previously being followed. An attempt is made to not repeat focal-follow sessions on the same individual during similar time-periods in a week.

Additional data are being collected *ad libitum* on dyadic agonistic interactions to establish a dominance-rank hierarchy between females in the troop. Dominance rank between the males is relatively straightforward and easy to deduce. However, dominance rank between females is more complicated to decipher, and therefore data collected are being analyzed with the aid of MatMan software 1.1 (Vries, 1995).

Parasite extraction and quantification

For parasite infection analysis, one fecal sample is collected for each focal-individual every 2 weeks, fixed with 10% formalin in 14-ml plastic test tubes, and brought to the lab for processing. Extraction of parasite eggs follows a modified formalin-ether method, or MGL (Ritchie, 1948; Hasegawa, Chapter 2, this volume), except that ether is substituted with ethyl acetate because it is less volatile (Truant *et al.*, 1981). The end product after the extraction process is homogenously suspended in 25 ml of 10% formalin stirred on a magnetic stir-plate, and then 5 separate aliquots are used to identify and count parasite eggs within the 0.15-ml grid of a MacMaster chamber slide. Count data are used to estimate the number of eggs for each parasite species per gram of feces from each individual monkey.

Results

A total of 192 h of observations have been recorded for individuals in Umi troop in three 4-week blocks. Results from a matrix with 113 total dyadic interactions suggest a linear dominance hierarchy between the 13 adult females in the troop (Landau's linearity index (h^1) = 0.563, P = 0.003). The number of one-way relationships is 47 (60.26%) and none are two-way. However, 31 (39.74%) of the interactions between all of the females are still unknown, especially those involving mid-ranking individuals, and more data are needed for better resolution of the dominance-rank hierarchy.

Preliminary results from one sample from each of the 16 macaques during the March 2007 sampling block show that individuals from Umi troop are

Table 19.1. *Population parameters for five species of nematode parasites found infecting Japanese macaques from Umi troop on Yakushima Island during March 2007. Prevalence is the percentage of hosts examined (n = 16) that are infected, and mean intensity of infection is the average number of eggs per gram of feces infected with each parasite species*

Parasite	Prevalence	Average eggs/g ± SD
Indirect life cycle		
Streptopharagus pigmentatus	87.5%	3800.5 ± 3716.9
Gongylonema pulchrum	6.25%	4.6 ± 9.9
Direct life cycle		
Oesophagostomum aculeatum	75%	66.7 ± 52.9
Trichuris trichiura	25%	135.9 ± 277.4
Strongyloides fulleborni	18.75%	1.5 ± 5.9

infected with *Streptopharagus pigmentatus*, *Gongylonema pulchrum*, *Trichuris trichiura*, *Oesophagostomum aculeatum*, and *Strongyloides fuelleborni* (Table 19.1). All of these nematode parasites were reported in the only previous parasite study on Japanese macaques from Yakushima (Gotoh, 2000). Our collaborator Hideo Hasegawa (Oita Medical University, Japan) has also confirmed identification of eggs from Umi individuals, and eggs and adults collected from a female macaque found dead in a troop neighboring Umi. Two of these species are trophically transmitted, and to complete their life cycle, they depend on a monkey's ingestion of insects (dung beetles or cockroaches) carrying juvenile stages encysted in muscle tissue. The other three species have a direct life cycle, which means that transmission from one host to the next occurs through a monkey's direct contact with parasite infective stages.

Prevalence of infection, or the proportion of hosts examined that harbor parasites, varies between parasite species. The two most prevalent eggs in fecal samples are of *S. pigmentatus*, which is trophically transmitted, and *O. aculeatum*, which has a direct life cycle (Table 19.1). However, all individual macaques are infected with at least one parasite species.

Mean intensity of infection, measured as the number of eggs/gram of feces from infected individuals, varies considerably between parasite species. The largest egg counts are for *S. pigmentatus* and *T. trichiura* (Table 19.1). A large coefficient of variance in each of the parasite species is typical of populations that have an aggregated distribution (Scott, 1987), and statistical comparisons against predicted random (Poisson) distributions show significant differences in four of the five species (Figure 19.1).

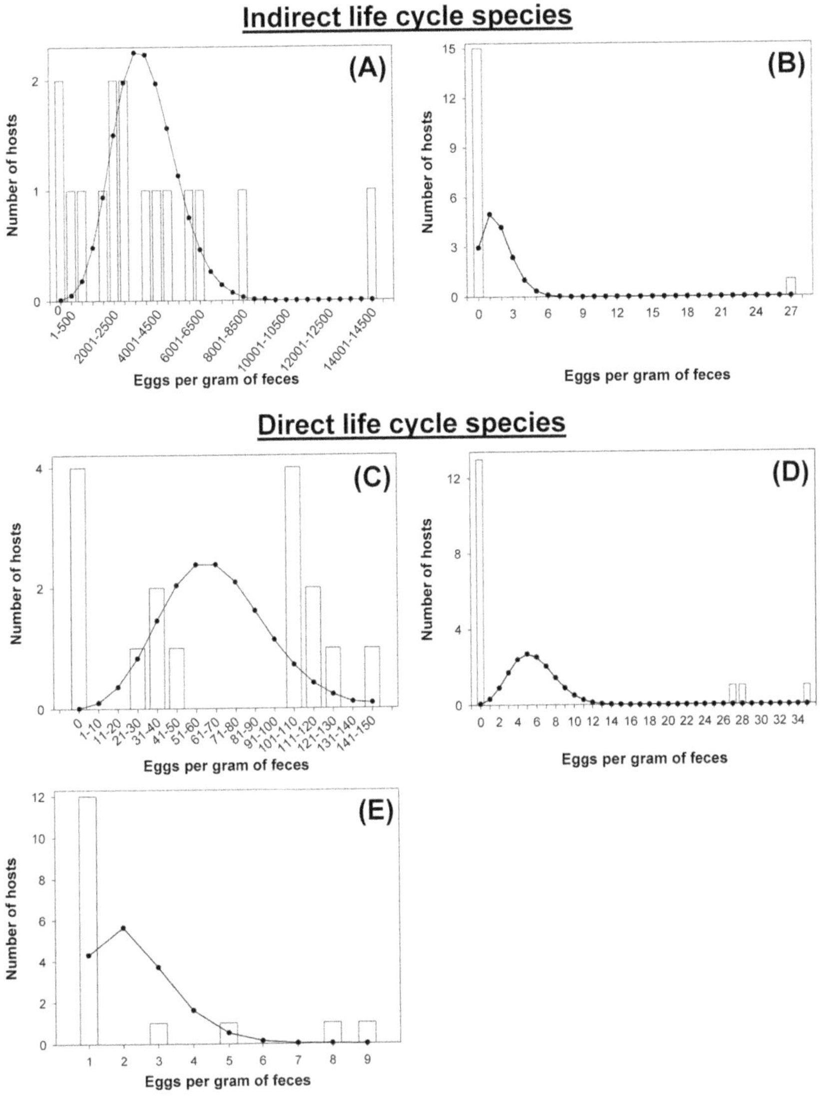

Figure 19.1. The observed (open bars) and the predicted random, or Poisson (dotted line) distributions of the parasite populations found infecting Japanese macaques on Yakushima Island during March 2007. Indirect life cycle, or trophically transmitted species, include: (A) *Streptopharagus pigmentatus* ($\chi^2 = 5.8$, DF = 1, P = 0.01) and (B) *Gongylonema pulchrum* ($\chi^2 = 12.4$, DF = 1, P < 0.001). Direct life cycle species include: (C) *Oesophagostomum aculeatum* ($\chi^2 = 14.9$, DF = 1, P < 0.001), (D) *Strongyloides fulleborni* ($\chi^2 = 16.9$, DF = 1, P < 0.001) and (E) *Trichuris trichiura* ($\chi^2 = 1.1$, DF = 1, P = 0.29).

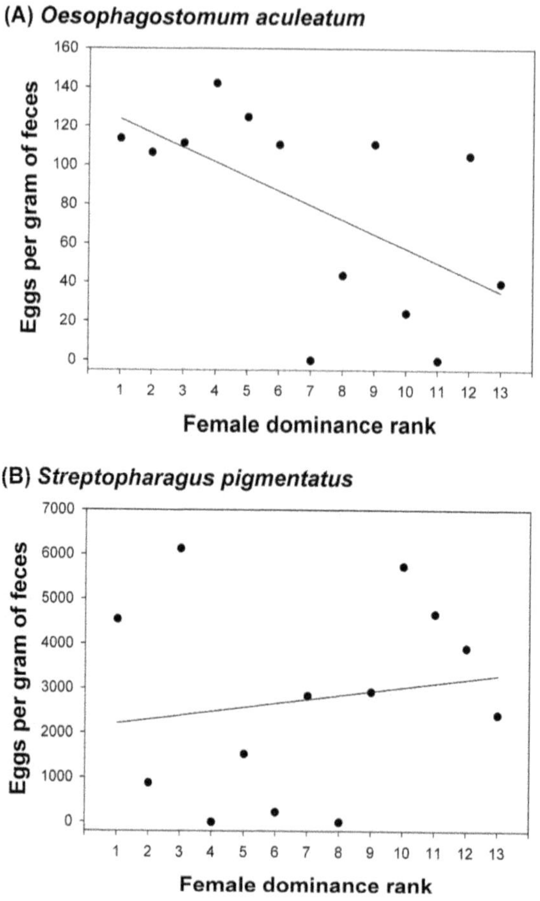

Figure 19.2. The observed relation between female dominance rank and the mean number of (A) *Oesophagostomum aculeatum* and (B) *Streptopharagus pigmentatus* eggs counted per gram of feces from Japanese macaques on Yakushima Island during March 2007.

Host social rank can explain variations in the number of eggs counted in the females of Umi troop for one of the parasite species. There is a significant negative relation between host social rank and the number of *O. aculeatum* eggs per individual (Figure 19.2A; ANOVA: $F_{1,13} = 5.45$, $P = 0.04$, $R^2 = 0.33$). However, social rank does not significantly explain variations in the number of *S. pigmentatus* eggs counted (Figure 19.2B; $F_{1,13} = 0.29$, $P = 0.6$, $R^2 = 0.03$). No comparison was made for the other three parasite species because they infected only one or two individual females.

Discussion

Data from only one field-sampling month show that parasite populations generally have an aggregated distribution in the troop of Japanese macaques studied on Yakushima Island. A review of the relationship between the variance and the mean number of parasites per host across 269 natural populations of metazoan parasites infecting vertebrate hosts shows that an almost perfect linear relation exists, with the mean explaining almost 90% of the variability in the variance (Shaw & Dobson, 1996). Thus, it is not surprising that this pattern emerges in parasite populations that infect macaques on Yakushima, and may have even been predicted with some degree of confidence before sampling ever began. However, the factors that lead to these aggregated parasite distributions can vary across host populations, and can be more elusive to decipher (Poulin, 2007a). One explanatory factor that is independent of the density of parasite populations, and thought to be important in the study of primate parasite ecology, is host behavior (Nunn & Altizer, 2006). Japanese macaques are highly social by nature with a stable linear dominance hierarchy among females in a troop, and testing the prediction that social dominance can be important to the aggregated distribution of their parasite populations on Yakushima was an obvious place to begin our analyses.

Dominance rank can explain some of the variation in the intensity of infection with *O. aculeatum*. Dominant individuals are expected to have more exposure to parasites because they may experience increased social contact (Nunn & Altizer, 2006), and this prediction may be applicable to those parasite species with transmission strategies that require direct contact between the hosts and infective stages, such as *O. aculeatum*. Indeed, the pattern for this parasite species shows that high-ranking female macaques have higher egg counts, but interestingly social rank does not explain the variation in infection with the other species with similar transmission biology. High-ranking individuals have also been predicted to contact more parasites that are trophically transmitted because they can dominate access to food, water, and other resources (Nunn & Altizer, 2006; Muelhenbein, Chapter 4, this volume), but infection patterns with *S. pigmentatus* show no relation with dominance rank. A lack of difference in infections with parasite species that require insects as intermediate hosts suggests females of all ranks in Umi troop are, at the very least, consuming similar numbers of infected beetles and/or cockroaches in their territory. Thus, the relation between social dominance and the different parasite species that infect Japanese macaques on Yakushima is far from clear, especially because dominance rank relations among Umi females is not yet complete. More importantly, what is clear is that we are seeing a reflection of parasite–host relationships at one particular point in time, and it is the dynamics of these relations across contiguous seasons that ultimately interest us to better answer a wider range of questions.

One of the topics that we believe can be addressed through our studies on Yakushima includes the infection probability after hosts have come in contact with parasites (Nunn & Altizer, 2006). For example, dominant individuals may experience immunosuppressive effects of testosterone and other stress hormones that could result in a positive association between high rank and parasitism (Muehlenbein, Chapter 4, this volume). This may explain why high-ranking females have higher levels of infection with *O. aculeatum*, but it does not explain why we do not see a similar pattern in the other parasite species. A different prediction suggests that dominant individuals may have better access to resources, which could strengthen their immune system and lower parasite levels. We are measuring corticosteroid hormones in Umi troop females to examine the relation between hormone and parasite levels, and test some of these predictions. The early patterns presented here suggest that there may be a relation between social rank and infection levels for at least one species of parasite in Yakushima macaques, and future results from analysis of stress hormones may provide some insight into the proximate determinants of this pattern.

Other ecological factors can also affect interactions between host and parasites. For example, diet is thought to be a critical component, and arguments exist for including the potential for encountering parasites into optimal foraging theory (Lozano, 1991). This idea suggests that animals may not only forage optimally for nutritional gains, but that it may also be advantageous for them to avoid ingesting items that are a source of parasites, or select a diet with anti-parasitic capabilities (Lozano, 1991; Huffman, 1997, 2006). Japanese macaques are dietary generalists and are able to alter their feeding behaviors as seasonal changes dictate the availability of food items (Maruhashi, 1980; Hill, 1997; Hanya, 2003, 2004; Agetsuma 1995a, b). That two species of trophically transmitted intestinal nematodes parasitize macaques on Yakushima is an indication that insects, and in particular dung beetles and/or cockroaches, comprise at least part of their diet. Indeed, animal matter (largely insects) made up nearly 8% of their annual diet in one study at Yakushima (Hill, 1997). Thus, according to predictions from the optimal foraging theory modified to include probabilities of infection with parasites (Lozano, 1991), the nutritional gains associated with consuming insects in Yakushima macaques must outweigh the risks of gaining parasites.

What is not known is to what degree the diet of the macaques reduces parasite infection. Extensive literature reviews on chimpanzee and gorilla diet have shown that they consume many plants, or the same plant parts used medicinally by humans for the treatment of parasites and symptoms related to infection (Huffman *et al.*, 1998; Cousins & Huffman, 2002). Preliminary results indicate that Yakushima macaques ingest at least 46 plant parts (38 species from 27 families) with known medicinal or bioactive properties (MacIntosh *et al.*, in prep.). This number constitutes roughly 17% of all plant parts consumed by these animals (see Maruhashi, 1980; Agetsuma, 1995a, b; Hill, 1997; Domingo-Roura & Yamagiwa, 1999; Hanya, 2003, 2004; L. Tarnaud, pers. comm.). Furthermore, at least 22 of these parts (19 species from 16 families) are reportedly used by humans in the treatment of gastrointestinal upset or parasite- and pathogen-related symptoms. Specifically, four parts (four species of different families) are anthelmintic, directly linked to the treatment of nematode infections. However, the mere presence of these plant species in the diet is insufficient to make any conclusions about their actual effectiveness. It will be necessary to demonstrate actual changes in health condition and parasite infection levels of individuals after ingestion of these potential medicinal plants (Huffman & Seifu, 1989; Huffman *et al.*, 1993, 1996; Huffman & Caton, 2001). Currently, work is underway in examining parasite infection levels of individual macaques in Umi group with relation to their specific dietary behaviors, and one goal is to isolate the effect of secondary plant metabolites on their intestinal parasites. Thus, if a significant effect of plant metabolites on infection levels can be demonstrated, it may provide one other explanation for the variation in the aggregated distribution of parasites in hosts, and even dampen any potential relation with social dominance.

Conclusions

The preliminary patterns presented in this chapter briefly highlight how we can test some early predictions in the fledgling field of primate parasite ecology by studying the ecology of parasites in Japanese macaques on Yakushima Island. There are few detailed studies on the ecology of parasites and their non-human primate hosts in any system, but we would argue that one easy way to begin such a research program is to explore patterns relating to the distribution of parasites in their host populations. One advantage to studying the ecology of parasites in Japanese macaques from Yakushima is the ease with which individuals can be identified to facilitate tracking variations in their infection levels across multiple seasons, and correlating these changes to behavioral interactions and social relationships that are largely influenced by kinship. This, of course, is not unique to Japanese macaques, but nonetheless, it could lead to exciting insights into the factors that generate the commonly observed distribution of parasites. One clear conclusion is that a lot of the detailed work on the ecology of this host species exists thanks to the tireless efforts of devoted primatologists, but more work needs to be done on the ecology of their parasites. Data from just one month has generated relatively simple patterns, but they emphasize the importance of long-term data to generate more robust analyses and conclusions. For example, seasonality can be very important to the dynamics of parasites and other infectious diseases (Altizer *et al.*, 2006; Lass & Ebert, 2006; Huffman *et al.*, Chapter 16, this volume), and any future endeavors in the study of primate parasite ecology must consider the relative importance of this factor to patterns measured in natural populations. Indeed, it is an exciting time for studying parasite ecology in non-human primates, and our hope is that testing predictions based on patterns from natural Japanese macaque populations will make exciting contributions to this fledgling field, just as they have already done to the general field of primatology.

Acknowledgements

We are grateful to the Japan Ministry of Education, Culture, Sports, Science and Technology (Monbukagakusho) and the Japan Society for the Promotion of Science (JSPS) for Grant-in-Aid funds to MAH and ADH, and the Kagoshima Prefectural Government and Yakushima World Heritage Office for permission to conduct our research on Yakushima Island. Logistic support from the Yakushima Field Station of Kyoto University is also greatly appreciated. ADH was generously supported with a JSPS Fellowship for Foreign Researchers, and AJM by a Monbukagakusho scholarship for foreign students.

References

Agetsuma, N. (1995a). Foraging synchrony in a group of Yakushima macaques (*Macaca fuscata yakui*). *Folia Primatologica*, **64**, 167–179.

Agetsuma, N. (1995b). Foraging strategies of Yakushima macaques (*Macaca fuscata yakui*). *International Journal of Primatology*, **16**, 595–609.

Altizer, S. Dobson, A., Hosseini, P. *et al.* (2006). Seasonality and the dynamics of infectious diseases. *Ecology Letters*, **9**, 467–484.

Anderson, R. M. & May, R. M. (1979). Population biology of infectious diseases: Part I. *Nature*, **280**, 361–367.

Chapman, C. A., Gillespie, T. R. & Goldberg, T. L. (2005). Primates and the ecology of their infectious diseases: how will anthropogenic change affect host–parasite interactions? *Evolutionary Anthropology*, **14**, 134–144.

Cheney, D. L., Seyfarth, R. M., Andelman, S. J. & Lee, P. C. (1988). Reproductive success in vervet monkeys. In *Reproductive Success*, ed. T. H. Clutton-Brock. Chicago, IL: The University of Chicago Press, pp. 227–239.

Combes, C. (2001). *Parasitism: The Ecology and Evolution of Intimate Interactions.* Chicago, IL: The University of Chicago Press.

Cousins, D. & Huffman, M. A. (2002). Medicinal properties in the diet of gorillas – an ethnopharmacological evaluation. *African Study Monographs*, **23**, 65–89.

Crofton, H. D. (1971). A model of host-parasite relationship. *Parasitology*, **63**, 343–364.

Domingo-Roura, X. & Yamagiwa, J. (1999). Monthly and diurnal variations in food choice by *Macaca fuscata yakui* during the major fruiting season at Yakushima, Japan. *Primates*, **40**, 525–536.

Eley, R. M., Strum, S. C., Muchemi, G. & Reid, G. D. F. (1989). Nutrition, body condition, activity patterns, and parasitism of free-ranging troops of olive baboons (*Papio anubis*) in Kenya. *American Journal of Primatology*, **18**, 209–219.

Freeland, W. J. (1983). Parasites and the coexistence of animal species. *American Naturalist*, **121**, 223–236.

Gillespie, T. R., Chapman, C. A. & Greiner, E. C. (2005). Effects of logging on gastrointestinal parasite infections and infection risk in African primates. *Journal of Applied Ecology*, **42**, 699–707.

Gotoh, S. (2000). Regional differences in the infection of wild Japanese macaques by gastrointestinal helminth parasites. *Primates*, **41**, 291–298.

Gouzoules, S. & Gouzoules, H. (1987). Kinship. In *Primate Societies*, ed. B. B. Smuts, D. L. Cheney, R. M. Seyfarth & R. W. Wrangham. Chicago, IL: The University of Chicago Press, pp. 299–305.

Hanya, G. (2003). Age differences in food intake and dietary selection of wild male Japanese macaques. *Primates*, **44**, 333–339.

Hanya, G. (2004). Seasonal variations in the activity budget of Japanese macaques in the coniferous forest of Yakushima: effects of food and temperature. *American Journal of Primatology*, **63**, 165–177.

Hausfater, G. & Watson, D. F. (1976). Social and reproductive correlates of parasite ova emissions by baboons. *Nature*, **262**, 688–689.

Hill, D. A. (1997). Seasonal variation in the feeding behavior and diet of Japanese macaques (*Macaca fuscata yakui*) in lowland forest of Yakushima. *American Journal of Primatology*, **43**, 305–322.

Hudson, P. J., Dobson, A. P. & Newborn, D. (1998). Prevention of population cycles and parasitism. *Science*, **282**, 2256–2258.

Huffman, M. A. (1997). Current evidence for self-medication in primates: a multidisciplinary perspective. *Yearbook of Physical Anthropology*, **40**, 171–200.

Huffman, M. A. (2001). Self-medicative behavior in the African Great Apes: an evolutionary perspective into the origins of human traditional medicine. *BioScience*, **51**, 651–661.

Huffman, M. A. (2006). Primate self-medication. In *Primates in Perspective,* ed. C. J. Campbell, A. Fuentes, K. C. MacKinnon, M. Panger & S. K. Bearder. Oxford: Oxford University Press, pp. 677–690.

Huffman, M. A. & Caton J. M. (2001). Self-induced increase of gut motility and the control of parasitic infections in wild chimpanzees. *International Journal of Primatology*, **22**, 329–346.

Huffman, M. A. & Seifu, M. (1989). Observations on the illness and consumption of a possibly medicinal plant *Vernonia amygdalina* by a wild chimpanzee in the Mahale Mountains, Tanzania. *Primates*, **30**, 51–63.

Huffman, M. A., Gotoh, S., Izutsu, D., Koshimizu, K. & Kalunde, M. S. (1993). Further observations on the use of the medicinal plant, *Vernonia amygdalina* (Del) by a wild chimpanzee, its possible effect on parasite load, and its phytochemistry. *African Study Monographs*, **14**, 227–240.

Huffman, M. A., Page, J. E., Sukhdeo, M. V. K. *et al.* (1996). Leaf-swallowing by chimpanzees, a behavioral adaptation for the control of strongyle nematode infections. *International Journal of Primatolology*, **17**, 475–503.

Huffman, M. A., Ohigashi H., Kawanaka, M. *et al.* (1998). African great ape self-medication: a new paradigm for treating parasite disease with natural medicines? In *Towards Natural Medicine Research in the 21st Century*, ed. Y. Ebizuka. Amsterdam: Elsevier Science B.V., pp. 113–123.

Iwano, T. (1983). Socioecology of Yakushima macaques (1973–1976). *Nihonzaru*, **5**, 86–95 (in Japanese).

Kawamura, S. & Itani, J. (1952). *Natural Society of Yakushima Macaques*. Primate Research Group of Kyoto University (in Japanese).

Koyama, N. (1970). Changes in dominance rank and division of a wild Japanese monkey troop in Arashiyama. *Primates*, **11**, 335–390.

Lafferty, K. D. & Kuris, A. M. (2002). Trophic strategies, animal diversity and body size. *Trends in Ecology and Evolution*, **17**, 507–513.

Lass, S. & Ebert, D. (2006). Apparent seasonality of parasite dynamics: analysis of cyclic prevalence patterns. *Proceedings of the Royal Society of London* B, **273**, 199–206.

Lozano, G. A. (1991). Optimal foraging theory: a possible role for parasites. *Oikos*, **60**, 391–395.

Maruhashi, T. (1980). Feeding behavior and diet of the Japanese monkey (*Macaca fuscat yakui*) on Yakushima island, Japan. *Primates*, **21**, 141–160.

Meade, B. J. (1984). Host–parasite dynamics among *Amboseli baboons*. PhD thesis, Virginia Polytechnic Institute and State University, Blacksburg, Virginia.

Melnick, D. J. & Pearl, M. C. (1987). Cercopithecines in multimale groups: genetic diversity and population structure. In *Primate Societies*, ed. B. B. Smuts, D. L. Cheney, R. M. Seyfarth & R. W. Wrangham. Chicago, IL: University of Chicago Press, pp. 121–134.

Minchella, D. J. & Scott, M. E. (1991). Parasitism: a cryptic determinant of animal community structure. *Trends in Ecology and Evolution*, **6**, 250–254.

Müller-Graf, C. D. M., Collins, D. A. & Woolhouse, M. E. J. (1996). Intestinal parasite burden in five troops of olive baboons (*Papio cynocephalus anubis*) in Gombe Stream National Park, Tanzania. *Parasitology*, **112**, 489–497.

Nunn, C. L. & Altizer, S. (2005). The global mammal parasite database: an online resource for infectious disease records in wild primates. *Evolutionary Anthropology*, **14**, 1–2.

Nunn, C. L. & Altizer, S. (2006). *Infectious Diseases in Primates: Behavior, Ecology and Evolution*. Oxford: Oxford University Press.

Pennycuick, L. (1971). Frequency distributions of parasites in a population of three-spined sticklebacks, *Gasterosteus aculeatus* L., with particular reference to the negative binomial distribution. *Parasitology*, **63**, 389–406.
Poulin, R. (2007a). *Evolutionary Ecology of Parasites*, 2nd edn. Princeton, NJ: Princeton University Press.
Poulin, R. (2007b). Are there general laws in parasite ecology? *Parasitology*, **134**, 763–776.
Ritchie, L. S. (1948). An ether sedimentation technique for routine stool examination. *Bulletin of the U.S. Army Medical Department*, **8**, 326.
Shaw, D. J. & Dobson, A. P. (1996). Patterns of macroparasite abundance and aggregation in wildlife populations : a quantitative review. *Parasitology*, **111**, S111–S133.
Smuts, B. B., Cheney, D. L., Seyfarth, R. M. & Wrangham, R. W. (1987). *Primate Societies*. Chicago, IL: University Chicago Press.
Sukhdeo, M. V. K. & Hernandez, A. D. (2005). Food web patterns and the parasite's perspective. In *Parasitism and Ecosystems*, ed. F. Thomas, J.-F. Guegan & F. Renaud. Oxford: Oxford University Press, pp. 54–67.
Sussman, R. W. (2006). A brief history of primate field studies. In *Primates in Perspective*, ed. C. J. Campbell, A. Fuentes, K. C. MacKinnon, M. Panger & S. K. Bearder. Oxford: Oxford University Press, pp. 6–10.
Thierry, B. (2006). The macaques. In *Primates in Perspective*, ed. C. J. Campbell, A. Fuentes, K. C. MacKinnon, M. Panger & S. K. Bearder. Oxford: Oxford University Press, pp. 224–239.
Truant, A. L., Elliott, S. H., Kelly, M. T. & Smith, J. H. (1981). Comparison of formalin-ether sedimentation, formalin-ethyl acetate sedimentation, and zinc sulfate flotation techniques for detection of intestinal parasites. *Journal of Clinical Microbiology*, **13**, 882–884.
Vries, H. (1995). An improved test of linearity in dominance hierarchies containing unknown or tied relationships. *Animal Behavior*, **50**, 1375–1389.
Wallis, J. & Lee, D. R. (1999). Primate conservation: the prevention of disease transmission. *International Journal of Primatology*, **20**, 803–826.
Wolfe, N. D., Escalante, A. A., Karesh, W. B. *et al.* (1998). Wild primate populations in emerging infectious disease research: the missing link? *Emerging Infectious Diseases*, **4**, 149–158.
Wrangham, R. W. (1980). An ecological model of female-bonded primate groups. *Behaviour*, **75**, 262–300.
Yamada, M. (1963). Five natural troops of Japanese monkeys in Shodoshima Island I: distribution and social organization. *Primates*, **34**, 419–430.
Yamagiwa, J. & Hill, D. A. (1998). Intraspecific variation in the social organization of Japanese macaques: past and present scope of field studies in natural habitats. *Primates*, **39**, 257–273.
Yamagiwa, J., Izawa, K. & Maruhashi, T. (1998). Long-term studies on wild Japanese macaques in natural habitats at Kinkazan and Yakushima: preface. *Primates*, **39**, 255–256.

20 *Crop raiding: the influence of behavioral and nutritional changes on primate–parasite relationships*

ANNA H. WEYHER

Photograph by Anna H. Weyher (*Papio anubis*)

Introduction

As infection and disease are important determinants of animal health and well-being (Scott, 1988), parasitological studies have become increasingly important for the conservation of primates (Stuart & Strier, 1995). Anthropogenic

Primate Parasite Ecology. The Dynamics and Study of Host–Parasite Relationships, ed. Michael A. Huffman and Colin A. Chapman. Published by Cambridge University Press.

influences may represent important parameters shaping parasite infection in wild primates. The expansion of human populations into primate habitats can increase the chance for new parasite–host interactions and may have a negative impact on primate species' survival (Wolfe *et al.*, 1998). This change in habitat dynamics can cause the introduction of foreign parasites into a population and, through the consequences of changing foraging strategies, may alter the relationship primates have with the parasites already present in their environment (Dobson, 1985; Grenfell & Gulland, 1995; Daszak *et al.*, 2000; Patz *et al.*, 2000; Chapman *et al.*, 2005a).

Although it is likely that human activity plays a role in primate–parasite interactions, relatively few studies have examined patterns of variation in parasite infection between populations of the same species that differ in their interaction with humans (Eley *et al.*, 1989; McGrew *et al.*, 1989; Appleton & Henzi, 1993; Muller-Graf *et al.*, 1996, 1997; Hahn *et al.*, 2003; Gillespie *et al.*, 2005). Understanding the causes and consequences of such variation may have important implications for primate conservation, as pathogens are known to play a key role in regulating population numbers (Anderson & May, 1978; Scott, 1988; Scott & Dobson, 1989; Gregory & Hudson, 2000).

Recent research has found that in the last 50 years humans have changed the earth's ecosystems more rapidly and extensively than at any other time period in human history (Millennium Ecosystem Assessment 2005, at http://www.maweb.org). In particular land has been converted into cropland at an alarming rate. It is estimated that more than one half of six of the earth's major biomes has been converted to agriculture and it is expected that this pattern will continue to increase in the future (Millennium Ecosystem Assessment 2005). This increase in the amount of land used for agriculture has affected the earth's ecosystems in a number of ways. In particular the encroachment of humans into wild habitat has forced numerous species, including primates, to alter their behavior in an attempt to adapt to these changes.

The planting of crops is one important way in which human habitat modification has affected the behavior and diet of primates. Human encroachment into wild primate habitat has led a number of primate species to change their behavior in order to raid human crops (Else & Lee, 1986; Patterson & Wallis, 2005). It is likely that these changes in diet and behavior associated with crop raiding have altered parasite–host dynamics in these species. There are few data however on the impact of crop raiding on primate–parasite relationships (but see Chapman *et al.*, 2005b). This represents a fundamental gap in our knowledge as over 70 species of primates raid human crops, raiding by primates has been observed in almost all range countries, and with the expansion of human populations, raiding is a phenomenon that will increase considerably in the future (Marsh & Mittermeier, 1987; Hill, 1997;

Warren, 2003; Chapman *et al.*, 2005b; Patterson & Wallis, 2005; Weyher *et al.*, 2006).

Baboons are an excellent species in which to study the effects of crop raiding on primate–parasite relationships. Their behavioral and environmental flexibility has enabled baboons to adapt to changing environments quite readily (Altmann *et al.*, 1993; Hahn *et al.*, 2003). In response to human-induced changes in habitat, baboons have learned to modify their behaviors in order to raid human crops (Fortham-Quick & Demment, 1988; Warren, 2003).

Baboon crop raiding

With the continued encroachment of humans and human settlements on natural baboon habitat, they have learned to exploit agricultural foods and human refuse. There are a number of benefits that baboons may incur from consuming human foods, including foraging, nutritional, and reproductive benefits, increased time for resting and social activity, lower mortality, increased adult survival, higher growth weight and final weight, better body condition, and possibly greater immune responses (Fortham-Quick & Demment, 1988; Warren, 2003).

Crop and rubbish raiding appear to create nutritional benefits for baboons. It has been found that baboons that crop raid and feed on human refuse spend more time resting and in social activity than do their wild feeding counterparts (Fortham-Quick, 1984; Fortham-Quick, 1986; Eley *et al.*, 1989; Altmann *et al.*, 1993; Warren, 2003). Two studies have measured the body condition of troops of baboons to determine if baboons, which are feeding on human foods, are in better physical condition than wild foraging baboons (Eley *et al.*, 1989; Altmann *et al.*, 1993). Physical body condition was measured qualitatively by individual appearance and quantitatively by subcutaneous fat and body weight (Eley *et al.*, 1989; Altmann *et al.*, 1993). These studies show that food-enhanced baboons seem to gain nutritional enhancement and better body condition by supplementing their diets with human food (Eley *et al.*, 1989; Altmann *et al.*, 1993).

Crop raiding may also create costs for individual baboons and baboon troops. Crop raiders are seen as pests to farmers and villagers who are affected by the baboon's raiding. Often baboons are snared, poisoned, or shot in an attempt to stop the pests (Warren, 2003). So although a benefit of crop raiding may be increased nutritional gain from a reliable food source, baboon crop raiders may have a higher mortality risk due to retaliation by farmers (Warren, 2003). In addition, it has been found that increased contact with humans can increase glucocorticoid stress levels in wildlife (Creel *et al.*, 2002; Chapman *et al.*,

2005b). The negative interactions that baboon crop raiders have with farmers may likely increase stress and cortisol in these animals which in turn may decrease their immune responses.

Another potential cost of crop raiding may be increased risk of disease transmission from humans, human food, human rubbish, and domesticated animals (Dittus, 1974; Fortham-Quick & Demment, 1988; Eley *et al.*, 1989). There are a number of parasite species in human populations that are present throughout baboon ranges and are possibly transmissible to baboons (see Gasser *et al.*, Chapter 3, this volume for the difficulty of verifying transmission has occurred). The genetic proximity of humans and non-human primates creates a potential for transmission of parasites between humans and baboons (Wolfe *et al.*, 1998). Because of poor sanitation conditions in many areas where baboons feed on human crops and human refuse, these animals are likely to come into contact with soil, water, and food contaminated with human feces. Crop raiders' increased contact with areas where livestock and dogs range may also increase their chances of contracting new parasites from these animals.

In addition to changes in contact with anthropozoonotic parasites from humans, changes in diet and behavior associated with crop raiding may alter primate–parasite relationships. Crops that baboons consume include maize, cassava, nuts, and fruit. These crops are easily digested and high in energy (Fortham-Quick & Demment, 1988). As mentioned above, regularly eating human foods allows baboons to increase their overall physical condition. These changes in food consumption are also likely to alter parasite dynamics in these animals. In particular, changes in diet from crop raiding may affect parasite establishment, rates of reproduction, and intensities of infection (Weyher *et al.*, 2006).

Two previous studies have examined parasite infection in baboons that supplement their diet with human food (Eley *et al.*, 1989; Hahn *et al.*, 2003). These studies compared baboon troops that enhance their diet by raiding human refuse with those that only forage on wild-grown foods. Eley *et al.* (1989) compared body condition, activity patterns, and parasitism among three troops of olive baboons in Amboseli, Kenya, which had differing access to human refuse dumps. This study allowed for a determination of relationships between body condition and parasitic load due to access to differing types of food. Two troops' main source of food was the garbage pits and vegetable gardens near human settlements. The third troop's diet was limited to natural forage. In their study, Eley *et al.* (1989) found that the troops provisioned with human refuse were in overall better body condition than the wild-feeding troop. Specifically, the rubbish raiders weighed more and had more subcutaneous fat. The provisioned troops' activity budgets also differed. Compared with the wild-foraging troop, one food-enhanced troop was observed to forage less and to rest and socialize more. The authors attributed these differences in activity budget to the enhanced nutrition of the rubbish-raiding troops. Consuming human foods can provide free-ranging primate groups with significant nutritional benefits (Else & Lee, 1986; Fa & Southwick, 1988) allowing them to attain their energy requirements quickly. This saves the baboons' time and energy, which is reflected in their physical activity and body condition (Eley *et al.*, 1989).

In addition to differences in activity budgets and physical condition associated with rubbish-raiding, Eley *et al.* (1989) found significant differences in strongyle parasite outputs between the rubbish-raiding troop for which this was quantified and the wild-foraging troop. The troop that only foraged on wild foods was significantly more parasitized with strongyle eggs than the rubbish-raiding troop. Eley *et al.* (1989) suggest that the wild-foraging troop may have a higher helminth worm burden because they may be moderately malnourished and therefore immunosuppressed (Eley *et al.*, 1989).

Hahn *et al.* (2003) compared helminth parasite infection in three troops of Amboseli baboons in an attempt to determine if foraging on human refuse resulted in greater exposure to human parasites. Two troops sampled were completely wild foraging. The third troop sampled supplemented their diet with human refuse from garbage dumps. Due to sampling techniques the authors only compared parasites in one wild-foraging troop with the rubbish-raiding troop. Hahn *et al.* (2003) found significant differences between the troops in number of individuals infected with three helminth species. The wild-foraging troop had significantly more individuals parasitized with *Trichuris* sp. and *Physaloptera* sp. The rubbish-raiding troop was significantly more parasitized with *Streptophargus* sp. There were no anthropozoonotic parasites discovered.

Physaloptera sp. and *Streptophargus* sp. are spuriud nematodes that involve an arthropod intermediate host (Hahn *et al.*, 2003). Baboons become infected with the parasite by eating these arthropod intermediate hosts (Hahn *et al.*, 2003). Why the rubbish-raiding troop had significantly more individuals parasitized with *Streptophargus* sp. and the wild-foraging troop had significantly more individuals parasitized with *Physaloptera* sp. cannot be determined as consumption of arthropod hosts was not measured, but the authors suggest that the differences in feeding behavior between the two troops may have affected the rates of infection of these parasites (Hahn *et al.*, 2003). Perhaps different arthropods are living in the two troops' environments. The authors were surprised with the results of the *Trichuris* sp. infections in the two troops. They assumed that the rubbish-raiding troop would be significantly more parasitized with *Trichuris* sp. as they live in a closed and contaminated environment that could likely lead to direct re-infection with this parasite. The opposite was true. The wild-foraging troop had a higher percentage of individuals with *Trichuris* sp. infections. The authors were attempting to determine if the baboons were being exposed to this parasite from humans. It is very difficult to determine from simple laboratory analysis of fecal specimens if parasites are being transmitted from one host species to another. Parasites must first be identified to species level to determine if the parasites infecting both human and non-human primates are the same species. Furthermore, recent research using DNA fingerprinting of parasites of the same species show that clear genetic groupings can be distinguished between human and non-human primates (Chapman *et al.*, 2005b; de Gruijter *et al.*, 2005; Gasser *et al.*, Chapter 3, this volume). In addition, light and scanning electron microscopy studies have shown that morphological differences of particular parasite species may also determine differences between parasites of the same species that infect human and non-human primates (Ooi *et al.*, 1993; Chapman *et al.*, 2005b). Instead of exchange of parasites from humans to baboons, perhaps nutritional differences between the two troops can explain the percentage of individuals infected with each parasite.

These studies measured different parasite parameters in rubbish-raiding and wild-foraging baboons. Eley *et al.* (1989) compared helminth egg outputs and Hahn *et al.* (2003) measured number of individuals infected with particular parasites in each troop. Although their results are not conclusive, both studies found differences in parasite infection between rubbish-raiding and wild-foraging baboons that appear to be linked to differences in diet between the troops (Eley *et al.*, 1989; Hahn *et al.*, 2003).

Despite the potential importance of anthropogenic effects on baboon parasite infections, the above studies are the only research that have compared parasites found in wild-feeding baboon groups with those found in groups that forage on human refuse (Eley *et al.*, 1989; Hahn *et al.*, 2003). Moreover, no previous study has compared parasites of wild-feeding baboons with those that crop raid (Weyher *et al.*, 2006). However, food-enhancement from human refuse is potentially very different from food-enhancement from human crops. Little is known about baboon troops that enhance their diet by raiding agricultural crops, but that do not also supplement their diet with human refuse. Human crops are very high in energy (Fortham-Quick & Demment, 1988) and baboons that crop raid may have a higher caloric intake than rubbish-raiding baboons. In addition, rates of contact with infected stages of human parasites may be very different for crop-raiding and rubbish-raiding baboons. At rubbish heaps baboons are likely to be in closer contact with one another and with parasite-contaminated soil, while crop-raiding baboons may come into contact less often with infective stages of parasites when raiding due to the larger size of agricultural fields compared with rubbish dumps; however, this remains to be quantified.

Discovering the parasites that crop-raiding baboons harbor in comparison to troops living in similar environments that do not enhance their diet with human crops will help us to discover what, if any, anthropozootic parasites may be affecting these troops that range in areas near to humans and domesticated animals. Furthermore, comparing the intensities of infection between crop-raiding and wild-foraging baboons may help us determine if and how changes in diet and behavior associated with crop raiding may alter parasite infection in primates.

In an attempt to understand how crop raiding may affect parasite infection in baboons, I compared the parasite output in a crop-raiding and a wild-foraging troop of olive baboons in Gashaka Gumti National Park, Nigeria (Weyher *et al.*, 2006). This study provides new insight into how individual parasite species may react differently to varying levels of nutrition in baboons.

Methods

Gashaka Gumti National Park (GGNP) is located in northeastern Nigeria (6° 55′–8° 05′N and 11° 11′–12° 13′E) and shares its eastern border with the Republic of Cameroon (Dunn, 1999). This study was undertaken in and around the southern sector of GGNP at two research sites. The first, Gashaka, lies outside the park near the village of Gashaka. The second, Kwano, lies 13 km from Gashaka inside the park. Vegetation is similar at both sites and is typified by a mixture of Guinea savanna woodland and lowland forest (Warren, 2003).

One troop at each of the two sites was studied. The two troops studied here range in very similar habitats, have comparable home-range sizes with similar amounts of rainfall seasonally and annually, forage on similar wild foods (Warren, 2003), and at the time of this study had comparable group sizes. The major difference between the ecology of the two troops lies in crop raiding. The crop-raiding troop ranges just outside of the national park, south of the Gam-Gam River, above Gashaka village. This troop crop raids throughout the year; however, peak times of raiding occur in February/March, and in July when crops are ripe (Warren, 2003). The main crops taken include maize, guinea corn, and cassava (Warren, 2003). The wild-foraging troop lives within GGNP, 10 km from the range of the crop-raiding troop, and does not consume any food from anthropogenic sources.

Two long-term studies have been conducted on these baboons (Warren, 2003; Higham *et al.*, 2005) and therefore all adult individuals could be reliably identified in both troops. The crop-raiding troop consisted of four adult females and six adult males, the wild-foraging troop consisted of seven adult females and six adult males. Three fecal samples were collected from each adult individual in both troops over 6 weeks from April to May 2004 (total = 69 samples). Samples were preserved in 30 ml Meridian Para-Pak™ tubes pre-filled with 15 ml SAF (sodium-acetate-acetic acid) fixative. After thoroughly mixing the sample, 5 ml of fecal matter was added to each tube (measured by a fill-to line on each tube). Each sample from each individual was analyzed separately. One milliliter of each sample was analyzed individually using the modified Ridley formal-ether concentration technique (Allen & Ridley, 1970). All parasitic helminth eggs and larvae and protozoan cysts were quantified. For each parasite taxon found, the total egg/cyst/larva counts were summed to obtain a relative count for each parasite per sample slide (Weyher *et al.*, 2006).

For each individual sampled, the parasite count from each of the three samples was summed and averaged to determine a relative egg/cyst/larva output for each parasite found per sample slide. To normalize the data all mean parasite counts were Log (x + 1) transformed before statistical tests were performed. Two-way analysis of variance (SPSS™ 11.5 for Windows) was employed to look for variation in parasite output between troops. For each ANOVA test, parasite species was the dependent variable and troop was the fixed factor. For all tests, the level of significance was set at $P < 0.05$. For greater detail of sampling and statistical techniques see Weyher *et al.* (2006).

Results

All parasite taxa recovered in this study have been recorded in other baboon parasite studies and all have been found in olive baboons (Hausfater & Watson, 1976; Goldsmid & Rogers, 1978; Appleton *et al.*, 1986, 1991; Eley *et al.*, 1989; McGrew *et al.*, 1989; Appleton & Henzi, 1993; Appleton & Brain, 1995; Ghandour *et al.*, 1995; Muller-Graf *et al.*, 1996; Muriuki *et al.*, 1998; Murray *et al.*, 2000; Hahn *et al.*, 2003). Nine parasite taxa were recovered – seven helminth taxa and two protozoan taxa. *Balantidium coli* was the only protozoan parasite used in statistical analysis as all other protozoa recovered could only be identified to the order "Amoebida." One parasite could only be identified to phylum level and is referred to as "unidentified nematode." Only *Balantidium coli* and *Schistosoma mansoni* could be identified to species level. Baboons host a number of strongylid species including *Necator americanus* and *Ancylostoma duodenale* (Muller-Graf, 1994). As the eggs of these species are virtually indistinguishable, all were simply classified as being in the order Strongylida. All other parasites recovered were identified to genus level.

Of the 23 adult baboons examined all were infected with at least four parasite taxa (Figure 20.1). The crop-raiding troop had higher parasite species richness than the wild-foraging troop, though this difference is due to one female in the

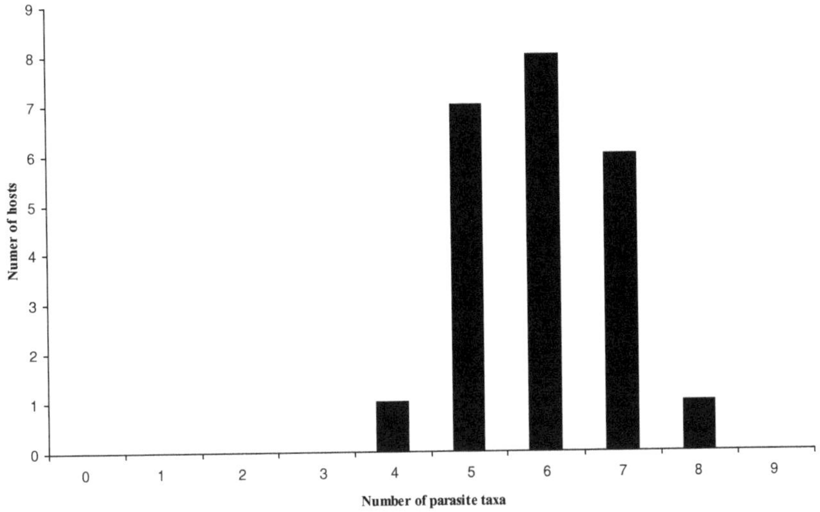

Figure 20.1. Parasite species richness across the two troops. Graph displays the number of individuals (y-axis) that harbored certain numbers of parasite taxa (x-axis).

crop-raiding troop, which was found to be parasitized with both *Schistosoma mansoni* and *Strongyloides* sp. These parasites were not recovered from the wild-foraging troop. Table 20.1 displays the range of each parasite egg/cyst count per sample slide, the median value, and the number of individuals in each troop which harbored each parasite. No statistical tests were carried out on prevalence of parasites between troops as only two troops were examined and the sample sizes were relatively small (Table 20.2).

The wild-foraging troop was found to be significantly more parasitized with *Physaloptera* sp. ($F = 15.235$, $df = 1$, $P = 0.001$) and *Trichuris* sp. ($F = 15.713$, $df = 1$, $P = 0.001$). The crop-raiding troop was found to be significantly more parasitized with *Balantidium coli* ($F = 11.787$, $df = 1$, $P = 0.003$). There were no significant differences between the crop-raiding and wild-foraging troops and any of the other parasite species recovered ($P > 0.129$ in all cases; Table 20.3). For results of parasite overdispersion, total helminth and protozoan loads, and variation in parasite output between sex, see Weyher *et al.* (2006).

Discussion

It has been suggested that crop raiding by wild animals is nutritionally beneficial because human food is a ready source of energy (Fortham-Quick & Demment,

Table 20.1. *Egg/cyst output for parasites recovered in the crop-raiding (CR) and wild-foraging (WF) baboon troops from baboons in Gashaka Gumti National Park, Nigeria (number of infected individuals are given in parentheses)*

	Mean egg/cyst output (per sample slide)		
Parasite	Range	Median	
CR troop (n = 10)			
Dicrocoelium sp.	1.00–4.33	1.34	(9)
Physaloptera sp.	0.67–20.67	3.67	(9)
Schistosoma mansoni	0.66	0.00	(1)
Strongylid sp.	1.33–71.67	4.17	(10)
Strongyloides sp.	0.33	0.00	(1)
Trichuris sp.	2.00–26.33	7.83	(10)
Unidentified nematode	0.67	0.00	(1)
Amoeba sp.	9.66–68.66	29.22	(10)
Balantidium coli	3.00–2089.67	4.67	(10)
NR troop (n = 13)			
Dicrocoelium sp.	0.33–1.67	0.67	(11)
Physaloptera sp.	2.00–331.67	43.00	(13)
Schistosoma mansoni	0.00	0.00	(0)
Strongylid sp.	0.33–71.67	4.67	(12)
Strongyloides sp.	0.00	0.00	(0)
Trichuris sp.	10.00–87.33	24.33	(13)
Unidentified nematode	0.33–14.33	0.00	(5)
Amoeba sp.	7.89–32.78	20.44	(13)
Balantidium coli	0.33–172.33	1.00	(10)

1988; Eley *et al.*, 1989; Altmann *et al.*, 1993; Warren, 2003; Chapman *et al.*, 2005b). Prior studies have found that the difference in diet between the crop-raiding and wild-foraging troops have affected the overall health of these baboons (Warren, 2003). The crop-raiding troop appears to be in markedly better physical condition and has a higher rate of pregnancy and infant survival than does the wild-foraging troop (Warren, 2003; Higham *et al.*, 2005). These benefits in nutrition and physical condition also appear to affect parasite infection in the crop-raiding troop.

This study found significant differences in fecal parasite outputs between the crop-raiding and wild-foraging baboons. However, the nature of the differences between troops in parasite infection varied according to the taxa considered. Two parasite taxa were more abundant in the wild-foraging troop, while one taxon was more abundant in the crop-raiding troop. Examination of the life cycles of each of these parasites helps us to understand why the outputs of particular parasite species were significantly different between the troops

Table 20.2. *Parasites found in samples analyzed in the crop-raiding and wild-foraging baboon troops in Gashaka Gumti National Park, Nigeria and prevalence of infection among adult individuals in each troop*

	Troop	
Parasite	CR (n = 10) No. infected individuals (%)	WF (n = 13) No. infected individuals (%)
Helminths		
Dicrocoelium sp.	90	85
Phylasoptera sp.	90	100
Schistosoma mansoni	10	0
Strongylid sp.	90	100
Strongyloides sp.	10	0
Trichuris sp.	100	100
Unidentified nematode	10	39
Protozoa		
Amoeba sp.	100	100
Balantidium coli	100	77

Table 20.3. *Mean egg/cyst outputs for the crop-raiding and wild-foraging baboon troops in Gashaka Gumti National Park, Nigeria and results from two-way ANOVA*

	Mean egg/cyst output			
Parasite	Wild-foraging troop	Crop-raiding troop	F	P
Amoeba sp.	20.57	32.30	1.776	0.198
Balantidium coli	43.41	520.87	11.787	**0.003**
Dicrocoeleum sp.	0.87	1.53	1.844	0.190
Physaloptera sp.	63.23	6.23	15.235	**0.001**
Schistosoma mansoni	0.00	0.03	2.180	0.156
Strongyloides sp.	0.00	0.03	2.180	0.156
Stongylid sp.	17.51	14.07	0.435	0.518
Trichuris sp.	32.69	9.57	15.713	**0.001**
Unidentified nematode	2.08	0.07	2.523	0.129

Note: The P values in bold indicate significant results.

Table 20.4. *Life cycles of parasites recovered from baboons in Gashaka Gumti National Park, Nigeria*

Parasite	Helminth or protozoan	Life cycle	Mode of transmission/reproduction
Amoeba sp.	Protozoan	Direct	Through ingestion of cysts, can reproduce within host
Balantidium coli	Protozoan	Direct	Through ingestion of trophozooites and cysts, can reproduce within host
Dicrocoeleum sp.	Helminth	Indirect	Through ingestion of intermediate host, eggs must be released into outside environment to develop into infective forms
Physaloptera sp.	Helminth	Indirect	Through ingestion of arthropod intermediate host, eggs must be released into outside environment to develop into infective forms
Schistosoma mansoni	Helminth	Indirect	Infection through penetration of skin by cercariae (snail intermediate host), eggs must be released into outside environment to develop into infective forms
Strongyloides sp.	Helminth	Direct	Through penetration by skin by larvae or ingestion of larvae, eggs must be released into outside environment to develop into infective forms
Strongylid sp.	Helminth	Direct	Through penetration of skin by larvae or ingestion of larvae, eggs must be released into outside environment to develop into infective forms
Trichuris sp.	Helminth	Direct	Through ingestion of eggs, must be released into outside environment to develop into infective forms
Unidentified nematode	Helminth	Unknown	Unknown

(Table 20.4). Different parasite species may be reacting very differently to the differing diets of the two troops.

The difference in parasite establishment and reproduction between helminth and protozoan parasites is extremely important here. In addition, the consideration of direct versus indirect transmission modes of these parasites helps us to understand differences in infection between the two troops. The increased level of nutrition from eating human crops appears to have decreased the intensities of infection of *Trichuris* sp. and *Physaloptera* sp., and increased the intensity of *Balantidium coli* in the crop-raiding troop.

Host nutritional status and parasite infection are known to be related (Dobson, 1985; Bundy & Golden, 1987; Nesheim, 1993; Coop & Holmes, 1996; Coop & Kyriazakis, 1999; Ezenwa, 2004). Nutrition of a host has been shown to be an important factor affecting parasite establishment, intensity of infection, and the internal environment of the host (Nesheim, 1993; Ezenwa, 2004). When parasites infect hosts, nutrients within the gut are taken away from normal body functioning for the survival and growth of the parasite (Coop & Kyriazakis, 1999). The health and diet of a host also play a role in decreasing or increasing susceptibility to parasitic infections and disease (Lozano, 1991) by directly affecting immune responses and acquisition of immunity to parasites (Coop & Holmes, 1996). In addition, a well-fed host is much more able to cope with parasitic infections already established (Hausfater & Sutherland, 1984). These relationships between host nutritional status and parasite infection are important to the present study.

A high-quality diet and/or consistent food availability may decrease a baboon group's susceptibility to certain helminth parasites (Eley *et al.*, 1989). The crop-raiding troop was found to have significantly lower outputs of the helminth parasites *Physaloptera* sp. and *Trichuris* sp. than the wild-foraging troop. It seems likely that crop-raiding baboons are able to fight off these helminth parasite infections more readily than the wild-foraging troop because they are in better physical condition from the crops that they consume. Interestingly, Hahn *et al.* (2003) found similar results between wild-foraging and rubbish-raiding baboons in Amboseli, Kenya. The wild-foraging troop had a significantly higher prevalence of both *Trichuris* sp. and *Physaloptera* sp. when compared with the rubbish-raiding baboons (Hahn *et al.*, 2003).

Previous studies of baboons have shown that food-enhancement from rubbish raiding improves physical condition and decreases intensities of helminth parasite infections (Eley *et al.*, 1989; Altmann *et al.*, 1993; Hahn *et al.*, 2003). These studies suggest that there may be a causal relationship between host nutrition and helminth parasite infection (Eley *et al.*, 1989). My results indicate that crop raiding similarly contributes to lower helminth loads.

The significant differences in output of *Physaloptera* sp. between the troops may be due to the life cycle of this parasite. *Physaloptera* sp. has an indirect life cycle. It is transmitted through the consumption of an intermediate arthropod host (Ivens *et al.*, 1978; Muller-Graf, 1994; Foreyt, 2001). It is possible that the wild-foraging troops' higher level of output of *Physaloptera* sp. is due to the fact that they are consuming the intermediate host of this parasite at higher rates than the crop-raiding troop. This study did not examine arthropod ingestion in the two troops of baboons.

In contrast to the differences found for the helminth species discussed above, the crop-raiding troop displayed a significantly higher *Balantidium coli* cyst output rate than did the wild-foraging animals. *Balantidium coli* is a protozoan parasite common in both humans and non-human primates in temperate and warm climates (Ash & Orthiel, 1997). The significant differences in *B. coli* output discovered between the two troops may be attributable to the crop raiders being exposed to higher levels of pathogens in their environment through their increased levels of anthropogenic contact. However, it is unknown if the exact species of this parasite infect both non-human and human primates. Consideration of the biology of *B. coli* suggests that the differences in nutritional status between the crop-raiding and wild-foraging troops may also be important here and that the improved nutritional status of crop-raiding animals actually leads to an increased intensity of infection of this parasite.

The difference between troops in *B. coli* output being of the opposite direction to that found for the helminth taxa may be due to a major reproductive difference between parasitic protozoa and parasitic helminths. Most parasitic protozoa reproduce within their host and their offspring are able to mature into adult forms within the same host, while parasitic helminths living in a host must release their eggs into the outside environment to develop into infective forms (Stephenson, 1987; Maizels *et al.*, 1993). Because of this ability to directly reproduce in the host, intensities of protozoan infections have been shown to increase rapidly even after a single infection event (Wakelin, 1996).

To survive and reproduce, protozoan trophozooites feed on organic material within their host (Noble *et al.*, 1989). Interestingly, it has been found that a host diet rich in starch favors the growth of *B. coli* (Noble *et al.*, 1989). This observation may therefore explain the difference in mean cyst output rates of *B. coli* between the crop-raiding and wild-foraging troops. The crops that the raiding troop consumes include starch-rich foods like maize, guinea corn, and cassava (Warren, 2003), which would be favorable to the growth and reproduction of *B. coli*. It is understood that the agricultural crops raided by the crop-raiding troop are much heavier in readily digestible starches than any single wild food that either of the troops consume (Fortham-Quick & Demment, 1988; Warren, 2003). *Balantidium coli* may therefore be benefiting from the increased starch intake of the crop-raiding troop and reproducing at higher rates within these hosts.

This study found significant differences in the output of three parasite species between a crop-raiding and wild-foraging troop of olive baboons that appear to be linked to differences in the troops' nutritional levels. The increased nutrition that the crop-raiding troop is gaining from supplementing their diet with human crops appears to affect their parasite infections. The crop-raiding troops' infections of the helminth parasites *Trichuris* sp. and *Physaloptera* sp. are significantly lower than the wild-foraging troop and it is likely that this is due to the fact that the crop-raiding troop is in better physical condition and therefore able to fight off these infections better than the wild-foraging troop. On the other hand, the crop-raiding troop displayed a higher *Balantidium coli* cyst output than the wild-foraging troop. This study suggests that the diet of the crop-raiding troop has also affected the infection of this parasite. The increase in starch intake of the crop-raiding troop has favored the increased intensity of *B. coli* in these animals.

It is clear that parasite transmission is very complex and a thorough understanding of the infective stages and transmission processes of each of the parasites found in analysis is important to fully understand the complex relationship baboon hosts are having with their environment. Foods that are eaten, physical condition of the animals, and rate of contact with contaminated soil, water, and infective hosts will all play a role in the rate of transmission of parasites in a baboon troop's habitat. In addition, an understanding of the level of infection and the particular pathogenicity of each parasite species within a host is extremely important as certain parasite infections are more likely than others to negatively impact a primate population.

Conclusion

This study was a snapshot in time in that it measured the outputs of parasites between the two troops over a short time period. It is significant however in that it suggests that different parasite species may be reacting in very different ways to changes in nutrition between two otherwise similar troops of baboons. To understand if these differences in parasite infection are constant it would be necessary to carry out a long-term study in which the baboons were sampled throughout the year in each season and at times when crops were and were not available. In addition, future research on the exact differences in caloric intake, energy expenditure, and nutrient ratios would help to determine if and what differences in diet and physical condition exist between the two troops.

This study only sampled two groups of baboons. As noted by previous authors, parasite fauna within primate groups are more similar than between groups (Freeland, 1979; Stuart *et al.*, 1998). Comparisons between troops using individuals as independent data points may therefore lead to pseudoreplication. Hence, a more statistically rigorous comparison of the differences in parasite fauna between crop-raiding and wild-foraging baboons will need more groups to be sampled, with each group being treated as an independent data point.

This study links consumption of human crops to changes in parasite infection in primates. These findings are important as it helps us to better understand how changes in primate behavior and diet as a result of crop raiding can affect primate parasite infection and health. With the continued encroachment of humans on primate habitat it is increasingly important that we understand how primates adapt and change to their environment. With crop raiding by primates already widespread (Naughton-Treves *et al.*, 1998; Warren, 2003) and likely to significantly increase in the future (Fortham-Quick & Demment, 1988; Warren, 2003), further research on the exact mechanisms underlying impacts of anthropogenic nutrition on primate–parasite dynamics are now urgently required.

This study is also important as it shows that when examining parasite infection in wild primates it is necessary to understand the biology of each individual parasite species. The life cycles and biology of particular parasites found within a host differ in a number of ways and are affected by multiple factors. Merely quantifying parasite infections between two groups or populations of primates does not answer specific questions about transmission rates and interactions between the parasites and their hosts. Numerous factors affect parasite–host relationships in primates including external environmental factors, behavior of the parasite and the host, host nutrition, parasite–parasite interactions, and host internal environment and immunity. As primatologists studying primate parasite ecology it is of utmost importance to understand the individual biology of the parasite species that infect the primates that we study. In essence, we must not only be primatologists, we must also become parasitologists.

Acknowledgements

I sincerely thank Dr. Colin Chapman and Dr. Michael Huffman for an invitation to contribute to this invaluable and much-needed volume. Special thanks to my professor and advisor Dr. Stuart Semple as this work could not have been carried out without his continual support. I thank the Nigerian authorities, the National Parks Service of Nigeria, through the Conservator General Alhaji Lawan B. Marguba, the staff and management of Gashaka Gumti National Park and NCF/WWF-UK for making this work possible. Fieldwork was supported by many people, including field assistants, Bobbo Buba and Nuhu. Volker Sommer (UCL) provided his help and support through the Gashaka Primate Project and James Higham gave invaluable assistance in the field. I thank them for all their work. I sincerely thank John Williams and Claire Rogers of the London School of Hygiene and Tropical Medicine for training in parasitological analysis. Without the support and expertise of all of these people, this project would not have been possible. I must also thank two anonymous reviewers for their constructive criticism of this chapter. Parts of this project were funded by grants from The Leakey Trust, the North of England Zoological Society, and Roehampton University.

References

Allen, A. V. H. & Ridley, D. S. (1970). Further observations in the formal-ether concentration technique for fecal parasites. *Journal of Clinical Pathology*, **23**, 545–547.

Altmann, J., Schoeller, D., Altmann, S. A., Muruthi, P. & Sapolsky, R. M. (1993). Body size and fatness of free-living baboons reflect food availability and activity levels. *American Journal of Primatology*, **30**, 149–161.

Anderson, R. M. & May, R. M. (1978). Regulation and stability of host-parasite population interactions. I. Regulatory processes. *Journal of Animal Ecology*, **47**, 219–247.

Appleton, C. C. & Brain, C. (1995). Gastro-intestinal parasites of *Papio cynocephalus ursinus* in the central Namib desert, Namibia. *African Journal of Ecology*, **33**, 257–265.

Appleton, C. C. & Henzi, S. P. (1993). Environmental correlates of gastrointestinal parasitism in montane and lowland baboons in Natal, South Africa. *International Journal of Primatology*, **14**, 623–635.

Appleton, C. C., Henzi, S. P., Whiten, A. & Byrne, R. (1986). The gastrointestinal parasites of *Papio ursinus* from the Drakensberg mountains, Republic of South Africa. *International Journal of Primatology*, **7**, 449–456.

Appleton, C. C., Henzi, S. P. & Whitehead, S. I. (1991). Gastro-intestinal helminth parasites of the chacma baboon, *Papio cynocephalus ursinus*, from the coastal lowlands of Zululand, South Africa. *African Journal of Ecology*, **29**, 149–156.

Ash, L. R. & Orthiel, T. C. (1997). *Atlas of Human Parasitology*, 4th edn. Chicago, IL: American Society of Clinical Parasitologists.

Bundy, D. A. P. & Golden, M. H. N. (1987). The impact of host nutrition on gastrointestinal helminth populations. *Parasitology*, **95**, 623–635.

Chapman, C. A., Gillespie, T. R. & Goldberg, T. L. (2005a). Primates and the ecology of their infectious diseases: how will anthropogenic change affect host-parasite interactions? *Evolutionary Anthropology*, **14**, 134–144.

Chapman, C. A., Speirs, M. L., Gillespie, T. R., Holland, T. & Austad, K. M. (2005b). Life on the edge: gastrointestinal parasites from the forest edge and interior primate groups. *American Journal of Primatology*, **68**, 1–12.

Coop, R. L. & Holmes, P. H. (1996). Nutrition and parasite interaction. *International Journal of Parasitology*, **26**, 951–962.

Coop, R. L. & Kyriazakis, I. (1999). Nutrition-parasite interaction. *Veterinary Parasitology*, **84**, 187–204.

Creel, S., Fox, J. E., Hardy, A. *et al.* (2002). Snowmobile activity and glucocorticoid stress responses in wolves and elk. *Conservation Biology*, **16**, 809–814.

Daszak, P., Cunningham, A. A. & Hyatt, A. D. (2000). Emerging infectious diseases of wildlife – threats to biodiversity and human health. *Science*, **287**, 443–449.

de Gruijter, J. M., Gasser, R. B., Polderman, A. M., Asigri, V. & Dijkshoorn, L. (2005). High resolution DNA fingerprinting by AFLP to study the genetic variation among *Oesophagostomum bifurcum* (Nematoda) from human and non-human primates from Ghana. *Parasitology*, **130**, 229–239.

Dittus, W. P. J. (1974). The sociological basis for conservation of the toque monkey (*Macaca sinica*) of Sri Lanka (Ceylon). In *Primate Conservation*, ed. HSH Prince Rainer III of Monaco and G. H. Bourne. New York, NY: Academic Press, pp. 238–267.

Dobson, A. P. (1985). The population dynamics of competition between parasites. *Parasitology*, **91**, 317–347.

Dunn, A. (1999). *Gashaka Gumti National Park: A Guide Book.* National Park Service of Nigeria, Lagos.

Eley, R. M., Strum, S. C., Muchemi, G. & Reid, G. D. F. (1989). Nutrition, body condition, activity patterns, and parasitism of free-ranging troops of olive baboons (*Papio anubis*) in Kenya. *American Journal of Primatology*, **18**, 209–219.

Else, J. G. (1991). Non-human primates as pests. In *Primate Responses to Environmental Change*, ed. H. O. Box. London: Chapman and Hall, pp. 155–156.

Else, J. G. & Lee, P. C. (eds.) (1986). *Primate Ecology and Conservation*, Vol. 2. Cambridge: Cambridge University Press.

Ezenwa, V. O. (2004). Interactions among host diet, nutritional status and gastrointestinal parasite infection in wild bovids. *International Journal of Parasitology*, **34**, 535–542.

Fa, J. E. & Southwick, C. H. (eds.) (1988). *Ecology and Behavior of Food Enhanced Primate Groups*. New York, NY: Alan R. Liss, Inc.

Foreyt, W. J. (2001). *Veterinary Parasitology: Reference Manual,* 5th edn. Ames, IA: Iowa State University Press.

Fortham-Quick, D. L. (1984). Effects of the consumption of human foods on the activity budgets of two troops of baboons, *Papio anubis* at Gilgil, Kenya. *International Journal of Primatology*, **5**, 339.

Fortham-Quick, D. L. (1986). Activity budgets and the consumption of human foods in two troops of baboons (*Papio anubis*) at Gilgil, Kenya. In *Primate Ecology and Conservation*, ed. J. G. Else & P. C. Lee. New York, NY: Cambridge University Press, pp. 221–228.

Fortham-Quick, D. L. & Demment, M. W. (1988). Dynamics of exploitation: differential energetic adaptation of two troops of baboons to recent human contact. In *Ecology and Behavior of Food Enhanced Primate Groups*, ed. J. E. Fa & C. H. Southwick, New York, NY: Alan R. Liss.

Freeland, W. J. (1979). Primate social groups as biological islands. *Ecology*, **60**, 719–728.

Ghandour, A. M., Zahid, N. Z., Banaja, A. A., Kamal, K. B. & Boug, A. I. (1995). Zoonotic intestinal parasites of hamadryas baboons *Papio hamadryas* in the western and northern regions of Saudi Arabia. *Journal of Tropical Medicine and Hygiene*, **98**, 431–439.

Gillespie, T. R., Chapman, C. A. & Greiner, E. C. (2005). Effects of logging on gastrointestinal parasite infections and infection risk in African primates. *Journal of Applied Ecology*, **42**, 699–707.

Goldsmid, J. M. & Rogers, S. (1978). A parasitological study on the chacma baboon (*Papio ursinus*) from the northern Transvaal. *Journal of the South African Veterinary Association*, **49**, 109–110.

Gregory, R. D. & Hudson, P. J. (2000). Parasites take control. *Nature*, **406**, 33–34.

Grenfell, B. T. & Gulland, F. M. D. (1995). Introduction: ecological impact of parasitism on wildlife host populations. *Parasitology*, **111**, S3–S14.
Hahn, N. E., Proulx, D., Muruthi, P. M., Alberts, S. & Altmann, J. (2003). Gastrointestinal parasites in free-ranging Kenyan baboons (*Papio cynocephalus* and *P. anubis*). *International Journal of Primatology*, **24**, 271–279.
Hausfater, G. & Sutherland, R. (1984). Little things that tick off baboons. *Natural History*, **2**, 54–61.
Hausfater, G. & Watson, D. F. (1976). Social and reproductive correlates of parasite ova emissions by baboons. *Nature*, **262**, 688–689.
Higham, J., Ross, C., Warren, Y. *et al.* (2005). Four years of rainforest baboons at Gashaka. *Primate Report*, **72**, 46–47.
Hill, C. M. (1997). Crop-raiding by wild vertebrates: the farmer's perspective in an agricultural community in western Uganda. *International Journal of Pest Management*, **43**, 77–84.
Ivens, V. R., Mark, D. L. & Levine, N. D. (1978). *Principal Parasites of Domestic Animals in the United States,* Special Publication 52. Urbana-Champaign, IL: College of Agriculture and Veterinary Medicine, University of Illinois.
Lozano, G. A. (1991). Optimal foraging theory: a possible role for parasites. *Oikos*, **60**, 391–395.
Maizels, R. M., Bundy, D. A. P., Selkirk, M. E., Smith, D. F., & Anderson, R. M. (1993). Immunological modulation and evasion by helminth parasites in human populations. *Nature*, **365**, 797–805.
Marsh, C. W. & Mittermeier, R. A. (1987). *Primate Conservation in the Tropical Rain Forest.* New York, NY: Alan R. Liss.
McGrew, W. C., Tutin, C. E. G., Collins, D. A. & File, S. K. (1989). Intestinal parasites of sympatric *Pan troglodytes* and *Papio* spp. at two sites: Gombe (Tanzania) and Mt. Assirik (Senegal). *American Journal of Primatology*, **17**, 147–155.
Muller-Graf, C. D. M. (1994). *Ecological Parasitism of Baboons and Lions.* PhD thesis, University of Oxford.
Muller-Graf, C. D. M., Collins, D. A. & Woolhouse, M. E. J. (1996). Intestinal parasite burden in five troops of olive baboons (*Papio cynocephalus anubis*) in Gombe Stream National Park, Tanzania. *Parasitology*, **112**, 489–497.
Muller-Graf, C. D. M., Collins, D. A., Packer, C. & Woolhouse, M. E. J. (1997). *Schistosoma mansoni* infection in a natural population of olive baboons (*Papio cynocephalus anubis*) in Gombe Stream National Park, Tanzania. *Parasitology*, **115**, 621–627.
Muriuki, S. M. K., Murugu, R. K., Munene, E., Karere, G. M. & Chai, D. C. (1998). Some gastro-intestinal parasites of zoonotic (public health) importance commonly observed in Old world non-human primates in Kenya. *Acta Tropica*, **71**, 73–82.
Murray, S., Stem, C., Boudreau, B. & Goodall, J. (2000). Intestinal parasites of baboons (*Papio cynocephalus anubis*) and chimpanzees (*Pan troglodytes*) in Gombe National Park. *Journal of Zoo and Wildlife Medicine*, **31**, 176–178.
Naughton-Treves, L., Treves, A., Chapman, A. C. & Wrangham, R. (1998). Temporal patterns of crop-raiding by primates: linking food availability in croplands and adjacent forest. *Journal of Applied Ecology*, **35**, 596–606.

Nesheim, M. C. (1993). Human nutrition needs and parasitic infections. *Parasitology*, **107**, S7–S18.

Noble, E. R., Noble, G. A., Schad, G. A. & MacInnes, A. J. (eds.) (1989). *Parasitology: The Biology of Animal Parasites*, 6th edn. Philadelphia, PA: Lea and Febiger.

Ooi, H. K., Tenora, F., Itoh, K. & Kamiya, M. (1993). Comparative study of *Trichuris trichura* from nonhuman primates and from man. *Journal of Veterinary Medical Science*, **55**, 363–366.

Patterson, J. D. & Wallis, J. (eds.) (2005). *Commensalism and Conflict: The Human-Primate Interface. Special Topics in Primatology*, Vol. 4. Norman, OK: American Society of Primatologists.

Patz, J. A., Graczyk, T. K., Geller, N. & Vittor, A. Y. (2000). Effects of environmental change on emerging parasitic diseases. *International Journal of Parasitology*, **30**, 1395–1405.

Scott, M. E. (1988). The impact of infection and disease on animal populations: implications for conservation biology. *Conservation Biology*, **2**, 40–56.

Scott, M. E. & Dobson, A. (1989). The role of parasites in regulating host abundance. *Parasitology Today*, **5**, 176–183.

Stephenson, L. S. (1987). *Impact of Helminth Infections on Human Nutrition*. London: Taylor & Francis.

Stuart, M. D. & Strier, K. B. (1995). Primates and parasites: a case study for a multidisciplinary approach. *International Journal of Primatology*, **16**, 577–593.

Stuart, M. D., Greenspan, L. L., Glander, K. E. & Clarke, M. R. (1990). A coprological survey of parasites of wild mantled howling monkeys, *Alouatta palliata palliata*. *Journal of Wildlife Disease*, **26**, 347–349.

Stuart, M., Pendergast, V., Rumfelt, S. *et al.* (1998). Parasites of wild howlers (*Alouatta* spp.). *International Journal of Primatology*, **19**, 493–512.

Wakelin, D. (1996). *Immunity to Parasites: How Parasitic Infections are Controlled*, 2nd edn. Cambridge: Cambridge University Press.

Warren, Y. (2003). Olive baboons (*Papio cynocephalus anubis*): behaviour, ecology, and human conflict in Gashaka Gumti National Park, Nigeria. PhD thesis, University of Surrey, Roehampton.

Weyher, A. H., Ross, C. & Semple, S. (2006). Gastrointestinal parasites in crop raiding and wild foraging *Papio anubis* in Nigeria. *International Journal of Primatology*, **27**, 1519–1534.

Wolfe, N. D., Escalante, A. A., Karesh, W. B. *et al.* (1998). Wild primate populations in emerging infectious disease research: the missing link? *Emerging Infectious Disease*, **4**, 149–158.

21 *Can parasite infections be a selective force influencing primate group size? A test with red colobus*

COLIN A. CHAPMAN, JESSICA M. ROTHMAN, AND
STACEY A. M. HODDER

Photograph by Colin A. Chapman (*Procolobus rufomitratus*)

Primate Parasite Ecology. The Dynamics and Study of Host–Parasite Relationships, ed. Michael A. Huffman and Colin A. Chapman. Published by Cambridge University Press.

Introduction

Identifying ecological factors underlying group size and social organization is a central focus of primatology (van Schaik, 1983; van Schaik & van Hooff, 1983; Chapman & Chapman, 2000b). Factors that might explain the variation in group size have been extensively evaluated in terms of costs and benefits. Animals that live in a group are thought to increase their fitness by avoiding predation (e.g. benefiting from group defense) or social pressures (e.g. infanticide). These factors favoring group living are countered by the costs of group living, such as competition for food or elevated travel costs as a result of patch depletion. The majority of primate socioecology models have focused on these advantages and disadvantages (Wrangham, 1980; van Schaik, 1989; Isbell, 1991; Sterck *et al.*, 1997; Chapman & Chapman, 2000b; Crockett & Janson, 2000; Chapman & Pavelka, 2005).

In contrast to these mainstream theories are a series of ideas about how parasite infections could be a selective force influencing primate social organization. In a series of intriguing papers published in the 1970s, Freeland proposed that aspects of primate social organization evolved to decrease the impact of parasite infections (Freeland, 1977a, 1976, 1977b, 1979, 1980). One of these predictions centers on the effects of parasitism on group size. Freeland (1976) suggested that as group size increases, there is an increased chance of mortality due to acquiring pathogenic levels of parasites. This could occur as a result of environmental contamination with infectious material caused by a large number of animals concentrated in a small area or because larger groups have more immigrants, which may introduce novel parasites. He documented that for mangabeys (*Lophocebus albigena*) the number of intestinal parasite species per individual was a function of group size (Freeland, 1979). Despite the potential significance for parasites to influence primate social organization, these ideas were generally not further empirically tested (but see recent reviews: Loehle, 1995; Heymann, 1999; Nunn *et al.*, 2000; Nunn & Altizer, 2006). This is illustrated by the fact that in the classic book on primate behavior and ecology by Smuts *et al.* (1987), disease is not listed in the index (Heymann, 1999), nor is parasite.

The objective of this chapter is to test predictions concerning how parasite infection could be a selective force influencing primate group size. We first examine the rationale behind various predictions and the conditions required for these predictions to operate. Subsequently, we discuss counter-strategies that primates could employ to avoid acquiring parasites, while still receiving the benefits associated with living in a group. Finally, we present data to explore the relationship between group size and parasite infection for two groups of

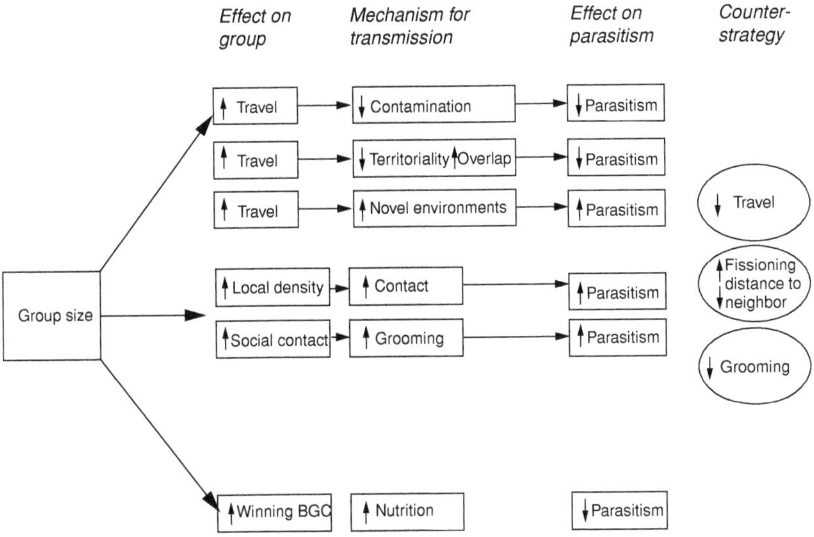

Figure 21.1. A consideration of how group size may affect primate groups, the mechanisms of parasite transmission, and the effect on parasitism. Also considered are possible counter-strategies that primates could use to decrease impacts of parasitism. BGC is between group contest competition. An example of the logic depicted here is that if group size increases social contact, including aggressive interactions, and grooming increases as a means of reconciling those interactions, this may lead to an increase in parasitism. If this does there would be selection for decreased grooming (or another counter-strategy could operate).

red colobus (*Procolobus rufomitratus*) living in Kibale National Park, Uganda that differ markedly in size.

Do parasite infections vary as a function of primate group size and can this select for groups of different sizes?

Parasite infections can influence fitness (Coop & Holmes, 1996; Murray *et al.*, 1998) and if the intensity of infections varies as a function of group size this could lead to selection for groups of different sizes. Parasite infections could be influenced by group size in a number of ways; some of which will produce very diametrically opposed outcomes (Figure 21.1). Here we consider how group size could lead to changes in four non-mutually exclusive factors that could influence the intensity of parasites: local density, social contact, travel, and the outcome of between group contest competition (BGC).

First, factors that influence host proximity and the number and duration of contacts among individuals can affect parasite transmission among hosts (Anderson, 1978; Altizer *et al.*, 2003). In host–parasite models with direct transmission, the probability that parasites will spread though the host population is an increasing function of parasite pathogenicity, host density, and host immune response; new infections depend on host contact rates and per contact probabilities of successful infection (Anderson & May, 1992). As a result, directly transmitted parasites may have increased transmission rates in denser populations (Freeland, 1976; Anderson & May, 1992; Loehle, 1995). Accordingly, corresponding increases in prevalence and a higher number of parasite species with direct transmission may be harbored by the host population (Anderson & May, 1992; Roberts *et al.*, 2000; Altizer *et al.*, 2003). This theory has been supported for primates by meta-analyses and empirical research. For example, Nunn *et al.* (2003) analyzed a comparative data set of 941 host–parasite combinations representing 101 anthropoid primate species and 231 parasite taxa, and demonstrated that host population density was positively associated with total parasite species richness. A higher prevalence of intestinal parasites was found in howler monkeys (*Alouatta palliata*) in a dense population than a less dense one (Stuart *et al.*, 1990). Similarly, Chapman *et al.* (2005b) demonstrated that the prevalence of *Trichuris* sp. in two colobine species increased when there was a dramatic increase in population density in a forest fragment associated with groups immigrating to the fragment (but see Stoner, 1996). These mathematical models, meta-analyses, and a handful of empirical studies suggest that high density, which would likely occur at a local scale with larger groups, increases the probability that parasites spread through a group.

Secondly, this expected density effect relies on one of two mechanisms operating: the number of social interactions and partners must increase with group size, or there must be an increase in the probability of contacting contaminated material with increasing group size (considered next). In a comparative analysis of 44 primate species, Dunbar (1991) demonstrated that the frequency of social grooming increased with group size. This increased contact is associated with the potential for increased parasite transmission through contact with infective stages on the fur, contact with other animals' saliva, and/or transfer of respiratory pathogens (Nunn & Altizer, 2006). As a consequence, through more frequent social interaction that involves physical contact, such as grooming, individuals in larger groups may acquire more severe parasite infections than individuals in smaller groups.

Thirdly, the way a group travels through its environment will influence the probability of contacting infected areas or individuals. There are three competing hypotheses concerning the way ranging patterns may be influenced by parasitism (Nunn & Dokey, 2006). (a) For parasite species transmitted by the fecal-oral route, transmission can be influenced by the number and duration of contacts with contaminated feces (Freeland, 1979, 1980; Anderson & May, 1992; Ezenwa, 2004; Vitone *et al.*, 2004). Thus, the greater the contact between hosts and infected feces (i.e. repeated use of the same contaminated area), the higher the parasite load and the more species-rich the parasite infection within a host. Accordingly, there is a greater potential for severe symptoms (Altizer *et al.*, 2003; Chapman *et al.*, 2005b). Freeland (1976) proposed that the need to avoid acquiring diseases could influence ranging behavior. In a test of the effect of movement patterns on infection risk, he noted that 40% of defecations hit branches that are used for mangabey locomotion before reaching the forest floor (Freeland, 1980). He also noted that the mangabeys remained in areas longer in the rainy season than in the dry season, and speculated that infection risk was lower in the rainy season due to the fact that rain washes infective stage parasites to the forest floor, reducing contamination levels in the treetops (but see Olupot *et al.*, 1997 for opposing data). (b) When intensive ranging is associated with territoriality, this could decrease home range overlap and reduce exposure to new parasites (Nunn & Dokey, 2006). (c) Alternatively, since the probability of contacting infected items may increase as groups travel further and larger groups tend to travel further than smaller groups (Isbell, 1991; Snaith & Chapman, 2007), there may be an increase in infections with group size. Furthermore, if large groups range further and encounter more novel habitats, the richness of their parasites may increase as a function of group size (Freeland, 1979; Loehle, 1995; Nunn & Altizer, 2006). This would particularly apply to species that have large degrees of home range overlap with conspecifics or with sympatric species capable of hosting the same parasites (but see Gasser *et al.*, Chapter 3, this volume for the difficulty in determining if interspecific transmission has occurred).

Fourthly, when assessing any proposed selective force influencing group size, it is assumed that the factor has a biologically meaningful effect relative to the advantages of larger groups (i.e. the benefits do not outweigh the costs). In cases where large group size increases the likelihood of winning between group contests for food (Koenig, 2002; C. A. Chapman, unpublished data), the nutritional gains from a larger group size may be substantial and may compensate for the costs of increased parasite infections associated with larger groups. How costly a parasitic infection will be parasite- and situation-dependent. For example, helminth and protozoan parasites can impact host survival and reproduction directly through pathological effects and indirectly by reducing host condition (Coop & Holmes, 1996; Murray *et al.*, 1998) and even up-regulation of host immunity can reduce breeding success (Ilmonen *et al.*, 2000). However, parasites do not necessarily induce negative health effects if hosts have adequate energy reserves or nutrient supplies (Munger & Karasov, 1989; Gulland, 1992; Milton, 1996), suggesting that the outcome of host–parasite associations may be contingent on host nutritional status, as well as on the severity of infection (Chapman *et al.*, 2006).

Counter-strategies that primates may employ to avoid parasite infections

Theory supports the idea that parasites with direct (oral-fecal) transmission strategies are more common in large groups relative to small groups because large groups have more social contacts and contact with contaminated items (Anderson & May, 1992). However, the concurrent increased probability of pathogenic effects on hosts is also expected to result in selection for host strategies to minimize the impact of parasite infections (Freeland, 1976). For example, if the level of parasite infection found in large groups has a fitness cost it can limit group size. If counter-strategies exist, we would expect individuals in larger groups to have mechanisms to decrease the amount of social contact or decrease the chances of contact with contaminated items. To decrease social contact, large social groups could fission into subgroups, they could increase group spread (increasing nearest-neighbor distance), or they could groom less. To avoid contact with contaminated items, large groups could travel further than smaller groups, they could revisit the same feeding or sleeping site less frequently, or again they could fission into subgroups.

Let us consider just two counter-strategies in more detail: fissioning of a large group into smaller subgroups, and an increase in travel during the day (day range). Chimpanzees (*Pan troglodytes*), spider monkeys (*Ateles* sp.), and gelada baboons (*Theropithecus gelada*) routinely fission into smaller units and reunite at later times (i.e. from days to weeks) (Stammbach, 1987; Chapman, 1990a; Mitani *et al.*, 2002) and this is considered a central part of their social organization. However, many species which are considered to have cohesive groups also occasionally divide for extended periods before reuniting, including long-tailed macaques (*Macaca fascicularis*) (van Schaik & van Noordwijk, 1988), red colobus (Chapman & Chapman, 2000a), and howler monkeys (Aureli *et al.*, in press; Chapman, 1990b). Dividing a group into small subgroups would both decrease social contact and decrease the distance they would need to travel to acquire adequate food resources, which may decrease disease transmission. The fact that many species form subgroups suggests that fissioning could be a widespread strategy.

Similarly, for a number of species, increases in day range have been documented to occur with increases in group size (reviewed by Isbell, 1991).

Again, by being in a small group animals would both decrease social contact and decrease the distance they would need to travel, thus decreasing disease transmission.

However, both the formation of subgroups and day range–group size relationships have been argued to be a function of feeding competition since with large groups the nutritional needs of all group members cannot be easily met as patches are depleted more rapidly. Consequently, the group is forced to travel longer distances or form subgroups (Chapman & Chapman, 2000b). Thus, investigating these possible counter-strategies will require finding instances where groups divide when food resources would not require it, or it will necessitate a multivariate approach where the variance can be partitioned between factors related to the density and distribution of food patches and to parasite infection risk. Furthermore, Freeland (1976) points out that if large groups increase distance traveled each day they may enter new habitats which could harbor novel parasites. An increase in parasite richness could result in greater morbidity and mortality. For example, in humans, *Schistosoma mansoni* has an increased effect on the development of malnutrition in the presence of *Trichuris trichiura* (Parraga *et al.*, 1996) and a variety of parasites demonstrate greatly elevated pathogenic effects in the presence of HIV (Kaplan *et al.*, 1996).

In closing, in a discussion concerning counter-strategies it is important to stress that if elaborate defenses have evolved to reduce parasite transmission or its impact on hosts then expected correlations between parasitism and group size may not be found (i.e. the "ghost of parasitism past"; Nunn & Altizer, 2006). As a result, we might expect to find that some populations or species might live in large groups, but have low parasite levels because they have evolved counter-strategies.

Case study: Red colobus of Kibale

We examined the effect of group size on gastrointestinal helminth (nematode) infections of red colobus monkeys (*Procolobus rufomitratus*) in Kibale National Park, Uganda. The existence of two neighboring red colobus groups that differ strikingly in size (i.e. ~36 versus 132 individuals) allowed us to evaluate the hypothesis that larger social groups have higher levels of parasite infection and more diverse parasitic faunas than smaller social groups. We examined fecal samples to determine indices of gastrointestinal parasite infections in these two groups over 23 months, and collected behavioral data on both groups to determine whether animals in the large group were employing counter-strategies to reduce the risk of infection. We contrasted the parasite infections of the large and small group of red colobus, and collected behavioral data to test five predictions concerning possible counter-strategies that could be employed to decrease parasite infection risk. Firstly, we examined whether the large group traveled more than the small group, with the expectation that the large group would travel further in search of non-contaminated areas. Secondly, we determined the home range size of both groups with the expectation that the large group would have a bigger home range than the small group and thus could more easily find non-contaminated areas. Thirdly, we expected the large group to fission into smaller subgroups more frequently than the small group, which would effectively decrease social contact and the likelihood of encountering contaminated material (but see potential conflicting factors discussed above). Fourth, we expected members of the large group to engage in grooming less than members of the small group to decrease social contact and reduce the chances of disease transmission. Finally, we predicted that the large group would increase group spread (indexed by an increasing nearest neighbor distance) relative to the small group and reduce the chances of disease transmission.

Methods

Fecal samples (n = 1033) were collected from two red colobus groups in the Kanyawara study area of Kibale between July 2004 and May 2006. Kibale (795 km^2; 0° 13′–0° 41′N and 30° 19′–30° 32′E) is a moist, mid-altitude (920–1590 m) evergreen forest intermixed with swamps, secondary forests, and grasslands (Struhsaker, 1997; Chapman & Lambert, 2000). Two groups were selected based on their size: the smaller group consisted of 36 individuals and the large group consisted of 132 individuals.

In total, 628 fecal samples were collected from the larger group (average per month = 30.0, range 17–76), and 405 samples were collected from the smaller group (average per month = 19.8, range 14–43). Fecal samples were collected directly after defecation, stored in 10% formalin solution, processed using sodium nitrate flotation (Sloss *et al.*, 1994), and examined using a Leica DM 2500 microscope with PL Fluotar objectives. Eggs and larvae of parasites were counted and identified on the basis of egg color, shape, contents, and size. Digital images were taken with an Infinity 1 (1.3 megapixel) camera of representative samples of each parasite or for parasites where identification was questionable, and measurements were made to the nearest 0.1 micron ± SD. These images were sent to taxonomic experts to verify identification. Coprocultures and necropsies were used to match parasite eggs to larvae for positive identification (Gillespie *et al.*, 2005a, b, Chapman *et al.*, unpublished data). The parasite infections were described in terms of prevalence of infection, species richness, load (eggs/g), and multiple infections. Prevalence is the proportion of individuals sampled that are infected with a particular parasite. Comparisons of parasite prevalence can be a useful indicator of parasites that may be impacting host populations (i.e. population declines have been correlated to increased infection prevalence). Since we could not individually recognize each animal in the study groups, repeat samples likely occurred and thus this should be viewed as an index of prevalence. In a quantitative evaluation of this issue, Huffman *et al.* (1997) contrasted incidences of infection based on the number of fecal samples obtained from chimpanzees (*Pan troglodytes*) from Mahale, Tanzania versus that based on the number of known individuals and documented that individual infection rates, the preferred unit of comparison, was statistically higher than rates based on all samples. This is because multiple sampling from the same individual may be required to detect an infection. The frequency of multiple-species infections (i.e. the proportion of a population with more than one species of parasite) can be another useful index, as multiple-species infections are associated with a greater potential for morbidity and mortality (Kaplan *et al.*, 1996; Parraga *et al.*, 1996). Parasite egg production, or load (egg/g), is highly variable and some researchers suggest that it may not be indicative of actual infection intensity. However, with *Trichuris* sp. from infected red colobus, we typically obtained $<$10 eggs/g, but during certain times, individuals in certain groups consistently had much higher loads (i.e. over 300 egg/g; all individuals in a group for an extended time; Chapman *et al.*, unpublished data). This suggests that load might be a useful index of parasite infection since high levels are temporally and spatially specific; however, we recommend that the results concerning load be viewed critically and only be considered of interest if in concordance with other indices of parasite infection.

Each group was observed for 6 days a month, with observations starting just after sunrise and ending around the time that the group became inactive for the day. During a 15-min block, five point samples were made of different individuals to determine their activity (e.g. grooming, traveling, feeding, resting). During each point sample the nearest neighbor was identified and its distance from the subject was visually estimated. The location of what was perceived to be the center of the group was plotted on detailed trail maps every 15 minutes. For the purpose of determining when a group subdivided, a group was only considered to be divided if two distinct units were separated by more than 100 m, and typically when the group did divide they were separated by more than 500 m.

For the purposes of determining the home range size of the two groups, a detailed map of the Kanyawara study area was divided into 1 ha grid cells by superimposing a 100 $\times$ 100 m grid over the trail system. The ranging patterns of the two groups were extracted from follows made between July 2004 and

May 2006. We used these maps to determine which grid cells were entered each day, and to establish the spatial extent of the group's home range during the study (considered to be all grid cells entered plus any cells that fell inside a minimum polygon formed by the outermost entered cells; Snaith & Chapman, unpublished manuscript).

Both two-tailed paired-sample t-tests (paired by month to eliminate seasonal variation in infections, e.g. average amount of time spent grooming each month) were used to determine if there were any differences between the small and large red colobus groups with respect to the following: (1) parasite prevalence, (2) species richness, (3) eggs/g, and (4) multiple infections. The paired t-test assumes that months are independent and none of the indices showed a correlation between successive months ($P > 0.1$ in all instances). Paired t-tests were used to evaluate behavioral differences between the two groups.

Results and discussion

Differences in parasite infections between large and small groups

The helminth parasite community described from the samples collected from both groups included the following nematodes: *Trichuris* sp. (Superfamily Trichuroidea), *Oesophagostomum* sp., an unidentified strongyle nematode (Superfamily Strongyloidea), *Strongyloides fulleborni* (Superfamily Rhabditoidea), and *Colobenterobius* sp. (Superfamily Oxyuroidea; Table 21.1). We also identified two protozoans, likely *Entamoeba coli* and *E. histolytica/dispar*. However, because of the small sample size for *Colobenterobius* sp., the unidentified strongyle nematode, and *Strongyloides fulleborni*, these parasites are not considered individually except in the analysis of richness and multiple infections. Further, *E. histolytica* and *E. dispar* have cysts that are morphologically indistinguishable and it was only recently that *E. dispar* was considered a distinct species (Gatti *et al.*, 2002). However, *E. histolytica* is pathogenic, while *E. dispar* is not. Thus, we do not consider the *E. histolytica/dispar* complex further with the exception of analyses of multiple infections and the richness of the parasite community, where they are considered as one species. The large and small group had the same community of parasites.

The paired t-test revealed no differences between the large and small group with respect to *Trichuris* eggs per gram of all samples (paired t-test $t = 0.262$, $P = 0.796$), *Trichuris* eggs per gram in infected individuals ($t = 0.754$, $P = 0.459$), *Oesophagostomum* eggs per gram of all samples (paired t-test $t = 1.000$, $P = 0.329$), *Oesophagostomum* eggs per gram in infected individuals ($t = 0.439$, $P = 0.666$), *Oesophagostomum* prevalence ($t = -0.963$,

Table 21.1. *A description of the parasites found in the large and small group of red colobus (*Procolobus rufomitratus*) in Kibale National Park, Uganda*

Parasite species	Mode of infection	Morbidity/mortality	Image
Oesophagostomum spp.	Larvae ingested	Severe diarrhea, weight loss, death	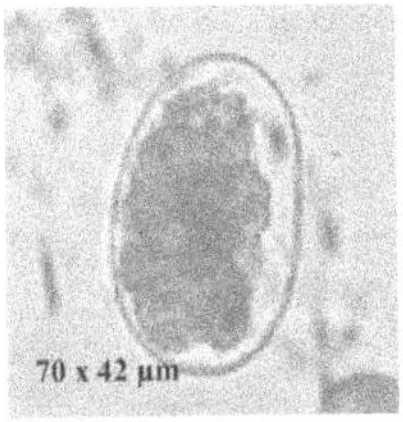
Entamoeba coli	Cyst or trophozoite ingested	Typically asymptomatic	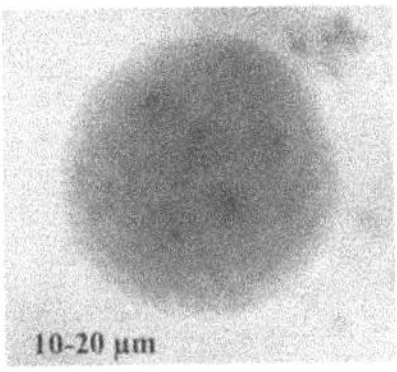
Entamoeba histolytica	Cyst or trophozoite ingested	Hepatic and gastric amoebiasis, dysentery, death	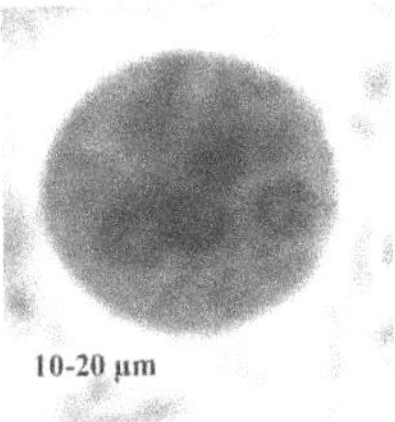
Entamoeba dispar	Cyst or trophozoite ingested	Asymptomatic	Same as *E. histolytica*
Strongyloides fulleborni	Larvae ingested, skin penetration	Mucosal inflammation, ulceration, death	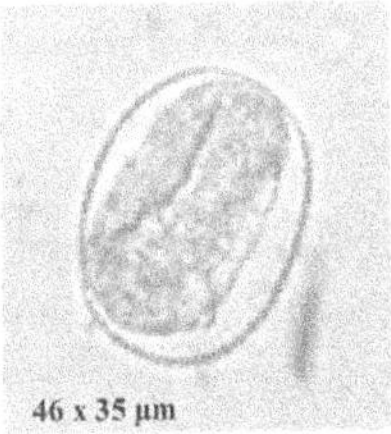
Trichuris sp.	Larvated egg ingested	Heavy infections cause abdominal pain and distension, bloody diarrhea, weight loss, nutritional deficiency, rectal prolapse, and reduced growth in human children	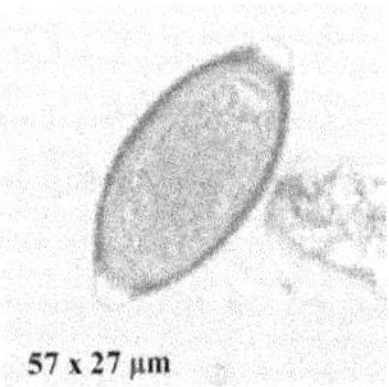

(*cont.*)

Table 21.1 (*cont.*)

Parasite species	Mode of infection	Morbidity/mortality	Image
Colobenterobius sp.	Ingestion of eggs	Unknown	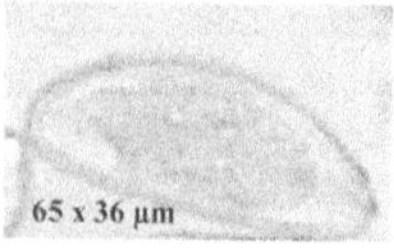
Strongyle (unidentified)	Larvae ingested	Unknown	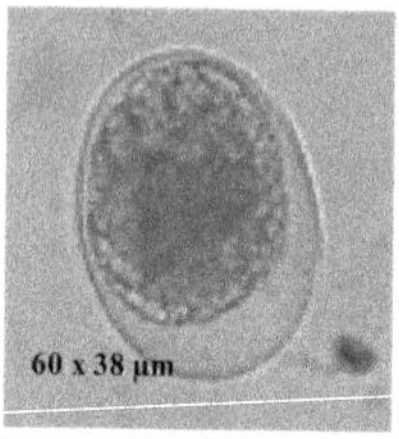

$P = 0.347$), multi-species infections ($t = 0.899$, $P = 0.380$), and species richness ($t = 1.420$, $P = 0.171$). However, the prevalence of *Trichuris* was found to be lower in the large group (mean monthly prevalence = 47.3%) when compared with the small group (57.0%), which is counter to what was predicted ($t = 2.394$, $P = 0.027$). If the logic behind the expectation that the large group would have greater parasite infections than the small group is correct, the fact that we do not find differences in parasite infections would suggest that behavioral strategies may be operating to decrease infections in the large group.

Counter-strategies and behavioral differences between large and small groups

Firstly, as predicted, the large group spent more of its time traveling (average percent time spent traveling a month = 3.31, SD = 0.835) than the small group (average = 1.32, SD = 0.241; paired t-test = 2.468, P = 0.022). Secondly, the home range size of the large group (125 ha) was larger that of the small group (52 ha). Both groups moved extensively through their home range and entered most regions of the home range every 3–4 months.

Third, as predicted, if the subgroup was employing counter-strategies to decrease parasite transmission, the larger group frequently divided into subgroups, while the smaller group did not. It was difficult to determine when the large group divided because it was difficult to obtain an accurate count of the subgroup that was being followed, or locate two or more subgroups of the large group that contained recognizable individuals. Consequently, our estimate of the rate of subgroup formation is likely an underestimate. The large group divided into subgroups on at least 37% of observation days (n = 132 days), while the small group never divided (n = 107 days). It may be more appropriate to consider each month as an independent unit for analysis. At this level of analysis the large group formed subgroups in 75% of the observation blocks, while the small group never formed subgroups.

Fourth, the prediction that individuals in large groups would decrease the likelihood of parasite transmission by decreasing the rate of grooming was not supported. The large group spent 5.77 ± 0.46% of the time grooming, while the small group spent 5.14± 0.43% of the time grooming (paired t-test, t = 1.179, $P = 0.251$, $n = 23$ months). It may be that the need to maintain social cohesion in large groups by social grooming (Dunbar, 1991) outweighs any benefits to reduce the potential of parasite transmission.

Finally, as predicted, nearest neighbor distances were greater in the large group (average distance per month = 3.05 m, SD = 0.466, $n = 23$ months) than the small group (average = 2.73 m, SD = 0.568; paired t-test = 3.08, $P = 0.006$, $n = 23$ months). Whether this distance is biologically relevant with regards to disease transmission depends on the disease and its mode of transmission; however, an average difference of only 32 cm is likely too short to have an effect.

Conclusions and future directions

There was no evidence that the large group of red colobus had more severe parasite infections than the small group. However, behavioral differences were found between the groups that suggest that the large group may be engaging in strategies to decrease the rate of transmission of parasites. Unfortunately, decisively concluding that the observed behavioral differences are a counter-strategy evolved to decrease parasite infections is not possible. This is an unfortunate reality for two reasons. First, for the notion that these behaviors act to decrease disease transmission to be accepted, researchers will have to demonstrate that ecological arguments that have been substantially empirically tested are inadequate in explaining the observation. For example, the tendency for the large group to fission may be a strategy to reduce travel costs that are associated with its size. The large group will deplete patches faster, which will necessitate greater travel. Forming smaller subgroups during periods of food scarcity may reduce the costs of travel. This argument follows logic presented in the ecological constraints model (Milton, 1984; Chapman & Chapman, 2000b), which has already received support for red colobus (Chapman & Chapman, 2000a). The second reason to be skeptical concerning claims that these behaviors represent counter-strategies to decrease parasite infection is that many of the predictions have multiple outcomes. For example, fissioning could reduce the local density and decrease the chance of coming into contact with contaminated material and thus decrease transmission. Alternatively, it could place group members in new habitats where they could encounter novel parasites which they could acquire and subsequently transmit back to other group members.

What will be required in the future are studies that are carefully designed to distinguish between alternative hypotheses. For example, Nunn & Dokey (2006) used a meta-analysis of helminth richness of 119 primate species to distinguish between two competing hypotheses concerning how ranging behavior influences parasitism. They demonstrated that helminth richness increased with an index of territoriality, thus the physical separation of groups associated with territoriality did not lead to declines in measures of parasitism, rather parasitism increased as would be expected if the more intensive use of a home range increased exposure to parasites.

Future studies should attempt to determine which of the mechanisms that propose that parasites influence group size are potentially operating. For example, it should be possible to determine if species benefit nutritionally from winning between-group competition and whether this is of the magnitude to compensate for any elevation in any indices of parasitism. Ideal subjects for such a study would be black-and-white and red colobus. In black-and-white colobus it is the small groups that tend to win in between-group contest (Harris, 2005), while in red colobus it is the larger group that wins (Chapman, unpublished data). Thus, one would predict that the difference in infection levels between large and small black-and-white colobus groups would be much greater than between large and small red colobus groups, since a large black-and-white colobus group is doubly disadvantaged as they do not have a nutritional gain from winning contests over food and they have an elevated risk of parasite infection.

In conclusion, it is an exciting time for studies of primate disease ecology and it is a time where we can re-examine some of the ideas that were proposed in the past. Today, we not only have accumulating data sets of primates from around the globe (Nunn & Altizer, 2005), we have information on how anthropogenic disturbance influences parasite–host relationships (Chapman *et al.*, 2005a), we can use new computational and modeling tools (see Nunn, Chapter 5, and Hasegawa *et al.*, Chapter 25, this volume), and we have a rebirth of interest from researchers from a variety of fields as demonstrated in this volume.

Acknowledgements

Funding for this research was provided by Canada Research Chairs Program, Wildlife Conservation Society, Natural Science and Engineering Research

Council of Canada, and National Science Foundation. Permission to conduct this research was given by the National Council for Science and Technology, and the Uganda Wildlife Authority. L. J. Chapman, E. Greiner, T. Snaith, and C. Wong provided helpful comments on this research, C. Wong and C. Walsh greatly assisted with the processing of samples, and E. Greiner aided in parasite identification.

References

Altizer, S., Nunn, C. L., Thrall, P. H. *et al.* (2003). Social organization and parasite risk in mammals: integrating theory and empirical studies. *Annual Review of Ecology and Systematics*, **64**, 517–547.

Anderson, R. M. (1978). The regulation of host population growth by parasitic species. *Parasitology*, **76**, 119–157.

Anderson, R. M. & May, R. M. (1992). *Infectious Diseases of Humans: Dynamics and Control.* Oxford: Oxford University Press.

Aureli, F., Shaffner, C. M., Boesch, C. *et al.* (in press). Fission-fusion dynamics: new frameworks for comparative research. *Current Anthropology.*

Chapman, C. A. (1990a). Association patterns of spider monkeys: the influence of ecology and sex on social organization. *Behavioral Ecology and Sociobiology*, **26**, 409–414.

Chapman, C. A. (1990b). Ecological constraints on group size in three species of neotropical primates. *Folia Primatologica*, **55**, 1–9.

Chapman, C. A. & Chapman, L. J. (2000a). Constraints on group size in red colobus and red-tailed guenons: examining the generality of the ecological constraints model. *International Journal of Primatology*, **21**, 565–585.

Chapman, C. A. & Chapman, L. J. (2000b). Determinants of group size in primates: the importance of travel costs. In *On the Move: How and Why Animals Travel in Groups*, ed. S. Boinski & P. A. Garber. Chicago, IL: University of Chicago Press, pp. 24–41.

Chapman, C. A. & Lambert, J. E. (2000). Habitat alteration and the conservation of African primates: case study of Kibale National Park, Uganda. *American Journal of Primatology*, **50**, 169–185.

Chapman, C. A. & Pavelka, M. S. M. (2005). Group size in folivorous primates: ecological constraints and the possible influence of social factors. *Primates*, **46**, 1–9.

Chapman, C. A., Gillespie, T. R. & Goldberg, T. L. (2005a). Primates and the ecology of their infectious diseases: how will anthropogenic change affect host-parasite interactions? *Evolutionary Anthropology*, **14**, 134–144.

Chapman, C. A., Speirs, M. L. & Gillespie, T. R. (2005b). Dynamics of gastrointestinal parasites in two colobus monkeys following a dramatic increase in host density: contrasting density-dependent effects. *American Journal of Primatology*, **67**, 259–266.

Chapman, C. A., Wasserman, M. D., Gillespie, T. R. *et al.* (2006). Do nutrition, parasitism, and stress have synergistic effects on red colobus populations living in forest fragments? *American Journal of Physical Anthropology*, **131**, 525–534.

Coop, R. L. & Holmes, P. H. (1996). Nutrition and parasite interaction. *International Journal of Parasitology*, **26**, 951–962.

Crockett, C. M. & Janson, C. H. (2000). Infanticide in red howlers: female group size, male membership, and a possible link to folivory. In *Infanticide by Males and its Implications*, ed. C. P. van Schaik & C. H. Janson. Cambridge: Cambridge University Press, pp. 75–98.

Dunbar, R. I. M. (1991). Functional significance of social grooming in primates. *Folia Primatologica*, **57**, 121–131.

Eisenberg, J. F., Muckenhirn, N. A. & Rudran, R. (1972). The relation between ecology and social structure in primates. *Science*, **176**, 863–874.

Ezenwa, V. O. (2004). Host social behavior and parasitic infection: a multifactorial approach. *Behavioral Ecology*, **15**, 446–454.

Freeland, W. (1977a). *The Dynamics of Primate Parasites*. Ann Arbor, MI: University of Michigan.

Freeland, W. J. (1976). Pathogens and the evolution of primate sociality. *Biotropica*, **8**, 12–24.

Freeland, W. J. (1977b). Blood-sucking flies and primate polyspecific associations. *Nature*, **269**, 801–802.

Freeland, W. J. (1979). Primate social groups as biological islands. *Ecology*, **60**, 719–728.

Freeland, W. J. (1980). Mangabey (*Cercocebus albigena*) movement patterns in relation to food availability and fecal contamination. *Ecology*, **61**, 1297–1303.

Gatti, S., Swiercynski, G., Robinson, F. *et al.* (2002). Amoebic infections due to the *Entamoeba histolytica–Entamoeba dispar* complex: a study of the incidence in a remote rural area of Ecuador. *American Journal of Tropical Medicine and Hygiene*, **67**, 123–127.

Gillespie, T. R., Chapman, C. A. & Greiner, E. C. (2005a). Effects of logging on gastrointestinal parasite infections and infection risk in African primates. *Journal of Applied Ecology*, **42**, 699–707.

Gillespie, T. R., Greiner, E. C. & Chapman, C. A. (2005b). Gastrointestinal parasites of the colobus monkeys of Uganda. *Journal of Parasitology*, **91**, 569–573.

Gulland, F. M. D. (1992). The role of nematode parasites in Soay sheep (*Ovis aries* L.) mortality during a population crash. *Parasitology Research*, **105**, 493–503.

Harris, T. R. (2005). Roaring, intergroup aggression, and feeding competition in black and white colobus monkeys (*Colobus guereza*) at Kanyawara, Kibale National Park, Uganda. New Haven, CI: Yale University.

Heymann, E. W. (1999). Primate behavioural ecology and diseases – some perspectives for a future primatology. *Primate Report*, **55**, 53–65.

Huffman, M. A., Gotoh, S., Turner, L. A., Hamai, M. & Yoshida, K. (1997). Seasonal trends in intestinal nematode infection and medicinal plant use among chimpanzees in the Mahale Mountains, Tanzania. *Primates*, **38**, 111–125.

Ilmonen, P., Taarna, T. & Hasselquist, D. (2000). Experimentally activated immune defense in female pied flycatchers results in reduced breeding success. *Proceedings of the Royal Society of London*, **267**, 665–670.

Isbell, L. A. (1991). Contest and scramble competition: patterns of female aggression and ranging behaviour among primates. *Behavioral Ecology*, **2**, 143–155.

Kaplan, J. E., Hu, D. U., Holmes, K. K. *et al.* (1996). Preventing opportunistic infections in human immunodeficiency virus-infected persons: implications for the developing world. *American Journal of Tropical Medicine and Hygiene*, **55**, 1–11.

Koenig, A. (2002). Competition for resources and its behavioral consequences among female primates. *International Journal of Primatology*, **23**, 759–783.

Loehle, C. (1995). Social barriers to pathogen transmission in wild animal populations. *Ecology*, **76**, 326–335.

Milton, K. (1984). Habitat, diet and activity patterns of free-ranging woolly spider monkeys (*Brachyteles arachnoides* E. Geoffroy 1806). *International Journal of Primatology*, **5**, 491–514.

Milton, K. (1996). Effects of bot fly (*Alouattamyia baeri*) parasitism on a free-ranging howler (*Alouatta palliata*) population in Panama. *Journal of the Zoological Society of London*, **239**, 39–63.

Mitani, J. C., Watts, D. P. & Muller, M. N. (2002). Recent developments in the study of wild chimpanzee behaviour. *Evolutionary Anthropology*, **11**, 9–25.

Munger, J. C. & Karasov, W. H. (1989). Sublethal parasites and host energy budgets: tapeworm infection in white-footed mice. *Ecology*, **70**, 904–921.

Murray, D. L., Keith, L. B. & Cary, J. R. (1998). Do parasitism and nutritional status interact to affect production in snowshoe hares? *Ecology*, **79**, 1209–1222.

Nunn, C. L. & Altizer, S. M. (2005). The global mammal parasite database: an online resource for infectious disease records in wild primates. *Evolutionary Anthropology*, **14**, 1–2.

Nunn, C. L. & Altizer, S. (2006). *Infectious Diseases in Primates: Behavior, Ecology and Evolution*. Oxford: Oxford University Press.

Nunn, C. L. & Dokey, A. T. W. (2006). Ranging patterns and parasitism in primates. *Biology Letters*, **2**, 351–354.

Nunn, C. L., Gittleman, J. L. & Antonovics, J. (2000). Promiscuity and the primate immune system. *Science*, **290**, 1168–1170.

Nunn, C. L., Altizer, S., Jones, K. E. & Sechrest, W. (2003). Comparative tests of parasite species richness in primates. *American Naturalist*, **162**, 597–614.

Parraga, I. M., Assis, A. O. & Parado, M. S. (1996). Gender differences in growth of school-aged children with schistosomiasis and geohelminth infection. *American Journal of Tropical Medicine and Hygiene*, **55**, 150–156.

Olupot, W., Chapman, C. A., Waser, P. M. & Isabirye-Basuta, G. (1997). Mangabey (*Cercocebus albigena*) ranging patterns in relation to fruit availability and the risk of parasite infection in Kibale National Park, Uganda. *American Journal of Primatology*, **43**, 65–78.

Roberts, M. G., Dobson, A. P., Ameberg, P., de Leo, G. A. & Krecek, R. C. (2000). Parasite community ecology and diversity. In *The Ecology of Wildlife Diseases*, ed. P. J. Hudson, A. Rizzoli, B. T. Grenfell, H. Heesterbeek & A. P. Dobson. Oxford: Oxford University Press, pp. 63–82.

Sloss, M. W., Kemp, R. L. & Zajac, A. M. (1994). *Veterinary Clinical Parasitology*, 6th edn. Ames, IA: Iowa State University Press.

Smuts, B. B., Cheney, D. L., Seyfarth, R. M., Wrangham, R. W. & Struhsaker, T. T. (1987). *Primate Societies*. Chicago, IL: The University of Chicago Press.

Snaith, T. V. & Chapman, C. A. (2007). Primate group size and socioecological models: do folivores really play by different rules? *Evolutionary Anthropology*, **16**, 94–106.

Stammbach, E. (1987). Desert, forest, and montane baboons: multilevel societies. In *Primate Societies*, ed. B. B. Smuts, D. L. Cheney, R. M. Seyfarth, R. W. Wrangham & T. T. Struhsaker. Chicago, IL: University of Chicago Press, pp. 112–120.

Sterck, E. H. M., Watts, D. P. & van Schaik, C. P. (1997). The evolution of female social relationships in nonhuman primates. *Behavioral Ecology and Sociobiology*, **41**, 291–309.

Stoner, K. (1996). Prevalence and intensity of intestinal parasites in mantled howling monkeys (*Alouatta palliata*) in northeastern Costa Rica: implications for conservation biology. *Conservation Biology*, **10**, 539–546.

Struhsaker, T. T. (1997). *Ecology of an African Rain Forest: Logging in Kibale and the Conflict between Conservation and Exploitation.* Gainesville, FL: University of Florida Press.

Stuart, M. D., Greenspan, L. L., Glander, K. E. & Clarke, M. R. (1990). A coprological survey of parasites of wild mantled howling monkeys, *Alouatta palliata palliata. Journal of Wildlife Diseases*, **26**, 547–549.

van Schaik, C. P. (1983). Why are diurnal primates living in groups? *Behaviour*, **87**, 120–144.

van Schaik, C. P. (1989). The ecology of social relationships amongst female primates. In *Comparative Socioecology: The Behavioural Ecology of Humans and Other Mammals*, ed. V. Standen & R. A. Foley. Boston, MA: Blackwell Scientific Publications, pp. 195–218.

van Schaik, C. P. & van Hooff, J. A. R. A. M. (1983). On the ultimate causes of primate social systems. *Behaviour*, **85**, 91–117.

van Schaik, C. P. & van Noordwijk, M. A. (1988). Scramble and contest feeding competition among female long-tailed macaques (*Macaca fasicularis*). *Behaviour*, **105**, 77–98.

Vitone, N. D., Altizer, S. & Nunn, C. L. (2004). Body size, diet and sociality influence the species richness of parasitic worms in anthropoid primates. *Evolutionary Ecology Research*, **6**, 183–189.

Wrangham, R. W. (1980). An ecological model of female-bonded primate groups. *Behaviour*, **75**, 262–300.

22 *How does diet quality affect the parasite ecology of mountain gorillas?*

JESSICA M. ROTHMAN, ALICE N. PELL, AND DWIGHT D. BOWMAN

Photograph by Jessica Rothman (*Gorilla beringei*)

Introduction

Host nutritional status affects host–parasite dynamics, but little is known about this relationship in free-ranging primate populations. Nutrition and parasitic infections interact in three important ways. Firstly, since nutritional deficits

Primate Parasite Ecology. The Dynamics and Study of Host–Parasite Relationships, ed. Michael A. Huffman and Colin A. Chapman. Published by Cambridge University Press.

compromise immune function, the ability of the host's immune response to resist parasitic infections and cope with the negative consequences of parasites depends on adequate nutritional supplies (Coop & Kyriazakis, 2001). Secondly, some parasites, gastrointestinal parasites in particular, may reduce the ability of the host to absorb nutrients, thereby altering digestive efficiency and further compromising the host's nutritional status (Koski & Scott, 2001). Third, because parasites have their own nutritional demands, they compete for the host's food supply and further intensify the effects of malnutrition. As a consequence, concurrent malnutrition and parasitic infection may initiate a downward spiral where each intensifies the effect of the other, leading to increased nutritional stress, reduced immunocompetence, and increased infection (Koski & Scott, 2001). Alternatively, good nutrition can increase the ability of the host to overcome parasitism by improving immune response, affecting parasite establishment, and decreasing the chance for re-infection (Coop & Kyriazakis, 2001).

Some dietary items may possess antiparasitic properties. Janzen (1978) speculated that folivorous primates lack the protozoan parasites seen in frugivores because foliage contains higher concentrations of secondary compounds than fruit. More recent studies have demonstrated that protozoan parasites are indeed found in folivorous species (Stuart *et al.*, 1998; Gillespie *et al.*, 2005), but there is support for the idea that secondary compounds may help to limit parasite infections in wild primate populations (reviewed in Huffman, 1997; Nunn & Altizer, 2006). Condensed tannins are among the most widespread secondary compounds in tropical foliage, and have been a focus of primate feeding ecology studies for decades (Milton, 1979; Oates *et al.*, 1980; Wrangham & Waterman, 1981; Glander, 1982). They have long been considered as a feeding deterrent because they reduce palatability and bind protein, minerals, and other macromolecules (Mueller-Harvey, 2006). In domestic animals, they decrease productivity by lowering digestibility, protein availability, and food intake, affecting weight gain and milk yield. However, tannins also possess properties that positively affect health. For example, they suppress intestinal parasites in domestic ruminants (Hoste *et al.*, 2006). One study suggested that sifakas (*Propithecus verreauxi verreauxi*) consume tannins intentionally for medicinal purposes (Carrai *et al.*, 2003). These authors proposed that female sifakas increased their tannin intake during the birthing season, and suggested that the tannins consumed may provide anti-abortive and anti-hemorrhagic effects, and stimulate milk secretion. To our knowledge, no studies have quantified the effects of tannin intake on the parasite infections of primates.

The objective of this study was to understand how diet quality affected parasite infections of a group of mountain gorillas (*Gorilla beringei*) inhabiting Bwindi Impenetrable National Park, Uganda. We predicted that if diet quality decreased, this would affect parasite infections. We considered two indices of diet quality: concentrations of protein and condensed tannins. Since the Bwindi gorillas eat diets that contain condensed tannins (Rothman *et al.*, 2006), we predicted that increased dietary concentrations of condensed tannins would decrease parasite infection. Since varying protein and tannin concentrations may not necessarily have an immediate effect on parasite infections, we predicted that weeks when the individuals consumed lower quality diets would be followed by weeks with more severe parasite infections. Accordingly, we explored this question on two time scales. We considered the diet eaten by the gorillas, and the parasite infections 2 and 4 weeks after ingestion of a particular diet. We chose these two time scales because experimental studies on livestock have demonstrated that changes in parasite infections are associated with the diet that an animal has obtained in the prior 2 weeks (Coop & Kyriazakis, 2001). Our data collection was facilitated by our ability to systematically observe identifiable individuals, determine their nutrient intake, and monitor their parasite infections almost daily for one year.

Methods

Study site

Located in southwestern Uganda, Bwindi Impenetrable National Park (BINP) (0° 53′–1° 08′S, 29° 35′–29° 50′E) is a 331 km^2 rugged mountainous rain forest at 1160–2600 m asl characterized by steep-sided hills and narrow valleys (Butynski & Kalina, 1993; McNeilage *et al.*, 2001). The study area is dominated by forest gaps (67%; Nkurunungi *et al.*, 2004), which are both natural and a result of pit sawing prior to 1991 (Babassa *et al.*, 2004). Other habitats include: mixed forest (29%), mature forest (2%), and swamps (2%) (Nkurunungi *et al.*, 2004). The annual rainfall during the study was 1646 mm and the mean annual rainfall at the study site is 1440 mm. Annual rainfall is bimodal: the dry months are typically December to February and June to August, while the wet months are September to November and March to May.

Study group

The Kyagurilo group of gorillas has been habituated to human observers since the mid-1990s through daily monitoring by park rangers and field assistants of the Uganda Wildlife Authority and the Institute of Tropical Forest Conservation. Each individual is identifiable based on facial characteristics. The group ranges in a 40 km^2 area of the interior of BINP at high altitudes of the park, between 2100–2500 m above sea level (Robbins & McNeilage, 2003). At the time of the study, August 2002–2003, the group included 12 independent individuals (two silverbacks, six females, and four juveniles) and two infants. One additional infant was born during the study. To our knowledge, the group does not range in areas inhabited by humans, and group members only encounter humans when park staff and/or researchers observe them from a distance of >5 m for 4 hours daily. Tourists do not have access to this group.

Estimation of diet quality

To evaluate the quality of the diets consumed by each independent gorilla (adults and juveniles; infants were not studied), focal sampling was used to determine the concentrations of protein and tannin in the diet (detailed in Rothman, 2006). We observed the animals on 319 days during August 2002–2003. We observed each independent animal in 30 min intervals. Although all animals were habituated to observers, the dense terrestrial vegetation often interfered with observations of specific animals. Therefore, focal individuals were followed to the extent possible until they were out of view. If the focal individual was out of view for more than 5 min, another animal was chosen. Since the goal of our focal sampling was to estimate nutrient intake, if the animal was out of view before the 30 min focal sample was complete, data on food intake rates were considered if they continued for more than 5 minutes. Missing intervals where the animal was out of view while feeding were not used in calculations of nutrient intake. When several animals could be observed simultaneously, opportunistic focal observations were made of the visible individuals. Our focal follows totaled 1318 hours.

During focal follows, we counted all food items ingested during timed feeding bouts. A feeding bout began whenever an animal first made contact with a food item and ended when the animal stopped feeding on any food item for 10 seconds. Multiple food items were frequently eaten during these feeding bouts. Observer biases were minimized by standardizing observational techniques by having all researchers observe the same animal on 3 days per month.

Because of variation in the weight and size of food items and how they are processed by different-sized gorillas (i.e. a leaf in the dry season may be smaller and lighter than a leaf in the wet season, and a juvenile may take smaller bites than an adult), specific food units were defined for each food item. A unit could be considered a single plant item (e.g. a leaf, or fruit), the approximate dimensions of the food item (for peel, bark, pith, wood) or in the case of small leaflets or fruits, the average number of item in a cluster (Chivers, 1998). To account for variation across time and space, we calibrated these units by weighing them immediately after collection (n = 50) to calculate the mean unit wet and dry weight for each food item regularly (at least once per month for each food that represented > 1% of the wet weight dietary intake).

Food items eaten during our focal observations were collected from the exact plant eaten by the gorillas when possible or from several adjacent plants of the same species. We attempted to process the selected food items similarly to the gorillas. For example, if the gorilla removed the outer peel of herbaceous stems, but ate the interior stem core, we selected the inner stem core for analysis. After collection, plants were immediately weighed in the field using a portable balance to the nearest 0.1 g. The samples were dried in a dark area of the field station (≤25 °C). When the samples achieved a constant weight, they were ground in a Wiley Mill through a 1-mm screen. Food species that were eaten often during the study were regularly sampled and reanalyzed to adjust for seasonal and environmental variation in nutritional composition (Chapman *et al.*, 2003). In total, we analyzed 336 plant samples.

Foods were analyzed by JMR in the Animal Nutrition laboratory at Cornell University. The amount of nitrogen in each food items was estimated using a Leco FP-528 combustion analyzer using standard procedures (AOAC, 1990). The measurement of total nitrogen provides an estimate of crude protein (protein levels = n × 6.25). However, a substantial portion of the nitrogen may be bound to the lignin in tropical plant parts and therefore unavailable to animals as a protein source (Milton & Dintzis, 1981; Conklin-Brittain *et al.*, 1999). For example, in many food items, the nitrogen bound to acid detergent fiber (mainly comprised of cellulose and lignin) exceeded 10% of the total nitrogen. We estimated this bound nitrogen by analyzing the nitrogen bound to the acid detergent fiber (cellulose + lignin) following the methods by Licitra *et al.* (1996).

To account for this indigestible nitrogen in our estimates of protein intake by gorillas, we calculated digestible protein in each plant part in the following way. First, we subtracted the amount of nitrogen bound to the acid detergent residue (ADIN) from the nitrogen to calculate the available nitrogen (National Research Council, 2003). Then we multiplied the amount of available nitrogen by 6.25 to estimate the amount of protein available to the gorillas (e.g. [N −ADIN] × 6.25 = available protein).

To estimate the amounts of condensed tannins in food items (% dry matter), we followed the methods of Hagerman & Butler (1994). The dried, ground plant samples were extracted in 70% (v/v) aqueous acetone in a sonicator containing ice water bath for 20 min. After 20 min, extracts were centrifuged at 2500 g for 10 min. The supernatant was collected and stored, covered in the dark at 4 °C. These extractions were repeated four times, and the supernatants were combined to reach a final concentration of 10 mg of crude plant material per ml of acetone extract. The acid butanol assay was used to estimate the amount of condensed tannin in each crude extract (Porter *et al.*, 1986) with controls for pigments in the crude plant extract (Watterson & Butler, 1983). Using Sephadex LH-20, internal standards were generated for each plant part representing at least 0.5% by weight of the diet (Hagerman & Butler, 1994). These purified tannins were used to develop standard curves which were used to estimate the amounts of tannin in the plant extracts of unknown quantities of condensed tannin.

Purifying tannin from the plant species of interest is important because there is a great deal of variation in the structure of tannins, which affects their reactivity in spectrophotometric assays. Many studies use a single standard to estimate tannin quantity in primate foods (e.g. quebracho: Remis *et al.*, 2002; Powzyk & Mowry, 2003; Norconk & Conklin-Brittain, 2004). However, this approach can cause ambiguous results (Giner-Chavez *et al.*, 1997; Schofield *et al.*, 2001; del Pino *et al.*, 2005). For example, a fruit commonly eaten by gorillas could misleadingly be estimated as anywhere between 0.3–11% DM in condensed tannin content, solely based on which of four external standards was chosen (Rothman *et al.*, 2006), when its actual tannin content is 8.9% DM.

To estimate the concentrations of protein and tannins in the diets of each individual, we multiplied the number of units consumed during a focal follow by the mean weight of that unit and then by the percent dry weight protein or tannin content of each food item. We summed this for all units consumed during each focal follow and multiplied this by the number of minutes spent feeding during that day. This result was divided by the total dry weight consumed during the day. Although we were only able to watch the gorillas for 4 hours, we considered this 4-hour period to be representative of the entire day and scaled the data accordingly. On 7 days additional observation time was permitted and the gorillas were followed from first contact until they built night nests (16:30–17:30 hrs). We confirmed that our 4-hour observations were representative of the remainder of the day (Rothman, 2006). The mean nutrient intake per follow per individual was used to calculate nutrient intake over a 2-week period.

Parasite analysis

Like other great apes, gorillas typically sleep each night in a newly built nest; infants nest with their mothers (Schaller, 1963), and before leaving its nest in the morning, a gorilla typically defecates on the nest's outer rim. Fecal samples were collected from the night nest of each group member and were assigned to age-sex classes by the diameter of the central bolus of the dung (infant < 2 cm; juvenile 2–4 cm; adult female 5–7 cm; silverback > 7 cm; Schaller, 1963); a technique commonly used in censuses (McNeilage *et al.*, 2001). The ability to differentiate fecal samples by age and sex was confirmed by opportunistically collecting samples when defecation was observed by identifiable animals for verification: males (n = 27, mean 7.1 ± 0.4 cm), females (n = 33, 6.0 ± 0.4 cm), and juveniles (n = 16, 3.9 ± 1.3 cm), and there was no overlap in size between any of the age-sex categories including juvenile males and adult females without infants. Since the gorillas did not have diarrhea or soft stool in their night nests during the study, the diameter of dung was considered to reliably correspond to a specific age-sex class.

In addition to classifying samples by age and sex, some fecal samples could be assigned to identifiable individuals including a female, a juvenile, and the two silverbacks. Because one female and her juvenile nested together, their nest was the only one in the group that contained feces of both a juvenile and female. The two silverback nests were differentiated on the basis of distance from the rest of the group. As in an earlier study (Schaller, 1963), the dominant silverback slept in the middle of the group and the subordinate silverback slept > 10 m from other group members (JMR, unpublished observation). Therefore, all dung samples collected from the nest >10 m from the center of the group were considered to be from the subordinate silverback and those samples collected from the silverback nest in the center of the group were considered to be from the dominant silverback male.

Fecal samples were collected once weekly from all group members and assigned to a particular age-sex category based on their dung diameter. For at least 4 of 7 days of each week, fecal samples from the dung that could be assigned to the four individually identifiable gorillas were collected: the dominant silverback (n = 306), the subordinate silverback (n = 256), a female (n = 239), and a juvenile (n = 239). Though every attempt was made to collect samples daily for these four animals, sometimes the gorilla nests were not located or the dung samples had been tampered with by the gorillas. This did not happen often and was not biased towards any one month or season, so we do not feel this is significant. A sample (~3 g) was taken from the interior of the fecal bolus and preserved in pre-weighed tubes containing 10% formalin for later examination.

At the College of Veterinary Medicine at Cornell University, feces were processed using the formalin-ethyl acetate sedimentation concentration technique (Garcia, 1999). Because specific identification of eggs in the feces would have required coproculture (in the case of the strongylid eggs) or details of worms present in these gorillas (based on necropsies), familial or generic determinations based on the appearance of the eggs or larvae are presented.

Data analysis

We explored the possibility that diet quality is related to indices of intestinal parasitism. We considered the percentage of available protein content of the diet, and the percentage of the diet comprised of condensed tannins (% dry matter basis) as indicators of diet quality, and the percentage of fecal samples positive for a particular parasite, and mean infection intensity (eggs per gram (epg) or larvae per gram (lpg) in all samples) as indices of parasitic infection. We considered these indices biweekly (every two weeks), so the results of all samples collected in a specific week for each age-sex class and identifiable individuals were pooled and the mean was taken. Since the amount of food eaten did not vary over the study period for each individual (Rothman, 2006), using diet concentrations is a valid estimation of diet quality.

Because we did not expect to see an effect on parasite infections immediately (Coop & Kyriazakis, 2001), we used a time lag of 2 weeks and 4 weeks to consider whether diet quality affected indices of parasite infections. Since our data were not normally distributed, we used Spearman rank correlations to assess relationships between diet quality and parasitic infection. We consider our analysis preliminary and exploratory, and so we did not correct for multiple comparisons using Bonferroni corrections (Nakagawa, 2004).

Results

Diet quality

Our assessment of diet quality revealed that intake of protein varied throughout the year, but diets were always well in excess of the protein requirements of humans, and never fell below a mean of 8% available protein (Table 22.1). This suggests that protein was not limiting and therefore may not have an effect on parasite infections.

Tannin content in the gorilla diet also varied throughout the year. During some weeks of the year, the gorillas ate a diet that contained minute concentrations of tannins (0.1% of the dry matter) and other weeks the diet contained appreciable concentrations of tannins (8.8%) (Figure 22.1).

Parasite community

The parasite community described from the 1408 fecal samples included the following: unidentified strongyle eggs, *Anoplocephala gorillae* (Superfamily

Table 22.1. *The diet quality (biweekly mean ± SD) and parasite infections (biweekly mean and range) of a group of mountain gorillas in Bwindi Impenetrable National Park, Uganda*

Infection index	Dominant silverback	Subordinate silverback	All females	All juveniles	Known female	Known juvenile
Percent of samples positive for						
Strongyle eggs	87.5 (67–100)	77.8 (36–100)	75.9 (16–100)	67.9 (25–100)	71.1 (0–100)	62.4 (0–100)
Probstmayria sp.	13.9 (0–43)	7.6 (0–38)	12.2 (0–37)	9.8 (0–38)	10.6 (0–71)	10.0 (0–86)
Anoplocephala gorillae	8.0 (0–21)	8.5 (0–33)	8.3 (0–33)	7.2 (0–50)	6.3 (0–43)	10.5 (0–50)
Strongyloides fuelleborni	2.0 (0–22)	0.83 (0–12.5)	0 (0)	0.7 (0–38)	0.38 (0–9.0)	0.33 (0–7.7)
Parasite burden						
Strongyles (egg/g)	89.0 (21–267)	29.9 (5–76)	40.1 (2–184)	35.8 (3–132)	27.0 (0–105)	19.7 (0–61)
Probstmayria sp. (larvae/g)	0.75 (0–4)	0.28 (0–1)	1.0 (0–12)	0.80 (0–10)	0.66 (0–9)	0.73 (0–12)
Anoplocephala gorillae (egg/g)	0.72 (0–4.9)	1.29 (0–12.1)	0.7 (0–10)	1.0 (0–16)	0.70 (0–13)	1.4 (0–19)
Strongyloides fuelleborni (egg/g)	0.15 (0–1.2)	0.04 (0–0.75)	0 (0)	0.7 (0–17)	0.01 (0–0.27)	0.01(0–0.23)
Diet quality						
Protein (% dry matter basis)	14.5 ± 3.3	13.7 ± 4.9	12.9 ± 2.7	12.9 ± 3.0	12.8 ± 5.5	11.5 ± 4.2
Tannin intake (% dry matter basis)	3.3 ± 2.6	3.0 ± 2.8	3.6 ± 2.7	3.8 ± 2.8	3.0 ± 3.7	3.2 ± 3.1

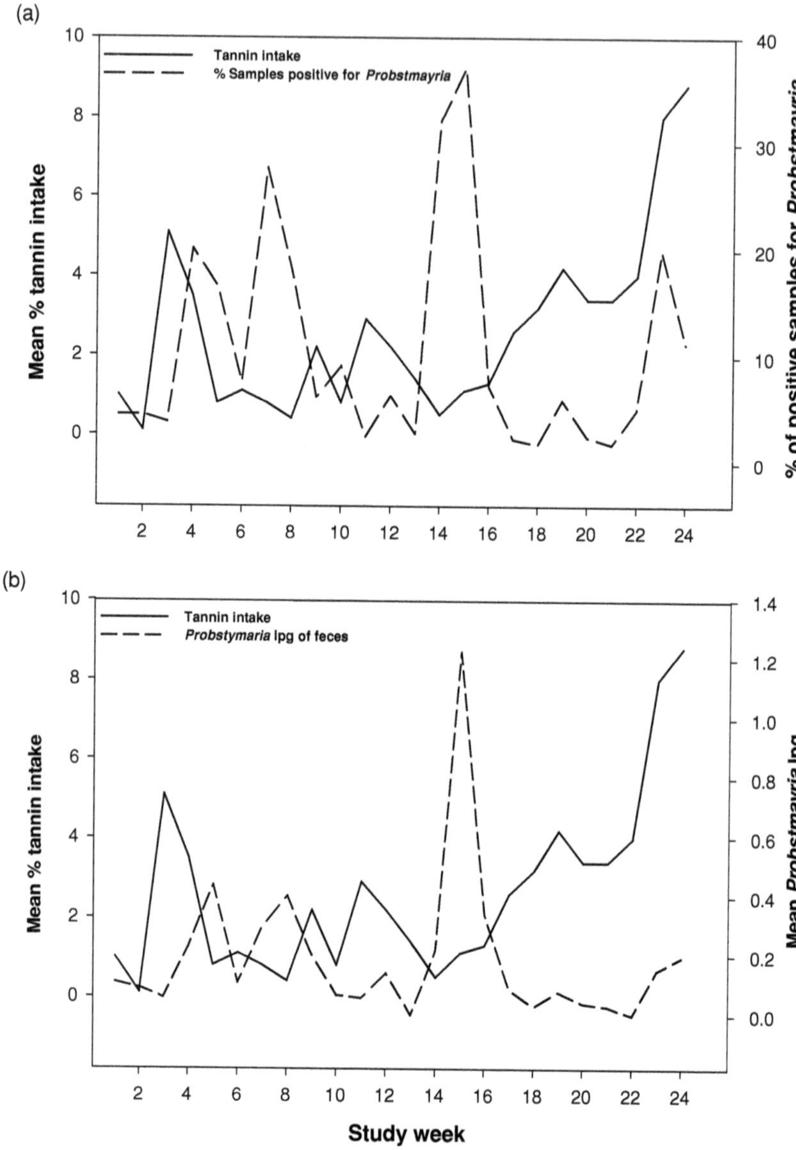

Figure 22.1. The relationship between the concentration of tannins in the diet of mountain gorillas and the (a) percentage of fecal samples positive for *Probstmayria* after a 2-week lag, and (b) mean number of *Probstmayria* larvae per gram (lpg) of feces after a 2-week lag. The data represent the mean infection levels and tannin intake of the four identifiable individuals monitored.

Anoplocephalidae), *Probstmayria* sp. (Superfamily Cosmocercoidea), *Strongyloides fuelleborni* (Superfamily Rhabditoidea) (Rothman *et al.*, 2008a). In addition, on three occasions an unidentified trematode egg was found in the feces of three different individuals (the dominant silverback and two different juveniles). This trematode egg was not considered further. Based on coprocultures and a necropsy of a female gorilla in Bwindi (Durette-Dusset *et al.*, 1992; Ashford *et al.*, 1996), the strongyle eggs present in the majority of fecal samples could be: *Oesophagostomum* sp., *Murshidia devians*, *Paralibyostrongylus kalinae*, and/or *Hyostrongylus kigezensis*. *Paralibyostrongylus kalinae* and *H. kigeziensis* have only been described from gorillas in Bwindi (Durett-Dusset *et al.*, 1992). Since all of these parasites have similar eggs and the gorillas host multiple strongylid species, it would be difficult to differentiate the strongylid eggs found in the feces to genera or species without necropsy.

In weekly samples collected from all group members categorized by age-sex class and fecal samples attributed to specific individuals, we found all of the above parasites during the year (Table 22.1). During some weeks the samples from all individuals were positive for strongyles, *A. gorilla,* and *Probstmayria*, indicating that all animals were infected and the prevalence of these parasites in the group was 100%. Nematodes and tapeworms can live for a long time inside their hosts, which may result in chronic infection. Since we do not know anything about the lifespan of these worms in particular, it is unclear whether the variation in egg output and percentage of positive samples represents the acquisition of a new infection, or heterogeneity in egg output. Since we did not repeatedly examine the same dung sample, we do not know whether variation could be due to methods of examination, but our techniques to examine fecal samples are standard practice in parasitology (Garcia, 1999).

Relationships between diet quality and parasite infections

We compared the biweekly mean diet quality with the mean biweekly infection levels (at 2 and 4 weeks after consumption of a particular diet quality) of each age-sex class (females and juveniles) and the known individuals (two silverbacks, female, and juvenile). We contrasted two indices of parasite infection with protein and tannin concentrations in the diet: parasite burden (eggs or larvae per gram) and percentage of positive fecal samples.

Following a 2- and 4-week time lag, we found no correlations between protein intake and indices of parasite infection for any animal or age-sex class for all parasites found ($P > 0.20$). However, if individuals in the group had a higher tannin intake during the previous 2 weeks, then indices of parasitic infection were affected in some animals. Specifically, we found correlations between tannin intake and indices of one parasite, *Probstmayria* sp. (Figure 22.1), a pinworm that is likely of little clinical importance and is found in many species of mammals. When tannins were consumed, the percentage of positive samples for *Probstmayria* sp. infections decreased in the following 2-week period for females (Spearman rank coefficient: −0.42, $P = 0.05$) and juveniles (−0.54, $P < 0.01$), and the known juvenile (−0.72, $P < 0.01$). However, this relationship was not found for the dominant silverback (−0.28, $P = 0.20$), the subordinate silverback (0.03, $P = 0.89$), or the known female (−0.29, $P = 0.19$). Similarly, following periods of tannin consumption, the number of *Probstmaryia* larvae per gram of feces decreased for the females (−0.48, $P = 0.02$) and juveniles (−0.55, $P = 0.01$), the dominant silverback (−0.48, $P = 0.02$), and the known juvenile (−0.50, $P = 0.02$). This relationship was not found in the subordinate silverback (0.03, $P = 0.89$) or the known female (−0.11, $P = 0.61$). We found no significant correlations for any other indices of infection for any of the other parasites with a 2-week time lag. Following a 4-week time lag, we found no correlations between tannin intake and indices of parasitic infection for any members of the gorilla group ($P > 0.20$).

Discussion

Our results support predictions that the intake of condensed tannins could be affecting one parasite that infects the Bwindi mountain gorillas. We found trends suggesting that *Probstmayria* sp., a pinworm, was negatively affected by the level of condensed tannins in the diet, suggesting that the level of tannins may impact this parasite of the gorillas. Further long-term work, complementary experimental studies, and a larger sample size are needed to explore these trends further. The majority of tannins in the diet of the gorillas were provided by the staple foods eaten by the gorillas. The variation in dietary composition throughout the year, combined with the intraspecific variation in tannin content in staple foods, accounted for the varying concentrations of tannin intake. For example, *Urera hypselodendron* leaves, one of the most frequently eaten foods, was highly variable in tannin content. According to the season and growing conditions, tannin concentrations in the leaves varied from 0% to 15% dry weight in areas where the gorillas fed (JMR, unpublished data). The gorillas did not appear to alter their diet based on this intraspecific variation in the quality of this food item. Thus, it is our view that the gorillas do not intentionally adjust their diet to reduce their parasite loads (i.e. intentionally consume tannins for medicinal purposes), but that variation in tannin concentration in their diets is a consequence of the intraspecific and interspecific variation in nutritional content of their diets. Over the year, the gorillas ate diets that

(a)

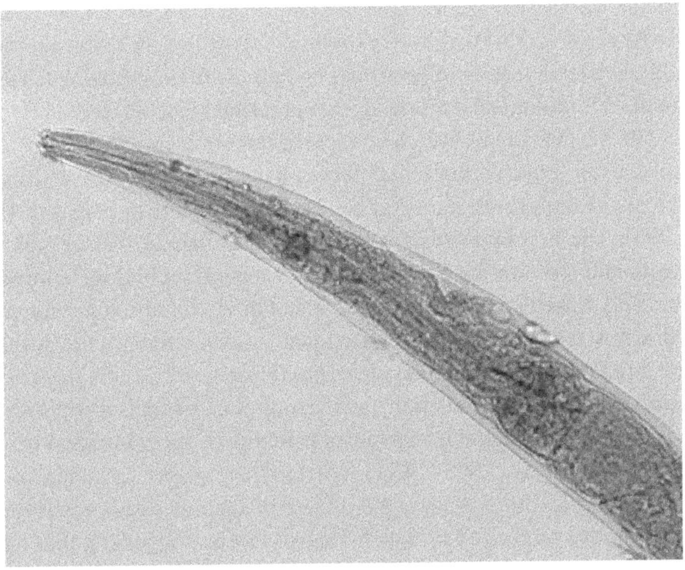

(b)

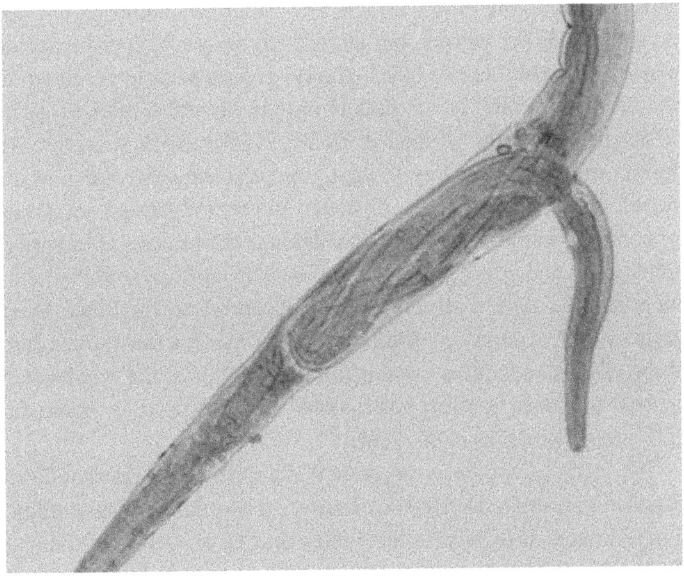

Figure 22.2. A female *Probstmayria* sp. giving birth. She was found in the fecal sample of an adult female mountain gorilla. (a) head (40 × magnification) (b) body (10×).

contained between 0.1% and 8.8% tannin on a dry weight basis (Figure 22.1). Hoste *et al.* (2006) reviewed experimental studies on domestic ruminants and suggested that threshold levels of condensed tannin should reach at least 3–4% tannin for biological activity against parasites to be observed.

The mechanism by which condensed tannins may affect *Probstmayria* sp. is not known, but two main hypotheses have been proposed with respect to the effects of tannins on parasitic nematodes of ruminant livestock (Hoste *et al.*, 2006). The first hypothesis is that condensed tannins act as a pharmaceutical agent and have antihelmintic properties that affect biological processes of the parasite. Based on a series of *in vitro* and in vivo studies on ruminants and their parasites, it has been demonstrated that tannins influence the development and establishment of early nematode stages (Hoste *et al.*, 2006). In vitro studies have consistently shown that crude tanniferous plant extracts affect the mobility and viability of infective stage larvae, or hatching of eggs (Molan *et al.*, 2000, 2002, 2003; Max *et al.*, 2005). When polyethylene glycol (a tannin binding agent) was added to these assays, the effects of tannins disappeared and nematode populations recovered to match control levels, suggesting that tannins play a key role. Condensed tannins have a high affinity for proline and hydroxyproline, two amino acids that are found on the cuticular coating of nematodes (Bahuaud *et al.*, 2006; Hoste *et al.*, 2006). The cuticular coating is not only found on the outside of the worms, but also covers internal reproductive and digestive anatomy. The number of free hydroxyl groups on condensed tannin is directly related to the inability of trichostrongyle larvae to remove its outer coating or sheath (exsheath) (Brunet & Hoste, 2006), which is a critical step in their development from the free-living to the parasitic stage. Lesions are evident on the internal structures of the digestive and reproductive tract of larvae exposed to tannins. Therefore, tannins may damage the biological integrity of cuticular membranes and affect parasite viability. The wide variation in tannin chemistry suggests that tannin structure and concentration modulate these effects on gastrointestinal parasites. Although there are only a few studies that have looked at the effects of tannins on non-ruminant animals, the results of these studies support that tannins effect gastrointestinal parasites (e.g. swine: Salajpal *et al.*, 2004; rodents: Rojas *et al.*, 2006).

The second hypothesis suggests that condensed tannins could improve host condition in ruminant livestock indirectly by binding high-quality dietary protein, allowing it to bypass the rumen and be absorbed in the small intestine. At low levels, condensed tannins could be beneficial in this way to ruminants because they could protect important amino acids from being used by ruminal microbes (Wang *et al.*, 1994). Instead, this high-quality protein is absorbed in the small intestine, improving nutritional status and alleviating any parasite-induced protein deficiency (Bown *et al.*, 1991). While this may be important for ruminants, it would not apply to gorillas because they are simple stomached animals with hindgut fermention. Therefore, this is not a plausible hypothesis as to why indices of *Probstmayria* sp. infections were reduced in these gorillas. Additionally, improving host resilience to parasitic infection would take time as nutritional condition improved; reduction of parasite infections has been fairly rapid after tannin intake in most studies (about 2 weeks), suggesting a direct antihelmintic effect (Coop & Kyriazakis, 2001).

Probstmayria sp. is a pinworm found in a variety of species including chimpanzees, baboons, domestic horses, zebra, tapirs, and swine. It has been reported as a parasite of gorillas at several research sites (Sleeman *et al.*, 2000; Lilly *et al.*, 2002; Rothman & Bowman, 2003; Freeman *et al.*, 2004). Despite its widespread prevalence, *Probstmayria* spp. are little studied. This is probably because they are of minimal clinical importance; pinworms in humans are more of a nuisance than a problem. They cause anal itching and restlessness, which could lead to insomnia and stress. Transmission is direct and occurs when infective larvae in the feces are ingested; coprophagy, which was observed rarely (three times in > 1300 hours of observation), may facilitate transmission (Graczyk & Cranfield, 2003). *Probstmayria* are viviparous and their entire life cycle takes place internally so that large numbers in the large bowel can accumulate rapidly. As a result, since gorillas are hindgut fermenters and food is probably retained in the colon for a longer time relative to other segments of their gastrointestinal tract, exposure of the worms to condensed tannins may be facilitated.

No other parasites appeared to be affected by condensed tannins. Heavy infections of strongylids are associated with dysentery, mucosal inflammation, ulceration, weight loss, and death, so these worms could cause substantial damage, severe pain, and pose a serious threat to the gorillas. Postmortems of mountain gorillas in the nearby Virunga region revealed nodular worm disease in three of eight gorillas (Mudakikwa *et al.*, 2001). Strongyle eggs were routinely found in the gorilla feces, and did not fluctuate in response to increased dietary tannin intake. It is not known whether eggs found in the feces represent more than one species, which could confound the results because different nematodes may have different responses to tannins. The results of several mountain gorilla necropsies suggest that multiple strongyle infections are common (Fossey, 1983; Durette-Dussette *et al.*, 1992; Ashford *et al.*, 1996; Mudakikwa *et al.*, 2001). Alternatively, the fact that we saw no variation in egg output with tannin intake could be a consequence of the location of the worms. Most of the strongyles identified from the Bwindi gorillas live in the small intestine and stomach. Since food likely moves more rapidly through these segments, reduced tannin–parasite interaction may reduce the impact of tannins. There was no evidence that tannins affected the tapeworm

Anoplocephala gorillae, which has been recovered from the small intestine of the necropsied gorillas (Mudakikwa *et al.*, 2001). It may be that the location of tapeworms does not allow for adequate exposure to tannin, or it is possible that cestodes are not affected by tannins in the same manner as nematodes. In a study of growing goats, increasing levels of a high tannin feed affected nematode eggs and coccidian oocysts in the feces, but had no effect on the number of cestode eggs (Dung *et al.*, 2005). Very few studies have investigated the effects of tannins on tapeworms and therefore our knowledge is minimal.

It is well-established that protein and energy malnutrition with accompanying parasitic infection affect susceptibility to parasites, and that parasitic infection further exacerbates nutritional stress due to the competition for resources (Smith *et al.*, 2005). The interactive effect of concomitant nutritional stress and parasitic infection has demonstrably led to population crashes in the wild (Gulland, 1992; Milton, 1996; Newey *et al.*, 2005). The synergy between nutritional stress and parasitism has been suggested to be the mechanism driving population declines in endangered red colobus monkey populations living in fragments of an African rain forest (Chapman *et al.*, 2006); within contiguous forest, where nutritional supplies were abundant, these interactive effects were not observed (Chapman *et al.*, 2007). Fluctuations in the dietary protein content did not affect the parasite ecology of the gorillas, likely because the gorillas did not appear to have a limiting amount of protein in their diets and many of their staple foods eaten daily contain > 20% crude protein (Rothman *et al.*, 2006a). Energy intake of these gorillas did not fluctuate seasonally and also did not appear to be limiting in the diet of the gorillas (Rothman *et al.*, 2008b). With sufficient amounts of dietary energy and protein, the effects of parasitism may be negligible (Munger & Karasov, 1989), and since these gorillas did not show evidence of nutritional stress, it is not surprising that fluctuations in protein content did not affect parasite infections (Murray *et al.*, 1998).

While the gorillas appeared to be in good clinical health throughout the study, illnesses may prompt self-medication (Huffman *et al.*, 1997). Huffman (1997) outlined the conditions required as evidence for intentional medicinal plant use: (1) The intake of plant species which are not normally consumed as part of the regular diet and do not provide a nutritional benefit. (2) The presence of an illness or parasite infection. (3) A positive change in the illness or parasite infection after ingestion of the proposed medicinal plant. Although the gorillas harbored parasite infections throughout the study, we did not see any signs of intestinal disease, such as diarrhea or blood in the stool, nor did we observe any evidence of self-medicative behaviors, such as leaf swallowing, which has been suggested to assist in the expulsion of parasites from the intestinal tract of chimpanzees, bonobos, and lowland gorillas (Huffman *et al.*, 1996; Dupain *et al.*, 2002; Huffman, 2003; Dupain *et al.*, Chapter 14, this volume). During the year, the gorillas ate a diverse diet including at least 158 food items from 107 food species; many are used by humans in surrounding villages for their medicinal properties (Rothman, unpublished data; Cousins & Huffman, 2002). Minute quantities of a diversity of foods may provide important prophylactic health benefits (Sherman & Billing, 1999; Arimond & Ruel, 2004). Although it did not appear that the gorillas were intentionally consuming plants for their medicinal value, we suggest that properties of their diet affect parasite infections.

Our study permitted an exploration of the effects of diet on the parasite infections of endangered mountain gorillas. While gorillas in Bwindi are protected and healthy, many other populations are being decimated by commercial hunting, habitat destruction through mechanized logging, and Ebola (Walsh *et al.*, 2003; Bermejo *et al.*, 2006). As humans invade the habitats of gorillas, the unfortunate reality is that we are even more pressed to understand baseline patterns of host–parasite relationships (Chapman *et al.*, 2005; Nunn & Altizer, 2006). This knowledge will allow us to more clearly identify the consequences of our actions and contribute to informed management plans that mitigate disease risks in endangered apes.

Acknowledgements

We thank Moses Akatorana and the assistants at the Institute of Tropical Forest Conservation for their hard work in the field. We are grateful to Kathy Duisenberre, James Robertson, Debbie Ross, Mike Van Amburgh, Peter Schofield, Susanne Pelton, Ralph Obendorf, and Ron Butler for assistance with analyses, technical laboratory training, and use of equipment. We thank Peter Van Soest, Debbie Cherney, Linda D'Anna, Ellen Dierenfeld, Skip Hintz, Araceli Lucio-Forster, Jan Liotta, Colleen McCann, John Bosco Nkurunungi, John Bosco Nizeyi, Eloy Rodriguez, and Colin Chapman for insightful discussions related to this project. We thank Tamaini Snaith for helpful comments that improved the quality of this chapter. Alastair McNeilage, William Olupot, Dennis Babaasa, Robert Bitariho, John Makombo, and Aventino Kasangaki provided logistical support in the field. Jane Engel and the Robert G. Engel Family Foundation, the Department of Animal Science at Cornell University, Mario Einaudi Foundation, Cornell University Graduate School, and the Institute of African Development provided funding for this research. We thank the Uganda Wildlife Authority and the Uganda Council for Science and Technology for permission to conduct this research. Finally, we thank Mike Huffman and Colin Chapman for inviting us to contribute to this volume.

References

AOAC (1990). *Official Methods of Analysis*. Arlington, VA: Association of Official Analytical Chemists.

Arimond, M. & Ruel, M. T. (2004). Dietary diversity is associated with child nutritional status: evidence from 11 demographic and health surveys. *Journal of Nutrition*, **134**, 2579–2585.

Ashford, R. W., Lawson, H., Butynski, T. M. & Reid, G. D. F. (1996). Patterns of intestinal parasitism in the mountain gorilla *Gorilla gorilla* in the Bwindi Impenetrable Forest, Uganda. *Journal of Zoology*, **239**, 507–514.

Babassa, D., Eily, G., Kasangaki, A., Bitariho, R. & McNeilage, A. (2004) Gap characteristics and regeneration in Bwindi Impenetrable National Park, Uganda. *African Journal of Ecology*, **42**, 217–224.

Bahuaud, D., De Montellano, C. M. O., Chauveau, S. *et al.* (2006). Effects of four tanniferous plant extracts on the *in vitro* exsheathment of third-stage larvae of parasitic nematodes. *Parasitology*, **132**, 545–554.

Bermejo, M., Rodriguez-Teijeiro, J. D., Illera, G. *et al.* (2006). Ebola outbreak killed 5000 gorillas. *Science*, **314**, 1564–1564.

Bown, M. D., Poppi, D. P. & Sykes, A. R. (1991). The effect of post-ruminal infusion of protein or energy on the pathophysiology of *Trichostrongylus colubriformis* infection and body composition in lambs. *Australian Journal of Agricultural Research*, **42**, 253–267.

Brunet, S. & Hoste, H. (2006). Monomers of condensed tannins affect the larval exsheathment of parasitic nematodes of ruminants. *Journal of Agricultural and Food Chemistry*, **54**, 7481–7487.

Butynski, T. M. & Kalina, J. (1993). Three new mountain national parks for Uganda. *Oryx*, **27**, 214–224.

Carrai, V., Borgognini-Tarli, S. M., Huffman, M. A. & Bardi, M. (2003). Increase in tannin consumption by sifaka (*Propithecus verreauxi verreauxi*) females during the birth season: a case for self-medication in prosimians? *Primates*, **44**, 61–66.

Chapman, C. A., Chapman, L. J., Rode, K. D., Hauck, E. M. & McDowell, L. R. (2003). Variation in the nutritional value of primate foods: among trees, time periods, and areas. *International Journal of Primatology*, **24**, 317–333.

Chapman, C. A., Gillespie, T. R. & Goldberg, T. L. (2005). Primates and the ecology of their infectious diseases: how will anthropogenic change affect host-parasite interactions? *Evolutionary Anthropology*, **14**, 134–144.

Chapman, C. A., Wasserman, M. D., Gillespie, T. R. *et al.* (2006). Do nutrition, parasitism, and stress have synergistic effects on red colobus populations living in forest fragments? *American Journal of Physical Anthropology*, **131**, 525–534.

Chapman, C. A., Saj, T. & Snaith, T. V. (2007). Temporal dynamics of nutrition, parasitism and stress in colobus monkeys: implications for population regulation and conservation. *American Journal of Physical Anthropology*, **134**, 240–250.

Chivers, D. J. (1998). Measuring food intake in wild animals: primates. *Proceedings of the Nutrition Society*, **57**, 321–332.

Conklin-Brittain, N. L., Dierenfeld, E. S., Wrangham, R. W., Norconk, M. & Silver, S. C. (1999). Chemical protein analysis: a comparison of Kjeldahl crude protein and total ninhydrin protein from wild, tropical vegetation. *Journal of Chemical Ecology*, **25**, 2601–2622.
Coop, R. L. & Kyriazakis, I. (2001). Influence of host nutrition on the development and consequences of nematode parasitism in ruminants. *Trends in Parasitology*, **17**, 325–330.
Cousins, D. & Huffman, M. A. (2002). Medicinal properties in diet of gorillas: an ethno-pharmacological evaluation. *African Study Monographs*, **23**, 65–89.
Dung, N. T., Mui, N. T. & Ledin, I. (2005). Effect of replacing a commercial concentrate with cassava hay (*Manihot esculenta* Crantz) on the performance of growing goats. *Animal Feed Science and Technology*, **119**, 271–281.
Dupain, J., Van Elsacker, L., Nell, C. *et al.* (2002). New evidence for leaf swallowing and *Oesophagostomum* infection in bonobos (*Pan paniscus*). *International Journal of Primatology*, **23**, 1053–1062.
Durette-Dusset, M. C., Chabaud, A. G., Ashford, R. W., Butynski, T. & Reid, G. D. F. (1992). Two new species of the Trichostrongylidae (Nematoda, Trichostrongyloidea), parasitic in *Gorilla gorilla beringei* in Uganda. *Systematic Parasitology*, **23**, 159–166.
Fossey, D. (1983). *Gorillas in the Mist*. New York, NY: Houghton Mifflin Company.
Freeman, A. S., Kinsella, J. M., Cipolletta, C., Deem, S. L. & Karesh, W. B. (2004). Endoparasites of western lowland gorillas (*Gorilla gorilla gorilla*) at Bai Hokou, Central African Republic. *Journal of Wildlife Diseases*, **40**, 775–781.
Garcia, L. S. (1999). *Practical Guide to Diagnostic Parasitology*. Washington, DC: American Society of Microbiologists.
Gillespie, T. R., Greiner, E. C. & Chapman, C. A. (2005). Gastrointestinal parasites of the colobus monkeys of Uganda. *Journal of Parasitology*, **91**, 569–573.
Giner-Chavez, B. I., Van Soest, P. J., Robertson, J. B. *et al.* (1997). A method for isolating condensed tannins from crude plant extracts with trivalent ytterbium. *Journal of the Science of Food and Agriculture*, **74**, 359–368.
Glander, K. E. (1982). The impact of plant secondary compounds on primate feeding behaviour. *Yearbook of Physical Anthropology*, **25**, 1–18.
Graczyk, T. K. & Cranfield, M. R. (2003). Coprophagy and intestinal parasites: implications to human-habituated mountain gorillas (*Gorilla gorilla beringei*) of the Virunga mountains and Bwindi Impenetrable Forest. *Primate Conservation*, **19**, 58–64.
Gulland, F. M. D. (1992). The role of nematode parasites in Soay sheep (*Ovis aries* L.) mortality during a population crash. *Parasitology*, **105**, 493–503.
Hagerman, A. E. & Butler, L. G. (1994). Assay of condensed tannins or flavonoid oligomers and related flavonoids in plants. *Oxygen Radicals in Biological Systems, Pt D*, **234**, 429–437.
Hoste, H., Jackson, F., Athanasiadou, S., Thamsborg, S. M. & Hoskin, S. O. (2006). The effects of tannin-rich plants on parasitic nematodes in ruminants. *Trends in Parasitology*, **22**, 253–261.
Huffman, M. A. (1997). Current evidence for self-medication in primates: a multidisciplinary perspective. *Yearbook of Physical Anthropology*, **40**, 171–200.

Huffman, M. A. (2003). Animal self-medication and ethno-medicine: exploration and exploitation of the medicinal properties of plants. *Proceedings of the Nutrition Society*, **62**, 371–381.

Huffman, M. A., Page, J. E., Sukhdeo, M. V. K. *et al.* (1996). Leaf-swallowing by chimpanzees: a behavioral adaptation for the control of strongyle nematode infections. *International Journal of Primatology*, **17**, 475–503.

Huffman, M. A., Gotoh, S., Turner, L. A., Hamai, M. & Yoshida, K. (1997). Seasonal trends in intestinal nematode infection and medicinal plant use among chimpanzees in the Mahale Mountains, Tanzania. *Primates*, **38**, 111–125.

Janzen, D. H. (1978). Complications in interpreting the chemical defenses of trees against tropical arboreal plant-eating vertebrates. In *The Ecology of Arboreal Folivores*, ed. G. G. Montgomery. Washington, DC: Smithsonian Institute, pp. 73–84.

Koski, K. G. & Scott, M. E. (2001). Gastrointestinal nematodes, nutrition and immunity: breaking the negative spiral. *Annual Review of Nutrition*, **21**, 297–321.

Licitra, G., Hernandez, T. M. & VanSoest, P. J. (1996). Standardization of procedures for nitrogen fractionation of ruminant feeds. *Animal Feed Science and Technology*, **57**, 347–358.

Lilly, A. A., Mehlman, P. T. & Doran, D. (2002). Intestinal parasites in gorillas, chimpanzees, and humans at Mondika research site, Dzanga-Ndoki National Park, Central African Republic. *International Journal for Parasitology*, **23**, 555–573.

Max, R. A., Wakelin, D., Craigon, J. *et al.* (2005). Effect of two commercial preparations of condensed tannins on the survival of gastrointestinal nematodes of mice and goats *in vitro*. *South African Journal of Animal Science*, **35**, 213–220.

McNeilage, A., Plumptre, A. J., Brock-Doyle, A. & Vedder, A. (2001). Bwindi Impenetrable National Park, Uganda: gorilla census 1997. *Oryx*, **35**, 39–47.

Milton, K. (1979). Factors influencing leaf choice by howler monkeys: a test of some hypotheses of food selection by generalist herbivores. *American Naturalist*, **114**, 362–378.

Milton, K. (1996). Effects of bot fly (*Alouattamyia baeri*) parasitism on a free-ranging howler monkey (*Alouatta palliata*) population in Panama. *Journal of Zoology*, **239**, 39–63.

Milton, K. & Dintzis, F. (1981). Nitrogen-to-protein conversion factors for tropical plant samples. *Biotropica*, **12**, 177–181.

Molan, A. L., Waghorn, G. C., Min, B. R. & McNabb, W. C. (2000). The effect of condensed tannins from seven herbages on *Trichostrongylus colubriformis* larval migration *in vitro*. *Folia Parasitologica*, **47**, 39–44.

Molan, A. L., Waghorn, G. C. & McNabb, W. C. (2002). Effect of condensed tannins on egg hatching and larval development of *Trichostrongylus colubriformis in vitro*. *Veterinary Record*, **150**, 65–69.

Molan, A. L., Meagher, L. P., Spencer, P. A. & Sivakumaran, S. (2003). Effect of flavan-3-ols on *in vitro* egg hatching, larval development and viability of infective larvae of *Trichostrongylus colubriformis*. *International Journal for Parasitology*, **33**, 1691–1698.

Mudakikwa, A. B., Cranfield, M. R., Sleeman, J. M. *et al.* (2001). Clinical medicine, preventive health care and research on mountain gorillas in the Virunga

Volcanoes region. In *Mountain Gorillas: Three Decades of Research at Karisoke*, eds. M. M. Robbins, P. Sicotte & K. J. Stewart. Cambridge: Cambridge University Press, pp. 341–360.
Mueller-Harvey, I. (2006). Unraveling the conundrum of tannins in animal nutrition and health. *Journal of the Science of Food and Agriculture*, **86**, 2010–2037.
Munger, J. C. & Karasov, W. H. (1989). Sublethal parasites and host energy budgets: tapeworm infection in white-footed mice. *Ecology*, **70**, 904–921.
Murray, D. L., Keith, L. B. & Cary, J. R. (1998). Do parasitism and nutritional status interact to affect production in snowshoe hares? *Ecology*, **79**, 1209–1222.
Nakagawa, S. (2004). A farewell to Bonferroni: the problems of low statistical power and publication bias. *Behavioral Ecology*, **15**, 1044–1045.
National Research Council (2003). *Nutrient Requirements of Primates.* Washington, DC: National Academies Press.
Newey, S., Shaw, D. J., Kirby, A. *et al.* (2005). Prevalence, intensity and aggregation of intestinal parasites in mountain hares and their potential impact on population dynamics. *International Journal for Parasitology*, **35**, 367–373.
Nkurunungi, J. B., Ganas, J., Robbins, M. M. & Stanford, C. B. (2004). A comparison of two mountain gorilla habitats in Bwindi Impenetrable National Park, Uganda. *African Journal of Ecology*, **42**, 289–297.
Norconk, M. A. & Conklin-Brittain, N. L. (2004). Variation on frugivory: the diet of Venezuelan white-faced sakis. *International Journal of Primatology*, **25**, 1–26.
Nunn, C. L. & Altizer, S. (2006). *Infectious Diseases in Primates: Behavior, Ecology and Evolution.* Oxford: Oxford University Press.
Oates, J. F., Waterman, P. G. & Choo, G. M. (1980). Food selection by the south Indian leaf monkey, *Presbytis johnii* in relation to leaf chemistry. *Oecologia*, **45**, 45–56.
Oates, J. F., Whitesides, G. H., Davies, A. G. *et al.* (1990). Determinants of variation in tropical forest biomass: new evidence from west Africa. *Ecology*, **71**, 328–343.
del Pino, M. C. A., Hervas, G., Mantecon, A. R., Giraldez, R. J. & Frutos, P. (2005). Comparison of biological and chemical methods, and internal and external standards, for assaying tannins in Spanish shrub species. *Journal of the Science of Food and Agriculture*, **85**, 583–590.
Porter, L. J., Hrstich, L. N. & Chan, B. G. (1986). The conversion of procyanidins and prodelphinidins to cyanidin and delphinidin. *Phytochemistry*, **25**, 223–230.
Powzyk, J. A. & Mowry, C. B. (2003). Dietary and feeding differences between sympatric *Propithecus diadema diadema* and *Indri indri*. *International Journal of Primatology*, **24**, 1143–1162.
Remis, M. J., Dierenfeld, E. S., Mowry, C. B. & Carroll, R. W. (2002). Nutritional aspects of western lowland gorilla (*Gorilla gorilla gorilla*) diet during seasons of fruit scarcity at Bai Hokou, Central African Republic. *International Journal of Primatology*, **22**, 807–836.
Robbins, M. M. & McNeilage, A. (2003). Home range and frugivory patterns of mountain gorillas in Bwindi Impenetrable National Park, Uganda. *International Journal of Primatology*, **24**, 467–491.
Rojas, D. K., Lopez, J., Tejada, I. *et al.* (2006). Impact of condensed tannins from tropical forages on *Haemonchus contortus* burdens in Mongolian gerbils

(*Meriones unguiculatus*) and Pelibuey lambs. *Animal Feed Science and Technology*, **128**, 218–228.

Rothman, J. M. (2006). *Nutritional Ecology and Parasite Dynamics of Mountain Gorillas.* Ithaca, NY: Cornell University.

Rothman, J. M. & Bowman, D. D. (2003). Endoparasites of mountain gorillas. In *Companion and Exotic Animal Parasitology*, ed. D. D. Bowman. Ithaca, NY: International Veterinary Information Service.

Rothman, J. M., Dierenfeld, E. S., Molina, D. O. *et al.* (2006). Nutritional chemistry of foods eaten by gorillas in Bwindi Impenetrable National Park, Uganda. *American Journal of Primatology*, **68**, 675–691.

Rothman, J. M., Pell, A. N. & Bowman, D. D. (2008b). Host–parasite ecology of the helminths of mountain gorillas. *Journal of Parasitology*, **94**, 834–840.

Rothman, J. M., Dierenfeld, E. S., Hintz, H. F. & Pell, A. N. (2008a). Nutritional quality of gorilla diets: consequences of age, sex and season. *Oecologia*, **155**, 111–122.

Salajpal, K., Karolyi, D. K., Beck, R. *et al.* (2004). Effect of acorn (*Quercus robur*) intake on faecal egg count in outdoor reared black Slavonian pig. *Acta Agriculturae Slovenica*, **S1**, 163–178.

Schaller, G. B. (1963). *The Mountain Gorilla: Ecology and Behavior.* Chicago, IL: Chicago University Press.

Schofield, P., Mbugua, D. M. & Pell, A. N. (2001). Analysis of condensed tannins: a review. *Animal Feed Science and Technology*, **91**, 21–40.

Sherman, P. W. & Billing, J. (1999). Darwinian gastronomy: why we use spices. *BioScience*, **49**, 453–463.

Sleeman, J. M., Meader, L. L., Mudakikwa, A. B., Foster, J. W. & Patton, S. (2000). Gastrointestinal parasites of mountain gorillas (*Gorilla gorilla beringei*) in the Parc National des Volcans, Rwanda. *Journal of Zoo and Wildlife Medicine*, **31**, 322–328.

Smith, V. H., Jones, T. P. & Smith, M. S. (2005). Host nutrition and infectious disease: an ecological view. *Frontiers in Ecology and the Environment*, **3**, 268–274.

Stuart, M., Pendergast, V., Rumfelt, S. *et al.* (1998). Parasites of wild howlers (*Alouatta* spp.). *International Journal of Primatology*, **19**, 439–512.

Wang, Y. X., Waghorn, G. C., Barry, T. N. & Shelton, I. D. (1994). The effect of condensed tannins in *Lotus corniculatus* on plasma metabolism of methionine, cystine and inorganic sulfate by sheep. *British Journal of Nutrition*, **72**, 923–935.

Watterson, J. J. & Butler, L. G. (1983). Occurrence of an unusual leukoanthocyanidin and absence of proanthocyanidins in sorghum leaves. *Journal of Agricultural and Food Chemistry*, **31**, 41–45.

Wrangham, R. W. & Waterman, P. G. (1981). Feeding behaviour of vervet monkeys on *Acacia tortilis* and *Acacia xanthophloea* with special reference to reproductive strategies and tannin production. *Journal of Animal Ecology*, **50**, 715–731.

23 *Host–parasite dynamics: connecting primate field data to theory*

COLIN A. CHAPMAN, STACEY A. M. HODDER, AND JESSICA M. ROTHMAN

Photograph by Colin A. Chapman (*Colobus guereza*)

Introduction

The study of non-human primates offers numerous opportunities to advance theory. Many of these opportunities are associated with the fact that for many primate species, animals can be individually recognized. This has often

Primate Parasite Ecology. The Dynamics and Study of Host–Parasite Relationships, ed. Michael A. Huffman and Colin A. Chapman. Published by Cambridge University Press.

encouraged the collection of extensive longitudinal data from field populations and thus considerable amounts of data are available on demographic and behavioral changes associated with natural environmental variation (Struhsaker, 1976) or anthropogenic change (e.g. logging or forest fragmentation; Chapman *et al.*, 2000). In addition, primates are an extremely diverse taxonomic group along a number of dimensions. Species like the potto (*Perodicticus potto*) live solitary lives, where animals only meet when they have territorial disputes or to mate (Nekaris & Bearder, 2007). Marmosets and gibbons typically live in social units composed of a monogamous pair and their offspring (Bartlett, 2007; Digby *et al.*, 2007). Many species are found in larger groups in which there is either one or many males associating with a number of females and sometimes these associations can be very spatially and temporally flexible, as seen in chimpanzees (*Pan troglodytes*) (Cords, 1987; Melnick & Pearl, 1987; Chapman *et al.*, 1995; Aureli *et al.*, in press). In terms of group size, primate species range from solitary species, like the potto, to mandrills (*Mandrillus sphinx*) where groups as large as 845 animals have been described (Abernethy *et al.*, 2002). Even within one species there can be tremendous variation in group size. For example, the size of red colobus monkey (*Procolobus rufomitratus*) groups in Kibale ranges from 11–152 animals (Chapman, unpublished data). With this variation in social organization and group size, behavior can be extremely variable. For example, Sussman *et al.* (2005) summarized the activity budgets of diurnal primates and pointed out that prosimians such as the ruffed lemur (*Varecia variegata*) spend < 1 % in social interactions, while Francois langurs (*Presbytis francois*) spend 28% of their time in such social interactions. Similarly, the ecology of species can differ dramatically. In Uganda, baboons (*Papio anubis*) in some areas of Kibale National Park are restricted to forest, while in Queen Elizabeth National Park immediately to the south, the same population of baboons are savanna dwelling. Such habitat shifts can correspond with large changes in diet. For example, the leaf component in the diet of redtail monkeys (*Cercopithecus ascanius*) varies from 7–74% leaves across its range (Chapman *et al.*, 2002a).

Explaining this variation has long fascinated primate behavioral ecologists and individually recognizable primates have been monitored over long periods to develop and test theory. In comparison, researchers interested in studying primates to advance theory of disease dynamics have generally not, as of yet, taken advantage of these traits of primate populations and so this offers a vast, largely unexplored potential research opportunity. For example, in the late 1970s, Freeland (1976, 1979b) proposed a number of stimulating ideas linking primate socioecology and disease risk; however to this day these hypotheses remain largely untested (Altizer *et al.*, 2003; Nunn & Altizer, 2006). In contrast, Wrangham (1980) and van Schaik (1983) proposed links between primate socioecology and competitive regimes or predation and these hypotheses have been extensively evaluated (Koenig, 2002; Snaith & Chapman, 2007). This difference cannot be easily explained by methodological differences, because it is not clear that there are large differences in the ease of examining disease risk, versus examining competitive regimes or predation risk.

Approaches to studying host–parasite disease dynamics

There are two approaches that can be used to take advantage of the social and ecological diversity in the primate order and of individual recognition of primates to study disease issues. The first employs the use of comparative methods to address either ecological or evolutionary questions. The types of questions that this method can ask are largely unlimited. This approach has recently been exemplified in the research by Charles Nunn and colleagues. For example, Nunn *et al.* (2003) investigated the determinants of parasite species richness in primates using a data set of 941 host–parasite combinations in 101 anthropoid primates. After testing for the importance of host body mass, life history, social contact, population density, diet, and habitat diversity, they found that host population density was associated with parasite species richness. Similarly, Nunn & Dokey (2006) used similar comparative methods to evaluate two competing hypotheses with respect to how ranging influences parasite infections: (1) more intensive ranging could result in greater exposure to parasites that have accumulated in the soil or on vegetation, or (2) more intensive ranging could reflect territoriality, which could reduce home range overlap and intergroup contact, thereby producing lower infection levels. They found that an increased intensity of range use resulted in increased parasitism, therefore their results did not support the hypothesis that territoriality was evolved to reduce infection risk.

These comparative methods depend on the quality of the data that primate studies offer. For example, questions of parasite community structure, coevolution, and competition within a host are fascinating areas of inquiry that can be addressed using comparative methods with other species (Bush & Holmes, 1986; Holmes, 1987). However, they can likely be studied in more detail in species where complete necropsies are easy to conduct on a large sample of individuals (e.g. regularly hunted species of North American ducks) and primate studies will likely not play a significant role in addressing such bodies of theory.

While the use of the comparative method can be a very powerful approach, in this chapter we focus on the second approach: namely empirically testing hypotheses concerning infectious diseases with data collected from a single

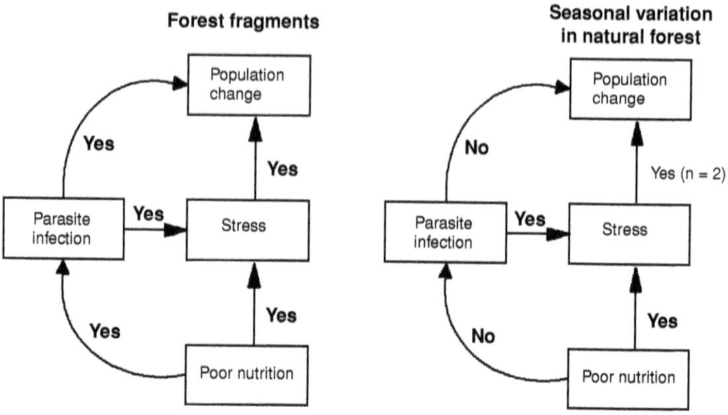

Figure 23.1. The relationship between nutrition, stress (fecal cortisol), parasite infection and population change in two systems where red colobus (*Procolobus rufomitratus*) were studied in Western Uganda.

field study or small set of field studies. An example of this approach is the research done to test the hypothesis that indices of parasite infections will increase when animals have a nutritionally poor diet (Figure 23.1) (Chapman *et al.*, 2006; Chapman *et al.*, 2007). To first address this hypothesis researchers monitored a series of forest fragments for 8 years and found that the change in colobus population size was correlated with both food availability and a number of indices of parasite infections. Changes in the quantity and quality of available food had a strong direct effect on population size, but it also had an indirect effect via parasite infections (i.e. if animals had a nutritionally poor diet they were more susceptible to parasitism, which led to population declines). Next the researchers asked the same question, but studied red colobus in continuous forest that were not nutritionally stressed. They found that ingesting a poor quality diet in one month corresponded with an increase in stress and parasite infection in the subsequent month; however, because the quality of the diet had little effect on parasite infections, there was no evidence of an amplification of the effect of poor diet due to parasite infections. This suggests that populations in forest fragments with very poor diets are impacted both by their diet and the fact that a poor diet facilitates more severe parasite infections; however, this combined effect is not seen in the environment where the animals are not nutritionally stressed.

Here we argue that field data exploiting the social and ecological variability across the primate order and the fact that primates can often be individually recognized can play an important role in testing theory related to host–parasite ecology. While there are a variety of theoretical ideas in the literature

(Anderson & May, 1992; Hudson *et al.*, 2001; Altizer *et al.*, 2003; Nunn & Altizer, 2006), we suggest that field studies of primate infectious diseases can contribute most to three theoretical areas:

(1) Disease transmission within and between social groups and potential counter-strategies.
(2) Individual variation in infection within groups.
(3) Ecological drivers of infection.

After presenting this argument, we illustrate the complexity of making clear predictions to test theory to demonstrate the need for carefully, well thought-out research. Finally, we demonstrate the value of developing such theories by outlining how they can guide construction of sound conservation and management plans. As a resource for researchers interested in developing these ideas, we attempt to present a compilation of studies that have presented predictions about primate infectious disease (Table 23.1).

Disease transmission within and between social group and potential counter-strategies

One of the strengths of primate field studies is that it is possible to quantify the movement and interactions of known individuals and groups. Once information is available on the diseases that infect these individuals, it is possible to examine questions related to disease transmission and spread. This provides an opportunity to test a number of hypotheses that have been proposed in the literature (Table 23.1). Thus, it allows one to investigate any social or ecological factors that influence contact between individuals, groups, or contaminated areas (e.g. frequently used food trees or sleeping sites).

In general, field research on primates has speculated on how behaviors might have evolved to avoid parasite transmission, but there is little quantitative information available to evaluate these speculations. However, if one turns to the agricultural and wildlife literature more information becomes available, which illustrates the potential for such research. For example, one of the most direct means of reducing parasite infection risk would be to evolve strategies to avoid contaminated areas and since primatologists typically follow known groups, knowledge of frequency of re-use of areas is readily available. Grazing mammalian herbivores cannot detect the presence of gastrointestinal parasites on their foods, but they can rely on the presence of feces as a means of avoiding areas where the potential for transmission is high. In both agricultural (Cooper *et al.*, 2000; Hutchings *et al.*, 2003) and wild systems (Hutchings *et al.*, 2002;

Table 23.1. *A variety of predictions that have been made with regards to (1) infection transmission and spread, (2) variance in infection probability and impact among types of individuals, and (3) ecological determinants of infection acquisition and loss, for which studies of primates are particularly suitable to address*

Prediction	References that address this prediction
Infection transmission and spread	
Parasite infections are expected to:	
Increase with population density	(Stuart *et al.*, 1990; Anderson & May, 1992; Milton, 1996; Nunn *et al.*, 2003; Chapman *et al.*, 2005b; Gillespie & Chapman, 2006)
Increase with group size	(Chapman *et al.*, Ch. 21, this volume; Freeland, 1976, 1979a; McGrew *et al.*, 1989; Loehle, 1995; Cote & Poulin, 1995; Nunn, 2002a, b; Nunn *et al.*, 2003; Vitone *et al.*, 2004)
Decrease with group size for arthropod parasites	(Davies *et al.*, 1991; Mooring & Hart, 1992; Nunn & Heymann, 2005)
Increase with increased contact rate (grooming, etc.)	(Moller *et al.*, 1993; Altizer *et al.*, 2003; Vitone *et al.*, 2004; Nunn & Altizer, 2006)
Decrease with group size if substructuring occurs	(Wilson *et al.*, 2003)
Increase with the frequency of formation of mixed species groups	(Freeland, 1977)
Richness will increase with large home ranges	(Freeland, 1980; Nunn *et al.*, 2003; Nunn & Dokey, 2006)
Intensity and prevalence may increase with increased intensity of home range use	(Freeland, 1980; Hausfater & Meade, 1982; Stoner, 1996; Nunn *et al.*, 2003; Nunn & Dokey, 2006)
Increased home range overlap	(Loehle, 1995; Tutin, 2000)
Decrease with increased group spread	(Nunn & Altizer, 2006)
Increase with the repeated use of sleeping sites or specific deposition sites	(Hausfater & Meade, 1982; Heymann, 1995; Gilbert, 1997; Di Bitetti *et al.*, 2000; Nunn & Heymann, 2005)
Variance in infection probability and impact among types of individuals	
High-ranking individuals exhibit elevated infections	(Hausfater & Watson, 1976; Muller-Graf *et al.*, 1997; Sapolsky, 2005; Muehlenbein, 2006)
Low-ranking individuals exhibit elevated infections	(Cheney *et al.*, 1988; Eley *et al.*, 1989; Sapolsky, 2005)
Dominant animals suffer immunosuppressive effect of testosterone	(Muehlenbein, Ch. 4, this volume; Sapolsky, 1992; Bercovitch & Ziegler, 2002)

Females avoid mating with parasitized males	(Ayers, 1986 in Hamilton & Zuk, 1982; Waitt *et al.*, 2003; Nunn & Altizer, 2006)
Younger individuals have less parasitic infections	(Nunn & Altizer, 2004)
Parasite infections vary between the sexes	(Hausfater & Watson, 1976; Milton, 1996; Stoner, 1996; Muller-Graf *et al.*, 1996, 1997; Fedigan & Zohar, 1997; Nunn & Altizer, 2004)
Ecological determinants of infection acquisition and loss	
Terrestriality leads to elevated infections	(Dunn, 1968; Muller-Graf *et al.*, 1997; Nunn *et al.*, 2000; Nunn, 2002a, b)
Territoriality leads to increased isolation and decreased parasite infections	(Loehle, 1995; Nunn & Dokey, 2006)
Frugivores consume more types of foods (e.g. intermediate hosts) and have elevated infections	(Dunn, 1968)
Use of medicinal plants decreases infections	(Huffman, 1997; Huffman *et al.*, 1997)
Drinking water from standing sources	(Hamilton *et al.*, 1976; McGrew *et al.*, 1989)
Dietary stress	(Chapman *et al.*, 2005b, 2006)
Elevated cortisol	(Chapman *et al.*, in press; Chapman *et al.*, 2006)
Seasonality	(Freeland, 1980; Hausfater & Meade, 1982; Milton, 1996; Nunn *et al.*, 2005)
Habitat type	(Bezjian *et al.*, in review; Stuart & Strier, 1995; Stoner, 1996; Stuart *et al.*, 1998; Gillespie *et al.*, 2005)

Ezenwa, 2004b) studies have demonstrated that herbivores avoid areas contaminated with feces of both their own and other species (Hutchings & Harris, 1997; Ezenwa, 2004b). In situations where it is not possible to avoid contaminated areas, herbivores shift the foraging strategy and graze on only the upper parts of the grass blades (Hutchings *et al.*, 2003). Further, this avoidance of potentially contaminated areas and the selective foraging on the upper parts of grass blades are exaggerated in parasitized animals relative to nonparasitized animals (Hutchings *et al.*, 1999), which could reflect the increased risk of mortality from further parasite intake (Fox, 1997). With respect to primates, Gilbert (1997) similarly suggested that red howlers (*Alouatta seniculus*) avoided contaminating areas where they could feed in the future by defecating in non-feeding sites that were free of underlying vegetation. Huffman (2007) suggests that hamadryas baboons (*Papio hamadryas*) avoid contaminated water by digging holes next to tainted water sites of livestock and wait for the water to filter through sand (see Huffman, 2007 for strategies involving self-medication).

Such examples lead one to suspect that many of the hypotheses regarding behavioral strategies primates have evolved to avoid parasite transmission may be valid (Table 23.1). This encourages efforts to conduct rigorous research to investigate some of these suggested strategies to decrease the risk of disease transmission and spread.

Individual variation in infection within groups

The detailed behavioral data that are often possible to obtain in primate studies allow researchers to classify individuals with respect to a variety of qualities and subsequently their parasite infections can be contrasted between groups of individuals that share particular traits. A number of studies have contrasted individuals by age, sex, and dominance rank and these investigations have generated a number of hypotheses that can be further tested in different primate populations and species (Table 23.1). To date these studies have primarily been limited to primates inhabiting African savanna; however, despite this limited sample size, results have differed, with some studies suggesting that higher-ranking individuals are more parasitized (Hausfater & Watson, 1976), while others have suggested that lower-ranking individuals are more strongly affected (Cheney *et al.*, 1988; Eley *et al.*, 1989), and still others have not found an effect of dominance on parasitism (Muller-Graf *et al.*, 1996). Physiological mechanisms creating these patterns have been suggested, such as relationships between dominance, testosterone, immune system efficiency, and infection level (Bercovitch & Ziegler, 2002; Muehlenbein, Chapter 4, this volume).

If one again turns to the agricultural or wildlife literature, a number of examples are available that illustrate the potential evolutionary pressures that different sorts of individuals possessing particular traits may face. For example, Ezenwa (2004a) demonstrated that among bovids territorial hosts are more likely to be infected by strongyle nematodes than non-territorial hosts. Further, she demonstrated that among gazelles the intensity of parasite infections was higher among territorial males compared with bachelor males, females, or juveniles, suggesting that territoriality either exposes the males to a higher parasite risk, or that these males become more susceptible to parasitism. The possibility that such patterns exist in primates raises some intriguing questions. For example, given that predation risk is viewed to be so intense that it strongly favors group living (van Schaik & van Hooff, 1983), it is puzzling that solitary males of these group-living species are so common (Butynski, 1990; Struhsaker & Pope, 1991). It may be that solitary males reduce their parasite infections, and that this facilitates growth and improved health, which makes them more competitive when the opportunity arises for them to enter a social group (Ezenwa, 2004a).

Ecological drivers of infection

Primate studies can also accurately assess diet and a variety of spatial and ecological aspects of daily life of many species (e.g. habitat use). Correspondingly, it is possible to contrast individuals, groups, populations, or species with respect to these ecological variables and their corresponding diseases and to understand how the spatial arrangement of the landscape influences disease transmission. For example, it is possible to evaluate whether the nature of the infection varies among habitats occupied by the same species (Huffman *et al.*, Chapter 16, this volume) or how the degree of habitat fragmentation affects infection risk.

There is a new and rapidly growing field of study that has been coined spatial epidemiology (Ostfeld *et al.*, 2005). This area of research investigates spatial variation in disease risk or incidence and considers how factors such as spatial localization of pathogens, host, or regions of elevated transmission probability influence the occurrence and spread of diseases. For example, 50% of the variation in the prevalence of eyeworm (*Loa loa*) between villages in west and central Africa can be explained by a simple set of environmental variables (forest, land cover, rainfall, temperature, topography, and soil (Thomson *et al.*, 2000). Or a more spatially explicit example is that Lyme disease (*Borrelia burgdorferi*) in New York State has been shown to be spatially aggregated on a scale of approximately 120 km. This scale appears to be related to the dispersal distance of key tick hosts (Glavanakov *et al.*, 2001).

Primate studies are just initiating research in the area of spatial epidemiology (Nunn *et al.*, 2005; Walsh *et al.*, 2005; Biek *et al.*, 2006; Hopkins & Nunn, 2007). For example, Walsh *et al.* (2005) quantified the spatial arrangement of genetic variance of the Ebola virus (ZEBOV) and found evidence of an advancing wave of outbreak originating from Yambuku, Democratic Republic of Congo after 1976. The distance among outbreaks suggests that the wave is advancing approximately 50 km per year.

Complexity of making predictions

Early studies proposing how infectious diseases placed selective pressures on primate behavioral traits were often relatively straightforward and suggested simple causal relationships. However, as the field developed researchers discovered complexities that illustrated that the situation was often not as simple as first perceived. It is now clear that when testing any specific hypothesis that infers that infectious diseases are selecting for a particular behavioral pattern, it is critical to eliminate possible alternative hypotheses (Nunn & Altizer, 2006; Nunn & Dokey, 2006; Chapman *et al.*, Chapter 21, this volume). A study ignoring this runs the risk of readers discounting the significance of the study because they view that the conclusions drawn represent just one possible interpretation, but that other interpretations are equally as feasible.

Effectively connecting primate field data to theory on host–parasite ecology will require studies that understand the complexity of the situation and consider alternative hypotheses. As a result, below we present two issues that illustrate the complexity of the situation and the need for such an approach.

Sociality: Does living in groups increase or decrease disease risk?

Dating back to the earliest publications detailing the advantages of group living, it has typically been assumed that living in groups increases local densities of animals (Alexander, 1974). Since parasite transmission is usually density dependent (Anderson & May, 1979), group-living animals were predicted to have greater parasite infection levels than solitary animals (Freeland, 1979b; Cote & Poulin, 1995). However, recent research has not supported this expectation and alternative hypotheses have been proposed. For example, two primate studies tested the prediction that infection should be positively related to group size by building on the assumption that increased infection levels should be indicated by increased leukocyte counts; however, neither study supported the prediction (Nunn, 2002a; Semple *et al.*, 2002). Thus, it appears that the relationship between group size and infection risk may not be a simple one. Countering predictions made in the earlier studies, some authors have recently argued that sociality could lower the risk of parasite transition if increased clustering of individuals into permanent groups with limited dispersal effectively quarantines parasites into discrete host "patches" (Freeland, 1979b; Wilson *et al.*, 2003; Ezenwa *et al.*, 2006; Nunn & Altizer, 2006). This idea has been supported by recent modeling efforts of disease transmission (Watve & Jog, 1997; Wilson *et al.*, 2003).

As a result of these conflicting ideas, some studies now suggest that living in groups increases infection risk, while other studies suggest that it decreases infection risk. This illustrates the complexity of advancing our understanding of the effect of sociality. In this case, it is valuable for a researcher studying a particular species or population to document whether or not there is a relationship between sociality/group size and infection level. This will start to provide a comparative database to evaluate the generality of one hypothesis versus the other. However, it is much more valuable to provide information that would allow future researchers to say under what conditions one or the other relationship would be expected and the other not. In the case of primates, this would involve documenting contact rates between groups or areas contaminated by neighboring groups, and dispersal rates of individuals among groups (see Ezenwa, 2004a for a study of this nature with ungulates).

Diet: What are the links between diet and disease risk?

Many of the commonly studied parasites of primates are transmitted orally, typically either when the primate ingests contaminated water, plant foods, or an intermediate insect host containing the infective stage of a parasite. Since the route of transmission is oral, one might expect differences in the richness or prevalence of parasite infections among individuals, groups, populations, or species eating different types of diets or having different digestive strategies (Nunn & Altizer, 2006). Since animals with more intricate digestive systems would have a greater habitat diversity for parasites (i.e. more microhabitats in the gut), one might expect that folivores would harbor richer parasite communities than frugivores (Kennedy *et al.*, 1986; Watve & Sukumar, 1995). On the other hand, folivores are thought to typically eat a diet higher in plant secondary compounds and these compounds may kill parasites before they can establish (Janzen, 1978; Huffman, 1997, 2007). However, frugivores are thought to eat a more varied diet and often consume invertebrates as a protein source. These invertebrates may be intermediate hosts for parasites with complex life cycles, and thus frugivores may be expected to have a higher species richness of parasites (Nunn & Altizer, 2006). With this set of hypotheses we have an "anatomical complexity" hypothesis suggesting folivores have higher infections than frugivores, a "secondary compound" hypothesis suggesting that folivores will have lower infections, and a "varied diet" hypothesis suggesting that frugivores should harbor richer parasite communities.

Two studies which addressed these hypotheses in primates (Nunn *et al.*, 2003; Vitone *et al.*, 2004) found only limited evidence of an association between parasite diversity and dietary categories, with the exception that the percentage of leaves in the diet was positively associated with the diversity of helminth species. However, it is possible that these studies fail to identify subtle patterns because there is a great deal of intraspecific variation in both primate diet (Chapman *et al.*, 2002a) and parasite biology (Holmes, 1987) that cannot be easily considered when conducting broad-scale contrasts.

It would appear that the situation is complex, and a major direction for future research should be to distinguish among these alternative hypotheses. There are at least three competing hypotheses and reason to be concerned that comparative studies cannot be conducted at a sufficiently fine scale to provide resolution among hypotheses. The way forward may be to design tests of these hypotheses which take advantage of the extensive variation in primate diets or individual recognition and age/sex differences in diet. With respect to the "varied diet" hypothesis, researchers could take advantage of extensive small-scale dietary variation that has been documented for a number of species. For example, within single geographical regions there are often populations of the same species where one population has a very monotonous diet, while the other population has a much more varied diet. In Kibale National Park, Uganda, red colobus at one site (Kahunge) spend 92% of their time feeding from one species (*Acacia hockii*), while at the main study site (Kanyawara) red colobus eat over 80 different plant species and no single food represents >10% of the foraging time (Chapman *et al.*, 2002b). Similarly, to evaluate the "secondary compound" hypothesis researchers could quantify the intake of specific secondary compounds known to negatively affect gastrointestinal parasites (e.g. steroid glucosides and sesquiterpene lactones: Huffman *et al.*, 1993; Ohigashi *et al.*, 1994; tannins: Carrai *et al.*, 2003; Hoste *et al.*, 2006; Rothman *et al.*, Chapter 22, this volume) and compare individuals or groups that are known to consume differing levels of these secondary compounds. If there is variability in secondary compound consumption, possibly due to seasonal or spatial variation in secondary compound availability (McKey *et al.*, 1978), it would be possible to evaluate whether secondary compounds drive differences in parasite infections amongst populations. Further research into the spatial and temporal variation of secondary compounds will likely reveal significant variation on finer spatial scales (Chapman *et al.*, 2003, Rothman *et al.*, Chapter 22, this volume), providing further opportunities to test this hypothesis.

Conservation application of these theoretical issues

The examination of these theoretical ideas and hypotheses are not simply addressing areas of academic interest; they can easily be applied to critically important conservation issues. At the present time the academic and conservation communities are both simply describing patterns of how anthropogenic change alters disease dynamics. However, to meet conservation goals it will be necessary to advance beyond description, to a stage where it is possible to predict what types of alteration to habitats will create what types of changes in disease patterns (Gillespie & Chapman, 2006). To do this the academic community must once again be able to set up testable hypotheses that distinguish between various alternatives.

To encourage research and debate on this topic we have outlined a series of predictions that deal with how different types of disturbances might alter parasite infections for primates (Table 23.2). For example, fragmentation may reduce parasite infections by effectively isolating populations, or it could increase infections because animals are restricted to small areas where they frequently revisit contaminated areas, or it could lead to decreased group size that would decrease interaction levels and decrease parasite infections. To resolve the importance of ideas such as these will, yet again, require careful research that tests specific hypotheses, while eliminating the alternatives. It is worth emphasizing that the field of spatial epidemiology may contribute greatly to conservation efforts. As has been demonstrated in other studies, this area may be able to predict sets of environmental variables (forest, land cover, rainfall, temperature, topography, and soil (Thomson *et al.*, 2000; Gillespie & Chapman, 2006) that can predict disease emergence, proportion of the communities most likely to transmit diseases (Goldberg *et al.*, 2007), or patterns of spread (Walsh *et al.*, 2005; Biek *et al.*, 2006).

The devastating impacts diseases like Ebola have had on both human and wildlife communities (Walsh *et al.*, 2003; Leroy *et al.*, 2004), and the immense social and economic costs created by viruses like HIV have made researchers and the general public acutely aware that disease issues are paramount in human and wildlife societies (Chapman *et al.*, 2005a). It is also clear that within the last several decades, humans have been responsible for considerable irrevocable changes to primate habitats. Unfortunately, most primates today live in disturbed habitat mosaics of farmland, human settlements, forest fragments,

Table 23.2. *Conservation threats and possible changes in parasite infections that result from particular changes caused by that threat. Note some predictions are driven by multiple factors*

Threat factors predicted to change in parasite infections	Predicted effect on parasite risk
Fragmentation	
Populations are physically isolated in patches	Negative
Change in social structure and decreased interaction rates	Negative
Decreased group size	Negative
Edge microclimate – typically drier	Negative
Animals obtain a nutritionally poor diet	Positive
Vector activity increased by increased numbers of humans and domestic animals	Positive
Because of reduced range, increase in revisitation rates to the same areas	Positive
Habitat degradation (e.g. logging)	
Change in social structure and decreased interaction rates	Negative
Decreased group size	Negative
Edge microclimate – typically drier	Negative
Increase in revisitation rates to the same area	Positive
Hunting	
Decrease density and lower transmission probability	Negative
Lower density results in improved nutrition	Negative
Increased stress	Positive
Social disruption and increased dispersal rates	Positive
Human contact	
Access to crops and increased nutrition	Negative
Direct transmission of anthropozoonotic parasites (genetic tests required)	Positive
Induces stress, which increases susceptibility (e.g. harassment by people or dogs)	Positive

and isolated protected areas (Chapman & Peres, 2001). An understanding of how disease shapes primate populations in these modified habitats will be critical in guiding future conservation plans.

Acknowledgements

Funding for this research was provided by Canada Research Chairs Program, Wildlife Conservation Society, Natural Science and Engineering Research Council of Canada, and National Science Foundation. Permission to conduct this research was given by the National Council for Science and Technology, and the Uganda Wildlife Authority. Lauren J. Chapman, Mike Huffman, Charles Nunn, and Tamaini Snaith provided helpful comments on this research.

References

Abernethy, K. A., White, L. J. T. & Wickings, E. J. (2002). Hordes of mandrills (*Mandrillus sphinx*): extreme group size and seasonal male presence. *Journal of Zoology (London)*, **258**, 131–137.

Alexander, R. D. (1974). The evolution of social behaviour. *Annual Review of Ecology and Systematics*, **5**, 325–382.

Altizer, S., Nunn, C. L., Thrall, P. H. *et al.* (2003). Social organization and parasite risk in mammals: integrating theory and empirical studies. *Annual Review of Ecology and Systematics*, **64**, 517–547.

Anderson, R. M. & May, R. M. (1979). Population biology of infectious disease: Part 1. *Nature*, **280**, 361–367.

Anderson, R. M. & May, R. M. (1992). *Infectious Diseases of Humans: Dynamics and Control.* Oxford: Oxford University Press.

Aureli, F., Shaffner, C. M., Boesch, C. *et al.* (in press). Fission-fusion dynamics: new frameworks for comparative research. *Current Anthropology*.

Bartlett, T. Q. (2007). The Hylobatidae: small apes of Asia. In *Primates in Perspective*, ed. C. J. Campbell, A. Fuentes, K. C. MacKinnon, M. Panger & S. K. Bearder. Oxford: Oxford University Press, pp. 274–289.

Bercovitch, F. B. & Ziegler, T. E. (2002). Current topics in primate socioendrocrinology. *Annual Review of Anthropology*, **31**, 45–67.

Bezjian, M., Gillespie, T. R., Chapman, C. A. & Greiner, E. C. (in review). Gastrointestinal parasites of forest baboons, *Papio anubis*, in Kibale National Park, Uganda.

Biek, R., Walsh, P. D., Leroy, E. M. & Real, L. A. (2006). Recent common ancestry of Ebola Zaire visus found in a bat reservoir. *Plos Pathogens*, **2**, e90.

Bush, A. O. & Holmes, J. C. (1986). Intestinal helminths of lesser scaup ducks: patterns of association. *Canadian Journal of Zoology*, **64**, 132–141.

Butynski, T. M. (1990). Comparative ecology of blue monkeys (*Cercopithecus mitis*) in high- and low-density sub-populations. *Ecological Monographs*, **60**, 1–26.

Carrai, V., Borgognini-Tarli, S. M., Huffman, M. A. & Bardi, M. (2003). Increase in tannin consumption by sifaka (*Propithecus verreauxi verreauxi*) females during the birth season: a case for self-medication in prosimians? *Primates*, **44**, 61–66.

Chapman, C. A. & Peres, C. A. (2001). Primate conservation in the new millennium: the role of scientists. *Evolutionary Anthropology*, **10**, 16–33.

Chapman, C. A., Wrangham, R. W. & Chapman, L. J. (1995). Ecological constraints on group size: an analysis of spider monkey and chimpanzee subgroups. *Behavioral Ecology and Sociobiology*, **36**, 59–70.

Chapman, C. A., Balcomb, S. R., Gillespie, T. R., Skorupa, J. P. & Struhsaker, T. T. (2000). Long-term effects of logging on African primate communities: a 28-year

comparison from Kibale National Park, Uganda. *Conservation Biology*, **14**, 207–217.
Chapman, C. A., Chapman, L. J., Cords, M. *et al.* (2002a). Variation in the diets of *Cercopithecus* species: differences within forests, among forests, and across species. In *The Guenons: Diversity and Adaptation in African Monkeys*, ed. M. Glenn & M. Cords. New York, NY: Plenum Press, pp. 319–344.
Chapman, C. A., Chapman, L. J. & Gillespie, T. R. (2002b). Scale issues in the study of primate foraging: red colobus of Kibale National Park. *American Journal of Physical Anthropology*, **117**, 349–363.
Chapman, C. A., Chapman, L. J., Rode, K. D., Hauck, E. M. & McDowell, L. R. (2003). Variation in the nutritional value of primate foods: among trees, time periods and areas. *International Journal of Primatology*, **24**, 317–333.
Chapman, C. A., Gillespie, T. R. & Goldberg, T. L. (2005a). Primates and the ecology of their infectious diseases: how will anthropogenic change affect host-parasite interactions? *Evolutionary Anthropology*, **14**, 134–144.
Chapman, C. A., Speirs, M. L., Gillespie, T. R., Holland, T. & Austad, K. (2005b). Life on the edge: gastrointestinal parasites from forest edge and interior primate groups. *American Journal of Primatology*, **68**, 1–12.
Chapman, C. A., Wasserman, M. D., Gillespie, T. R. *et al.* (2006). Do nutrition, parasitism, and stress have synergistic effects on red colobus populations living in forest fragments? *American Journal of Physical Anthropology*, **131**, 525–534.
Chapman, C. A., Saj, T. L. & Snaith, T. V. (2007). Temporal dynamics of nutrition, parasitism, and stress in colobus monkeys: implications for population regulation and conservation. *American Journal of Physical Anthropology*, **134**, 240–250.
Cheney, D. L., Seyfarth, R. M., Andelman, S. J. & Lee, P. C. (1988). Reproductive success in vervet monkeys. In *Reproductive Success*, ed. T. H. Clutton-Brock. Chicago, IL: The University of Chicago Press, pp. 384–402.
Cooper, J., Gordon, I. J. & Pike, A. W. (2000). Strategies for avoidance of faeces by grazing sheep. *Applied Animal Behaviour Science*, **69**, 15–33.
Cords, M. (1987). Forest guenons and patas monkeys: male-male competition in one-male groups. In *Primate Societies*, ed. B. B. Smuts, D. L. Cheney, R. M. Seyfarth, R. Wrangham & T. T. Struhsaker. Chicago, IL: The University of Chicago Press, pp. 98–111.
Cote, I. M. & Poulin, R. (1995). Parasitism and group size in social animals: a meta-analysis. *Behavioral Ecology*, **6**, 159–165.
Davies, C. R., Ayres, J. M., Dye, C. & Deane, L. M. (1991). Malaria infection rate of Amazonian primates increases with body weight and group size. *Functional Ecology*, **5**, 655–662.
Di Bitetti, M. S., Vidal, E. M. L., Baldovino, M. C. & Benesovsky, V. (2000). Sleeping site preferences in tufted capuchin monkeys (*Cebus apella nigritus*). *American Journal of Primatology*, **50**, 257–274.
Digby, L. J., Ferrari, S. F. & Saltzman, W. (2007). Callitrichines: the role of competition in cooperatively breeding species. In *Primates in Perspective*, ed. C. J. Campbell, A. Fuentes, K. C. MacKinnon, M. Panger & S. K. Bearder. Oxford: Oxford University Press, pp. 85–106.

Dunn, F. L. (1968). The parasites of *Saimiri*: in the context of platyrrhine parasitism. In *The Squirrel Monkey*, ed. L. A. Rosenblum & R. W. Cooper. New York, NY: Academic Press, pp. 31–68.

Eley, R. M., Strum, S. C., Muchemi, G. & Reid, G. D. F. (1989). Nutrition, body condition, activity patterns, and parasitism of free-ranging troops of olive baboons (*Papio anubis*) in Kenya. *American Journal of Primatology*, **18**, 209–219.

Ezenwa, V. O. (2004a). Host social behavior and parasitic infection: a multifactorial approach. *Behavioral Ecology*, **15**, 446–454.

Ezenwa, V. O. (2004b). Selective defecation and selective foraging: antiparasite behavior in wild ungulates? *Ethology*, **110**, 851–862.

Ezenwa, V. O., Price, S. A., Altizer, S., Vitone, N. D. & Cook, K. C. (2006). Host traits and parasite species richness in even and odd-toed hoofed mammals, Artriodactyla and Perissodactyla. *Oikos*, **115**, 526–536.

Fedigan, L. M. & Zohar, S. (1997). Sex differences in mortality of Japanese macaques: twenty-one years of data from the Arashiyama West population. *American Journal of Physical Anthropology*, **102**, 161–175.

Fox, M. T. (1997). Pathophysiology of infection with gastrointestinal nematodes in domestic ruminants: recent developments. *Veterinary Parasitology*, **72**, 285–308.

Freeland, W. J. (1976). Pathogens and the evolution of primate sociality. *Biotropica*, **8**, 12–24.

Freeland, W. J. (1977). Blood-sucking flies and primate polyspecific associations. *Nature*, **269**, 801–802.

Freeland, W. J. (1979a). Mangabey (*Cercocebus albigena*) social organization and population density in relation to food use and availability. *Folia Primatologica*, **32**, 108–124.

Freeland, W. J. (1979b). Primate social groups as biological islands. *Ecology*, **60**, 719–728.

Freeland, W. J. (1980). Mangabey (*Cercocebus albigena*) movement patterns in relation to food availability and fecal contamination. *Ecology*, **61**, 1297–1303.

Gilbert, K. A. (1997). Red howling monkey use of specific defecation sites as a parasite avoidance strategy. *Animal Behaviour*, **54**, 451–455.

Gillespie, T. R. & Chapman, C. A. (2006). Prediction of parasite infection dynamics in primate metapopulations based on attributes of forest fragmentation. *Conservation Biology*, **20**, 441–448.

Gillespie, T. R., Chapman, C. A. & Greiner, E. C. (2005). Effects of logging on gastrointestinal parasite infections and infection risk in African primates. *Journal of Applied Ecology*, **42**, 699–707.

Glavanakov, S., White, D. J., Caraco, T. *et al.* (2001). Lyme disease in New York State: spatial pattern at a regional scale. *American Journal of Tropical Medicine and Hygiene*, **65**, 538–545.

Goldberg, T. L., Gillespie, T. R., Rwego, I. B. *et al.* (2007). Patterns of gastrointestinal bacterial exchange between chimpanzees and humans involved in research and tourism in western Uganda. *Biological Conservation*, **135**, 527–533.

Hamilton, W. D. & Zuk, M. (1982). Heritable true fitness and bright birds: a role for parasites? *Science*, **218**, 384–387.

Hamilton, W. J., Buskirk, R. E. & Buskirk, W. H. (1976). Defense of space and resources by chacma (*Papio ursinus*) baboon troops in an African desert and swamp. *Ecology*, **57**, 1264–1272.

Hausfater, G. & Meade, B. J. (1982). Alternation of sleeping groves by yellow baboons (*Papio cynocephalus*) as a strategy for parasite avoidance. *Primates*, **23**, 287–297.

Hausfater, G. & Watson, D. F. (1976). Social and reproductive correlates of parasite ova emissions by baboons. *Nature*, **262**, 688–689.

Heymann, E. W. (1995). Sleeping habits of tamarins, *Saguinus mystax and Saguinus fuscicollis* (Mammalia: Primates; Callitrichidae), in north-eastern Peru. *Journal of Zoology (London)*, **237**, 211–226.

Holmes, J. C. (1987). The structure of helminth communities. *International Journal of Parasitology*, **48**, 97–100.

Hopkins, M. E. & Nunn, C. L. (2007). A global gap analysis of infectious agents in wild primates. *Diversity and Distribution*, **13**, 561–572.

Hoste, H., Jackson, F., Athanasiadou, S., Thamsborg, S. M. & Hoskin, S. O. (2006). The effects of tannin-rich plants on parasitic nematodes in ruminants. *Trends in Parasitology*, **22**, 253–261.

Hudson, P. J., Rissoli, A., Grenfell, B. T., Heesterbeek, H. & Dobson, A. P. (eds.) (2001). *The Ecology of Wildlife Diseases*. Oxford: Oxford University Press.

Huffman, M. A. (1997). Current evidence for self-medication in primates: a multidisciplinary perspective. *Yearbook of Physical Anthropology*, **40**, 171–200.

Huffman, M. A. (2007). Primate self-medication. In *Primates in Perspective*, ed. C. J. Campbell, A. Fuentes, K. C. MacKinnon, M. Panger & S. K. Bearder. Oxford: Oxford University Press, pp. 677–690.

Huffman, M. A., Gotoh, S., Izutsu, D., Koshimizu, K. & Kalunde, M. S. (1993). Further observations on the use of the medicinal plant, *Vernonia amygdalina* (Del) by a wild chimpanzee, its possible effect on parasite load, and its phytochemistry. *African Study Monographs*, **14**, 227–240.

Huffman, M. A., Gotoh, S., Turner, L. A., Hamai, M. & Yoshida, K. (1997). Seasonal trends in intestinal nematode infection and medicinal plant use among chimpanzees in the Mahale Mountains, Tanzania. *Primates*, **38**, 111–125.

Hutchings, M. R. & Harris, S. (1997). The effects of farm management practices on cattle grazing behaviour and the potential for transmission of bovine tuberculosis from badgers to cattle. *Veterinary Journal*, **153**, 149–162.

Hutchings, M. R., Kyriazakis, I., Gordon, I. J. & Jackson, F. (1999). Tradeoffs between nutrient intake and faecal avoidance in herbivore foraging decisions: the effect of parasite status, level of feeding motivation and sward nitrogen content. *Journal of Animal Ecology*, **68**, 310–323.

Hutchings, M. R., Milner, J. M., Gordon, I. J., Kyriazakis, I. & Jackson, F. (2002). Grazing decisions of Soay sheep, *Ovis aries*, on St. Kilda: a consequence of parasite distribution. *Oikos*, **96**, 235–244.

Hutchings, M. R., Athanasiadou, S., Kyriazakis, I. & Gordon, I. J. (2003). Can animals use foraging behaviour to combat parasites? *Proceedings of the Nutrition Society*, **62**, 361–370.

Janzen, D. H. (1978). Complications in interpreting the chemical defenses of trees against tropical arboreal vertebrates. In *The Ecology of Arboreal Folivores*, ed. G. G. Montgomery. Washington, DC: Smithsonian Institution Press, pp. 73–94.

Kennedy, C. R., Bush, A. O. & Aho, J. M. (1986). Patterns in helminth communities: why are birds and fish different? *Parasitology*, **93**, 205–215.

Koenig, A. (2002). Competition for resources and its behavioral consequences among female primates. *International Journal of Primatology*, **23**, 759–783.

Leroy, E. M., Rouguet, P., Formenty, P. *et al.* (2004). Multiple Ebola virus transmission events and rapid decline of Central African Wildlife. *Science*, **303**, 387–390.

Loehle, C. (1995). Social barriers to pathogen transmission in wild animal populations. *Ecology*, **76**, 326–335.

McGrew, W. C., Tutin, C. E. G., Collin, D. A. & File, S. K. (1989). Intestinal parasites of sympatric *Pan troglodytes* and *Papio* spp. at two sites Gombe (Tanzania) and Mt. Assirik (Senegal). *American Journal of Primatology*, **17**, 147–155.

McKey, D. B., Waterman, P. G., Mbi, C. N., Gartlan, J. S. & Struhsaker, T. T. (1978). Phenolic content of vegetation in two African rain forests: ecological implications. *Science*, **202**, 61–64.

Melnick, D. J. & Pearl, M. C. (1987). Cercopithecines in multimale groups: genetic diversity and population structure. In *Primate Societies*, ed. B. B. Smuts, D. L. Cheney, R. M. Seyfarth, R. Wrangham & T. T. Struhsaker. Chicago, IL: University of Chicago Press, pp. 121–134.

Milton, K. (1996). Effects of bot fly (*Alouattamyia baeri*) parasitism on a free-ranging howler (*Alouatta palliata*) population in Panama. *Journal of Zoological Society of London*, **239**, 39–63.

Moller, A. P., Dufva, R. & Allander, K. (1993). Parasites and the evolution of host social behaviour. *Advances in the Study of Behaviour*, **22**, 65–102.

Mooring, M. S. & Hart, B. L. (1992). Animal grouping for protection from parasites: selfish herd and encounter-dilution effects. *Behaviour*, **123**, 173–193.

Muehlenbein, M. P. (2006). Intestinal parasite infections and fecal steroid levels in wild chimpanzees. *American Journal of Physical Anthropology*, **130**, 546–550.

Muller-Graf, C. D. M., Collias, D. A. & Woolhouse, M. E. J. (1996). Intestinal parasite burden in five troops of olive baboons (*Papio cynocephalus anubis*) in Gombe Stream National Park, Tanzania. *Parasitology Research*, **112**, 489–497.

Muller-Graf, C. D. M., Collins, D. A., Packer, C. & Woolhouse, M. E. J. (1997). *Schistosoma mansoni* infection in a natural population of olive baboons (*Papio cynocephalus anubis*) in Gombe Stream National Park, Tanzania. *Parasitology*, **115**, 621–627.

Nekaris, A. & Bearder, S. K. (2007). The lorisiform primates of Asia and Mainland Africa. In *Primates in Perspective*, ed. C. J. Campbell, A. Fuentes, K. C. Mackinnon, M. Panger & S. K. Bearder. Oxford: Oxford University Press, pp. 24–45.

Nunn, C. L. (2002a). A comparative study of leukocyte counts and disease risk in primates. *Evolution*, **56**, 177–190.

Nunn, C. L. (2002b). Spleen size, disease risk and sexual selection: a comparative study in primates. *Evolutionary Ecology Research*, **4**, 91–107.

Nunn, C. L. & Altizer, S. (2004). Sexual selection, behaviour, and sexually transmitted diseases. In *Sexual Selection in Primates: New and Comparative Perspectives*, ed. P. M. Kappeler & C. P. van Schaik. Cambridge: Cambridge University Press, pp. 117–130.

Nunn, C. L. & Altizer, S. (2006). *Infectious Diseases in Primates: Behavior, Ecology and Evolution*. Oxford: Oxford University Press.

Nunn, C. L. & Dokey, A. T. W. (2006). Ranging patterns and parasitism in primates. *Biology Letters*, **2**, 351–354.

Nunn, C. L. & Heymann, E. W. (2005). Malaria infection and host behaviour: a comparative study of neotropical primates. *Behavioural Ecology and Sociobiology*, **59**, 30–37.

Nunn, C. L., Gittleman, J. L. & Antonovics, J. (2000). Promiscuity and the primate immune system. *Science*, **290**, 1168–1170.

Nunn, C. L., Altizer, S., Jones, K. E. & Sechrest, W. (2003). Comparative tests of parasite species richness in primates. *American Naturalist*, **162**, 597–614.

Nunn, C. L., Altizer, S. M., Sechrest, W. & Cunningham, A. A. (2005). Latitudinal gradients of parasite species richness in primates. *Diversity and Distributions*, **11**, 249–256.

Ohigashi, H., Huffman, M. A., Izutsu, D. *et al.* (1994). Toward the chemical ecology of medicinal plant use in chimpanzees: the case of *Vernonia amygdalina*, a plant used by wild chimpanzees possibly for parasite-related diseases. *Journal of Chemical Ecology*, **20**, 541–553.

Ostfeld, R. S., Glass, G. E. & Keesing, F. (2005). Spatial epidemiology: an emerging (or re-emerging) discipline. *Trends in Ecology and Evolution*, **20**, 328–336.

Sapolsky, R. M. (1992). Neuroendocrinology of the stress-response. In *Behavioral Endocrinology*, ed. J. B. Becker, S. M. Breedlove & D. Crews. Cambridge, MA: MIT Press, pp. 287–566.

Sapolsky, R. M. (2005). The influence of social hierarchy on primate health. *Science*, **308**, 648–652.

Semple, S., Cowlishaw, G. & Bennett, P. (2002). Immune system evolution among anthropoid primates: parasites, injuries, and predators. *Proceedings of the Royal Society of London Series B, Biological Sciences*, **269**, 1031–1037.

Snaith, T. V. & Chapman, C. A. (2007). Primate group size and socioecological models: do folivores really play by different rules? *Evolutionary Anthropology*, **16**, 94–106.

Stoner, K. (1996). Prevalence and intensity of intestinal parasites in mantled howling monkeys (*Alouatta palliata*) in northeastern Costa Rica: implications for conservation biology. *Conservation Biology*, **10**, 539–546.

Struhsaker, T. T. (1976). A further decline in numbers of Amboseli vervet monkeys. *Biotropica*, **8**, 211–214.

Struhsaker, T. T. & Pope, T. R. (1991). Mating system and reproductive success – a comparison of two African forest monkeys (*Colobus badius* and *Cercopithecus ascanius*). *Behaviour*, **117**, 182–205.

Stuart, M. D. & Strier, K. B. (1995). Primates and parasites – a case for a multidisciplinary approach. *International Journal of Primatology*, **16**, 577–593.

Stuart, M. D., Greenspan, L. L., Glander, K. E. & Clarke, M. R. (1990). A coprological survey of parasites of wild mantled howling monkeys, *Alouatta palliata palliata*. *Journal of Wildlife Diseases*, **26**, 547–549.

Stuart, M. D., Greenspan, L. L., Glander, K. E. & Clarke, M. R. (1998). Parasites of wild howlers (*Alouatta* spp.). *International Journal of Primatology*, **19**, 493–512.

Sussman, R. W., Garber, P. A. & Cheverud, J. M. (2005). Importance of cooperation and affiliation in the evolution of primate sociality. *American Journal of Physical Anthropology*, **128**, 84–97.

Thomson, M. C., Obsomer, V., Dunne, M., Connor, S. J. & Molyneux, D. H. (2000). Satellite mapping of *Loa loa* prevalence in relation to ivermectin use in west and central Africa. *Lancet*, **356**, 1077–1078.

Tutin, C. E. G. (2000). Ecologie et organisation des primates de la foret tropicale Africaine: aide a la comprehension de la tranmission des retrovirus. *Foret Tropicales et Emergences Virales*, **93**, 157–161.

van Schaik, C. P. (1983). Why are diurnal primates living in groups? *Behaviour*, **87**, 120–144.

van Schaik, C. P. & van Hooff, J. A. R. A. M. (1983). On the ultimate causes of primate social systems. *Behaviour*, **85**, 91–117.

Vitone, N. D., Altizer, S. & Nunn, C. L. (2004). Body size, diet and sociality influence the species richness of parasitic worms in anthropoid primates. *Evolutionary Ecology Research*, **6**, 183–189.

Waitt, C., Little, A. C., Wolfensohn, S. *et al.* (2003). Evidence from rhesus macaques suggests that male coloration plays a role in female primate mate choice. *Proceedings of the Royal Society of London Series B, Biological Sciences*, **270**, S144–S146.

Walsh, P. D., Biek, R. & Real, L. A. (2005). Wave-like spread of Ebola Zaire. *PLoS Biology*, **3**, 1946–1953.

Walsh, T. D., Abernethy, K. A., Bermejo, M. *et al.* (2003). Catastrophic ape decline in western equatorial Africa. *Nature*, **422**, 611–617.

Watve, M. G. & Jog, M. M. (1997). Epidemic diseases and host clustering: an optimum cluster size ensures maximum survival. *Journal of Theoretical Biology*, **184**, 167–171.

Watve, M. G. & Sukumar, R. (1995). Parasite abundance and diversity in mammals: correlates with host ecology. *Proceeding of the National Academy of Sciences USA*, **92**, 8945–8949.

Wilson, K., Knell, R., Boots, M. & Koch-Osborne, J. (2003). Group living and investment in immune defense: an interspecific analysis. *Journal of Animal Ecology*, **72**, 133–143.

Wrangham, R. W. (1980). An ecological model of female-bonded primate groups. *Behaviour*, **75**, 262–300.

Part IV
Conclusions

24 *Ways forward in the study of primate parasite ecology*

COLIN A. CHAPMAN, MICHAEL A. HUFFMAN, SADIE J. RYAN, RAJA SENGUPTA, AND TONY L. GOLDBERG

Photograph by Colin A. Chapman (*Cercopithecus ascanius*)

Introduction

The devastating impacts that diseases like Ebola have had on both human and wildlife communities (Walsh *et al.*, 2003; Leroy *et al.*, 2004) and the immense

Primate Parasite Ecology. The Dynamics and Study of Host–Parasite Relationships, ed. Michael A. Huffman and Colin A. Chapman. Published by Cambridge University Press.

social and economic costs associated with viruses like HIV (Piot *et al.*, 2004) underscore the practical and societal necessity of understanding infectious diseases in primates and the dynamics of their zoonotic transmission. Given that monkeys and apes often share parasites with humans, understanding the ecology of infectious diseases in non-human primates is of paramount importance to human health planning. This is well illustrated by the HIV viruses, the causative agents of human AIDS, which evolved recently from related SIV viruses of chimpanzees (*Pan troglodytes*) and sooty mangabeys (*Lophocebus atys*; Hahn *et al.*, 2000), as well as by outbreaks of Ebola hemorrhagic fever in African communities, which trace their origins to zoonotic transmissions of Ebola virus from local apes (Formenty *et al.*, 1999). Another example that has a deeper evolutionary origin is malaria. Di Fiore *et al.* (Chapter 7, this volume) review the impacts of malaria on human health and report that there were 396 million cases of human malaria and 1.1 million deaths in 2001 due to infection by *Plasmodium falciparum* alone.

Infectious disease also poses significant conservation risks to non-human primate populations, many of which are already threatened or endangered by habitat loss and/or hunting. For example, evidence indicates that Ebola virus outbreaks have contributed to the reduction of ape population densities by more than 50% over a broad geographic region between 1983 and 2000 (Walsh *et al.*, 2003, 2005). Such examples clearly illustrate the importance of understanding disease dynamics from a conservation perspective. In addition, although humans have always shared habitats with non-human primates, the dynamics of human–primate interactions are changing radically (Chapman & Peres, 2001). Within the last several decades, humans have been responsible for irreversible changes to primate habitats. Most primates today live in anthropogenically disturbed habitat mosaics of farmland, human settlements, forest fragments, and isolated protected areas. As anthropogenic habitat change forces humans and primates into closer and more frequent contact, the risks of interspecific disease transmission increase (Daszak *et al.*, 2000, see also Wolfe and Switzer, Chapter 17, this volume).

When we (Huffman and Chapman) started work on this edited volume, we began with an appreciation that information about diseases from conservation and human-health planning perspectives had stimulated a considerable amount of recent research and the field of primate disease ecology was rapidly gaining momentum. To verify this impression we conducted a simple, but telling, analysis using the Web of Science. We simply put the terms "parasite" and "primate" into the topic search and quantified the number of articles published each year containing these terms. Figure 24.1 illustrates that our impression that the field was gaining momentum was correct. If one reads the titles of these articles, it is also clear that growth in the field has been dominated by

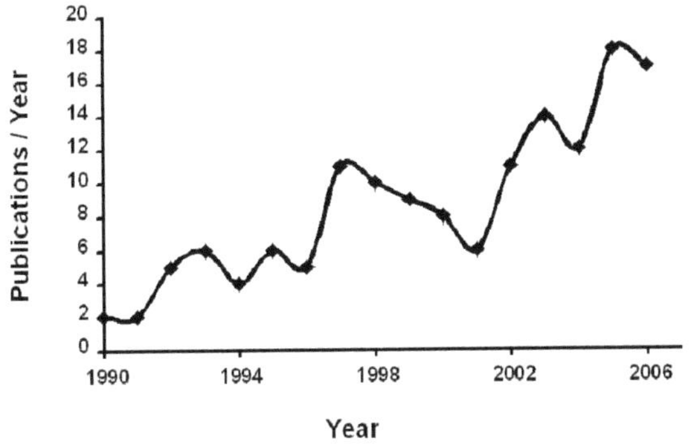

Figure 24.1. The number of publications per year revealed by a search using the Web of Science, with the search terms "parasite" and "primate."

studies of disease ecology and evolution, and that taxonomic research has continued to grow steadily, but at a much slower rate. The complexity of taxonomic identification, as illustrated by the chapters of Gasser *et al.* (Chapter 3) and Hasegawa (Chapter 2), indicates that taxonomic studies are as needed as ecological and evolutionary ones.

While authors of this contributed volume were not asked to identify the agenda of the next generation of research, their results clearly cannot help but do so. To build on or crystallize this agenda, rather than reviewing what has been presented in each chapter or presenting an overview, we would like to build on the impression that this area of study will soon see great advances. Thus, we will present some ideas/approaches that we hope will guide an insightful development of studies in this area. To do this, the editors of this volume (Huffman and Chapman) brought together three additional colleagues who we knew would bring yet more diversity to the perspectives and approaches and asked them to help to illustrate the ways forward for investigations into the field of primate disease ecology.

Methods and molecules

A number of chapters in this volume speak to a vital need to improve the "tool kit" that we use when asking questions in primate disease ecology. First, in the chapter by Greiner and McIntosh, it becomes clear that a variety of traditional methods can be used to characterize parasite infections, but that the selection of methods depends on the species of primate, the parasites in question, and logistical issues. This notion is amplified in Hasegawa's Chapter 2. It becomes clear in later chapters (Dupain *et al.* (14), Vitazkova (18), Weyher (20), and

Chapman *et al.* (23)) that there is little consistency in the methods used and that the field in general needs to be able to answer some specific methodological questions concerning how to characterize infections in such a way as to make comparable cross-study comparisons, particularly when comparisons are made at the population level.

To illustrate the complexity of this issue, let us consider one question: "How many samples should be examined to contrast two groups or populations?" We need to be able to answer this question for many of the inquiries being repeatedly raised, such as how does a specific type of anthropogenic habitat disturbance influence the parasite infections of a particular primate or what is the effect of seasonality across populations of any given parasite infection. If all animals are individually recognizable then this question becomes somewhat simpler and there is research suggesting how many samples of a single individual are needed to describe an infection (Muehlenbein *et al.*, 2004). But when populations are being compared and individual identification is not possible, the way forward is more challenging. It is always important to avoid sample bias, using the same methodology over similar periods of the year for similar durations for each population being compared (Huffman *et al.*, Chapter 16, this volume). Where individual identification is possible, the rules for sampling single individuals need to be applied, doing ones best to monitor all individuals of a population consistently over time. It is common for comparative studies to take a "slice of the pie approach," sampling group(s) for a short period. This runs into difficulties, especially because seasonality of infection can bring highly contrasting results from the same population depending upon when you sample it (Hernandez *et al.*, Chapter 19, this volume; Huffman *et al.*, 1997) or from different populations if the sampling period is too short, and or the periods sampled significantly differ with respect to season.

On the surface the question of how many samples to collect would appear to be a simple one and the answer would be that one would attempt to obtain as many samples as logistically possible. However, the situation is not so simple. There must be a compromise between increasing the sample size and doing assessments of infections using different techniques (e.g. sedimentation, flotation). In the Chapman lab at McGill University, performing both sedimentation (all of the sediment) and flotation takes 6 to 15 times longer than performing flotation alone. A means of increasing the chances of detecting differences among groups or populations would be to decrease variance in sampling unrelated to the question being asked. One might choose only to collect samples at a specific time of the day if diurnal variation in egg output were high, as is the case for many protozoan species (Ezenwa, 2003), or one might sample only from specific age/sex classes if variation were high along this dimension. Unfortunately, little information exists on the magnitude of diurnal variation or variation among age/sex groups for most primate or parasite species. Providing better guidelines as to what methods are best in what situations will greatly increase our ability to compare studies and make generalizations.

A number of chapters in this volume also illustrate the improvement that comes from adding molecular approaches to our "tool kit." This is nicely illustrated by Gasser *et al.*'s Chapter 3 describing how molecular methods were needed to determine that *Oesophagostomum bifurcum* in humans is genetically distinct from that harbored by some non-human primates, that these genetic variants have distinct transmission patterns, and that non-human primates are not a source of human oesophagostomiasis. Thus, it is clear that molecular studies have contributed in substantial ways to our understanding of infectious disease in primates, and the past utility of molecular laboratory and analytical tools bodes well for their future in the field. Genetic data have, to date, been used in studies of primate disease ecology primarily for two related purposes: molecular subtyping and molecular epidemiology. Both uses have improved our understanding of primate infectious disease transmission dynamics.

The use of genetic data to subtype primate pathogens has not been merely an exercise in molecular systematics. Rather, studies examining the intraspecific molecular taxonomy of primate pathogens have shed light on key hypotheses relevant to the transmission of microbes between non-human primates and other species (including humans). For example, mountain gorillas in Bwindi Impenetrable National Park, Uganda can be infected with the gastrointestinal protozoa *Cryptosporidium parvum* and *Giardia duodenalis* (Nizeyi *et al.*, 1999, 2002). Both of these pathogens have genetically distinguishable subtypes (traditionally "genotypes" in the case of *C. parvum* and "assemblages" in the case of *G. duodenalis*) that are associated with transmission among ecologically distinct sets of hosts (Ey *et al.*, 1997; Peng *et al.*, 1997). Graczyk and colleagues (Graczyk *et al.*, 2001, 2002) used molecular subtyping methods to demonstrate that Bwindi gorillas are infected with "genotype 2" *C. parvum* and "Assemblage A" *G. duodenalis*, both of which are associated with zoonotic and anthroponotic transmission (as opposed to other subtypes that are considered to be more host restricted). Molecular subtyping of these protozoa in gorillas therefore strengthens the hypothesis that humans, and perhaps their livestock, represent a source of pathogen transmission for wild gorillas. Molecular data can also weaken hypotheses about the importance of interspecific transmission. Case-in-point are studies by Gasser and colleagues of the clinically relevant "nodule worm," *Oesophagostomum bifurcum*, in Togo and Ghana (Gasser *et al.*, 2006, Gasser *et al.*, Chapter 3, this volume).

Molecular epidemiology can be viewed as an extension of molecular subtyping, in that it attempts not only to classify pathogens, but also to infer patterns of transmission from degrees of genetic similarity among isolates and/or their phylogenetic history (see Di Fiore *et al.*, Chapter 7, this volume). In studies of primate pathogens, the molecular epidemiological approach has been applied most widely to viruses. Phylogenetic analyses based on nucleotide sequences, for example, led to the surprising discovery that both HIV-1 and HIV-2 originated from separate (and, in the case of HIV-2, multiple) cross-species transmissions of simian immunodeficiency virus from non-human primates to humans (Gao *et al.*, 1992, 1999). Wolfe and colleagues examined other primate retroviruses (simian foamy viruses and simian T-cell lymphotropic viruses) and noted substantial "mixing" of human and non-human primate retroviral sequences on phylogenetic trees – a key observation supporting the concept of "viral chatter," or the continuous introduction of endemic primate retroviruses into "high risk" humans, such as hunters (Wolfe *et al.*, 2005; Wolfe & Switzer, Chapter 17, this volume). In combination with careful analyses of geographic associations among outbreaks, molecular sequence data from Ebola virus isolates in Gabon and Republic of Congo demonstrated that localized epidemics in people probably originated through contact with different infected ape or duiker carcasses (Leroy *et al.*, 2004).

Viruses with RNA genomes lend themselves well to molecular epidemiological studies of transmission, due to their high evolutionary rates (Domingo *et al.*, 1996). However, molecular epidemiological methods for inferring transmission among pathogen isolates need not be restricted to such pathogens. Goldberg *et al.* (2007) used a PCR-based DNA fingerprinting approach in combination with population genetic and phylogenetic analyses to infer transmission of the common gastrointestinal bacterium *Escherichia coli* between humans and chimpanzees. The study inferred high rates of bacterial transmission between chimpanzees and people employed in chimpanzee research and tourism in Kibale National Park, Uganda, and showed that chimpanzees harbor bacteria that are resistant to multiple antibiotics used by local people. Novel laboratory and analytical methods may need to be developed for studies of bacteria and other microbes that evolve more slowly than viruses (e.g. Goldberg, 2003; Goldberg *et al.*, 2006).

Molecular methods themselves are evolving at a remarkable rate. Techniques such as real-time quantitative PCR (Higuchi *et al.*, 1993), for example, now significantly speed the diagnostic process and can indicate not only whether a specific pathogen subtype is present in a clinical sample, but also at what concentration. The rapid and cost-effective sequencing of entire eukaryotic genomes is now becoming a reality (Margulies *et al.*, 2005). Refinement of molecular laboratory methods and associated analytical strategies should only enhance future studies of the transmission dynamics of a broad taxonomic range of primate pathogens.

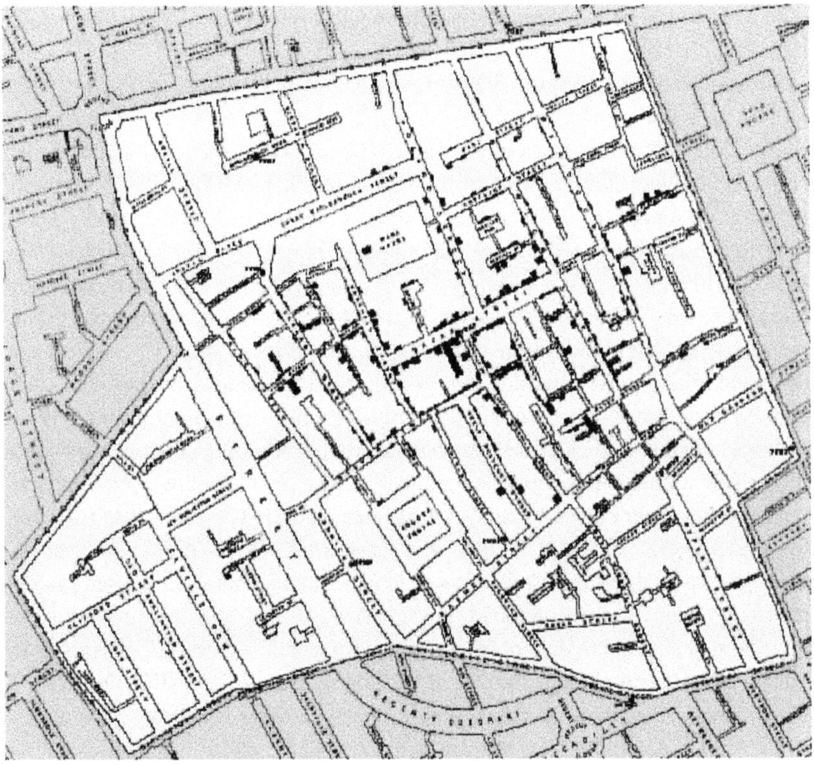

Figure 24.2. Modified map from John Snow's (1855) map of cholera incidences (each case is shown as a bar) and location of hand pumps.

Combining spatial and temporal considerations

The collection of research chapters in this volume illustrated that while disease-transmission and agent-based modeling frameworks (Nunn, Chapter 5, this volume) are exciting trends in the study of infectious diseases, researchers often appear unintentionally to exclude the spatial dimension (Dietz & Hadeler, 1988; Blower *et al.*, 1995; Oli *et al.*, 2006). Perhaps because temporality is implicit in modeling, space is not (even typically within agent-based models). Yet, very early studies of infectious diseases recognized the importance of space. For example, John Snow was critical in tracing the origins of a cholera outbreak to a hand pump and is considered a precursor of statistical mapping methods (Figure 24.2; Snow, 1855). Here we suggest that one way forward that will bring new insights into primate disease ecology will involve spatial epidemiology, with its emphasis on space and spatially explicit models of disease vectors and transmission (Elliot & Wartenburg, 2004).

Two parallel developments are likely to impact the growth of spatially explicit models in epidemiology significantly: (1) the ubiquitous availability of both spatial data and digital earth technologies (Gore, 1998; Masser, 1999; Fonseca *et al.*, 2002), and (2) a growing awareness of basic spatial operations available within traditional GIS software (e.g. overlays and buffers; Xiang, 1993).

A rich repertoire of spatially explicit data is made available in the USA (and worldwide) through publicly funded data-gathering and dissemination initiatives. The 2000 Shuttle Radar Topography Mission (SRTM), for example, provided high-resolution topographic information (90 m × 90 m) for the terrestrial surface worldwide. Similarly, the Landsat series of satellites have obtained large amounts of multi-spectral information since 1972, having accumulated 1.7 million scenes and over 630 terabytes of data (http://www.landimaging.gov/about.html). These data sets are complemented with satellite data collected by other nations, such as Canada's RADARSAT, Europe's SPOT, and India's IRS series satellites. In addition, many countries have available large socioeconomic and biophysical data sets and many are freely available for download. Digital earth technologies are proving to be increasingly popular in the public domain, with technologies such as Mapquest, Yahoo Maps!, and Google Earth fast becoming part of daily life. The growth of such technologies, in combination with the existing repertoire of spatially explicit data, has the power to transform knowledge about the etiology of an infectious disease. For example, Butler (2007), a reporter for *Nature*, generated an animated sequence of H5N1 outbreaks using Google Earth, which helped illustrate the spatiotemporal dynamics of disease occurrence between 2003 and 2006 (Figure 24.3). Indeed, numerous blogs on the internet refer to this map, an indication of its power to influence both the scientific and non-scientific audience.

The next natural step is the use of analysis that spatially integrates environmental (e.g. elevation, land use land cover) and socioeconomic (e.g. income, population density) factors to understand causal mechanisms of infectious disease transmission. The two common GIS-based tools of "overlay" (that identifies relationships between two variables in the same spot) and "buffer" (that identifies relationships between a variable and another (or itself) as a function of distance) can play an important role in explaining disease etiology. Figure 24.4 is a reproduction from the journal *Emerging Infectious Diseases* that uses the "overlay" functionality of a standard GIS package (ArcGIS 9.2) to identify the predominance of certain ecoregions as hotspots for occurrence of bird flu (H5N1) (Sengupta *et al.*, 2007). Only 25 ecoregions, representing 8.8% of the terrestrial surface area, accounted for 2407 (76.8%) of the reported H5N1 cases

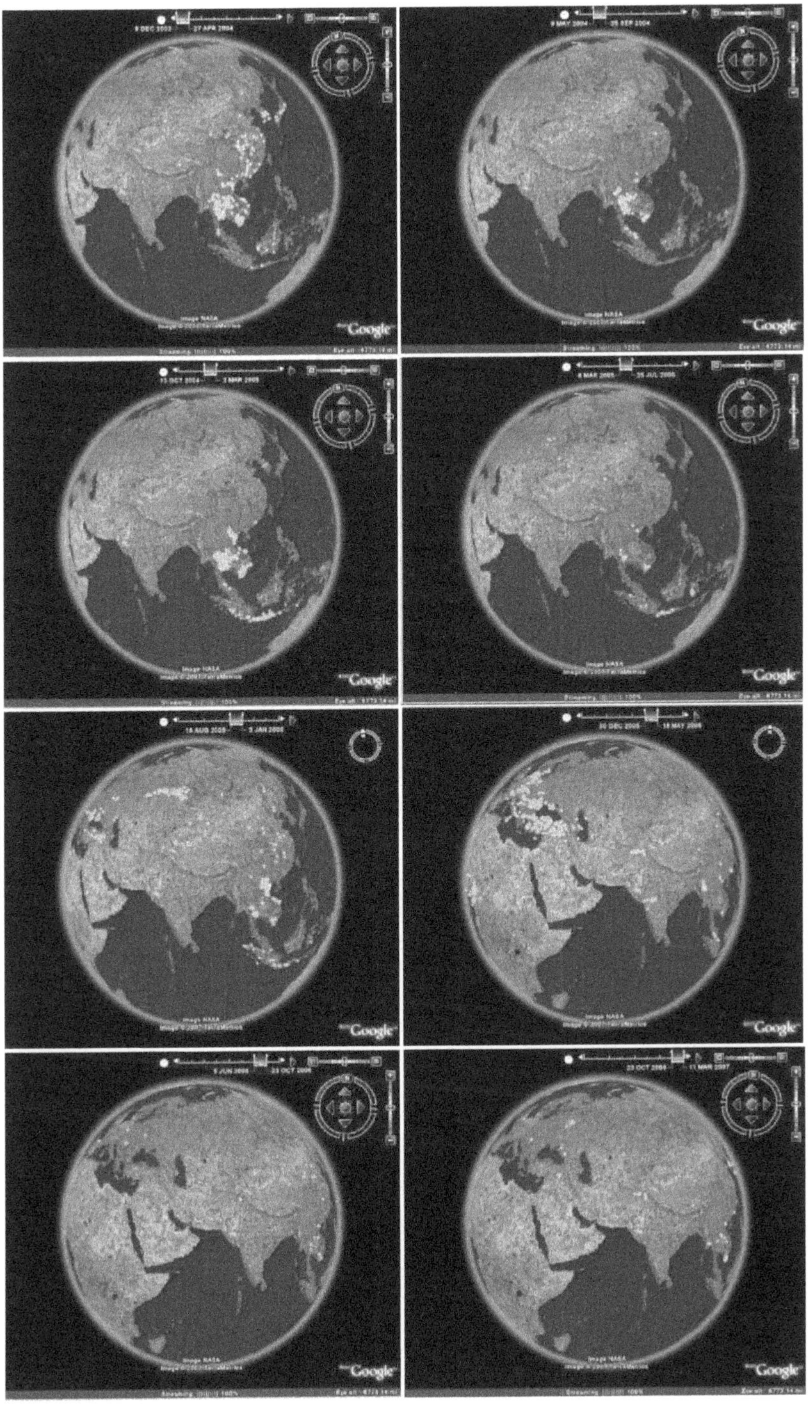

Figure 24.3. Snapshots of Declan Butlers' maps of bird flu occurrences between 2003–2006.

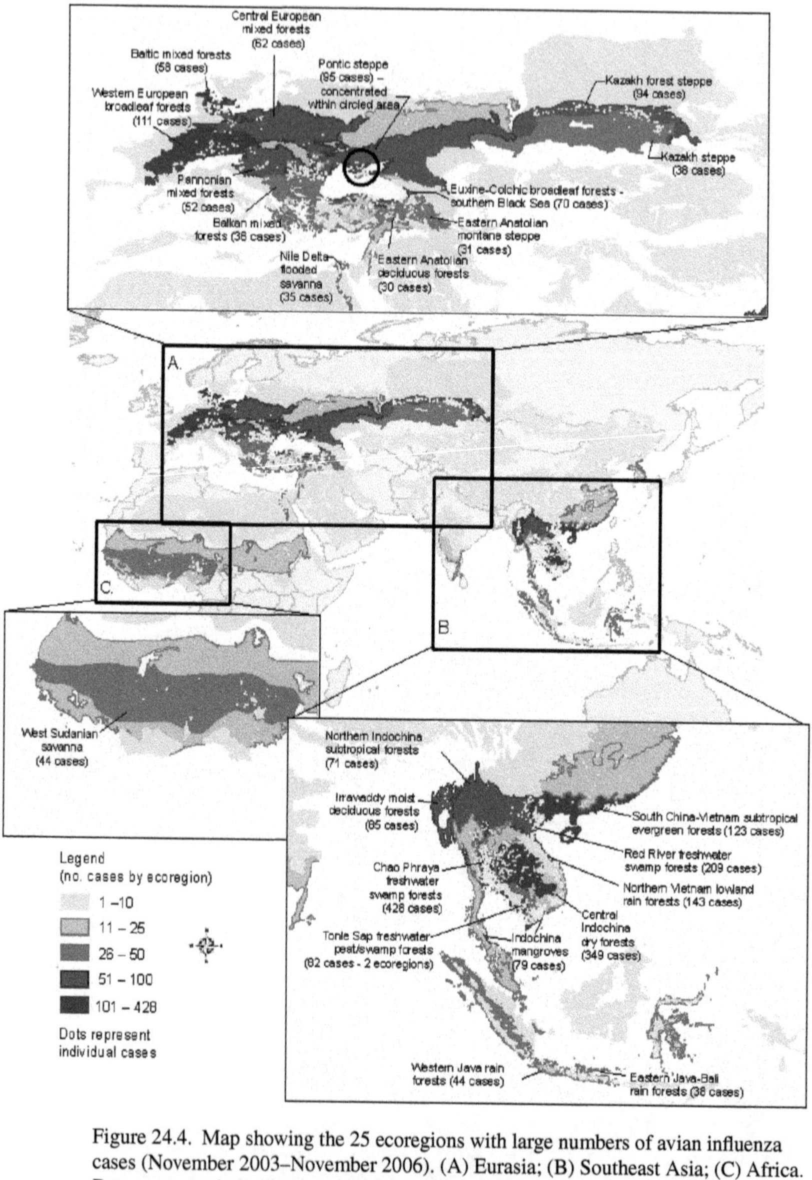

Figure 24.4. Map showing the 25 ecoregions with large numbers of avian influenza cases (November 2003–November 2006). (A) Eurasia; (B) Southeast Asia; (C) Africa. Dots represent individual cases (figure reprinted with permission © Centers for Disease Control and Prevention. *Emerging Infectious Diseases*, **13**, 1269–1271).

between 2003 and 2006. Greater understanding of such techniques could prove useful for locating causes of disease transmissions that may have ecological underpinnings (Wolfe *et al.*, 2005; Despommier *et al.*, 2006). In effect, the consideration of multiple variables at the same location, or the values of the same variable across several locations, can be seen as the first step towards developing a true spatial process model to represent a phenomenon that varies through space and time.

Most infectious diseases do indeed vary in this manner. Infection between primate populations, for example, are a function of the presence of parasites within host populations, its prevalence, the interaction of infected individual(s) with other members of the population (or other populations), and a number of spatially explicit environmental variables (Thomson *et al.*, 2000). An agent-based model, such as the one described by Nunn (Chapter 5, this volume) is able to capture the spatial and temporal dimensions of infectious disease transmission if and only if the spatial dimension has been explicitly specified within the model. Further, most models are univariate with respect to the environmental variables. This is likely not true for most infectious diseases. Research by Thomson *et al.* (2000) on eyeworm identified that variation in the prevalence of eyeworm among villages in west and central Africa can be explained by a number of environmental variables that included land cover, rainfall, temperature, topography, and soil (although one has to be careful here as some of these variables may co-vary).

Finally, one cannot underestimate the importance of integrating the temporal and spatial dimensions. Jacques (2000) correctly criticizes GIS for its static view and lack of process-based disease models. The development of temporal data storage and retrieval methodologies, and models that translate spatio-temporal data into meaningful health outcomes, are both necessary and legitimate demands to enable GIS to properly support spatial epidemiology (Kaur *et al.*, 2004). However, just as temporality is a legitimate demand, so too is the need to incorporate the ability of process-based epidemiological models to access, retrieve, manipulate, and identify relationships between multiple environmental variables stored in spatially explicit data sets. This will truly put space alongside time in epidemiological models and help advance the field forward.

Scaling up: the need for models

To understand the role of disease in the persistence of primate populations, we believe that it will be extremely insightful to create appropriate and flexible

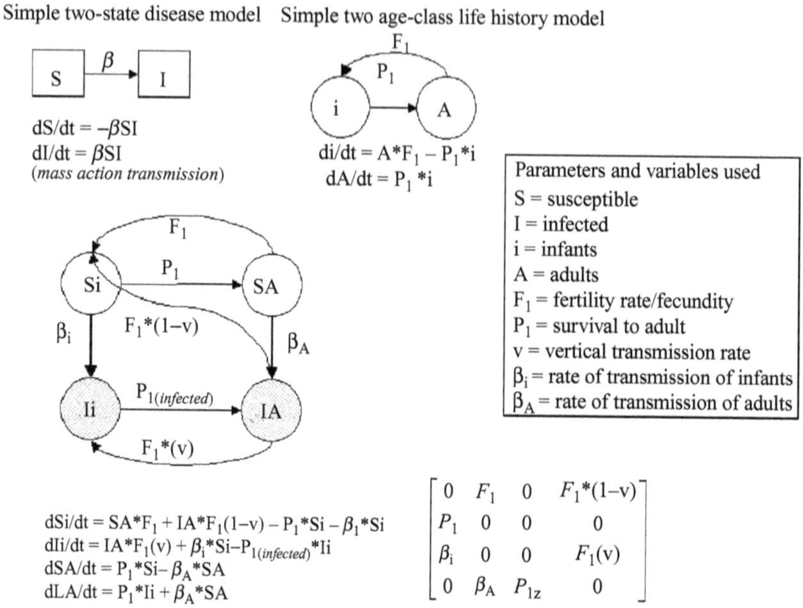

Figure 24.5. In this example, we demonstrate how an age-class model can be overlaid with a two-state transmission model to generate a framework for describing the transmission dynamics of a simple directly transmitted disease within a demographically explicit population model. The scaling up of this type of model to multiple age classes and multiple disease states is straightforward.

quantitative frameworks. Integrating population processes into models of disease impacts can help not only to guide data collection, but also lets the user explore how disease might interact with other forces, such as environmental fluctuations or climate change.

Simple process models of population increase and decrease (discrete or continuous equations) can be modified to include additional mortality due to disease impact, or adjusted to examine resource-dependent responses. However, for a taxon in which social interactions, group living, and population structure are often quite well known and definitely of interest, slightly more complex (but not complicated) models may be more appropriate. For example, a simple age-based demographic model of a primate population can be used to project the growth or decline of a population (Figure 24.5). Age-specific rates of survival and reproduction are the key components in traditional life tables, which for many primates are available in the literature (e.g. Kappeler & Pereira, 2003). These can be used in a deterministic matrix model for demographic projection to estimate whether the population is growing, declining, or remaining stable (Caswell, 1989; Ebert, 1999).

Matrix life-history models also allow the examination of which age classes are having the most impact on the trajectory of the population. This is at the core of many population viability analyses (PVAs) that inform conservation actions and policies. The sensitivity of the population trajectory to certain age classes can help reveal which parts of the population's structure needs protection to improve the recovery of the whole population. One highly cited example of the utility of a PVA is in the protection of loggerhead sea turtles (*Caretta caretta*), whose population was threatened both by beach use, endangering the eggs and hatchlings, and by adults drowning in fish nets. In their classic papers, Crouse and Crowder and colleagues (Crouse *et al.*, 1987; Crowder *et al.*, 1994) showed that adult survival was in fact more important to long-term population dynamics than protection of the younger life stages, and this led mandatory turtle excluder devices in fishing nets – which has been adopted globally for many turtle populations. In developing models for primates this will be important in the cases where diseases hit one component of a population and not others. In the development of population models for viability analyses it became apparent that stochasticity in both demographic and extrinsic inputs can drive a population dramatically in the short term, relative to longer-term effects in larger populations. This is important when considering primate populations, as many remnant populations are small or fragmented, meaning that population biology of small numbers is at play.

The impact of disease on a primate population can be accessed through models at many different levels. If the state of an infected individual is identifiable (observations of physical manifestations of infection, such as high parasite loads, lesions, lethargy) and can be associated with differential survival, it is easy to see how this can be incorporated into a sensitivity analysis in a PVA type of framework. This is particularly interesting if the disease impact changes with age and/or sex class of the individuals. For example, a disease causing high infant mortality will create a very different population signal than one that affects dispersing age animals (Oli *et al.*, 2006). Even in the absence of empirical data, exploring potential theoretical impacts of disease by examining the sensitivities of different age classes can help inform the potential risks to that population.

It is also important to consider the potential impact of a disease due to its intrinsic traits. Some diseases, such as viral hemorrhagic fevers (e.g. Ebola and Marburg), cause mortality quickly, which may not allow rapid spread through a population, due to a short infectious period and a quick removal of the infected individuals from the population. In contrast to the "fast" dynamics of directly transmitted diseases with high mortality impacts, it is also important to consider the more prolonged role of chronic disease and of indirect transmission. Chronic diseases tend to lead to issues not of mortality, but of morbidity; these "slower" diseases may render individuals weaker, or less resilient in the face of other extrinsic shocks to the system. One characteristic of chronic weak states is the reduced ability to reproduce, or provide sufficient resources for offspring. Another factor is that weak or frail individuals are less likely to survive during low resource periods, and are less likely to rebound from short-term environmental shocks. The role of these life-history and resilience impacts can be modeled quite simply, using state-dependent frameworks, or simple simulations in matrix frameworks.

Since primates characteristically live in groups, the structure of the population can act to "slow down" disease spread, if individuals get infected and die before emigrating from the group and thus there is no spread. If a disease has a long chronic phase or is infectious for a long time, the structure of a population becomes less relevant, as the between-group travel time becomes shorter than the duration of infectiousness and the disease can spread. The relationship between population structure and direct disease spread has been explored by Cross *et al.* (2005) within a theoretical framework, and the ramifications of different structure on disease due to social system differences in primates and other animals has been explored (Thrall *et al.*, 2000; Altizer *et al.* 2003; Ryan *et al.*, in prep.). In particular, Thrall *et al.* (2000) demonstrate that the role of social structure and the sexual transmission mode for disease has varying effects on the prevalence and extent of an outbreak for different sexes in the population. Social structure and metapopulation structure can be modeled implicitly in deterministic frameworks, simply by continuing the "compartmental" class structure into larger dimensions of a matrix. However, it will rapidly become apparent to the user that for primates, more realistic depictions of spatial structure may provide better information. Overlaying disease impacts on the individual and agent-based models as described by Charles Nunn in this volume is another step, ranging in complexity from individual states (diseased or not) to explicit dynamics of the parasites themselves.

Explicit modeling of host–parasite dynamics wherein external phases of parasites are included is fairly well established (Hudson *et al.*, 1992); however, the inclusion of multiple hosts and multiple parasites into models leads to quite complicated and data-hungry frameworks. The value of agent-based models with high levels of complexity is the potential to reveal novel properties of the system, commonly referred to as "emergent properties." These properties can demonstrate unexpected drivers in the system because of complex interactions that are not captured by more simple descriptions of the system. Importantly, the flexibility of parameters – wherein the user can simulate a range of values – provides a means of sensitivity analysis, and the ability to explore which components of the system are contributing to the drivers. This is a less precise means of examining sensitivity than the more analytically tractable nature of eigen value methods in matrix models (Mesterton-Gibbons, 2000; Yearsley, 2004); nonetheless, it is similarly useful.

Conclusions

Primate parasite ecology is in a period of rapid growth and development and it is our opinion, and we hope that the chapters in this edited volume provide a convincing case, that research in this field offers a great deal to enhance our understanding of this fascinating group of mammals and this diverse group of parasites. We view that by employing the tools and approaches such as those outlined in this last chapter, unique insights will be derived moving this field forward dramatically.

Given the plight of primates in the tropics, we wish to end our volume by reminding the reviewer of the typical state of affairs of primate conservation and emphasizing the role parasitism can play in enhancing the effects of other factors. The tropical forests that most primate species occupy are undergoing rapid anthropogenic transformation. Cumulatively, countries with primate populations are losing approximately 125 000 km^2 of forest annually (Chapman & Peres, 2001). Other populations are being affected by forest degradation (logging and fire) and hunting (Oates, 1996). This is the typical means of considering conservation threats: in terms of readily apparent factors such as habitat loss/degradation and over-harvesting. However, when situations are considered in more depth it often becomes apparent that population declines are associated with a complex set of interactions among the environment and biota, which in combination overwhelm the populations' ability to withstand change. It is clear that primate pathogens are important members of this complex set of interactions. For example, Ebola has clearly been related to the dramatic decline in gorilla populations (Walsh *et al.*, 2003; Walsh *et al.*, Chapter 8, this volume; Chapman *et al.*, 2005), yet there is also evidence that Ebola outbreaks occur in habitats currently undergoing fragmentation in areas that are a mosaic of forests and open habitats (Morvana *et al.*, 1999). Similarly, in forest fragments, nutritional stress is correlated with high gastrointestinal parasite infections and population declines (Chapman *et al.*, 2006).

Conservation biologists and managers interested in protecting specific species or ecosystems benefit from understanding the interactions that characterize those ecosystems, because by doing so it becomes possible to predict population or ecosystem changes that may not be intuitively obvious (e.g. a population will decline if it is over-harvested). The field of disease ecology is an exciting area where by understanding how changes in the ecosystem, like climate change or habitat fragmentation, can alter host–disease interactions we can provide conservation biologists with such non-intuitive tools for species protection. Furthermore, by understanding how changes in host–disease interactions can cause population declines, we need not simply respond to change, but we can be proactive and manage the system to prevent declines.

References

Altizer, S., Nunn, C. L., Thrall, P. H. *et al.* (2003). Social organization and parasite risk in mammals: integrating theory and empirical studies. *Annual Review of Ecology and Systematics*, **64**, 517–547.

Blower, S. M., McLean, A. R., Porco, T. C. *et al.* (1995). The intrinsic transmission dynamics of tuberculosis epidemics. *Nature Medicine*, **1**, 815–821.

Butler, D. (2007). Cheaper approaches to flu divide researchers. *Nature*, **448**, 976–977.

Caswell, H. (1989). *Matrix Population Models.* Sunderland, MA: Sinauer Associates, Inc.

Chapman, C. A. & Peres, C. A. (2001). Primate conservation in the new millennium: the role of scientists. *Evolutionary Anthropology*, **10**, 16–33.

Chapman, C. A., Gillespie, T. R. & Goldberg, T. L. (2005). Primates and the ecology of their infectious diseases: how will anthropogenic change affect host–pathogen interactions? *Evolutionary Anthropology*, **14**, 134–144.

Chapman, C. A., Wasserman, M. D., Gillespie, T. R. *et al.* (2006). Do nutrition, parasitism, and stress have synergistic effects on red colobus populations living in forest fragments? *American Journal of Physical Anthropology*, **131**, 525–534.

Cross, P. C., Lloyd-Smith, J. O., Johnson, P. L. F. & Getz, W. M. (2005). Dueling timescales of host movement and disease recovery determine invasion of disease in structured populations. *Ecology Letters*, **8**, 587–595.

Crouse, D. T., Crowder, L. B. & Caswell, H. (1987). A stage-based population model for loggerhead sea turtles and implications for conservation. *Ecology*, **68**, 1412–1423.

Crowder, L. B., Crouse, D. T., Heppelll, S. S. & Martin, T. H. (1994). Predicting the impact of turtle excluder devices on loggerhead sea turtle populations. *Ecological Applications*, **4**, 445–455.

Daszak, P., Cunningham, A. A. & Hyatt, A. D. (2000). Wildlife ecology – emerging infectious diseases of wildlife – threats to biodiversity and human health. *Science*, **287**, 443–449.

Despommier, D., Ellis, B. R. & Wilcox, B. A. (2006). The role of ecotones in emerging infectious diseases. *EcoHealth*, **3**, 281–289.

Dietz, K. & Hadeler, K. P. (1988). Epidemiological models for sexually-transmitted diseases. *Journal of Mathematical Biology*, **26**, 1–25.

Domingo, E., Escarmís, C., Sevilla, N. *et al.* (1996). Basic concepts in RNA virus evolution. *FASEB Journal*, **10**, 859–864.

Ebert, T. A. (1999). *Plant and Animal Populations: Methods in Demography.* San Diego, CA: Academic Press.

Elliot, P. & Wartenburg, D. (2004). Spatial epidemiology: current approaches and future challenges. *Environmental Health Perspectives*, **112**, 998–1006.

Ey, P. L., Mansouri, M., Kulda, J. *et al.* (1997). Genetic analysis of *Giardia* from hoofed farm animals reveals artiodactyl-specific and potentially zoonotic genotypes. *Journal of Eukaryotic Microbiology*, **44**, 626–635.

Ezenwa, V. O. (2003). The effect of time of day on the prevalence of coccidian oocysts in antelope faecal samples. *African Journal of Ecology*, **41**, 192–193.

Fonseca, F. T., Egenhofer, M. J., Agouris, P. & Câmara, G. (2002). Using ontologies for integrated geographic information systems. *Transactions in GIS*, **6**, 231–257.

Formenty, P., Hatz, C., Le Guenno, B. *et al.* (1999). Human infection due to Ebola virus, subtype Cote d'Ivoire: clinical and biologic presentation. *Journal of Infectious Diseases*, **179**, S48–S53.

Gao, F., Yue, L., White, A. T. *et al.* (1992). Human infection by genetically diverse SIVSM-related HIV-2 in west Africa. *Nature*, **358**, 495–499.

Gao, F., Bailes, E., Robertson, D. L. *et al.* (1999). Origin of HIV-1 in the chimpanzee *Pan troglodytes troglodytes*. *Nature*, **397**, 436–441.

Gasser, R. B., de Gruijter, J. M. & Polderman, A. M. (2006). Insights into the epidemiology and genetic make-up of *Oesophagostomum bifurcum* from human and non-human primates using molecular tools. *Parasitology*, **132**, 453–460.

Goldberg, T. L. (2003). Application of phylogeny reconstruction and character-evolution analysis to inferring patterns of directional microbial transmission. *Preventive Veterinary Medicine*, **61**, 59–70.

Goldberg, T. L., Gillespie, T. R. & Singer, R. S. (2006). Optimization of analytical parameters for inferring relationships among *Escherichia coli* isolates from repetitive-element PCR by maximizing correspondence with multilocus sequence typing data. *Applied and Environmental Microbiology*, **72**, 6049–6052.

Goldberg, T. L., Gillespie, T. R., Rwego, I. B. *et al.* (2007). Patterns of gastrointestinal bacterial exchange between chimpanzees and humans involved in research and tourism in western Uganda. *Biological Conservation*, **135**, 527–533.

Gore, A. (1998). The digital earth: understanding our planet in the 21st century. *Australian Surveyor*, **43**, 89–91.

Graczyk, T. K., DaSilva, A. J., Cranfield, M. R. *et al.* (2001). *Cryptosporidium parvum* genotype 2 infections in free-ranging mountain gorillas (*Gorilla gorilla beringei*) of the Bwindi Impenetrable National Park, Uganda. *Parasitology Research*, **87**, 368–370.

Graczyk, T. K., Bosco-Nizeyi, J., Ssebide, B. *et al.* (2002). Anthropozoonotic *Giardia duodenalis* genotype (assemblage) A infections in habitats of free-ranging human-habituated gorillas, Uganda. *Journal of Parasitology*, **88**, 905–909.

Hahn, B. H., Shaw, G. M., De Cock, K. M. & Sharp, P. M. (2000). AIDS as a zoonosis: scientific and public health implications. *Science*, **287**, 607–617.

Higuchi, R., Fockler, C., Dollinger, G. & Watson, R. (1993). Kinetic PCR analysis: real-time monitoring of DNA amplification reactions. *Biotechnology*, **11**, 1026–1030.

Hudson, P. J., Dobson, A. P. & Newborn, D. (1992). Do parasites make prey vulnerable to predation: red grouse and parasites. *Journal of Animal Ecology*, **61**, 681–692.

Huffman, M. A., Gotoh, S., Turner, L. A., Hamai, M. & Yoshida, K. (1997). Seasonal trends in intestinal nematode infection and medicinal plant use among chimpanzees in the Mahale Mountains, Tanzania. *Primates*, **38**, 111–125.

Jacques, G. M. (2000). Spatial analysis in epidemiology: nascent science or a failure of GIS? *Journal of Geographical Systems*, **2**, 91–97.

Kappeler, P. M. & Pereira, M. E. (eds.). (2003). *Primate Life History and Socioecology*. Chicago, IL: University of Chicago Press.

Kaur, T., Singh, J., Huffman, M. A., Moscovice, L. R. & Nelson, P. (2004). CyberCHIMPP: a mobile field data collection system to facilitate chimpanzee research and conservation. *Folia Primatologica*, **75** (Suppl.), 288.

Leroy, E. M., Rouguet, P., Formenty, P. *et al.* (2004). Multiple Ebola virus transmission events and rapid decline of Central African Wildlife. *Science*, **303**, 387–390.

Margulies, M., Egholm, M., Altman, W. E. *et al.* (2005). Genome sequencing in microfabricated high-density picolitre reactors. *Nature*, **437**, 376–380.

Masser, I. (1999). All shapes and sizes: the first generation of national spatial data infrastructures. *International Journal of Geographical Information Science*, **13**, 67–84.

Mesterton-Gibbons, M. (2000). A consistent equation of ecological sensitivity in matrix population analysis. *Trends in Ecology and Evolution*, **15**, 115.

Morvana, J. M., Deubelb, V., Gounonc, P. *et al.* (1999). Identification of Ebola virus sequences present as RNA or DNA in organs of terrestrial small mammals of the Central African Republic. *Microbes and Infection*, **1999**, 1193–1201.

Muehlenbein, M. P., Watts, D. P. & Whitten, P. L. (2004). Dominance rank and fecal testosterone levels in adult male chimpanzees (*Pan troglodytes schweinfurthii*) at Ngogo, Kibale National Park, Uganda. *American Journal of Primatology*, **64**, 71–82.

Nizeyi, J. B., Mwebe, R., Nanteza, A. *et al.* (1999). *Cryptosporidium* sp. and *Giardia* sp. infections in mountain gorillas (*Gorilla gorilla beringei*) of the Bwindi Impenetrable National Park, Uganda. *Journal of Parasitology*, **85**, 1084–1088.

Nizeyi, J. B., Sebunya, D., Dasilva, A. J. *et al.* (2002). Cryptosporidiosis in people sharing habitats with free-ranging mountain gorillas (*Gorilla gorilla beringei*), Uganda. *American Journal of Tropical Medicine and Hygiene*, **66**, 442–444.

Oates, J. F. (1996). Habitat alteration, hunting, and the conservation of folivorous primates in African forests. *Australian Journal of Ecology*, **21**, 1–9.

Oli, M. K., Venkataraman, M., Klein, P. A., Wendland, L. D. & Brown, M. B. (2006). Population dynamics of infectious diseases: a discrete time model. *Ecological Modelling*, **198**.

Peng, M., Xiao, M. L., Freeman, A. R. *et al.* (1997). Genetic polymorphism among *Cryptosporidium parvum* isolates: evidence of two distinct human transmission cycles. *Emerging Infectious Diseases*, **3**, 567–573.

Piot, P., Feachem, R. G.A., Jong-Wook, L. & Wolfensohn, J. D. (2004). The global response to AIDS: lessons learned, next steps. *Science*, **304**, 1909–1910.

Ryan, S. J., Nunn, C. L. & Dobson, A. P. (in prep.). Socially directed disease transmission in primates: a modeling approach.

Sengupta, R., Rosenshein, L., Gilbert, M. & Weiller, C. (2007). Ecoregional dominance in spatial distribution of avian influenza (H5N1) outbreaks. *Emerging Infectious Diseases*, **13**, 1269–1271.

Snow, J. (1855). *On the Mode of Communication of Cholera*, 2nd edn. London: John Churchill.

Thomson, M. C., Obsomer, V., Dunne, M., Connor, S. J. & Molyneux, D. H. (2000). Satellite mapping of *Loa loa* prevalence in relation to ivermectin use in West and Central Africa. *Lancet*, **356**, 1077–1078.

Thrall, P. H., Antonovics, J. & Dobson, A. P. (2000). Sexually transmitted diseases in polygynous mating systems: prevalence and impact on reproductive success. *Proceedings of the Royal Society of London Series B, Biological Sciences*, **267**, 1555–1563.

Walsh, P. D., Biek, R. & Real, L. A. (2005). Wave-like spread of Ebola, Zaire. *PLoS Biology*, **3**, 1946–1953.

Walsh, T. D., Abernethy, K. A., Bermejo, M. *et al.* (2003). Catastrophic ape decline in western equatorial Africa. *Nature*, **422**, 611–617.

Wolfe, N. D., Daszak, P., Kilpatrick, A. M. & Burke, D. S. (2005). Bushmeat hunting, deforestation, and prediction of zoonoses emergence. *Emerging Infectious Diseases*, **11**, 1822–1827.

Xiang, W. N. (1993). Application of a GIS-based stream buffer generation model to environmental policy evaluation. *Environmental Management*, **17**, 817–827.

Yearsley, J. M. (2004). Transient population dynamics and short-term sensitivity analysis of matrix population models. *Ecological Modelling*, **177**, 245–258.

25 *Useful diagnostic references and images of protozoans, helminths, and nematodes commonly found in wild primates*

HIDEO HASEGAWA, COLIN A. CHAPMAN, AND MICHAEL A. HUFFMAN

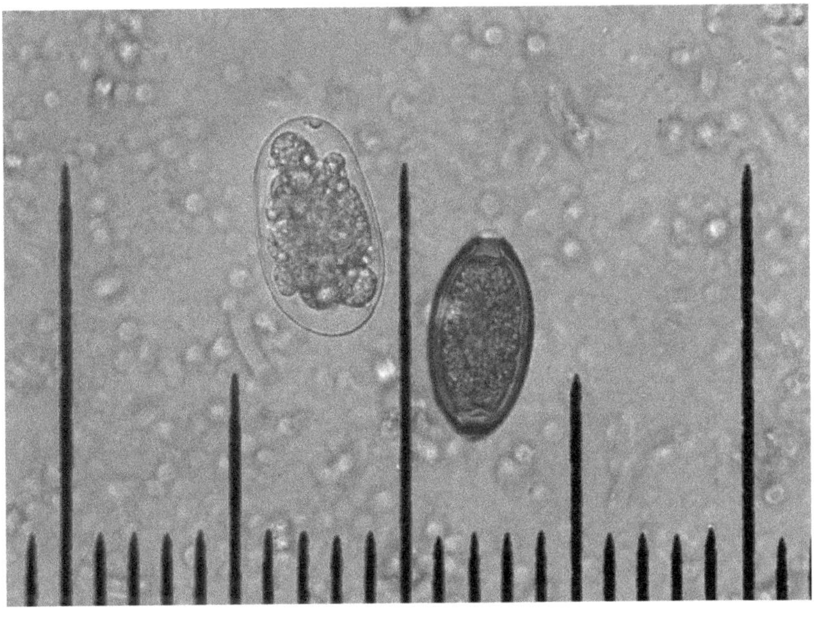

Photograph by Stacey Hodder (*Oesophagostomum* sp., left; *Trichuri* sp., right)

Primate Parasite Ecology. The Dynamics and Study of Host–Parasite Relationships, ed. Michael A. Huffman and Colin A. Chapman. Published by Cambridge University Press.

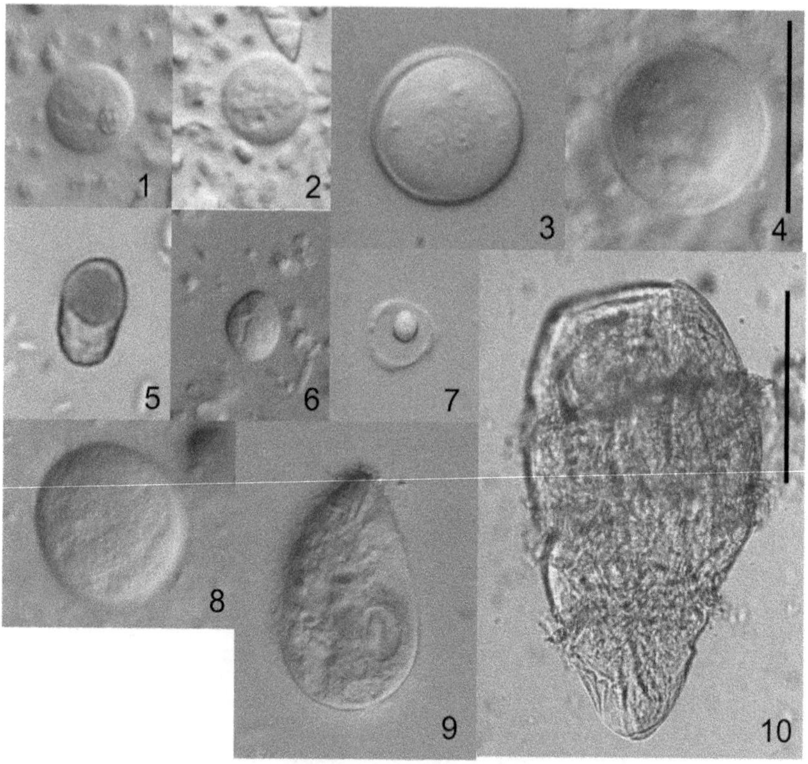

Figure 25.1. Protozoans that have appeared in the feces of primates. 1–2. *Entamoeba* sp. cyst (host: *Macaca fuscata*). 3. *Entamoeba coli* cyst (host: *M. fuscata*). 4. *Entamoeba dispar* cyst (host: *M. fuscata*). 5. *Iodamoeba buetschlii* (host: *Pan troglodytes*). 6. *Chilomastix mesnili* cyst (host: *Pan troglodytes*, captive). 7. *Blastocyctis hominis* cyst (host: *P. troglodytes*, captive). 8. *Balantidium coli* cyst (host: *M. fuscata*). 9. *B. coli* trophozoite (host: *P. troglodytes*, captive). 10. *Troglodytella abrassarti* trophozoite (host: *P. troglodytes*, captive). (Scale 25 μm for 1–7, 50 μm for 8–10). Photos/material: T. Dagg and A. D. Hernandez (4, 8), K. J. Petrzelkova (3, 6, 7, 9, 10).

This chapter provides a collection of useful references and diagnostic images of protozoans, helminths, and nematodes commonly found in primates from the wild and captivity. The host species noted after the taxa name in the plates denote the host species from which each particular image was taken, but does not necessarily present the full range of host species in which these parasites are found. We refer the reader to the Global Mammal Parasite Database at http://www.mammalparasites.org/ for a more comprehensive list of host–parasite associations of these and other parasites.

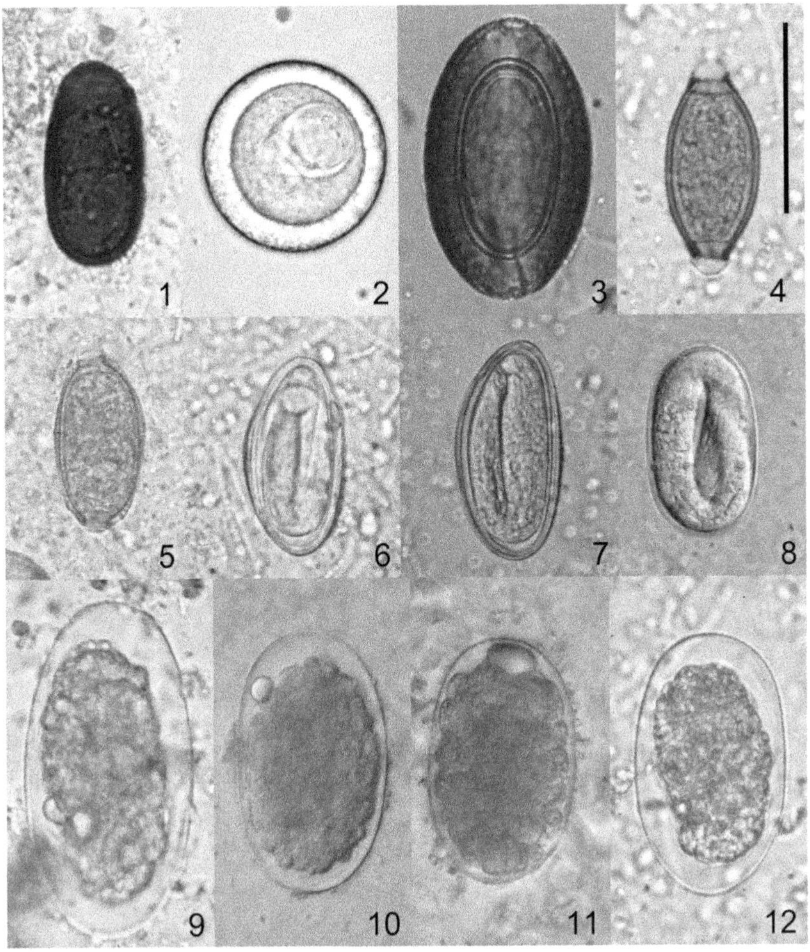

Figure 25.2. Eggs of some common helminths parasitic in primates (I).
1. Dicrocoeliidae sp. (host: *Pan paniscus*). 2. *Bertiella sturdi* (host: *Macaca fuscata*). 3. *Prosthenorchis elegans* (host: *Saimiri sciureus*). 4. *Trichuris* cf. *trichiura* (host: *P. paniscus*). 5. *Capillaria brochieri* (host: *P. paniscus*). 6. *Enterobius anthropopitheci* (host: *P. paniscus*). 7. *Enterobius vermicularis* (host: *Pan troglodytes*, captive). 8. *Strongyloides fuelleborni* (host: *M. fuscata*). 9. *Oesophagostomum* cf. *stephanostomum* (host: *P. paniscus*). 10, 11. *Oesophagostomum aculeatum* (host: *M. fuscata*). 12. Strongylyda sp. (hookworm) (host: *P. paniscus*). (Scale: 50 μm). Photos courtesy of K. Ando (2) and H. Sato (3).

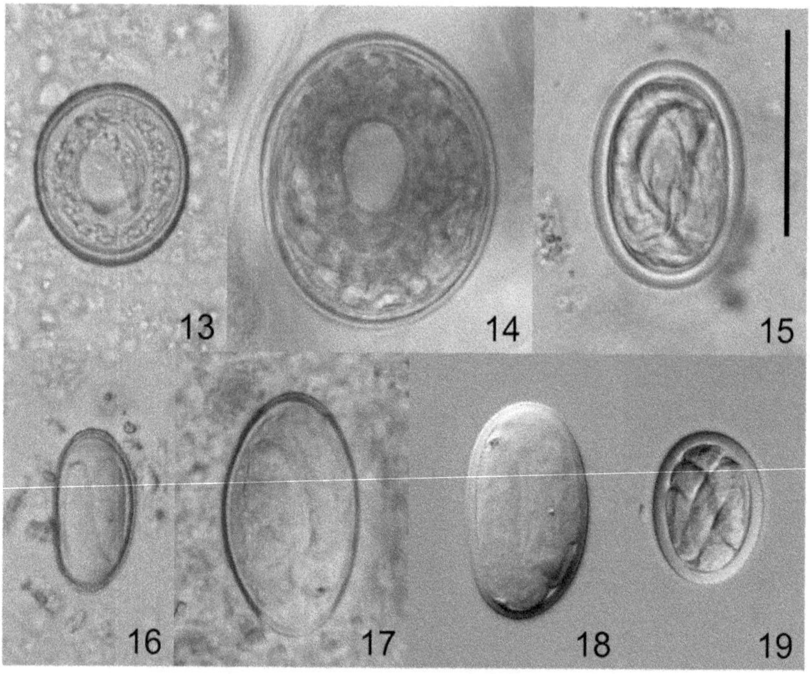

Figure 25.3. Eggs of some common helminths parasitic in primates (II). 13. *Subulura* sp. (host: *Pan troglodytes*, courtesy of IJP). 14. *Subulura* sp. (host: *Cercopithechus mitis kolbi*). 15. *Protospirura muricola* (host: *Pan troglodytes*). 16. *Streptopharagus pigmentatus* (host: *Macaca fuscata*), 17. *Gongylonema pulchrum* (host: *M. fuscata*). 18. *Spirura guianensis* (host: *Saimiri sciureus*, captive). 19. *Pterygodermatites nycticebi* (host: *Nycticebus coucang*, captive). (Scale: 50 μm). Photos/material courtesy of United States National Parasite collection (14), A. D. Hernandez and T. Dagg (17), H. Sato (18).

These images are meant to be a preliminary guide to identification. Care must be taken, as some eggs and larvae look similar to others, and definitive species identification by adult worms is always preferable (Greiner & McIntosh, Chapter 1, this volume; Hasegawa, Chapter 2, this volume). For the non-specialist, close collaboration with a parasitologist experienced in primates is highly recommended.

The bibliography contains recommended resources for identification of the parasite images provided and for parasites in general. An important source of verification is via the comparison of one's specimens with voucher specimens and/or photo material that can often be borrowed on inter-institutional loan from leading depositories of parasite material such as the USNPC (http://www.ars.usda.gov/is/np/systematics/animalpar.htm). Another option is to contact an experienced parasitologist willing to verify or make positive IDs

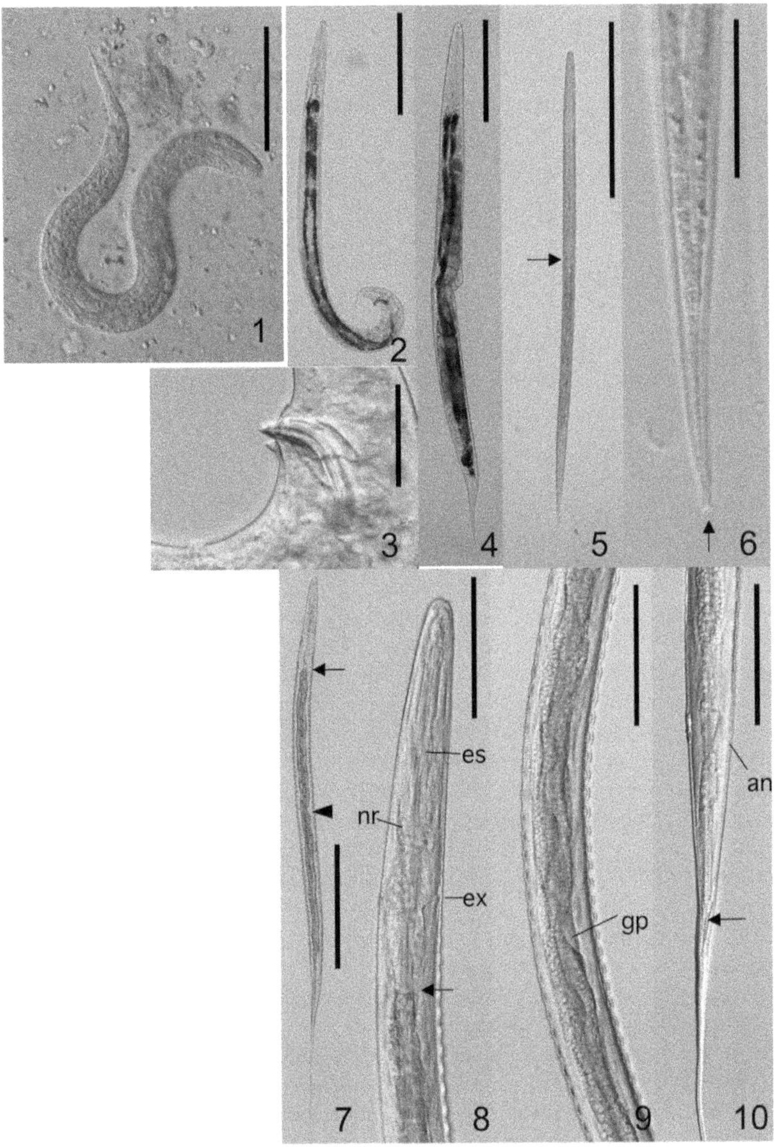

Figure 25.4. Common nematodes appearing in feces and fecal cultures of primates. 1. Rhabditoid larva of *Strongyloides stercoralis* (host: *Pan troglodytes*, captive). 2–6. *Strongyloides fuelleborni* (host: *Macaca fuscata*). 2. Free-living male, 3. Spicule and gubernaculum of free-living male. 4. Free-living female. 5. Filariform larva (arrow indicates esophago-intestinal junction). 6. Posterior end of filariform larva. 7–10. Infective larva of *Oesophagostomum aculeatum* (host: *Macaca fuscata*). 7. Entire worm (arrow and arrowhead indicate esophago-intestinal junction and gonadal primordium, respectively). 8. Anterior portion (arrow indicates esophago-intestinal junction). 9. Middle intestine region showing zigzagged lumen. 10. Posterior portion showing level of tail tip (arrow). Abbreviations: es. esophagus, ex. excretory pore, gp. gonadal primordium, nr. nerve ring. (Scale: 200 μm for 2, 4, 7; 50 μm for 1, 5, 8–10; 25 μm for 3, 6). Material: K. Petrzelkova (1).

for you. For this, it is often necessary to send worm specimens preserved in formaldehyde or alcohol for morphological and genetic identification (Gasser *et al.*, Chapter 3, this volume).

Acknowledgements

We wish to thank the following individuals for generously providing materials and/or photographic contributions: Katsuhiko Ando, Topher Dagg, Alexander D. Hernandez, Klara J. Petrzelkova, and Hiroshi Sato.

Bibliography

Anderson, R. C. (2000). *Nematode Parasites of Vertebrates. Their Development and Transmission*, 2nd edn. Wallingford, UK: CAB International.

Anonymous. *DPDx Laboratory Identification of Parasites of Public Concern*. Center for Disease Control and Prevention. http://www.dpd.cdc.gov/dpdx/.

Ando, K., Sato, Y., Miura, K. *et al.* (1994). The occurrence of *Bertiella sturdi* (Cestoda: Anoplocephalidae) in Japanese macaque, *Macaca fuscata*, from Mie Prefecture, Japan. *Japanese Journal of Parasitology*, **43**, 211–218.

Blanchard, J. L. & Eberhard, M. L. (1985). Case report: esophageal *Spirura* infection in a squirrel monkey (*Saimiri sciureus*). *American Journal of Primatology*, **10**, 279–282.

Bowman, D. D. (2003). *Georgis' Parasitology for Veterinarians*, 8th edn. Philadelphia, PA: Saunders Company.

Brack, M. (1987). *Agents Transmissible from Simians to Man*. Göttingen: Springer-Verlag.

Garcia, L. S. (2007). *Diagnostic Medical Parasitology*, 5th edn, Brooklyn. NY: American Microbiology Society Press.

Gasser, R. B., Woods, W. G., Huffman, M. A., Blotkamp, J. & Polderman, A. M. (1999). Molecular separation of *Oesophagostomum stephanostomum* and *Oesophagostomum bifurcum* (Nematoda: Strongyloidea) from non-human primates. *International Journal of Parasitology*, **29**, 1087–1091.

Hasegawa, H. & Udono, T. (2007). Chimpanzee pinworm, *Enterobius anthropopitheci* (Nematoda: Oxyuridae), maintained for more than twenty years in captive chimpanzees in Japan. *Journal of Parasitol*ogy, **93**, 850–853.

Hasegawa, H., Kano, T. & Mulavwa, M. A. (1983). A parasitological survey on the feces of Pygmy chimpanzees, *Pan paniscus*, at Wamba, Zaire. *Primates*, **24**, 419–423.

Hasegawa, H., Ikeda, Y., Fujisaki, A. *et al.* (2005). Morphology of chimpanzee pinworms, *Enterobius (Enterobius) anthropopitheci* (Gedoelst, 1916) (Nematoda: Oyuridae), collected from chimpanzees, *Pan troglodytes*, on Rubondo Island, Tanzania. *Journal of Parasitology*, **91**, 1314–1317.

Ikeda, Y., Fujisaki, A., Murata, K. & Hasegawa, H. (2003). Redescription of *Pterygodermatites (Mesopectines) nycticebi* (Mönnig, 1920) (Nematoda: Rictulariidae), a parasite of slow loris *Nycticebus coucang* (Mammalia: Primates). *Folia Parasitologica*, **50**, 115–120.

Itoh, K., Oku, Y., Okamoto, M. *et al.* (1988). Helminth parasites of the Japanese monkey, *Macaca fuscata fuscata* in Ehime Prefecture, Japan. *Japanese Journal of Veterinary Research*, **36**, 235–247.

Justine, J. L. (1988). *Capillaria brocheri* n. sp. (Nematoda: Capillariinae) parasite intestinal du chimpanzé *Pan paniscus* Zaïre. *Annales de Parasitologie humaine et comparée.*, **63**, 420–438.

Petrzelkova, K. J., Hasegawa, H., Moscovice, L. R. *et al.* (2006). Parasitic nematodes in the chimpanzee population on Rubondo Island, Tanzania. *International Journal of Primatology*, **27**, 767–777.

Rothman, J. & Bowmann, D. D. (2003). A review of the endoparasites of mountain gorillas. In *Companion and Exotic Animal Parasitology*, ed. D. D. Bowman. Ithaca, NY: International Veterinary Information Service, http://www.ivis.org.

Spear, R. (1989). Identification of species of *Strongyloides*. In *Strongyloidiasis: A Major Roundworm Infection of Man*, ed. D. I. Glove. Philadelphia, PA: Taylor & Francis, Ch. 2.

Tachibana, H., Cheng, X. J., Kobayashi, S., Fujita, Y. & Udono, T. (2000). *Entamoeba dispar*, but not *E. histolytica*, detected in a colony of chimpanzees in Japan. *Parasitology Research*, **86**, 537–541.

Tachibana, H., Cheng, X. J., Kobayashi, S. *et al.* (2001). High prevalence of infection with *Entamoeba dispar*, not *E. histolytica*, in captive macaques. *Parasitology Research*, **87**, 14–17.

Uni, S., Abe, M., Harada, K. *et al.* (1992). New record of *Gongylonema pulchrum* Molin, 1857 from a new host, *Macaca fuscata*, in Japan. *Annales de Parasitologie humaine et comparée*, **67**, 221–223.

Index

For EU product safety concerns, contact us at Calle de José Abascal, 56–1°, 28003 Madrid, Spain or eugpsr@cambridge.org.

www.ingramcontent.com/pod-product-compliance
Ingram Content Group UK Ltd.
Pitfield, Milton Keynes, MK11 3LW, UK
UKHW022114130426
469895UK00017B/210
* 9 7 8 1 1 0 8 8 2 9 4 0 3 *